T0320628

Resource Allocation for Wireless Networks

Merging the fundamental principles of resource allocation with the state of the art in research and application examples, Han and Liu present a novel and comprehensive perspective for improving wireless system performance. Cross-layer multiuser optimization in wireless networks is described systematically. Starting from the basic principles, such as power control and multiple access, coverage moves to the optimization techniques for resource allocation, including formulation and analysis and game theory. Advanced topics, such as dynamic resource allocation and resource allocation in antenna-array processing and in cooperative, sensor, personal-area, and ultrawide-band networks, are then discussed. Unique in its scope, timeliness, and innovative author insights, this invaluable work will help graduate students and researchers understand the basics of wireless resource allocation while highlighting modern research topics and will help industrial engineers improve system optimization.

Zhu Han is currently an assistant professor in the Electrical and Computer Engineering Department at Boise State University, Idaho. In 2003, he was awarded his Ph.D. in electrical engineering from the University of Maryland, College Park. Zhu Han has also worked for a period in industry, as an R & D Engineer for JDSD. Dr. Han is PHY/MAC Symposium Vice Chair of the IEEE Wireless Communications and Networking Conference, 2008.

K. J. Ray Liu is a Distinguished Scholar–Teacher of the University of Maryland, College Park. Dr. Liu is the recipient of numerous honors and awards, including best paper awards from IEEE Signal Processing Society (twice), IEEE Vehicular Technology Society, and EURASIP, as well as recognitions from University of Maryland, including the university-level Invention of the Year Award and the college-level Poole and Kent Company Senior Faculty Teaching Award.

Resource Allocation for Wireless Networks

Basics, Techniques, and Applications

ZHU HAN

Boise State University, Idaho

K. J. RAY LIU

University of Maryland, College Park

CAMBRIDGE
UNIVERSITY PRESS

University Printing House, Cambridge CB2 8BS, United Kingdom

Cambridge University Press is part of the University of Cambridge.

It furthers the University's mission by disseminating knowledge in the pursuit of
education, learning and research at the highest international levels of excellence.

www.cambridge.org
Information on this title: www.cambridge.org/9780521873857

© Cambridge University Press 2008

First published 2008

A catalogue record for this publication is available from the British Library

Library of Congress Cataloguing in Publication data

Han, Zhu, 1974–
Resource allocation for wireless networks : basics, techniques, and applications / Zhu Han, K.J. Ray Liu.
    p. cm.
Includes bibliographical references and index.
ISBN 978-0-521-87385-7 (hbk.)
1. Wireless communication systems.   2. Resource allocation.   I. Liu, K. J. Ray, 1961–   II. Title.
TK5103.2.H332   2006
621.384 – dc22   2008007048

ISBN 978-0-521-87385-7 Hardback

To my wife as well as my parents and sister in China.
– Zhu Han

To Lynne Liu.
– K. J. Ray Liu

Contents

Preface

Because of fading channels, user mobility, energy/power resources, and many other factors, cross-layer design and multiuser optimization are the keys to ensuring overall system performance of wireless networks. And resource allocation is one of the most important issues for implementing future wireless networks.

In the past decade, we have witnessed significant progress in the advance of resource allocation over wireless networks. It is not only an important research topic, but is also gradually becoming an integral teaching material for graduate-level networking courses.

Yet there are few books available to date that can serve such a purpose. Why? Because the field of resource allocation is such a versatile area that covers a broad range of issues, it is not easy to develop a comprehensive book to cover them all. For instance, resource allocation across various networking layers encounters different design constraints and parameters; different networking scenarios have different performance goals and service objectives; and different formulations of resource allocations need to employ different optimization tools.

To respond to the need of such a book for graduate students, researchers, and engineers, we try to tackle the difficulties by bringing together our research in resource allocation over the past decade and the basic material of resource allocation and optimization techniques to form the foundation of this book. Its intent is to serve either as a textbook for advanced graduate-level courses on networking or as a reference book for self-study by researchers and engineers.

This book covers three main parts. In Part I, the basic principles of resource allocation is discussed. Part II provides the background of optimization tools needed to conduct research and development in resource allocation. And in Part III, examples of advanced topics in resource allocation for different networking scenarios are the focus, to illustrate what one may encounter in different applications.

We would like to thank many for their contributions to some of the material presented in this book, including Thanongsak Himsoon, Ahmed Ibrahim, Zhu Ji, Andres Kwasinski, Xin Liu, Charles Pandana, Farrokh Rashid-Farrokhi, Javad Razavilar, Ahmed Sadek, Wipawee Siriwongpairat, Guan-Ming Su, Weifeng Su, Lee Swindlehurst, Beibei Wang, and Min Wu. We also would like to thank Greg Heinzman for his editing assistance.

Zhu Han
K. J. Ray Liu

1 Introduction

Over the past decade, there has been a significant advance in the design of wireless networks, ranging from physical-layer algorithm development, medium-access control (MAC) layer protocol design, to network- and system-level optimization. Many wireless standards have been proposed to suit the demands of various applications. Over time, researchers have come to the realization that, for wireless networks, because of fading channels, user mobility, energy/power resources, and many other factors, one cannot optimize wireless communication systems as has been traditionally done in wired networks, in which one can simply focus on and optimize each networking layer without paying much attention to the effects of other layers. For wireless networks, cross-layer optimization is a central issue to ensure overall system performance. Yet resource allocation is one of the most important issues for cross-layer optimization of wireless networks.

For instance, across different layers, one cannot design physical-layer coding, modulation, or equalization algorithms by assuming that the MAC layer issues are completely perfect, and vice versa. There are also user diversities—different users at different times and locations may suffer different channel conditions, and therefore may have different demands and capability. Fixing and allocating bandwidths and resources without considering such user diversity can simply waste system resources, and thus performances. In addition, in wireless networks there are space, time, and frequency diversities as well. Taking advantage of those diversities can significantly improve communication performance. All those factors contribute to the need of careful consideration of resource allocations.

We have witnessed the advance of resource allocation in recent years with tremendous progress. As one can imagine, because of the number of degrees of freedom and many different parameters, resource allocation is a broad issue covering a wide range of problems. Therefore, the optimization tools employed also vary a lot. Besides the commonly used convex optimization in communication system design, many resource allocation problems are nonlinear and nonconvex in nature. When it comes to channel allocation and scheduling, sometimes the problems become integer, combinatorial, or both. If one takes into account time-varying conditions, then the problem evolves into one of dynamic optimization. When cooperation among distributed and autonomous users is considered, game theory can be employed to find the optimal strategy and solution. It is fair to say that there is no single optimization tool available to solve all resource-allocation problems at once.

What makes resource allocation more challenging is that, in fact, when it comes to the applications, different wireless networks aim at different service goals, and therefore have different design specifications. One network can be severely energy sensitive and power constrained, whereas the other can be bandwidth limited or throughput hungry. In some situations, a network may have a high degree of mobility with opportunistic access, whereas in other cases a network has an ultrawide bandwidth to share with others but little mobility.

As such, different networks face different resource-allocation problems, different characteristics of problems employ different optimization techniques, and joint considerations of different layers encounter different constrained optimization issues.

This book aims at providing a comprehensive view to answer the preceding challenges in the hope of allowing readers to be able to practice and optimize the allocation of scant wireless resources over assorted wireless network scenarios. Given the nature of the topic, this book is interdisciplinary in that it contains concepts in signal processing, economics, decision theory, optimization, information theory, communications, and networking to address the issues in question. In addition, we try to provide innovative insight into the vertical integration of wireless networks through the consideration of cross-layer optimization.

The goals of this book are for readers to have a basic understanding of wireless resource-allocation problems, to be equipped with an adequate optimization background to conduct research on or design wireless networks, and to be well informed of state-of-the-art research developments. To achieve these goals, this book contains three parts:

- Part I: Basic Principles
  We will study the basic principles of resource allocation for multiple users to share the limited resources in wireless networks under different practical constraints. In addition to the explanation of the basic principles, we will also illustrate the limitations and trade-offs of different approaches.
- Part II: Optimization Techniques
  We will consider various optimization techniques that can be applied to wireless resource-allocation problems. These techniques will be categorized and then compared for their advantages and disadvantages. Some applications for different network scenarios are given as examples.
- Part III: Advanced Topics
  Through the use of some state-of-the-art design examples of different wireless networks, we will illustrate the wide varieties of topics and their potential future design directions. By considering the technical challenges of a variety of networks, we will show how to employ different techniques for different scenarios such as cellular networks, wireless local-area networks, ad hoc/sensor networks, ultrawide-band networks, and collaborative communication networks.

In Part I, the chapters that cover the basics of resource allocation are as follows:

2. Wireless Networks: An Introduction
   In this chapter, we first consider different wireless channel models such as large-scale propagation-loss models and small-scale propagation-loss models. Then, according to

the decreasing order of the coverage areas, we discuss four types of wireless networks: cellular networks, WiMax networks, WiFi networks, and wireless personal-area networks. In the wireless ad hoc networks without a network infrastructure, autonomous users should be able to establish the basic network functions in a distributed way. Wireless sensor networks can detect the events and transmit the information to the data-gathering point with a low consumption of energy. Finally, to cope with the limited spectrum, a cognitive radio can detect the spectrum hole and utilize the unused spectrum.

3. Power Control

Power control is an effective resource-allocation method to combat fading channel and cochannel interference. The transmitted power is adjusted according to the channel condition so as to maintain the received signal quality. Power control is not a single user's problem, because a user's transmit power causes other users' interferences. We describe the basics of power control first. Then we classify the power-control schemes and discuss the centralized, distributed, and statistical schemes in details. Finally, code-division multiple-access power control is highlighted.

4. Rate Adaptation

Rate adaptation is one of the most important resource-allocation issues, because the system can adapt the users' rates so that the limited radio resources can be efficiently utilized. In this chapter, we give an overview of the rate-adaptation system. Rate controls over different layers are discussed, such as source rate control, rate control for network/MAC layers, channel-coding rate control, and joint source–channel coding.

5. Multiple Access and Spectrum Access

The multiple-access scheme is a general strategy to allocate limited resources, such as bandwidth and time, to guarantee the basic quality of services, improve the system performances, and reduce the cost for the network infrastructures. In this chapter, we first study some fixed multiple-access methods such as frequency-division, time-division and code-division multiple access. Then scheduling and random-access protocols are investigated. A third-generation multiple-access system is given as an example. Although multiple access considers the problem of allocating limited radio resources to multiple users, spectrum access decides whether an individual user can access a certain spectrum. We study channel allocation and opportunity spectrum access. Finally, handoff and admission control are illustrated.

In Part II, the chapters cover optimization techniques commonly used in resource allocation:

6. Optimization Formulation and Analysis

In this chapter, we discuss how to formulate the wireless networking problem as a resource-allocation optimization issue. Specifically, we study what the resources are, what the parameters are, what the practical constraints are, and what the optimized performances across the different layers are. In addition, we address how to perform resource allocation in multiuser scenarios. The trade-offs between the different optimization goals and different users' interests are also investigated. The goal is to provide readers with a new perspective from the optimization point of view for wireless networking and resource-allocation problems.

7. Mathematical Programming

If the optimization problem is to find the best objective function within a constrained feasible region, such a formulation is sometimes called a mathematical program. Many real-world and theoretical problems can be modeled within this general framework. In this chapter, we discuss the four major subfields of mathematical programming: linear programming, convex programming, nonlinear programming, and dynamic programming. Finally, a wireless resource-allocation example using programming is illustrated.

8. Integer/Combinatorial Optimization

Discrete optimization is the problem in which the decision variables assume discrete values from a specified set. Combinatorial optimization problems, on the other hand, are problems of choosing the best combination out of all possible combinations. Most combinatorial problems can be formulated as integer programs. In wireless resource allocation, many variables have only integer values, such as the modulation rate, and other variables, such as channel allocation, have a combinatorial nature. Integer optimization is the process of finding one or more best (optimal) solutions in a well-defined discrete problem space. The major difficulty with these problems is that we do not have any optimality conditions to check whether a given (feasible) solution is optimal. We list several possible solutions such as relaxation and decomposition, enumeration, and cutting planes. Finally, a resource-allocation example is formulated and solved as a Knapsack problem.

9. Game Theory

Game theory is a branch of applied mathematics that uses models to study interactions with formalized incentive structures ("games"). It studies the mathematical models of conflict and cooperation among intelligent and rational decision makers. "Rational" means that each individual's decision-making behavior is consistent with the maximization of subjective expected utility. "Intelligent" means that each individual understands everything about the structure of the situation, including the fact that others are intelligent, rational decision makers. In this chapter, we discuss four different types of games, namely, the noncooperative game, repeated game, cooperative game, and auction theory. The basic concepts are listed, and simple examples are illustrated. The goal is to let the readers understand the basic problems and basic approaches. As a result, we hope the readers can formulate the problems and find solutions in their research areas.

In Part III, we consider some network-aware advanced topics to illustrate the versatility of resource allocation:

10. Resource Allocation with Antenna-Array Processing

For spatial diversity, transceivers employ antenna arrays and adjust their beam patterns such that they have good channel gain toward the desired directions, while the aggregate interference power is minimized at their output. Antenna-array processing techniques such as beamforming can be applied to receive and transmit multiple signals that are separated in space. Hence, multiple cochannel users can be supported in each cell to increase the capacity by exploring the spatial diversity. We investigate two examples. First, joint power control, beamforming, and base-station assignment are studied. Second, if the channel information is not available, blind beamforming can be employed to control multiple users' power and beam pattern to achieve the desired link qualities.

11. Dynamic Resource Allocation

A general strategy to combat detrimental effects, such as fading, is the dynamic allocation of resources such as transmitted power, modulation rates, channel assignment, and scheduling based on the channel conditions. Several design challenges need to be overcome: To optimize radio resource utilization, an important trade-off exists between system performances and fairness among users. To satisfy the growing demands for heterogeneous applications of wireless networks, it is critical to deliver flexible, variable-rate services with high spectral efficiencies to provide a different quality of service. Finally, if the dynamics of channels is known, each user can calculate the optimal dynamic strategies to maximize the long-term benefits.

12. Game-Theoretic Approaches for Resource Allocation

Some wireless networks, such as ad hoc networks, consist of a collection of radio transceivers without requiring centralized administration or a prearranged fixed network infrastructure. As a result, ensuring cooperation among selfish users becomes an important issue for designing wireless networks. Game theory is an effective method for analyzing and designing the distributed resource allocation. In this chapter, for noncooperative game theory, we study three examples for power control, multicell orthogonal frequency-division multiple-access channel allocation, and source–relay resource allocation for cooperative communications. For repeated game theory, we study a punishment-based approach for rate control and a self-learning-based approach for packet forwarding. Finally, for cooperative game theory, we use a negotiation-based approach for single-cell orthogonal frequency-division multiple-access resource allocation and opportunistic spectrum access for a cognitive radio.

13. Resource Allocation for Cooperative Networks

Cooperative communications have gained attention as an emerging transmit strategy for future wireless networks. Cooperative communications efficiently take advantage of the broadcasting nature of wireless networks. The basic idea is that users or nodes in a wireless network share their information and transmit cooperatively as a virtual antenna array, thus providing diversity that can significantly improve system performance. In this chapter, we investigate the impact of cooperative communications on the design of different layers.

14. Ad Hoc/Sensor/Personal-Area Networks

Over the past few decades, the increasing demands from military, national security, and commercial customers have been driving the large-scale deployment of ad hoc networks, sensor networks, and personal-area networks, which have no sophisticated infrastructures such as base stations. In these scenarios, the mobile users have to set up the network functionality on their own. For ad hoc networks, we investigate the connectivity problem. For sensor networks, we study how to prolong the lifetime.

Finally, for personal-area networks, we employ resource allocation to extend the coverage area.

15. Resource Allocation for Wireless Multimedia

    With the advancement of multimedia compression technology and wide deployment of wireless networks, there is an increasing demand especially for wireless multimedia communication services. To overcome many potential design challenges, dynamic resource allocation is a general strategy used to improve the overall system performance and ensure individual quality of service. Specifically, in this chapter, we consider two aspects of design issues: *cross-layer optimization* and *multiuser diversity*. We study how to optimally transmit multiuser multimedia streams, encoded by current and future multimedia codecs, over resource-limited wireless networks such as third-generation cellular systems, wireless local-area networks, fourth-generation cellular systems, and future wireless local-area networks and wireless metropolitan-area networks.

Part I

Basics Principles

2 Wireless Networks: An Introduction

2.1 Introduction

"Wireless network" refers to a telecommunications network whose interconnections between nodes is implemented without the use of wires. Wireless networks have seen unprecedent growth during the past few decades and will continuously evolve in the future. Seamless mobility and coverage ensure that various types of wireless connections can be made anytime, anywhere. In this chapter, we introduce some basic types of wireless networks and give the readers some preliminary backgrounds for the current state-of-the-art development.

Wireless networks use electromagnetic waves, such as radio waves, for carrying information. Therefore the performance is greatly influenced by randomly fluctuating wireless channels. To understand the channels, in Section 2.2, we will study the existing wireless channel models used for different network scenarios.

There are many existing wireless standards. We consider them according to the order of coverage area, and start with cellular wireless networks. The third-generation (3G) wireless cellular network standards have been enhanced to offer significantly increased performance for data and broadcast services through the introduction of high-speed downlink packet access, enhanced uplink, and multimedia broadcast multicast services. In Section 2.3, we provide an overview of the key elements and technologies. Specifically, we discuss WCDMA, CDMA2000, TD/S CDMA, and 4G and beyond.

WiMax, based on the IEEE 802.16 standard for a wireless metropolitan-area network (WMAN), is expected to enable true broadband speeds over wireless networks at a cost that enables mass-market adoption. WiMAX has the ability to deliver true broadband speeds and help make the vision of pervasive connectivity a reality. We discuss some techniques and the standard in Section 2.4.

A wireless local-area network (WLAN) is a network in which a mobile user can connect to a local-area network (LAN) through a wireless connection. The IEEE 802.11 group of standards specifies the technologies for a WLAN. Based on IEEE 802.11, WiFi is a brand originally licensed by the WiFi Alliance to describe WLAN technology. WiFi provides a low-cost and relatively simple way to gain high-speed access to the Internet. In Section 2.5, we study some specifications in IEEE 802.11 standards.

A wireless personal-area network (WPAN) is a personal-area network (PAN) for wireless interconnecting devices centered around an individual person's workspace. Typically, a WPAN uses a certain technology that permits communication within about

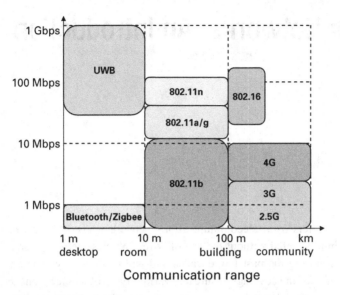

Figure 2.1 Standards comparison.

10 m; in other words, a very short range. IEEE 802.15 standards specify some technologies used in Bluetooth, Zigbee, and Ultra Wide Band. We investigate these technologies in Section 2.6.

We list different standards in Figure 2.1 for different communication rates and different communication ranges. Those standards will fit different needs of various applications, and we also discuss the techniques that can utilize multiple standards in different situations, so that a connection can be made anytime and anywhere.

Finally, in the last three sections in this chapter, we discuss some wireless networks without standards. Specifically, we study wireless ad hoc networks, wireless sensor networks, and cognitive radios, respectively. The motivations for deploying such networks, the design challenges to maintain basic functionality, and recent developments in real implementation are explained in detail.

2.2   Wireless Channel Models

Unlike the wired channels that are stationary and predictable, wireless channels are extremely random and hard to analyze. Models of wireless channels are one of the most difficult challenges for wireless network design. Wireless channel models can be classified as large-scale propagation models and small-scale propagation models, relative to the wavelength.

Large-scale models predict behavior averaged over distances much greater than the wavelength. The models are usually functions of distance and significant environmental features, and roughly frequency independent. The large-scale models are useful for modeling the range of a radio system and rough capacity planning. Some theoretical models (the first four) and experimental models (the rest) are listed as follows:

- Free-Space Model
Path loss is a measure of attenuation based on only the distance from the transmitter to the receiver. The free-space model is valid only in the far field and only if there is no interference and obstruction. The received power $P_r(d)$ of the free-space model as a function of distance d can be written as

$$P_r(d) = \frac{P_t G_t G_r \lambda^2}{(4\pi)^2 d^2 L},$$
(2.1)

where P_t is the transmit power, G_t is the transmitter antenna gain, G_r is the receiver antenna gain, λ is the wavelength, and L is the system-loss factor not related to propagation. Path-loss models typically define a "close-in" point d_0 and reference other points from the point. The received power in decibel form can be written as

$$P_d(d) \text{ dBm} = 10 \log \left[\frac{P_r(d_0)}{0.001\text{W}} \right] + 20 \log \left(\frac{d_0}{d} \right).$$
(2.2)

- Reflection Model
Reflection is the change in direction of a wave front at an interface between two dissimilar media so that the wave front returns into the medium from which it originated. A radio propagation wave impinges on an object that is large compared with the wavelength, e.g., the surface of the Earth, buildings, and walls.

A two-ray model is one of the most important reflection models for wireless channels. An example of reflection and the two-ray model is shown in Figure 2.2. The two-ray model considers a model in which the receiving antenna sees a direct-path signal as well as a signal reflected off the ground. Specular reflection, much like light off of a mirror, is assumed, and, to a very close approximation, the specular reflection arrives with strength equal to that of the direct-path signal. The reflected signal shows up with a delay relative to the direct-path signal and, as a consequence, may add constructively (in phase) or destructively (out of phase). The received power of the two-ray model can be written as

$$P_r = P_t G_t G_r \frac{h_t^2 h_r^2}{d^4},$$
(2.3)

where h_t and h_r are the transmitter height and receiver height, respectively.

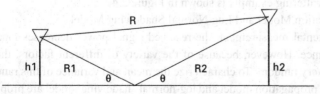

Figure 2.2 Reflection and two-ray model.

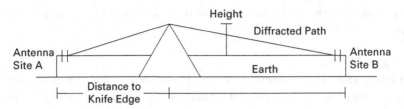

Figure 2.3 Diffraction and knife-edge model.

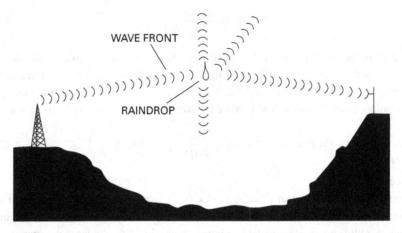

Figure 2.4 Scattering.

- Diffraction Model
  Diffraction occurs when the radio path between transmitter and receiver is obstructed by a surface with sharp irregular edges. Radio waves bend around the obstacle, even when a line of sight (LOS) does not exist. In Figure 2.3, we show a knife-edge diffraction model, in which the radio wave of the diffraction path from the knife edge and the radio wave of the LOS are combined together at the receiver. Similar to the reflection, the radio waves might add constructively or destructively.
- Scattering Model
  Scattering is a general physical process whereby the radio waves are forced to deviate from a straight trajectory by one or more localized nonuniformities in the medium through which it passes. In conventional use, this also includes deviation of reflected radiation from the angle predicted by the law of reflection. The obstructing objects are smaller than the wavelength of the propagation wave, e.g., foliage, street signs, and lamp posts. One scattering example is shown in Figure 2.4.
- Log-Scale Propagation Model and Log-Normal Shadowing Model
  From the experimental measurement, the received signal power decreases logarithmically with distance. However, because of the variety of different factors, the decreasing speed is very random. To characterize the mean and variance of this randomness, the log-scale propagation model and log-normal shadowing model are proposed, respectively.

The log-scale propagation model generalizes path loss to account for other environmental factors. The model chooses a distance d_0 in the far field and measures the path loss $PL(d_0)$. The propagation path-loss factor α indicates the rate at which the path loss increases with the distance. The path loss of the log-scale propagation model is given by

$$PL(d) \text{ (dB)} = PL(d_0) + 10\alpha \log \left(\frac{d}{d_0} \right). \qquad (2.4)$$

Shadowing occurs when objects block the LOS between the transmitter and the receiver. A simple statistical model can account for unpredictable "shadowing" as

$$PL(d) \text{ (dB)} = PL(d) + X_0, \qquad (2.5)$$

where X_0 is a zero-mean Gaussian random variable with variance typically from 3 to 12. The propagation factor and the variance of log-normal shadowing are usually determined by experimental measurement.

- Outdoor Propagation Models
  In outdoor models, the terrain profile of a particular area needs to be taken into account for estimating the path loss. Most of the following models are based on a systematic interpretation of measurement data obtained in the service area. Some typical outdoor propagation models are the Longley–Rice model, ITU terrain model, Durkins model, Okumura model, Hatas model, PCS extension of the Hata model, Walfisch and Bertoni model, and wideband PCS microcell model.

- Indoor Propagation Models
  For indoor applications, the distances are much smaller than those of the outdoor models. The variability of the environment is much greater, and key variables are the layout of the building, construction materials, building type, and antenna location. In general, indoor channels may be classified either as LOS or obstruction with a varying degree of clutter. The losses between floors of a building are determined by the external dimensions and materials of the building, as well as the type of construction used to create the floors and the external surroundings. Some of the available indoor propagation models are the Ericsson multiple-breakpoint model, ITU model for indoor attenuation, log-distance path-loss model, attenuation-factor model, and Devasirvathamòs model.

Small-scale (fading) models describe signal variability on a scale of wavelength. In fading, multipath effects and Doppler effects dominate. The fading is frequency dependent and time variant. The focus is on modeling "fading," which is the rapid change in signal over a short distance or length of time.

Multipath fading is caused by interference between two or more versions of the transmitted signal, which arrive at slightly different times. Multipath fading causes rapid changes in signal strength over a small travel distance or time interval, random frequency modulation that is due to varying Doppler shifts on different multipath signals, and time dispersion caused by multipath propagation delays.

To measure the time dispersion of multiple paths, the power-delay profile and root mean square (RMS) are the most important parameters. Power-delay profiles are generally represented as plots of relative received power as functions of excess delay with respect to a fixed time-delay reference. The mean excess delay is the first moment of the power-delay profile and is defined as

$$\bar{\tau} = \frac{\sum_k a_k^2 \tau_k}{\sum_k a_k^2}, \tag{2.6}$$

where τ_k is the delay of the kth multipath and a_k is its corresponding amplitude. The RMS is the square root of the second central moment of the power-delay profile:

$$\sigma_\tau = \sqrt{\bar{\tau}^2 - (\bar{\tau})^2}, \tag{2.7}$$

where

$$\bar{\tau}^2 = \frac{\sum_k a_k^2 \tau_k^2}{\sum_k a_k^2}. \tag{2.8}$$

Typical values of RMS delay spread are of the order of microseconds in outdoor mobile radio channels and of the order of nanoseconds in indoor radio channels.

Analogous to the delay spread parameters in the time domain, coherent bandwidth is used to characterize the channel in the frequency domain. Coherent bandwidth is the range of frequencies over which two frequency components have a strong potential for amplitude correlation. If the frequency correlation between two multipaths is above 0.9, then the coherent bandwidth is

$$B_c = \frac{1}{50\sigma}. \tag{2.9}$$

If the correlation is above 0.5,

$$B_c = \frac{1}{5\sigma}. \tag{2.10}$$

Coherent bandwidth is a statistical measure of the range of frequencies over which the channel can be considered flat.

Delay spread and coherent bandwidth describe the time-dispersive nature of the channel in a local area. But they do not offer information about the time-varying nature of the channel caused by relative motion of transmitter and receiver. Next we define Doppler spread and coherence time, which describe the time-varying nature of the channel in a small-scale region.

Doppler frequency shift is due to the movement of the mobile users. Frequency shift is positive when a mobile moves toward the source; otherwise, the frequency shift is negative. In a multipath environment, the frequency shift for each ray may be different, leading to a spread of received frequencies. Doppler spread is defined as the maximum Doppler shift,

$$f_m = \frac{v}{\lambda}, \tag{2.11}$$

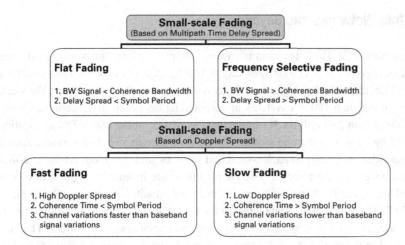

Figure 2.5 Classification of small fading.

where v is the mobile user's speed and λ is the wavelength. If we assume that signals arrive from all angles in the horizontal plane, the Doppler spectrum can be modeled as Clarke's model [247].

Coherence time is the time duration over which the channel impulse response is essentially invariant. Coherence time is defined as

$$T_c = \frac{C}{f_m},\tag{2.12}$$

where C is a constant. The coherence time definition implies that two signals arriving with a time separation greater than T_C are affected differently by the channel. If the symbol period of the baseband signal (reciprocal of the baseband signal bandwidth) is greater than the coherence time, then the signal will distort, because the channel will change during the transmission of the signal.

Based on the transmit signal's bandwidth and symbol period relatively to the multipath RMS and coherent bandwidth, small-scale fading can be classified as flat fading and frequency-selective fading. The classification means a bandlimited transmit signal sees a flat frequency channel or a frequency-selective channel. Based on coherence time that is due to Doppler spread, small-scale fading can be classified as fast fading and slow fading. The classification means that, during each signal symbol, the channel changes or does not change. The details are shown in Figure 2.5.

Multipath and Doppler effects describe the time and frequency characteristics of wireless channels. But further analysis is necessary for statistical characterization of the amplitudes. Rayleigh distributions describe the received signal envelope distribution for channels, in which all the components are non-LOS. Ricean distributions describe the received signal envelope distributions for channels in which one of the multipath components is the LOS component. Nakagami distributions are used to model dense scatters. Nakagami distributions can be reduced to Rayleigh distributions, but they give more control over the extent of the fading.

2.3   3G Cellular Networks and Beyond

Third-generation (3G) mobile communication systems based on wideband code-division multiple-access (WCDMA) and CDMA2000 (CDMA stands for code-division multiple-access) radio-access technologies have seen widespread deployment around the world. There are more than 160 3G systems in commercial operation in 75 countries with a total of more than 230 million 3G subscribers as of December 2005. The applications supported by these commercial systems range from circuit-switched services such as voice and video telephony to packet-switched services such as videostreaming, e-mail, and file transfer. As more packet-based applications are invented and put into service, the need increases for better support for different quality of service (QoS), higher spectral efficiency, and higher data rates for packet-switched services, in order to further enhance user experience while maintaining efficient use of system resources. This need has resulted in the evolution of 3G standards, as shown in Figure 2.6. For 3G cellular systems, there are two camps: the 3G Partnership Project (3GPP) [386] and the 3G Partnership Project 2 (3GPP2) [387], which is based on different second-generation (2G) technologies.

The development of 3G will follow a few key trends, and the evolution following these trends will continue as long as the physical limitations or backward compatibility

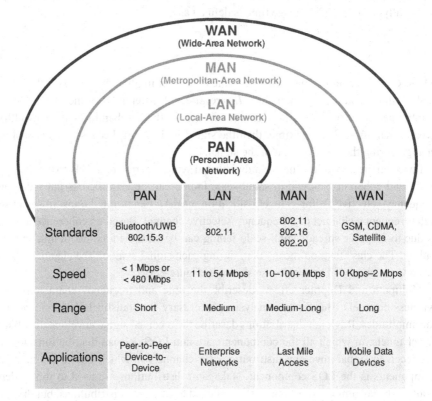

	PAN	LAN	MAN	WAN
Standards	Bluetooth/UWB 802.15.3	802.11	802.11 802.16 802.20	GSM, CDMA, Satellite
Speed	< 1 Mbps or < 480 Mbps	11 to 54 Mbps	10–100+ Mbps	10 Kbps–2 Mbps
Range	Short	Medium	Medium-Long	Long
Applications	Peer-to-Peer Device-to-Device	Enterprise Networks	Last Mile Access	Mobile Data Devices

Figure 2.6 Comparison of different wireless networks.

Table 2.1 Comparison of 3G standards

	CDMA2000	WCDMA	TD-SCDMA
Carrier bandwidth	1.25/3.75 MHz	5 MHz	1.6 MHz
Multiple access	DS/MC-CDMA	DS-CDMA	TDMA/DS-CDMA
Chip rate	1.2288/3.6864M chips/s	3.84M chips/s	1.28M chips/s
Frequency reuse	1	1	1
Channel coding	Convol./turbo code	Convol./turbo code	Convol./turbo code
Spreading code	Walsh, pseudo noise	OVSF	OVSF
Spreading factor	4-256	4-256	1,2,4,8,16
Data modulation	DL:QPSK;UL:BPSK	DL:QPSK;UL:BPSK	QPSK, 8-PSK
Frame length	5 ms, 20 ms	5 ms, 20 ms	10 ms
No. of slots/frame	16	16	7
Max. data rate	2.4 Mbps	2 Mbps	2 Mbps
Spectrum efficiency	1.0	0.4	1.25
Power control	Open/close 800 Hz	Open/close 1600 Hz	open/close 200 Hz
Receiver	Rake	Rake	MUD, Rake
Inter-BS timing	GPS synchronous	Asynch./synch.	Synchronous

requirements do not force the development to move from evolution to revolution. The key trends include the following:

- Voice services will also stay important in the foreseeable future, which means that capacity optimization for voice services will continue.
- Along with increasing use of Internet-protocol-based (IP-based) applications, the importance of data as well as simultaneous voice and data will increase.
- Increased need for data means that the efficiency of data services needs to be improved.
- When more and more attractive multimedia terminals emerge in the markets, the usage of such terminals will spread from office, homes, and airports to roads, and finally everywhere. This means that high-quality high-data-rate applications will be needed everywhere as well.
- When the volume of data increases, the cost per transmitted bit needs to decrease in order to make new services and applications affordable for everybody. The other current trend is that in the 3G evolution path very high data rates are achieved in hot spots with WLAN rather than with cellular-based standards.

In Table 2.1, we compare some of the technical parameters for the major 3G standards. In the following subsections, we discuss these standards in detail.

2.3.1   CDMA2000

The CDMA family of cellular networks grew out of work undertaken by Qualcomm, a California-based company. Working on direct-sequence spread-spectrum (DSSS) techniques, by using different spreading codes, a large number of users could occupy the

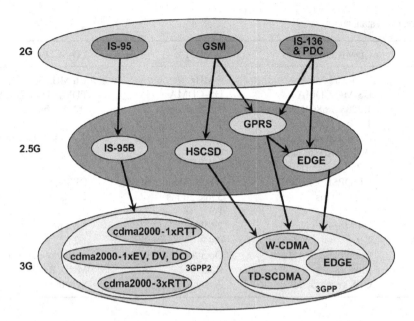

Figure 2.7 Evolution of 2G to 3G cellular networks [247].

same channel at the same time, which could provide a multiple-access scheme for cellular telecommunications. The first standard was the IS-95, and first network was launched in Hong Kong in 1996 under the brand name CDMAOne.

The CDMA system also has the following standards in its developmental stages, as shown in Figure 2.7: IS-95, IS-95A, IS-95B, CDMA2000(1x/EV-DO, 1xEV-DV, 1xRTT and 3xRTT). The first version of system standard IS-95 has never been launched for commercial purposes because of its prematurity. The IS-95A was originally applied for business and is still used widely nowadays. IS-95B was a short version because CDMA2000 standard was announced six months after it came out. The original IS-95A standard allowed for only circuit-switched data at 14.4 kbps, and IS-95B provided up to 64-kbps data rates as well as a number of additional services. A major improvement came later with the development of 3G services. The first 3G standard was known as CDMA2000 1x, which initially provided data rates up to 144 kbps. With further developments, the systems promise to allow a maximum data rate of 307 kbps.

In terms of the technology used, CDMAOne used a bandwidth of a 1.25-MHz channel so as to fit in with existing band plans and channel allocations. When using the system, different users are allocated different Walsh codes, the orthogonal spreading codes. Quadrature-phase-shift keying (QPSK) was used as a modulation form on the forward channel and offset QPSK (OQPSK) on the reverse channel. For CDMA2000, the Walsh code length was changed from 64 bits to 128 and new channels were introduced. Additionally, turbocodes were initially used for error correction.

The packet data MAC functions for IS-95 have only two states, active and dormant, as previously described. This simple approach works well for fairly low-speed data services with relatively low occupancy for any given user. However, this MAC model is inadequate to meet the aggressive requirements for very-high-speed data services, compared with many competing users in 3G systems. This is due to the excessive interference caused by idle users in the active state and the relatively long time and high system overhead required for transitioning from the dormant to the active state. To address these requirements, the CDMA2000 system incorporates a sophisticated MAC mechanism that includes two intermediate states (control hold state and suspended state) between the IS-95 active and dormant states.

To be backward compatible with IS-95 networks, the CDMA2000 radio interface retains many of the attributes of the IS-95 air interface design. In IS-95-B, higher data rates are provided through code aggregation. In CDMA2000, higher rates are achieved through either reduced spreading or multiple-code channels. In addition, there are a number of major enhancements in the CDMA2000 physical layer that facilitate advanced data services with higher rates and improved capacity. As a result, the upgrade cost of IS-95 to CDMA2000 is relatively low.

The designation "1xRTT" (1 times radio transmission technology) is used to identify the version of CDMA2000 radio technology that operates in a pair of 1.25-MHz radio channels (1×1.25 MHz, as opposed to 3×1.25 MHz in 3xRTT). 1xRTT almost doubles voice capacity over IS-95 networks. Although capable of higher data rates, most deployments limit the peak data rate to 144 kbps. Although 1xRTT officially qualifies as 3G technology, 1xRTT is considered by some to be a 2.5G (or sometimes 2.75G) technology. This has allowed it to be deployed in a 2G spectrum in some countries that limit 3G systems to certain bands. At this point, the development of CDMA2000 diverged. One development known as CDMA2000 1xEV-DO (data only) provided a data-only or data-optimized evolution, whereas another development known as CDMA2000 1xEV-DV (data and voice) provided data and voice evolutions.

- EV-DO
CDMA2000 1xEV-DO was originally not on the development roadmap for CDMA2000. As a result, EV-DO was defined under IS-856 rather than under IS-2000 and carries data only at broadband speed. The first commercial CDMA2000 1xEV-DO network was deployed by a Korea-based company, SK Telecom, in January, 2002. The system is becoming more widespread with 30 networks live across Asia, America, and Europe and 37 more networks deployed in 2006.

  With CDMA2000 1x and EV-DO release 0 well established, further development is looking to provide greater data speeds, better spectral efficiency, and improved network manageability. The evolution continues through EV-DO. It aggregates multiple channels for higher peak rates and supports up to 20-MHz bandwidth with peak rates increasing linearly. In this way, CDMA2000 1xEV-DO would be able to compete with WLANs while maintaining the mobility.
- EV-DV
The original roadmap for the development of CDMA2000 1x was to adopt the data and voice system, but it is unlikely to be deployed, although it could be capable of providing data at speeds up to 3.09 Mbits/s in the forward direction. To meet the requirements

of 1xEV-DV, there are a number of new features to be implemented. These included the addition of new channels, an adaptive modulation and coding scheme, the addition of an automatic repeat-request (ARQ) to the physical layer, and cell switching. 1xEV-DV has been developed to exploit the delay tolerance and diversity of multiuser packet data traffic via numerous air interface innovations. It has been developed in a cooperative fashion with the support and effort of the companies involved in the 3GPP2.

2.3.2   WCDMA/UMTS

WCDMA was developed by NTT DoCoMo as the air interface for their 3G network, FOMA. Later NTT DoCoMo submitted the specification to the International Telecommunication Union (ITU) as a candidate for the international 3G standard known as IMT-2000. The ITU eventually accepted WCDMA as part of the IMT-2000 family of 3G standards, as an alternative to CDMA2000, EDGE (enhanced data rate for GSM evolution), and the short-range, Digital Enhanced Cordless Telecommunication (DECT) system. Later, WCDMA was selected as the air interface for the Universal Mobile Telecommunications System (UMTS), the 3G successor to the GSM (Global System for Mobile Communication). During the evolution from 2G to 3G for WCDMA, there are some 2.5G technologies (shown in Figure 2.7) as follows:

- HSCSD: High-speed circuit-switched data (HSCSD) is an enhancement to circuit-switched data, the original data transmission mechanism of the GSM mobile phone system. One innovation in HSCSD is to allow different error-correction methods to be used for data transfer. The other innovation in HSCSD is the ability to use multiple time slots at the same time. Using the maximum of four time slots, this can provide an increase in the maximum transfer rate of up to 57.6 kbps (4 × 14.4 kbps).
- GPRS: General packet radio service (GPRS) is a mobile data service available to users of GSM and IS-136 mobile phones. GPRS data transfer is typically charged per megabyte of transferred data, whereas data communication via traditional circuit switching is billed per minute of connection time, independently of whether the user actually has transferred data or has been in an idle state. 2G cellular systems combined with GPRS are often described as "2.5G," that is, a technology between the second generation (2G) and third generation (3G) of mobile telephony. It provides moderate-speed data transfer by using unused time-division multiple-access (TDMA) channels in, for example, the GSM system.
- EDGE: Enhanced data rates for GSM evolution (EDGE), or enhanced GPRS (EGPRS), is a digital mobile phone technology that allows for increased data transmission rate and improved data transmission reliability. It is generally classified as a 2.75G network technology. EDGE has been introduced into GSM networks around the world since 2003, initially in North America. It can be used for any packet-switched applications such as an Internet connection. High-speed data applications such as video services and other multimedia benefit from EGPRS's increased data capacity. Circuit-switched EDGE is a possible future development. The highest rate is 384 kbps.

The rapid widespread deployment of WCDMA and an increasing uptake of 3G services are raising expectations with regard to new services such as web surfing and file transfer. Release 6 of WCDMA brought support for broadcast services through multimedia broadcast multicast services (MBMS), enabling applications such as mobile TV. WCDMA has been evolving to meet the increasing demands for high-speed data access and broadcast services. These two types of services have different characteristics, which influence the design of the enhancements. For high-speed data access, data typically arrive in bursts, posing rapidly varying requirements on the amount of radio resources required. The transmission is typically bidirectional, and low delays are required for a good end-user experience. As the data are intended for a single user, feedback can be used to optimize the transmission parameters. Broadcast/multicast services carry data intended for multiple users. Consequently, user-specific adaptation of the transmission parameters is cumbersome, and diversity not requiring feedback is crucial. Because of the unidirectional nature of broadcasted data, the low delay for transmission is not as important as for high-speed data access.

Release 5 of WCDMA introduced improved support for downlink packet data, often referred to as high-speed downlink packet access (HSDPA). In release 6, finalized early in 2005, the packet data capabilities in the uplink (enhanced uplink) were improved. In WCDMA, the shared downlink resource consists of transmission power and channelization codes in node B (the base station), whereas in the uplink the shared radio resource is the interference at the base station. A key characteristic of HSDPA is the use of shared-channel transmission. This implies that a certain fraction of the total downlink radio resources available within a cell, channelization codes and transmission power, is seen as a common resource that is dynamically shared between users, primarily in the time domain. The use of shared-channel transmission in WCDMA implemented through the high-speed downlink shared channel (HSDSCH), enables the possibility of rapidly allocating a large amount of the downlink resources to a user when needed. Fast scheduling is used to control allocation of the shared resource among users on a rapid basis. Additionally, a fast hybrid ARQ with soft combining enables fast retransmission of erroneous data packets. To meet the requirement on low delays and rapid resource (re)allocation, the corresponding functionality must be located close to the air interface. In WCDMA this has been solved by locating the enhancements in the base station as part of additions to the MAC layer.

- Link Adaptation

  Link adaptation is implemented by adjusting the channel-coding rate, and selecting between QPSK and 16 quadrature amplitude modulation (QAM). Higher-order modulation, such as 16 QAM, makes more efficient use of bandwidth than QPSK, but requires greater received Eb/N0. Consequently, 16 QAM is mainly useful in advantageous channel conditions. In addition, the data rate also depends on the number of channelization codes assigned for HSDSCH transmission in a transmission time interval (TTI). The data rate is selected independently for each 2-ms TTI by node B, and the link-adaptation mechanism can therefore track rapid channel variations.

- Scheduling
  The scheduler is a key element and to a large extent determines the overall downlink performance, especially in a highly loaded network. A practical scheduler strategy exploits the short-term variations (e.g., there that are due to multipath fading and fast interference variations) while maintaining some degree of long-term fairness between the users.
- Hybrid ARQ
  The third key feature of HSDPA is a hybrid ARQ with soft combining, which allows the terminal to rapidly request retransmission of erroneously received transport blocks, essentially fine-tuning the effective code rate and compensating for errors made by the link-adaptation mechanism. The terminal attempts to decode each transport block it receives and reports to node B its success or failure 5 ms after the reception of the transport block.

2.3.3   TD-SCDMA

Transmit diversity (TD) is one of the key contributing technologies to defining the ITU-endorsed 3G systems WCDMA and CDMA2000. Spatial diversity is introduced into the signal by transmitting through multiple antennas. The antennas are spaced far enough apart that the signals emanating from them can be assumed to undergo independent fading. In addition to diversity gain, antenna gain can also be incorporated through channel-state feedback. This leads to the categorization of TD methods into open-loop and closed-loop methods. Several methods of transmit diversity in the forward link have been either under consideration or adopted for the various 3G standards.

China has fully embraced the remarkable growth and unprecedented penetration of mobile services, and has become the world's largest mobile cellular market. TD-SCDMA was proposed by the China Wireless Technology Standard (CWTS) Group in 1998, approved as one of the 3G standards by ITU in May 2000, and joined 3GPP in March 2001. China puts a major effort into advancing its mobile communication systems and facilitating its own technological development in this critical area. TD-SCDMA, a combination of TDD and synchronous CDMA, offers several unique advantages over its alternatives, WCDMA and CDMA2000, such as flexible spectrum allocation, low-cost implementation, and easier migration from GSMs.

For TD-SCDMA, the channel includes three carriers using a low-chip-rate mode of 1.28 Mchips/s that corresponds to a carrier bandwidth of 1.6 MHz. This helps provide high flexibility in spectrum usage and network design, especially in densely populated areas. In addition, each TDMA frame of 5-ms duration is divided into seven time slots, which can be flexibly assigned to either multiple users or a single user that might require multiple time slots. In addition to the TDMA/TDD principle, TD-SCDMA uses a different CDMA mode from that of CDMA2000 and WCDMA systems, in which TD-SCDMA limits the number of codes for each time slot to a maximum of 16. This helps to reduce multiple-access interference (MAI) and to increase system capacity. Relying on a combination of TDD and synchronous CDMA, TD-SCDMA offers a number of attractive features, including unpaired frequencies, suitability for IP services, and capability to support asymmetric services in uplink/downlink. In addition,

Table 2.2 Comparison of 3G and 4G

Parameter	3G	4G
Major requirement driving architecture	Voice-driven data add on	Data/voice over IP
Network architecture	Wide-area cell based	Hybrid with WiFi and WPAN
Speed	384 kbps–2 Mbps	20–10 Mbps
Frequency band	1.8–2.4 GHz	2–8 GHz
Bandwidth	1.25, 5, or 20 MHz	100 MHz
Switching design	Circuit and packet	Packet
Access	DS-CDMA	OFDM/MC-CDMA
FEC	Convolution/turbo code	Concatenated coding
Component design	Antenna, multiband adapter	Smart antennas, software radios

TD-SCDMA systems also incorporate some new or unique technologies such as joint detection, adaptive antenna-array processing, dynamic channel allocation, and baton handover.

2.3.4   4G and Beyond

Looking at developments in the Internet and applications, we clearly see that the complexity of the transferred content is rapidly increasing and will increase further in the future. Generally it can be said that the more bandwidth that is available, the more bandwidth applications will consume. To justify the need for a new air interface, goals need to be set high enough to ensure that the system will be able to serve us long into the future. A reasonable approach would be to aim at 100-Mbps full-mobility wide-area coverage and 1-Gbps low-mobility local-area coverage with a next-generation cellular system in about 2010 in standards fora. Also, the future application and service requirements will bring new requirements to the air interface and new emphasis on air interface design. One such issue, which already has had a strong impact on 3G revolution is the need to support IP and IP-based multimedia. If both technology and spectra to meet such requirements cannot be found, the whole discussion of 4G may become obsolete. In Table 2.2, we compare key parameters of 4G with those of 3G.

For 4G standards, it is worth mentioning Flash-OFDM (fast low-latency access with seamless handoff orthogonal frequency-division multiplexing), a system that is based on OFDM and also specifies higher protocol layers. It was developed and is marketed by Flarion, which was acquired by Qualcomm. Flash-OFDM has generated interest as a packet-switched cellular bearer. Flash-OFDM competes with GSM and 3G networks. Flarion system is the first truly IP-based broadband cellular network designed for data, and it outperforms 3G in all critical areas of performance. For example, the system is capable of sustaining 12 Mbps of throughput per cell in a three-carrier three-sector configuration and has peak user data rates up to 3 Mbps, full cellular mobility, less than 20 ms of latency, and full QoS.

Next-generation wireless involves the concept of a major move toward ubiquitous wireless communications systems and seamless high-quality wireless services. 4G mobile communications involve a mix of concepts and technologies in the making.

Some can be recognized as derived from 3G and are called evolutionary (e.g., evolutions of WCDMA and CDMA2000), whereas others involve new approaches to wireless mobile and are sometimes labeled revolutionary, like OFDM/ WCDMA. What is important, though, is the common understanding that technologies beyond 3G are of fundamental relevance in the movement toward a new wireless world that is a total convergence of wireless mobile and wireless access communications. Any of these terms are meant to signify fundamentally better wireless mobile communications in the future.

2.4 WiMAX Networks

Wireless metropolitan-area network (WMAN) technology is a relatively new field that was started in 1998. From that time a new standard has emerged to handle the implementation, IEEE 802.16. The equivalent of 802.16 in Europe is HIPERMAN. The WiMAX Forum is working to ensure that 802.16 and HIPERMAN interoperate seamlessly. This standard has helped to pave the way for WMAN technology globally and since its first inception has now received six expansions onto the standards. WMAN differs from other wireless technologies in that it is designed for a broader audience, such as a large corporation or an entire city. The overall scheme is illustrated in Figure 2.8.

There are two main applications of WiMAX today: Fixed WiMAX applications are point-to-multipoint enabling broadband access to homes and businesses, whereas mobile WiMAX offers the full mobility of cellular networks at true broadband speeds. Both fixed and mobile applications of WiMAX are engineered to help deliver ubiquitous, high-throughput broadband wireless services at a low cost.

Next, we break up how WMAN technology works by frequency range (top to bottom) and available options in each range.

- 10GHz-66GHz range:
  In this range, a "WirelessMAN-SC" technique is used, which employs a single carrier for modulation. For allowing multiple users onto the network, WMAN uses either time-division duplexing (TDD) or frequency-division duplexing (FDD). TDD allows for variable asymmetry in both uplink and downlink connections and acts just like

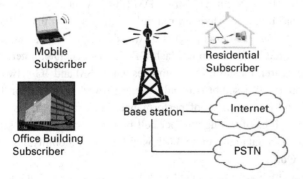

Figure 2.8 Basic implementation of a WMAN from the original IEEE 802.16 standard [372].

time-division multiplexing. TDD allows two users (in this case receiver/transmitter) with a two-way connection on a single frequency. FDD places the uplink and downlink signals on separate subbands and works best for symmetrical traffic (i.e., when the amount of uplink and downlink traffic is the same). For the base station, time-division multiplexing is utilized and for uplink, TDMA is employed.

- 2GHz-11GHz band:

For the licensed 2–11-GHz band, there are three different ways for a WMAN to work. The first way is "WirelessMAN-SCa," which works the same way as it did in the 10–66-GHz range. The second way is "WirelessMAN-OFDM," which employs an OFDM method. This method is similar to a multiple-carrier modulation, in which it divides a high bit stream into several low bit streams across subcarriers that are orthogonal to each other. The last way is "WirelessMAN-OFDMA." OFDMA is a multiple-user version of OFDM, in which the subcarriers are broken down even further into subsets, and each subset is the representative of a different user.

Another option when using the licensed 2–11-GHz band is the kind of acknowledgement system to use. For example, one of the acknowledgement systems that can be used in the 2–11-GHz band is ARQ. ARQ is an automatic repeat-request algorithm that has a transmitter send a packet. If the receiver retrieves the data with no error, it will send back an acknowledge (ACK) message saying it is ready for the next packet. If it sends a no-acknowledge (NACK) message or does not send a message, the transmitter will resend the data again.

In the 2–11-GHz license-exempt band, WMAN employs the "WirelessHUMAN" method. The WirelessHUMAN (wireless high-speed unlicensed metropolitan-area network) method employs the same OFDM method that was previously discussed for WirelessMAN-OFDM.

The 802.16 MAC uses a scheduling algorithm for which the subscriber station needs to compete once (for initial entry into the network). After the competition, the subscriber station is allocated an access slot by the base station. The time slot can enlarge and contract, but remains assigned to the subscriber station, which means that other subscribers cannot use it. The 802.16 scheduling algorithm is stable under overload and oversubscription (unlike 802.11). It can also be more bandwidth efficient. The scheduling algorithm also allows the base station to control QoS parameters by balancing the time-slot assignments among the application needs of the subscriber stations. Moreover, the MAC layer is also in charge of protocol data unit (PDU) assembly and disassembly. A detailed illustration for different layer protocols of 802.16 is shown in Figure 2.9.

The operation standards for WMANs are regulated under IEEE standard 802.16 [377]. WMANs are allowed the operating frequency range of 10–66 GHz. With such a broad spectrum to work with, WMANs have the ability to transmit over previous wireless frequencies such as IEEE 802.11b/g, causing less interference with other wireless products. The only downside to using such high frequencies is that WMAN needs a LOS between the transmitters and receivers, much like a directional antenna. Using a LOS, however, will decrease multipath distortion, allowing higher bandwidths to be achieved,

Table 2.3 Comparison of 802.16 standards

Parameter	802.16	802.16a/802.16d	802.16e
Date	Dec. 2001	Jan. 2003/Q3 2004	Q3, 2004
Spectrum	10–66 GHz	<11 GHz	<6 GHZ
Channels	LOS only	Non-LOS	Non-LOS
Modulation	QPSK,16 QAM, 64 QAM	OFDM256, QPSK, 16 QAM, 64 QAM	Same as 802.11a
Mobility	Fixed	Fixed	Pedestrian mobility, regional roaming
Bandwidth	20.25.28 MHZ	1.25–20 MHz	Same as 802.16a
Throughput	Up to 75 Mbps	Up to 75 Mbps	Up to 30 Mbps
Cell radius	1–3 miles	3–5 miles	1–3 miles

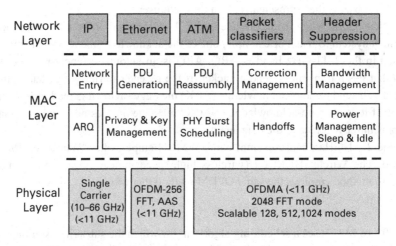

Figure 2.9 WMAX protocol stacks [383].

and can attain up to 75 Mbps for both uplink and downlink on a single channel [372]. Some extensions of 802.16 standards are listed as follows and in Table 2.3:

- IEEE 802.16a: The IEEE has developed 802.16a for use in licensed and license-exempt frequencies from 2 to 11 GHz. Most commercial interest in IEEE 802.16 is in these lower-frequency ranges. At the lower ranges, the signals can penetrate barriers and thus do not require a LOS between transceiver and antenna. This enables more flexible WiMAX implementations while maintaining the technologys data rate and transmission range. IEEE 802.16a supports mesh deployment, in which transceivers can pass a single communication on to other transceivers, thereby extending basic 802.16s transmission range.
- IEEE 802.16b: This extension increases the spectrum the technology can use in the 5- and 6-GHz frequency bands and provides QoS. WiMAX provides QoS to ensure priority transmission for real-time voice and video and to offer differentiated service levels for different traffic types.

- IEEE 802.16c: IEEE 802.16c represents a 10–66-GHz system profile that standardizes more details of the technology. This encourages more consistent implementation and, therefore, interoperability.
- IEEE 802.16d: IEEE 802.16d includes minor improvements and fixes to 802.16a. This extension also creates system profiles for compliance testing of 802.16a devices.
- IEEE 802.16e: This technology will standardize networking between carrier-fixed base stations and mobile devices, rather than just between base stations and fixed recipients. IEEE 802.16e would enable the high-speed signal handoffs necessary for communications with users moving at vehicular speeds.

In addition to IEEE 802.16, IEEE 802.20 (IEEE802.20) or the Mobile Broadband Wireless Access (MBWA) Working Group aims to prepare a formal specification for a packet-based air interface designed for IP-based services. The goal is to create an interface that will allow the creation of low-cost, always-on, and truly mobile broadband wireless networks, nicknamed Mobile-Fi. IEEE 802.20 will be specified according to a layered architecture, which is consistent with other IEEE 802 specifications. The scope of the working group consists of the physical (PHY), medium-access control (MAC), and logical-link control (LLC) layers. The air interface will operate in bands below 3.5 GHz and with a peak data rate of over 1 Mbps. The goals of 802.20 and 802.16e, the so-called "mobile WiMAX," are similar. A draft 802.20 specification was balloted and approved on January 18, 2006.

WiMAX can be viewed as "last-mile" connectivity at high data rates. This could result in lower pricing for both home and business customers as competition lowers prices. In areas without preexisting physical cable or telephone networks, WiMAX may be a feasible alternative for broadband access that has been economically unavailable. Prior to WiMAX, many operators were using proprietary fixed wireless technologies for broadband services. For this reason, WiMAX has its significant markets in rural areas and developing countries.

2.5   WiFi Networks

IEEE 802.11 denotes a set of WLAN standards developed by Working Group 11 of the IEEE LAN/MAN Standards Committee (IEEE 802). WiFi is a brand originally licensed by the WiFi Alliance to describe the underlying technology of WLAN based on the IEEE 802.11 specifications. It was developed to be used for mobile computing devices, such as laptops, in LANs, but is now increasingly used for more services, including Internet and VoIP (voiceover IP) phone access, gaming, and basic connectivity of consumer electronics, such as televisions, DVD players, or digital cameras.

In the PHY layer, 802.11b operates within the 2.4-GHz industrial, scientific, and medical (ISM) band. The original 802.11b defines data rates of 1 and 2 Mbps via radio waves using a frequency-hopping spread spectrum (FHSS) or a direct-sequence spread spectrum (DSSS). For FHSS, 2.4-GHz band is divided into 75 1-MHz subchannels. The sender and receiver agree on a hopping pattern, and data are sent over a sequence of the subchannels. Each conversation within the 802.11 network occurs over a different hopping pattern. Because of FCC regulations that restrict subchannel bandwidth to 1 MHz, FHSS techniques are limited to speeds of no higher than 2 Mbps. DSSS divides the 2.4-GHz band into 14 22-MHz channels. Adjacent channels overlap one another partially, with 3 of the 14 being completely nonoverlapping. The spreading code is the 11-bit Barker sequence. Binary-phase-shift keying (BPSK) and quadrature-phase-shift keying (QPSK) are used to provide different rates.

To increase the data rate to 5.5 Mbps and 11 Mpbs in the 802.11b standard, advanced coding technique, complementary code keying (CCK) is employed. A complementary code contains a pair of finite-bit sequences of equal length, such that the number of pairs of identical elements (1 or 0) with any given separation in one sequence is equal to the number of pairs of unlike elements having the same separation in the other sequence. A network using CCK can transfer more data per unit time for a given signal bandwidth than a network using the Barker code, because CCK makes more efficient use of the bit sequences. CCK consists of a set of 64 8-bit code words. The 5.5-Mbps rate uses CCK to encode 4 bits per carrier, whereas the 11-Mbps rate encodes 8 bits per carrier. Both speeds use QPSK as the modulation technique and signal at 1.375 Mps. Table 2.4 shows the differences rates for 802.11b.

802.11a adopts OFDM at 5.15–5.25 GHz, 5.25–5.35 GHz, and 5.725–5.825 GHz to support multiple data rates up to 54 Mbps. 802.11g utilizes the 2.4-GHz band with OFDM modulation and is also backward compatible with 802.11b. For OFDM, the fast Fourier transform (FFT) has 64 subcarriers. There are 48 data subcarriers and 4 carrier pilot subcarriers for a total of 52 nonzero subcarriers defined in IEEE 802.11a, plus 12 guard subcarriers. The IEEE 802.11a/g PHY layer provides eight PHY modes with different modulation schemes and different convolutional coding rates, and can offer various data rates. The configurations of these eight PHY modes are listed in Table 2.5.

To achieve higher data rates in the PHY layer, in January 2004, IEEE announced that it had formed a new 802.11 Task Group (TGn) to develop a new amendment to the 802.11 standard for WLANs. 802.11n builds on previous 802.11 standards by adding MIMO (multiple-input multiple-output). MIMO uses multiple transmitter and receiver antennas to allow for increased data throughput through spatial multiplexing and increased range by exploiting the spatial diversity. There are several proposal groups named TGnSync, WWiSE (short for "World-Wide Spectrum Efficiency"), and MITMOT ("MAC and MIMO Technologies for More Throughput"). All proposals occupy the 2.5-GHz frequency band with 20- or 40-MHz bandwidth so as to support the communication

Table 2.4 802.11b rates

Data rate	Code length	Modulation	Symbol rate	Bits/symbol
1 Mbps	11(DSSS)	BPSK	1 Mps	1
2 Mbps	11(DSSS)	QPSK	1 Mps	2
5.5 Mbps	8(CCK)	QPSK	1.375 Mps	4
11 Mbps	8(CCK)	QPSK	1.375 Mps	8

Table 2.5 PHY layer mode for IEEE 802.11a/g

Mode	Modulation	Channel coding	Data rate
1	BPSK	1/2	6 Mbps
2	BPSK	3/4	9 Mbps
3	QPSK	1/2	12 Mbps
4	QPSK	3/4	18 Mbps
5	16-QAM	1/2	24 Mbps
6	16-QAM	3/4	36 Mbps
7	64-QAM	2/3	48 Mbps
8	64-QAM	3/4	54 Mbps

Table 2.6 Comparison of 802.11 standards

Parameter	802.11b	802.11a/g	802.11n
Air Rate	11 Mbps	54 Mbps	200+ Mbps
MAC SAP Rate	5 Mbps	25 Mbps	100 Mbps
Range	30 m	30 m	50 m
Frequency	2.4 GHz	5.25,5.6,5.8 GHz/2.4 GHz	2.4 GHz
Bandwidth	20 MHz	20 MHz	20 or 40 MHz
Modulation	DSSS/CCK	DSSS/CCK/OFDM	DSSS/CCK/OFDM with MIMO
Special Streams	1	1	1,2,3,4

speed of more than 200 Mbps. 802.11n is backward compatible with 802.11b and 802.11g. In Table 2.6, we compare the parameters for the three 802.11 standards.

The IEEE 802.11 MAC protocol supports two kinds of access methods, namely, the distributed coordination function (DCF) and the point coordination function (PCF). In both mechanisms, only one user occupies all the bandwidth at each time slot. The PCF is based on polling, controlled by a point coordinator like an access point (AP), to communicate with a node listening and to see if the airwaves are free. The PCF seems to be implemented in only very few hardware devices as it is not part of the WiFi Alliance's interoperability standard.

In contrast, the DCF is an access mechanism using carrier-sense multiple access with collision avoidance (CSMA/CA). DCF mandates a station wishing to transmit to listen for the channel status for a DCF interframe space (DIFS) interval. If the channel is found busy during the DIFS interval, the station defers its transmission or proceeds otherwise. In a network in which a number of stations contend for the multiaccess channel, if multiple stations sense the channel is busy and defer their access, they will find that the channel is released virtually simultaneously and then try to seize the channel again at the same time. As a result, collisions may occur. To avoid such collisions, the DCF also specifies random backoff, which forces a station to defer its access to the channel for an extra period. The DCF also has an optional virtual carrier-sense mechanism that exchanges short request-to-send (RTS) and clear-to-send (CTS) frames between the source and destination stations before the long data frame is transmitted. The details of RTS/CTS will be given in a later chapter.

To take full advantage of the future market opportunity for WiFi, several key challenges must be overcome. In the following, we list some near-future design topics and their possible solutions.

- Security:

  Most concentration for WiFi is on free public access. However, eavesdroppers and hackers can take full advantage of WiFi systems. Currently, all 802.11a, b, and g devices support a WEP (wired equivalent privacy) encryption that has had flaws.

  IEEE 802.11i, also known as WiFi Protected Access 2 (WPA2), is an amendment to the 802.11 standard specifying security mechanisms for wireless networks. The 802.11i specification defines two classes of security algorithms: Robust Security Network Association (RSNA), and Pre-RSNA. Pre-RSNA security consists of WEP and 802.11 entity authentication. RSNA provides two data confidentiality protocols, called the temporal key integrity protocol (TKIP) and the countermode/CBC-MAC Protocol (CCMP). The RSNA establishment procedure includes 802.1X authentication and key management protocols. Beyond IEEE 802.11i, it is worth mentioning that WAPI (WLAN Authentication and Privacy Infrastructure) is a Chinese National Standard for WLANs (GB 15629.11-2003).

- Mobility:

  Mobility is an important attribute of wireless networks. Current WLAN standards provide mobility through roaming capabilities. IEEE 802.11p, also referred to as wireless access for the vehicular environment (WAVE), defines enhancements to 802.11 required to support intelligent transportation systems (ITSs) applications. This includes data exchange between high-speed vehicles and between the vehicles and the roadside infrastructure in the licensed ITS band of 5.9 GHz (5.85–5.925 GHz). 802.11p will be used as the groundwork for DSRC (Dedicated Short Range Communications), a U.S. Department of Transportation project, which will be emulated elsewhere, looking at vehicle-based communication networks, particularly for applications such as toll collection, vehicle safety services, and commerce transactions via cars. The ultimate vision is a nationwide network that enables communications between vehicles and roadside APs or other vehicles.

- QoS support:

  802.11e is the first wireless standard that spans home and business environments. It adds QoS features and multimedia support to the existing 802.11 wireless standards, while maintaining full backward compatibility with these standards. QoS and multimedia support are critical to wireless home networks where voice, video, and audio will be delivered. Broadband service providers view QoS and multimedia-capable home networks as an essential ingredient to offering residential customers video on demand, audio on demand, VoIP, and high-speed Internet access.

  802.11e introduces two enhancements, enhanced DCF (EDCF) and hybrid coordination function (HCF). In EDCF, a station with high-priority traffic waits a little less before it sends its packet, on average, than a station with low-priority traffic, so that high-priority traffic has a higher chance of being sent than low-priority traffic. In addition, each priority level is assigned a transmit opportunity (TXOP) that is a bounded time interval during which a station can send as many frames as possible. The HCF works more like PCF. With the PCF, QoS can be configured with great precision. QoS-enabled stations have the ability to request specific transmission parameters (data rate, jitter, etc.) that should allow advanced applications like VoIP and video streaming to work more effectively on a WiFi network.

- Integration of 3G and WLAN
  The 3G cellular networks and 802.11 WLANs possess complementary characteristics. 3G cellular networks promise to offer always-on, ubiquitous connectivity and mobility with relatively low data rates. 802.11 offers much higher data rates, comparable with those of the cellular networks, but can cover only smaller areas without mobility, suitable for hot-spot applications in hotels and airports. The performance and flexibility of wireless data services would be dramatically improved if users could seamlessly roam across the two networks. By offering integrated 802.11/3G services, 3G operators and Wireless Internet Service Providers (WISPs) can attract a wider user base and ultimately facilitate the ubiquitous introduction of high-speed wireless services. Users can also benefit from the enhanced performance and lower overall cost of such a combined service. For a network node changing the type of connectivity between 3G cellular phone and WLAN, the concept of vertical handoff will be discussed in a later chapter.

2.6   Wireless Personal-Area Networks

In this section, we first briefly review the 802.15 wireless personal-area standards. Then, for low data rate, we study the Bluetooth and Zigbee standards. Finally, we investigate the high-speed ultrawide-band standard.

A WPAN is a computer network used for wireless communication among devices [including telephones and personal digital assistants (PDAs)] close to one person. The devices may or may not belong to the person in question. The reach of a WPAN is typically a few meters. WPANs can be used for communication among the personal devices themselves (intrapersonal communication) or for connecting to a higher-level network and the Internet (an uplink). 802.15 is a communications specification that was approved in early 2002 by the IEEE Standards Association (IEEE-SA) for WPANs. Specifically, we list the following three substandards.

The IEEE Standard 802.15.1 was approved as a new standard for Bluetooth by the IEEE-SA Standards Board on 15 April 2002. The Bluetooth standard enables wireless communication between multiple electronic devices within 10 m of each other. Bluetooth devices are organized in piconets, which include one master device and up to seven slave devices. The Bluetooth devices communicate in the 2.4-GHz radio frequency (RF) band, enabling devices to communicate without LOS spacing, such as through walls or through a person's body. Bluetooth piconets utilize a FHSS in 79 1-MHz bands, reducing the likelihood of interference with other Bluetooth piconets.

802.15.3 is the IEEE standard for a high-data-rate WPAN designed to provide QoS for real-time distributions of multimedia content, like video and music. It is ideally suited for a home multimedia wireless network. The original standard uses a "traditional" carrier-based 2.4-GHz radio as the PHY layer. A follow-on standard, 802.15.3a, defines an alternative PHY layer; current candidate proposals are based on an ultrawide band (UWB), that will provide in excess of 110 Mbps at a 10-m distance and 480 Mbps at 2 m. This will allow applications requiring streaming of high-definition video between media servers and flat-screen high-definition (HD) monitors and extremely fast transfer of media files between media servers and portable media devices.

IEEE 802.15.4-2003 (Low-Rate WPAN) deals with a low data rate but a very long battery life (months or even years) and very low complexity. The first edition of the 802.15.4 standard was released in May 2003. In March 2004, after Task Group 4b was formed, Task Group 4 put itself in hibernation. The ZigBee set of high-level communication protocols is based on the specification produced by the IEEE 802.15.4 Task Group.

2.6.1   Bluetooth/Zigbee

Bluetooth is a standard for wireless communications that uses short-range radio frequencies to enable communication among multiple electronic devices. Bluetooth technology is envisioned as a replacement of the interconnection cables between personal devices such as notebook computers, cellular phones, PDAs, and digital cameras. Some typical applications are shown in Figure 2.10. If widely adopted, Bluetooth would enable a uniform interface for accessing data services. Thus calendars, address books, and business cards stored in personal devices could be automatically synchronized by use of push-button synchronization and proximity operation.

The Bluetooth Special Interest Group (SIG) was founded by Ericsson, IBM, Intel, Nokia, and Toshiba in February 1998 to develop an open specification for short-range wireless connectivity. The SIG offered all of the intellectual property explicitly included in the Bluetooth specification royalty-free to adopter members to facilitate the widespread acceptance of the technology. The SIG now includes thousands of companies. To use the intellectual property in the Bluetooth specification, adopter members must qualify any Bluetooth products they intend to bring to market through the Bluetooth qualification program. The Bluetooth qualification program includes radio and protocol conformance testing, profile conformance testing, and interoperability testing.

Bluetooth is a RF technology utilizing the unlicensed 2.4-GHz ISM band. Bluetooth enables wireless connections up to 10 m under standard transmitter power, and, because of the use of RFs, devices need not be within LOS of each other and may connect through walls or other nonmetal objects. In the active mode, Bluetooth devices typically consume 0.1 W of active power for class 1 with a range of 100 m, 2.5 mW for class 2 with a range of 10 m, and 1 mW for class 3 with a range of 1 m. The modulation technique utilized in Bluetooth technology is binary Gaussian frequency-shift keying,

Figure 2.10 Typical Bluetooth applications.

and the baud rate is 1 Msymbol/s. Thus the bit time is 1 μs and the raw transmission speed is 1 Mbps.

The baseband signals used in Bluetooth devices, which are typically 1 MHz in bandwidth, cannot directly be transmitted on the wireless medium. Modulation of the 1-MHz baseband signals into the 2.4-GHz band is difficult to achieve in one step because complementary metal-oxide semiconductor (CMOS) transistors do not operate at these frequencies. Bluetooth radio devices solve this problem by modulating the baseband signal onto an intermediate frequency, such as 3 MHz, and then use a frequency mixer to increase the frequency of the signal to the 2.4-GHz band.

Because the unlicensed ISM band in which Bluetooth operates is often cluttered with signals from other devices, such as garage door openers, baby monitors, cordless phones, and microwave ovens, Bluetooth utilizes a FHSS for security and to avoid interference with the signals from other devices. Frequency hopping also allows multiple piconets to exist within range of each other with minimal interference. Frequency hopping typically involves generating a frequency-shift-keyed signal, and then shifting the frequency of the frequency-shift-keyed signal by an amount determined by a pseudo-noise code. The pseudo-noise code is random in that it appears to be unpredictable to an outsider, but it is generated by deterministic means. The pseudo-noise code is unique to the piconet and is determined by the master device.

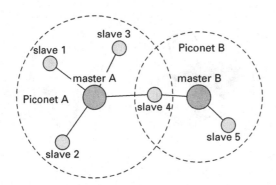

Figure 2.11 Piconets.

Bluetooth utilizes a slow hopping scheme, hopping in a pseudo-random fashion through 79 1-MHz channels. The frequency channels are located at $(2, 402 + k)$ MHz, with $k = 0, 1, \ldots, 78$. A Bluetooth piconet hops through 1600 different frequencies per second. Each frequency hop corresponds to one slot, with each slot lasting $1/1600 = 625$ μs. Each packet may be one, three, or five slots long. A frame consists of two packets, one packet being a transmit packet and the other packet being a receive packet.

A packet consists of an access code, a header, and a payload. The access code is 72 bits long and is used for clock synchronization, DC offset compensation, identification, and signaling. The header is 54 bits long and is used for addressing, identifying the packet type, controlling flow, sequencing to filter retransmitted packets, and verifying header integrity (ensuring that the header was not altered by another source). The payload is between zero and 2744 bits, depending on the type of packet. In packets that are one slot long, the payload is 240 bits long. In packets that are three slots long, the payload is 1500 bits long. In packets that are five slots long, the payload is 2744 bits long.

Each Bluetooth device includes a unique IEEE-type 48-bit address, called a Bluetooth device address, assigned to each Bluetooth device at manufacture, and a 28-bit clock. The clock ticks once every 312.5 μs, which corresponds to half the residence time in a frequency band when the radio hops at the rate of 1600 hops per second.

Bluetooth devices that are in communication with each other are organized into groups of two to eight devices called piconets, as shown in Figure 2.11. A piconet consists of a single master device and between one and seven slave devices. A device may belong to more than one piconet, but may be the master in no more than one piconet; thus a device may be a slave in two piconets or a master in one piconet and a slave in another piconet.

The slaves utilize the Bluetooth clock of the master to maintain time synchronization. The pseudo-random-hopping sequence is determined by the 48-bit Bluetooth device address of the master. The Bluetooth clock of the master clock determines the phase in the hopping pattern, thereby determining the particular frequency to be used at a particular time slot. Thus the communications channel in a particular piconet is fully identified by the master, and this communications channel serves to distinguish one piconet from another.

The master and the slaves alternate transmit opportunities according to a TDD scheme. According to this scheme, the master transmits on even-numbered time slots, as defined by the master's Bluetooth clock, while the slaves transmit on odd-numbered slots. A given slave may transmit only if the master has just transmitted to this slave.

To determine the presence and identities of other Bluetooth devices, Bluetooth devices engage in inquiry and page processes. The inquiry process is performed without knowledge of the identity or presence of other Bluetooth devices, whereas the paging process is performed with knowledge of the identity and presence of other Bluetooth devices.

During an inquiry process, a prospective master device makes its presence known by transmitting inquiry messages. Devices that are searching for inquiry messages respond with inquiry messages that contain their Bluetooth device addresses. After the master has acquired knowledge of the Bluetooth device address and presence of other Bluetooth devices within range, the master explicitly pages the other Bluetooth devices to join its piconet. Devices responding to the page will provide additional information, such as their clock phases, to the master.

Bluetooth devices have three low-power modes in which they reside when they are not in active communication. In sniff mode, a slave agrees with its master to listen for master transmissions periodically. In hold mode, a device agrees with another device in the piconet to remain silent for a given amount of time. A device that has gone into hold mode does not relinquish its temporary member address within the piconet. In park mode, a slave agrees with its master to park until further notice. In park mode, the slave device relinquishes its temporary member address within the piconet and periodically listens to transmissions from the master. The slave may be invited back to active communications by the master or may send a request to the master to be unparked.

Bluetooth devices typically provide link-layer security between any two Bluetooth radios. A challenge/response system, such as an E1 algorithm, is used for authentication. The authentication is based on a link key, which is a 128-bit shared secret between the two Bluetooth devices. The link key is generated by a challenge and response process between the two Bluetooth devices. Data sent between two Bluetooth devices may also be encrypted and may be ciphered with an E0 algorithm. An encryption key may be between 8 and 128 bits long and may be derived from the link key. The Bluetooth devices may use a configuration encryption key 0 to 16 bytes in length for key management and usage. The authentication and encryption keys may be generated with E2–E3 algorithms.

The specifications were formalized by the Bluetooth SIG. The SIG was formally announced on May 20, 1998. Today over 6000 companies worldwide are part of the SIG. It was established by Ericsson, Sony Ericsson, IBM, Intel, Toshiba and Nokia, and later joined by many other companies as associate or adopter members. Bluetooth is also known as IEEE 802.15.1 and the standards have the following versions:

- Bluetooth 1.0 and 1.0B
  Versions 1.0 and 1.0B had many problems, and the various manufacturers had great difficulties in making their products interoperable. 1.0 and 1.0B also had a mandatory Bluetooth Hardware Device Address (BD_ADDR) transmission in the handshaking

process, rendering anonymity impossible at a protocol level, which was a major setback for services planned to be used in Bluetooth environments.

- Bluetooth 1.1
  Many errors found in the 1.0B specifications were fixed. Support for nonencrypted channels and new features like the received signal-strength indicator (RSSI) were added.
- Bluetooth 1.2
  This version is backward compatible with 1.1 and the major enhancements include the adaptive frequency-hopping (AFH) spread spectrum, which improves resistance to RF interference by avoiding the use of crowded frequencies in the hopping sequence; higher transmission speeds in practice; extended synchronous connections (eSCOs), which improve the voice quality of audio links by allowing retransmissions of corrupted packets; host controller interface (HCI) support for three-wire universal asynchronous receiver/transmitter (UART), and HCI access to timing information for Bluetooth applications.
- Bluetooth 2.0
  This version is backward compatible with 1.x. The main enhancement is the introduction of an enhanced data rate (EDR) of 3.0 Mbps. This has the following effects: three times faster transmission speed up to 10 times in certain cases (up to 2.1 Mbps); 100-m range (depending on the class of the device); lower power consumption through a reduced duty cycle; simplification of multilink scenarios because of more available bandwidth; and further improved BER (bit error rate) performance.
- Bluetooth 2.1
  A draft version of the Bluetooth Core Specification Version 2.1 + EDR is now available from the Bluetooth website [384].

For a WPAN, besides Bluetooth technology, ZigBee is the name of a specification for a suite of high-level communication protocols using small, low-power digital radios based on the IEEE 802.15.4 standard for WPANs. ZigBee operates in the ISM radio bands; 868 MHz in Europe, 915 MHz in the United States, and 2.4 GHz in most jurisdictions worldwide. The technology is intended to be simpler and cheaper than other WPANs such as Bluetooth. The specification supports data transmission rates of up to 250 Kbps at a range of up to 30 m. ZigBee's technology is slower than 802.11b (11 Mbps) and Bluetooth (1 Mbps) but it consumes significantly less power.

2.6.2   Ultrawide Band

UWB is a technology for transmitting information spread over a large bandwidth (>500 MHz) that is able to share a spectrum with other users. In 2002, the FCC authorized the unlicensed use of UWB in 3.1 10.6 GHz. The intention is to provide an efficient use of scarce radio bandwidths while enabling high-data-rate personal-area network (PAN) wireless connectivity. Deliberations in the International Telecommunication Union Radiocommunication Sector (ITU-R) resulted in a report and recommendation on UWB in November of 2005.

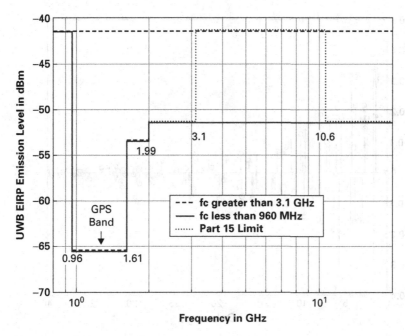

Figure 2.12 FCC UWB mask.

The FCC power spectral-density emission limit for UWB emitters operating in the UWB band is −41.3 dBm/MHz. This is the same limit that applies to unintentional emitters in the UWB band, the so-called Part 15 limit [371]. However, the emission limit for UWB emitters can be significantly lower (as low as −75 dBm/MHz) in other segments of the spectrum to prevent interference with other applications such as GPS. The FCC UWB spectrum mask is shown in Figure 2.12.

As in the IEEE 802.15.3a standard [77], the channel impulse response of the UWB can be modeled as the Saleh–Valenzuela (S-V) model [264]:

$$h(t) = \sigma^2 \sum_{c=0}^{C} \sum_{l=0}^{L} \alpha(c, l) \delta[t - T(c) - \tau(c, l)], \qquad (2.13)$$

where σ^2 represents total multipath energy, $\alpha(c, l)$ is the gain of the lth multipath component in the cth cluster, $T(c)$ is the delay of the cth cluster, and $\tau(c, l)$ is the delay of the lth path in the cth cluster relative to the cluster-arrival time. The cluster arrivals and the path arrivals within each cluster are modeled as a Poisson distribution with rate Λ and rate λ (where $\lambda > \Lambda$), respectively. $\alpha(c, l)$ are modeled as zero-mean, complex Gaussian random variables with variances [77] $\Omega(c, l) = E[|\alpha(c, l)|^2] = \Omega(0, 0) \exp\left(-\frac{T(c)}{\Gamma} - \frac{\tau(c,l)}{\gamma}\right)$, where $E[\cdot]$ is the expectation operation, $\Omega(0, 0)$ is the mean energy of the first path of the first cluster, Γ is the cluster-decay factor, and γ is the ray-decay factor. The total energy contained in terms $\alpha(c, l)$ is normalized to unity, i.e., $\sum_{c=0}^{C} \sum_{l=0}^{L} \Omega(c, l) = 1$. The channel parameters corresponding are specified in [77] for four environment categories, CM1–CM4:

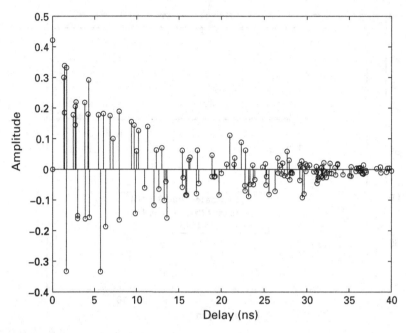

Figure 2.13 Typical UWB channel.

- Category CM1 (for Channel Model 1) consists of LOS paths in residences with $0 < d < 4$ m (RMS delay spread 5 ns);
- CM2 consists of NLOS paths in residences with $0 < d < 4$ m (RMS delay spread 8 ns);
- CM3 consists of NLOS paths in residences with 4 m $< d < 10$ m (RMS delay spread 14 ns);
- CM4 consists of NLOS paths with extreme delay spreads (RMS delay spread 25 ns).

In Figure 2.13, we show a typical UWB channel response over time.

The ability of UWB technology to provide significantly high data rates within short ranges has made it an excellent alternative for Bluetooth for the PHY layer of the IEEE 802.15.3a standard for WPANs. However, as with 802.11 standards, two opposing groups of UWB developers are competing over the IEEE standard. The two competing technologies are single-band UWB and multiband UWB. The single-band technique, backed by Motorola/XtremeSpectrum, supports the idea of an impulse radio that occupies a wide spectrum. The multiband approach divides the available UWB frequency spectrum into multiple smaller and nonoverlapping bands with bandwidths greater than 500 MHz to obey the FCC's definition of UWB signals. The multiband approach is supported by several companies, including Staccato Communications, Intel, Texas Instruments, General Atomics, and Time Domain Corporation.

For the single-band UWB, the most popular proposal is the direct-sequence (DS)-UWB, which uses a combination of a single-carrier spread-spectrum design and wide coherent bandwidth. Unlike conventional wireless systems, which use narrowband-modulated carrier waves to transmit information, DS-UWB transmits data by pulses of

Table 2.7 Rate-Dependent Parameters

Data rate (Mbps)	Modulation	Coding rate	Conjugate symmetric inputs to IFFT	Time-spreading factor
53.3	QPSK	1/3	Yes	2
55	QPSK	11/32	Yes	2
80	QPSK	1/2	Yes	2
106.7	QPSK	1/3	No	2
110	QPSK	11/32	No	2
160	QPSK	1/2	No	2
200	QPSK	5/8	No	2
320	QPSK	1/2	No	1
400	QPSK	5/8	No	1
480	QPSK	3/4	No	1

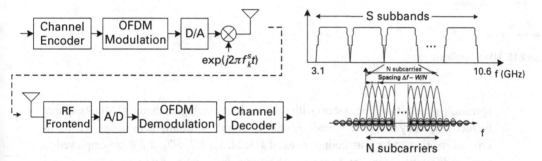

Figure 2.14 Multiband UWB system.

energy generated at very high rates: in excess of 10^9 pulses per second, providing support for data rates of 28, 55, 110, 220, 500, 660, and 1320 Mbps. A fixed UWB chip rate in conjunction with variable-length spreading codewords enables this scalable support.

For the multiband UWB, as shown in Figure 2.14, the available UWB spectrum, from 3.1 to 10.6 GHz, is divided into $S = 14$ subbands. Each subband occupies a bandwidth of at least 500 MHz in compliance with the FCC regulations. The UWB system employs OFDM with $N = 128$ subcarriers, which are modulated by QPSK. At each OFDM symbol period, the modulated symbol is transmitted over one of the S subbands. These symbols are time interleaved across subbands. Different bit rates are achieved by use of different channel-coding, frequency-spreading, or time-spreading rates. Frequency-domain spreading is obtained by choosing conjugate symmetric inputs to the IFFT (inverse fast Fourier transformation), whereas time-domain spreading is achieved by repeating the same information in an OFDM symbol on two different subbands [18]. The receiver combines the information transmitted via different times or frequencies to increase the signal-to-noise ratio (SNR) of received data.

As listed in Table 2.7, the multiband UWB system provides data rates ranging from 53.3 to 480 Mbps. For rates no higher than 80 Mbps, both time and frequency spreadings are performed, yielding an overall spreading gain of four. In the case of rates between 106.7 and 200 Mbps, only time-domain spreading is utilized, which results in an overall

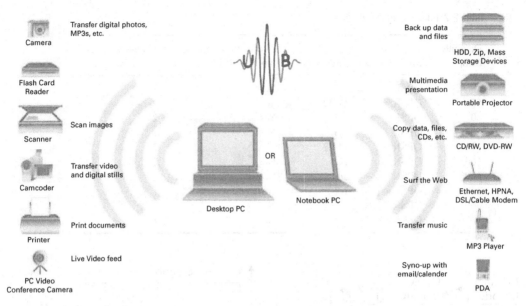

Camera — Transfer digital photos, MP3s, etc.

Flash Card Reader

Scanner — Scan images

Camcoder — Transfer video and digital stills

Printer — Print documents

PC Video Conference Camera — Live Video feed

Desktop PC   OR   Notebook PC

Back up data and files — HDD, Zip, Mass Storage Devices

Multimedia presentation — Portable Projector

Copy data, files, CDs, etc. — CD/RW, DVD-RW

Surf the Web — Ethernet, HPNA, DSL/Cable Modem

Transfer music — MP3 Player

Syno-up with email/calender — PDA

Figure 2.15 UWB applications.

spreading gain of two. The system with information rates higher than 200 Mbps exploits neither frequency nor time spreading, and its overall spreading gain is one. Forward-error-correction codes with coding rates of 1/3, 11/32, 1/2, 5/8, or 3/4 are employed to provide different channel protections with various information data rates.

Because of the extremely low emission levels currently allowed by regulatory agencies, UWB systems tend to be short range and high speed. High-data-rate UWBs can enable wireless monitors, the efficient transfer of data from digital camcorders, wireless printing of digital pictures from a camera without the need for an intervening personal computer, and the transfer of files among cell phone handsets and other handheld devices like personal digital audio and video players. Some applications are shown in Figure 2.15. UWB is also used in "see-through-the-wall" precision radar imaging technology, precision positioning and tracking (using distance measurements between radios), and precision time-of-arrival-based localization approaches.

2.7   Wireless Ad Hoc Networks

An ad hoc network is an autonomous collection of mobile users that communicate over bandwidth-constrained wireless links. The network is decentralized, in which all network activity, including discovering the topology and delivering messages, must be executed by the nodes themselves. Ad hoc networks need efficient distributed algorithms to determine network organization, link scheduling, and routing. For a special case of an ad hoc network, Mobile Ad Hoc Networks (MANETs), because the nodes are mobile, the network topology may change rapidly and unpredictably over time.

The first generation of ad hoc networks started about 1970, when Packet Radio Networks (PRNETs) were proposed by the U.S. Defense Advanced Research Projects Agency (DARPA) for multihop networks in a combat environment, and Areal Locations of Hazardous Atmospheres (ALOHA) was proposed in Hawaii for distributed channel access management. The second generation of ad hoc networks emerged in the 1980s, when the ad hoc network systems were further enhanced and implemented as a part of the Survivable Adaptive Radio Networks (SURAN) program. SURAN provided a packet-switched network to the mobile battlefield in an environment without infrastructure so as to be beneficial in improving the performance of radios by making them smaller, cheaper, and resilient to electronic attacks. In the 1990s, the concept of commercial ad hoc networks arrived with notebook computers and other feasible communications equipment. For example, the IEEE 802.11 subcommittee adopted the term "ad hoc networks."

The advantages of ad hoc networks are the easiness and speed of deployment, which are important requirements for military applications. For civil applications, ad hoc networks decrease dependence on expensive infrastructures. The set of applications for ad hoc networks is diverse, ranging from small, static networks that are constrained by power sources, to large-scale, mobile, highly dynamic networks. Some typical applications are PAN emergency operations such as policing and fire fighting, civilian environments such as taxi networks, and military use on the battlefields. One example of an ad hoc network is shown in Figure 2.16.

In contrast to the traditional wireless network with an infrastructure, an ad hoc network needs its own design requirements so as to be functional. We list some important aspects as follows:

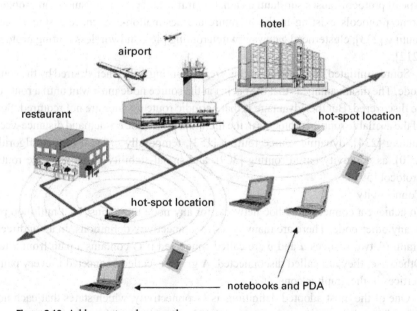

Figure 2.16 Ad hoc network example.

- Distributed Operation and Self-Organization

  No node in an ad hoc network can depend on a network in the background to support the basic functions like routing. Instead, these functions must be implemented and operated efficiently in a distributed manner. Moreover, in events such as topology changes that are due to mobility, the network can be self-organized to adapt to the changes.

  In addition, because the ad hoc nodes might belong to different authorities, they might not be necessary to cooperate to fulfill the network functions. However, this non-cooperation can cause severe network breakdown. Motivating distributed autonomous users is an important research and design topic. Traditionally, pricing anarchy is employed by use of the distributed control theory. Later in this book, we study how to explore the other methods, such as game theory, to motivate users' cooperative behaviors.

- Dynamic Routing

  For a MANET, the routing problem between any pair of nodes is challenging because of the mobility of the nodes. The optimal source-to-destination route is time variant. Moreover, compared with the traditional network in which the routing protocols are proactive, the ad hoc dynamic routing protocols are reactive. The routes are determined only when the source requests the transmission to the destination. There are two types of ad hoc dynamic routing protocols: table-driven routing protocols and source-initiated on-demand routing protocols.

  The table-driven routing protocols require each node to maintain one or more tables to store routing information. The protocols rely on an underlying routing table update mechanism that involves the constant propagation of routing information. Packets can be forwarded immediately because the routes are always available. However, this type of protocol causes substantial signaling traffic and power-consumption problems. Some protocols existing in the literature are destination-sequenced distance-vector routing [233], clusterhead gateway switch routing [46], and wireless routing protocols [212].

  Source-initiated on-demand routing creates routing only when desired by the source node. The disadvantage is that the packet at the source node must wait until a route can be discovered. But the advantage is that periodic route updates are not required. Some of the available routing protocols in the literature are ad hoc on-demand distance-vector routing [234], dynamic source routing [153], temporally ordered routing algorithm [230], associativity-based routing [309], and signal-stability-based adaptive routing protocol [59].

- Connectivity

  To achieve a connected ad hoc network, for any node there must be a multihop path to any other node. There are many types of connectivity definitions. In an undirected graph G, two vertices u and v are called connected if G contains a path from u to v. Otherwise, they are called disconnected. A graph is called connected if every pair of vertices in the graph is connected.

  One of the most adopted definitions is k-connectivity, which states that each node can at least connect to the rest of network if $k - 1$ of its neighbor nodes are destroyed.

DEFINITION 1 *A graph G with edge set $V(G)$ is said to be k-connected if $G \backslash Y$ is connected for all $Y \subseteq V(G)$ with $|Y| < k$. In other words, a graph is k-connected if the graph remains connected when fewer than k vertices are deleted from the graph.*

If a graph G is k-connected, and $k < |V(G)|$, then $k \leq \Delta(G)$, where $\Delta(G)$ is the minimum degree of any vertex $v \in V(G)$.

The following theorem, Menger's theorem, a special case of the max-cut–min-flow theorem, states how to calculate k for k-connectivity:

THEOREM 1 *Let G be a finite undirected graph and i and j two nonadjacent vertices. Then the size of the minimum vertex cut for i and j (the minimum number of vertices whose removal disconnects i and j) is equal to the maximum number of pairwise vertex-independent paths from i to j.*

- Mobility

To test a new protocol for a MANET, it is important to use a mobility model that accurately represents the mobile users using the protocol, so that the performances of practical implementation match the simulation results. There are two types of mobility models, traces and synthetic models. For traces, the mobility patterns are observed in real-life systems, so that accurate information is provided. However, traces are limited only for an existing environment. For an unknown environment, accurate synthetic models are necessary.

A synthetic model tries to simulate the real movement of mobile users. In [40], several mobility models are discussed and explained. If the different mobile users are moving randomly to each other, some mobility mobiles are listed as follows:

1. Random-walk mobility model: a simple model based on random directions and speeds.
2. Random-waypoint mobility model: a model including pause times between changes in directions and speeds.
3. Random-direction mobility model: a model forcing mobile users to travel to the edge of the simulation area before changing directions and speeds.
4. Boundless simulation area mobility model: a model converting a two-dimensional (2D) rectangular simulation area into a torus-shaped simulation area.
5. Gauss–Markov mobility model: a model using a set of parameters to change the degree of randomness in mobility patterns.
6. Probabilistic version of the random-walk mobility model: a model with a probability to determine the next position of mobile users.
7. City-section mobility model: a model in which movement is on the streets of a city.
8. Random-trip mobility model [177]: a model that contains as special cases the random waypoint on convex or nonconvex domains, random walk, billiards, city section, space graph, and other models.

If the mobile users are moved in groups, some mobility models are listed as follows:

1. Exponential correlated random mobility model: a model using a motion function to create movement.

2. Column mobility model: a model in which the mobile users form a line and are uniformly moving forward in a certain direction.
3. Nomadic community mobility model: a model in which mobile users move together from one position to the other.
4. Pursue mobility model: a model in which mobile users follow a given target.
5. Reference-point group mobility model: a model in which group movements are based on the path traveled by a logical center.

- Other issues such as lower power, security, and localization are discussed in the next section together with sensor networks.

Finally, we discuss two ad hoc standards in 802.11 for WLANs and 802.15 for WPANs. One future design goal of ad hoc networks is to let mobile users be able to form connections and perform basic functions. These mobile users belong to different types of networks, as shown in Figure 2.1, such as cellular, WiFi, and WPAN.

Most installed WLANs today utilize the "infrastructure" mode that requires the use of one or more APs. With this configuration, the AP provides an interface to a distribution system (e.g., Ethernet), which enables wireless users to utilize corporate servers and Internet applications. As an option, the 802.11 standard specifies the "ad hoc" mode, which allows the radio network interface card (NIC) to operate in what the standard refers to as an independent basic service set (IBSS) network configuration. With an IBSS, there are no APs. User devices communicate directly with each other in a peer-to-peer manner.

For WPAN applications such as Bluetooth, the ad hoc network is set up by forming piconets. Within each piconet, only one master device and possibly several slave devices form connections. A slave device can belong to different piconets and serve as a connection between piconets. The major functionality of the piconets are piconet forming and maintenance, packet forwarding, and intrapiconet and interpiconet scheduling.

2.8   Wireless Sensor Networks

A wireless sensor network (WSN) is a wireless network consisting of spatially distributed autonomous devices using sensors to cooperatively monitor physical or environmental conditions, such as temperature, sound, vibration, pressure, motion, or pollutants, at different locations. The goals and tasks of sensor networks are to determine the value of some parameter at a given location, to detect the occurrence of events of interest and estimate parameters of the detected events, to classify a detected object, or to track an object. The development of WSNs was originally motivated by military applications such as battlefield surveillance. However, WSNs are now used in many civilian application areas, including environment and habitat monitoring, healthcare applications, home automation, and traffic control. Some examples and applications are shown in Figure 2.17 and discussed in detail as follows:

- Military sensor networks to detect and gain as much information as possible about enemy movements, explosions, and other phenomena of interest. Sensor networks

(a) Military Sensor

(b) Traffic Sensor

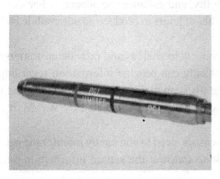

(c) Tsunami Sensor

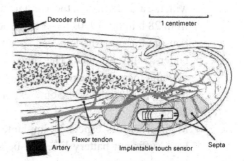

(d) Implanted Medical Sensor

Figure 2.17 Sensor examples.

to detect and characterize chemical, biological, radiological, nuclear, and explosive (CBRNE) attacks. For example, military sensors can track an enemy tank as it moves through the geographic area covered by the network.

- Sensor networks to detect and monitor environmental changes such as in plains, forests, and oceans. For example, for the nations that are exposed to seismic and tsunami risk, it is a critical task of designing, validating, and implementing cost-effective sensor networks for monitoring and transmitting scientific information regarding tsunami risk to action agencies.

- Wireless traffic sensor networks to monitor vehicle traffic on highways or in congested parts of a city. By far the most common technique for a traffic sensor is the inductive loop that is simply a coil of wire embedded in the road's surface. The sensor constantly tests the inductance of the loop in the road, and, when the inductance rises, there is a car waiting. In the traffic sensor network, the sensors can also detect a vehicle moving through an intersection and estimate the speed and direction of the vehicle. Another application is wireless parking lot sensor networks to determine which spots are occupied and which are free.

It is worth mentioning the unmanned driving supported by sensor networks in the future. The DARPA Grand Challenge since 2004 is a prize competition for driverless cars, sponsored by DARPA, the central research organization of the U.S. Department of Defense. Some laser-radar (LADAR) sensors exist today that have successfully been implemented and demonstrated to provide somewhat reliable obstacle detection that can be used for path planning and path selection. The sensors implanted into the highway can also help unmanned ground vehicles.

- Wireless surveillance sensor networks provide security in shopping malls, parking garages, and other civil facilities. The surveillance tasks can include maintaining constant sensor coverage of a given area and informing operators of changes within the area; discerning targets and identifying the potential target from the viewable region; and maintaining targets.

- Manufacturing sensors can facilitate the the monitoring and control process, which can reduce the cost, improve the flexibility, and enhance the accuracy. For example, with the aid of sensors, the average number of hours to produce an automobile reduced from 28 h in the 1990s to 12 h in 2005.

- Sensors for supermarkets can speed products to shelves and provide customers with better-quality produce. The sensors not only can provide information regarding the sale and location of certain product, but also can maintain quality produce and fresh foods on store shelves. Some experiments have been conducted by Wal-Mart since 2006.

- Medical sensors, especially implanted sensors, need to constantly monitor the patients, have sufficient battery life, and be able to transmit the sensed information out of a body via wireless channels.

In Figure 2.18, we show the typical structure of sensor networks. In addition to one or more sensors, each node in a sensor network is typically equipped with a radio transceiver or other wireless communication device, a small microprocessor, some memory, and an energy source, usually a battery. For the sensors, different sensing applications can include temperature, light, humidity, pressure, accelerometers, magnetometers, chemical, acoustics, and image/video. The microprocessor has a significant constraint in

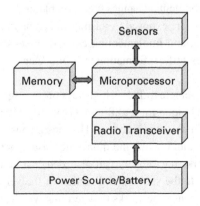

Figure 2.18 Sensor structure.

computational power. Currently, the devices typically have a component-based embedded operating system. The available memory is also very limited. Current radio transceivers for sensor networks are low rate and short range. Some sensors can be powered by a wired power source, whereas most of the widely deployed sensors are powered by a battery. Exchanging the energy-depleting sensors is a challenging job, so power saving is critical in the design of such WSNs.

The size of a single sensor node can vary from shoebox-sized nodes down to devices the size of a grain of dust. The cost of sensor nodes is similarly variable, ranging from hundreds of dollars to a few cents, depending on the size of the sensor network and the complexity required of individual sensor nodes. Size and cost constraints on sensor nodes result in corresponding constraints on resources such as energy, memory, computational speed, and bandwidth. Basically, to design a WSN, the following requirements should be considered [385]:

1. Large number of (mostly stationary) sensors: Aside from the use of mobile, unmanned, robotic sensors in military operations, most nodes in a smart sensor network are stationary. Networks of 10,000 or even 100,000 nodes are envisioned, so scalability is a major issue.

2. Low-energy use to extend sensor network lifetime: Because in many applications the sensor nodes will be placed in a remote area, service of a node may not be possible. In this case, the lifetime of a node may be determined by the battery life, thereby requiring the minimization of energy expenditure. There are many definitions of sensor network lifetime:
   - time of the first node failure;
   - time of a certain fraction of surviving nodes in a network;
   - mean expiration time;
   - in terms of packet-delivery rate;
   - in terms of the number of active flows;
   - time to the first loss of coverage;
   - time until the network is first partitioned;
   - time until the performance of a certain task falls below a criterion.

3. Network self-organization: Given the large number of nodes and their potential placement in hostile locations, it is essential that the network be able to self-organize. In other words, manual configuration is not feasible. Moreover, nodes may fail (either from lack of energy or from physical destruction), and new nodes may join the network. Therefore the network must be able to periodically reconfigure itself so that it can continue to function. Individual nodes may become disconnected from the rest of the network, but a high degree of connectivity must be maintained.

4. Collaborative signal processing: Yet another factor that distinguishes these networks from MANETs is that the end goal is detection/estimation of some events of interest, and not just communications. To improve the detection/estimation performance, it is often quite useful to fuse data from multiple sensors. This data fusion requires the transmission of data and control messages, and so it may put constraints on the network architecture.

5. Querying ability: A user may want to query an individual node or a group of nodes for information collected in the region. Depending on the amount of data fusion performed, it may not be feasible to transmit a large amount of the data across the network. Instead, various local sink nodes will collect the data from a given area and create summary messages. A query may be directed to the sink node nearest the desired location.

In practice, there are some other design issues [168], such as how to deploy the sensor networks, how to locate a specific sensor, how to shut down and reactivate a sensor to save energy, how to route the information back to data collecting points, how to reduce the packet forwarding loads by data fusion, and finally how to have secure sensor networks.

- Sensor network deployment
  The problem is to select the locations to place the sensor networks, give a particular application context, an operational region, and a set of wireless sensor devices. The sensors can be deployed in a structured sense or in a randomly scattering manner. The density of sensors is judged by the robustness and cost of the networks.

  Usually, the sensor networks have two types of sensors. The first type is the low-capability sensor that is in charge of collecting data, and the other type is the clusterhead or data sink that is more powerful in computation and data transmission. The network topology between these two types of sensors can be a star-connected single hop, a multihop mesh/grid, or a multiple-tier hierarchical cluster.

  Because the transmit power is bounded, the sensor can reach the other sensor with limited distance. This fact causes connectivity problems for sensor networks. Moreover, the sensing range is also restrained. The coverage of the sensing area is a function of the density of the sensors. There are different types of coverage metrics such as k-coverage, minimum coverage, and maximal breach distance [202].

- Localization
  Localization determines the location of a certain event sensed by the sensor. This location information can provide a location stamp over the event, track the monitored object, determine the coverage, form the cluster, facilitate routing, and perform efficient querying. Even though the information can be obtained by GPS, cost and indoor environments prohibit sensors from being equipped with GPS.

  Nevertheless, the task of localization captures multiple aspects of sensor networks: A PHY layer imposes measurement challenges that are due to multipaths, shadowing, sensor imperfections, changes in propagation properties. Extensive computation is necessary for many formulations of localization problems. Moreover, the problems sometimes should be solved in a distributed manner or on a memory-constrained processor. Next, for networking and coordination issues, sensor nodes have to collaborate and communicate with each other to know the topology of the whole networks. Finally, for system integration issues, it is challenging to integrate location services with other applications.

  There are several types of localization mechanisms. First, an active localization system sends signals to localize a target. The examples are radar or lidar (ladar).

Second, in cooperative localization, the target cooperates with the system. For example, the target emits a signal with known characteristics, and then the system deduces location by detecting a signal. Third, a passive localization system deduces location from observation of signals that are "already present." The example technique is to use the geometric methods to calculate the location by measuring the signal strength over the receivers in different locations. Finally, a blind localization system deduces the location of a target without a priori knowledge of its characteristics.

- Time synchronization

Localization provides the sensor networks with spatial information, whereas accurate time synchronization is also very essential as in the following examples: First, because the delay for the information to the sink is unpredictable, each sensor needs to have a consistent time stamp for the message. This is more important for some types of data such as tsunami alarms, because the time information provides many scientific clues. For localization, the transmitter and receiver need to have a synchronized time so that the time-of-flight can be calculated. For multiple access such as TDMA-based schemes, each sensor needs to transmit at the exact time slots. For sleep scheduling, the energy is saved by turning on and turning off the sensors at a certain time.

Accurate time can be obtained by a GPS signal. But this approach is very expensive. Quartz crystal oscillators can provide accuracy of about several microseconds. For better accuracy, some techniques such as a phase loop lock need to be implemented to synchronize the clocks.

- Sleeping mechanism

In most sensors, the primary source of power consumption is the radio for transmitting, receiving, and listening. If the sensors wake up only during the time when the radio is active and sleep in the remaining time, the limited energy can be conserved and the lifetime of sensor networks can be prolonged.

However, the sleep-and-wake-up mechanism causes other design problems. First, there is a trade-off between the delay of information and energy consumption. Moreover, the design of the MAC layer multiple-access protocol needs to consider the wake-up time. Further, the transmitter and receiver should be synchronized to wake up at the same time. Finally, the fairness issue needs to be considered so that some sensors will not be overloaded to be energy depleted too early.

- Energy-efficient routing

Because energy is a major concern for the design of WSNs, selecting energy-efficient routes from the data-collecting sensors to the data sink can significantly improve the network lifetime. In addition, when multiple routes are considered, individual optimal energy-efficient routes are not optimal, in the sense that some sensors on the critical paths might be depleted first. So joint optimization is necessary.

In addition to energy concern, routing protocols also should take into consideration the latency that is due to the sleeping mechanism of sensors. The routing protocols should also consider the data fusion/aggregation. Finally, for large sensor networks, scalability is an important issue. For the situations in which the sensors are mobile or can join/leave the network frequently, adaptive ability is also a design challenge.

- Fusion/aggregation

  Fusion/aggregation is a process dealing with the association, correlation, and combination of data and information from single and multiple sources to achieve refined position and identity estimates for observed entities and to achieve complete and timely assessments of situations and threats, and their significance. In WSNs, especially large ones, it is impossible and energy inefficient to gather the information to make the decision. Instead, along the path to the data sink, a data-fusion node collects the results from multiple nodes, it fuses the results with its own based on a decision criterion, and then it sends the fused data to another node/base station. By doing this, the traffic load can be greatly reduced and energy can be conserved.

  There are two forms of data fusion/aggregation. In the first form, data from different node measurements are combined together to form larger packets. It is simple to implement, but has a higher computational burden, a higher communication burden, and a larger training data requirement. The second type is decision fusion, in which the decisions (hard or soft) based on node measurements are combined. The decision fusion solves the problems of the first form. The decision can be made by mechanisms such as voting. For example, a fusion node arrives at a consensus by a voting scheme: majority voting, complete agreement, and weighted voting. Other fusion decision algorithms include the probability-based Bayesian model and stack generalization.

  For sensor networks, there are different fusion architectures from the data-collecting sensors to the data sink. In a centralized architecture, a central processor fuses the reports collected by all other sensing nodes. The centralized one has the advantage that an erroneous report can be easily detected and is simple to implement. On the other hand, it has the disadvantage that it is inflexible to sensor changes and the workload is concentrated at a single point. In a decentralized architecture, data fusion occurs locally at each node on the basis of local observations and the information obtained from neighboring nodes. There is no central processor node. The advantages are scalable and tolerant to the addition or loss of sensing nodes or dynamic changes in the network. In the hierarchical architecture, nodes are partitioned into hierarchical levels. The sensing nodes are at level 0 and the data sink is at the highest level. Reports move from the lower levels to higher ones. This architecture has the advantage that the workload is balanced among nodes.

- Security

  Because sensor networks may interact with sensitive data and/or operate in hostile unattended environments, like military sensors, it is important for security to be addressed from the system design. Moreover, because of inherent resource and computing constraints, security in sensor networks poses more challenges than traditional network/computer security. Possible security attacks include denial-of-service attack, Sybil attack, traffic-analysis attack, node replica attack, privacy attack, physical attack, and collusion attack. In the literature, there are some defensive mechanisms, such as key cryptography and trust management. The security issue is beyond the scope of this book. A good survey of security issues in WSNs can be found in [322].

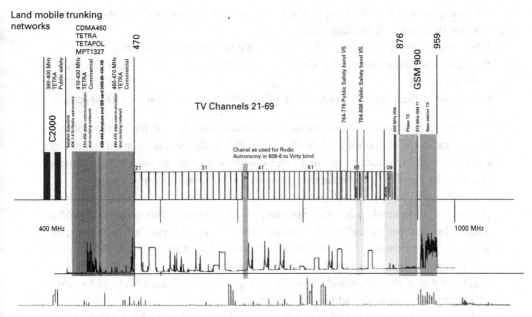

Figure 2.19 Spectrum usage.

2.9   Cognitive Radios

The Federal Radio Act under the U.S. FCC allows predetermined users the right to transmit at a given frequency. Nonlicensed users are regarded as "harmful interference," and in most cases sidebands were implemented to ensure that interference is not an issue. As technology advanced, higher-frequency bands were sold at auction, bringing considerable revenue to the government. For example since the 1994 PCS (personal communications service) auction, over U.S.$30 billion have been generated. As the demands for wireless communication become more and more pervasive, the wireless devices must find a way to obtain the right to transmit at frequencies in a limited radio band. However, there exist a large number of frequency bands that have considerable, and sometimes periodic, dormant time intervals. For example, TV stations do not work at night. As shown in Figure 2.19, there exist some spectrum holes at a given time over different spectrum bands. So there is a dilemma that, on the one hand, mobile users have no spectrum to transmit, whereas, on the other hand, some spectra are not fully utilized.

To cope with this dilemma, cognitive radio is a paradigm for wireless communication in which either a network or a wireless node changes its transmission or reception parameters to communicate efficiently without interfering with the licensed users. This alteration of parameters is based on the active monitoring of several factors in the external and internal radio environments, such as RF spectra, user behaviors, and network states.

Depending on the set of parameters taken into account in deciding on transmission and reception changes, and for historical reasons, we can distinguish certain types of cognitive radio:

- Full cognitive radio ("Mitola radio"): Every possible parameter observable by a wireless node or network is taken into account.
- Spectrum-sensing cognitive radio: Only the RF spectrum is considered.
- Licensed-band cognitive radio: Cognitive radio is capable of using bands assigned to licensed users.
- Unlicensed-band cognitive radio: Can utilize only unlicensed parts of RF spectrum.

Cognitive radio can be designed as an enhancement layer on top of the software-defined radio (SDR) concept. A SDR system is a radio communication system that can tune to any frequency band and receive any modulation across a large frequency spectrum by means of programmable hardware that is controlled by software.

A SDR performs significant amounts of signal processing in a general-purpose computer or a reconfigurable piece of digital electronics. The goal of this design is to produce a radio that can receive and transmit a new form of radio protocol just by running new software. The hardware of a SDR typically consists of a superheterodyne RF front end, which converts RF signals from (and to) analog intermediate-frequency (IF) signals, and analog-to-digital converters and digital-to-analog converters, which are used to convert a digitized IF signal from and to analog form, respectively.

Software radios have significant utility for military and cell phone services, both of which must serve a wide variety of changing radio protocols in real time. A SDR can currently be used to implement simple radio modem technologies. In the long run, the SDR is expected by its proponents to become the dominant technology in radio communication. It is the enabler of the cognitive radio.

One example of a cognitive structure is shown in Figure 2.20. The cognitive transmitter and receiver can adapt to 3G WiFi and WPAN networks. By sensing the available spectrum, the cognitive radio can adapt to the most suitable available communication links. For example, if the user is located at home, the user can communicate via Bluetooth; if the user travels to the airport, WiFi communication can be available; if the user drives on the highway, the cellular phone system can provide reliable communication links. Another analogous interpretation for cognitive radio is as follows: Licensed users can legally use a spectrum, much as high-occupancy vehicles (HOV) can drive on the HOV lanes. However, the HOV lanes are not always occupied. So, other than during rush hours, other vehicles can also drive on the HOV lanes. In this sense, the cognitive radios are similar to the other non-HOV vehicles.

Two major objectives of cognitive radio are to reliably communicate whenever and wherever and to efficiently utilize the radio spectrum. To achieve these objectives, three fundamental cognitive tasks [131] for cognitive radio must be fulfilled, as follows:

1. radio-scene sensing analyzes the interferences and detects the spectrum holes;
2. spectrum analysis, such as channel-state estimation and predictions of channel capacity;
3. transmitting power control and dynamic spectrum management.

In Figure 2.21, we show the cognitive cycle, in which three tasks interact with each to handle the outside world so that the best strategy can be calculated and implemented.

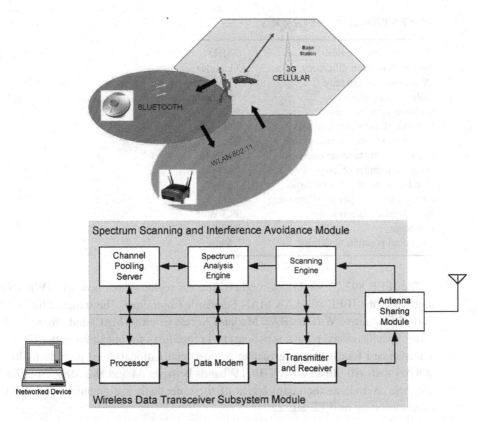

Figure 2.20 Cognitive radio structure.

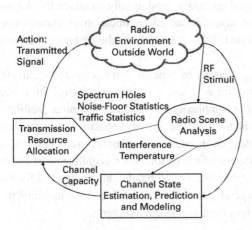

Figure 2.21 Cognitive cycle [4].

Table 2.8 WRAN coverage and capacity

RF channel bandwidth	6 MHz
Average spectrum efficiency	3 bit/s/Hz
Downlink user capacity	1.5 Mbps
Uplink user capacity	384 kbps
Oversubscription ratio	50
Number of users per downlink	600
Minimum number of users	90
Assumed early take-up rate	3 bit/s/Hz
Potential number of users	1800
Number of users per household	2.5
Number of users per coverage area	4500
WRAN base-station power	98.3 W
Coverage	30.7 km
Minimum population density	1.5/km^2

The IEEE 802.22 Working Group on wireless regional-area networks (WRANs) is a group of the IEEE 802 LAN/MAN Standards Committee. The standard for WRAN, Part 22 [Cognitive Wireless RAN Medium Access Control (MAC) and Physical (PHY) Layer Specifications] regulates policies and procedures for operation in the TV bands. The standard focuses on constructing a consistent, national, fixed point-to-multipoint WRAN that will utilize UHF/VHF TV bands between 54 and 862 MHz. Specific TV channels as well as the guard bands of these channels are planned to be used for communication in IEEE 802.22.

The IEEE, together with the FCC, is pursuing a centralized approach for available spectrum discovery. Specifically, each AP would be armed with a GPS receiver that would allow its position to be reported. This information would be sent back to centralized servers (in the United States these would be managed by the FCC), which would respond with the information about available free TV channels and guard bands in the area of the AP. Other proposals would allow local spectrum sensing only, in which the AP would decide by itself which channels are available for communication. A combination of these two approaches is also envisioned. Table 2.8 describes the coverage and capacity of IEEE 802.22 WRANs.

Overall, cognitive radio can bring a variety of benefits: For a regulator, it can significantly increase spectrum availability for new and existing applications. For a license holder, cognitive radio can reduce the complexity of frequency planning, facilitate the secondary spectrum market agreements, increase system capacity through access to more spectra, and avoid interference. For an equipment manufacturer, cognitive radio can increase demands for wireless devices. Finally, for a user, cognitive radio can bring more capacity per user, enhance interoperability and bandwidth-on-demand for Public Safety and Emergency Response operations, and provide ubiquitous mobility with a single-user device across disparate spectrum-access environments.

3   Power Control

3.1   Introduction

In wireless communications, transmission power is an important resource. Power control, also known as transmit power control, is a significant design problem in modern wireless networks. Power control comprises the techniques and algorithms used to manage and adjust the transmitted power of base stations and handsets. Power control also serves several purposes, including reducing cochannel interference (CCI), managing data quality, maximizing cell capacity, minimizing handset mean transmit power, etc. In this chapter, we illustrate the basic power-control problems and some possible solutions.

In wireless communication systems, two important detrimental effects that decrease network performance are the time-varying nature of the channels and CCI. The average channel gain is primarily determined by large-scale path-loss factors such as propagation loss and shadowing. The instant channel gain is also affected by small-scale fading factors such as multipath fading. Because the available bandwidth is limited, the channels are reused for different transmissions. This channel reuse increases the network capacity per area, but, on the other hand, it causes CCI. Because of these effects, the signal-to-interference-noise ratio (SINR) at a receiver output can fluctuate of the order of tens of decibels. Power control is an effective resource-allocation method to combat these detrimental effects. The transmitted power is adjusted according to the channel condition so as to maintain the received signal quality. Power control is no longer one user's problem, because a user's transmit power causes other users' interferences.

The *objective* of power control in wireless networks is to control the transmit power to guarantee a certain link quality and to reduce CCI. To maintain the link quality, it is necessary to keep the SINR above a threshold that is called the *minimum protection ratio*. In power control, transmitted power is constantly adjusted. Such a process improves the qualities of weak links, whereas, at the same time, it increases CCI during deep fading.

The existing power-control schemes can be classified as centralized or distributed, forward (down) link or reverse (up) link, open loop or close loop, etc. The power-control problem is not a trivial problem, in that there are trade-offs and practical constraints:

- There is a trade-off for each link's power. The increase in transit power will increase the link's SINR, but, on the other hand, the increased power interferes with other links and causes degradation of other links.

- For the uplink or multicell case, power control should be implemented in a distributed way. All users should use only their local information to control the power so that the limited power resources can be cooperatively utilized to improve the system performance while maintaining the users' QoS.
- It is necessary to have a simple implementation of the system without causing too much communication overhead and burden.
- The convergence speed for a power-control algorithm should be fast enough compared with the changing speed of the fading channels.
- The power-control scheme should be able to accommodate heterogeneous QoS requirements.

So the efficient management of the power resource has become an important research issue in recent years.

3.2   Basic Power-Control System Models

In this section, we discuss some different scenarios for power control such as single-cell case and multicell case, and uplink and downlink. We explain the basic problems and the differences among various scenarios. Then we discuss feasibility of power control. Finally, we give a two-user example to explain the basic ideas.

In Figure 3.1, we illustrate an example of multicell communication networks. A set of K transmitter–receiver pairs share the same channel. The link gain between transmitter i and receiver j is denoted by G_{ij}, and the ith transmitter power is P_i. For an isotropic antenna with unity gain in all directions, the signal power received at receiver i from transmitter j is $G_{ji} P_i$. It is assumed that transmitter i communicates with receiver i. Hence the desired signal at receiver i is equal to $G_{ii} P_i$, whereas the interfering signal power from other transmitters to receiver i is $I = \sum_{j \neq i} G_{ji} P_j$. Assume the thermal noise is white Gaussian noise with variance σ^2, and the SINR at the ith receiver is given by

$$\Gamma_i = \frac{G_{ii} P_i}{I + \sigma^2} = \frac{G_{ii} P_i}{\sum_{j \neq i} G_{ji} P_j + \sigma^2}. \tag{3.1}$$

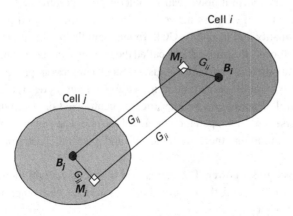

Figure 3.1 Example of multiple-cell link geometry and link-gain model.

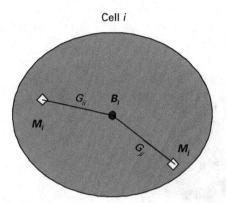

Figure 3.2 Example of single-cell link geometry and link-gain model.

The quality of the transmission link from transmitter i to receiver i depends on Γ_i. The quality is acceptable if Γ_i is above a certain threshold γ_i^{min}, which is also called the minimum protection ratio. γ_i^{min} is determined based on the link-quality requirement such as the targeted BER. Hence, for acceptable link quality,

$$\Gamma_i \geq \gamma_i^{min}, \forall i. \tag{3.2}$$

For a multicell case, the channel gain from one cell to the other (i.e., G_{ji}, $j \neq i$) is much smaller than the channel gain within each cell (G_{ii}). So the interferences from other users are relatively small. If the cochannel cells are separated far enough, the interferences term ($\sum_{j \neq i} G_{ji} P_j$) may be much less than the thermal-noise term (σ^2). Under this condition, the power-control problem becomes a single-user optimization problem and can be easily solved. However, if the cochannel cells are close to each other, we need to control the users' transmitted power in the different cells. Moreover, the channel estimations from one cell to the other (G_{ij}, $i \neq j$) are hard to obtain. So distributed power control using only local information is preferred for multicell power-control implementation.

In Figure 3.2, we show an example of a single-cell communication network. In contrast to the multicell case, there is only one base station in the network. The received SINR of the ith user is given by

$$\Gamma_i = \frac{\alpha_{ii} G_i P_i}{\sum_{j \neq i} \alpha_{ji} G_j P_j + \sigma^2}, \tag{3.3}$$

where α_{ij} is the orthogonal factor between user i and user j. For example, in a synchronized CDMA system, we have

$$\alpha_{ij} = \begin{cases} \text{processing gain,} & i = j. \\ 1, & i \neq j. \end{cases}$$

For the single-cell case, the interferences might be much larger compared with those in the multicell case. So power control is always necessary for all users every time. On the other hand, the channel gain G_j can be obtained easily. Consequently centralized power control can be possibly implemented.

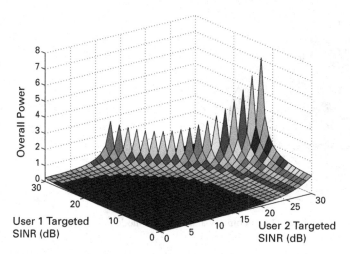

Figure 3.3 Overall transmitted power for different targeted SINRs.

From (3.1) and (3.3), we can see that users' SINR is linearly increased by its own power but nonlinearly constrained by other users' power. Because of this nonlinearity, the power-control problem is difficult to solve in wireless networks.

To study the behavior of the power-control problem, we use a two-user example to illustrate the basic ideas. In the multicell wireless network shown in Figure 3.1, we define the channel-gain matrix as $[\mathbf{G}]_{ij} = G_{ij}$. We assume $\sigma^2 = 0.001$ in (3.1). As an example, we assume

$$G = \begin{bmatrix} 4.984 & 0.067 \\ 0.0029 & 0.9580 \end{bmatrix}.$$

Obviously user 1 has a much better channel gain to its corresponding base station than that of user 2. Suppose each user's SINR is larger than a threshold γ_i. In Figure 3.3, we show the overall transmitted power versus the targeted SINR for both users. We can see that the overall power increases when the targeted SINR increases. Because user 1 has a better channel condition than that of user 2, the overall power increases slower when user 1 increases its targeted SINR than when user 2 increases its targeted SINR. When the targeted SINRs are large enough, the overall power begins to increase very quickly. Under this condition, in order to increase users' SINRs a little bit, the users have to transmit much more power. When the users' targeted SINRs are larger than some points, there is no sufficient power that can achieve the SINRs. We call the system infeasible under this condition. We define the feasibility in the following definition.

DEFINITION 2 *The power allocation is called* feasible *if there exists positive power allocation for all the users so that the targeted SINR can be achieved. More generally speaking, the power allocation is feasible if all other constraints, such as the maximal power from the transmitter, are also satisfied.*

The feasible range for all users might not be convex, especially when more constraints, such as the maximal power constraints, are considered. This nonconvex feasible range will cause many local optima for power-control optimization. The optimization algorithm can fall into these local optima and generate less desirable solutions. Moreover, even finding a solution within the feasible range may also be a difficult problem.

3.3   Classification of Power-Control Schemes

In this section, we discuss different types of power-control schemes according to how to measure the quantity, what the available measurements are, what the constraints are, and how much time delay can be accepted. Finally, we briefly review the literature of power control.

Based on the directions of communications, the power-control schemes can be classified as follows:

- Power control for reverse (up)link (from mobiles to base stations): The maximal transmit power is limited by the battery-based mobiles. Therefore power cannot be arbitrarily large. Moreover, the computational capability for mobiles is low. So it is impossible for mobiles to perform complex computing. Power control for reverse (up)link is one of the most important system requirements, especially for a DS-CDMA system with the near/far effect. The near/far effect means that, if there is no power control, the user near to the base station and with a good channel condition will have a very good channel condition, but its power will severely damage the link quality of the user that is far away from the base station and with a bad channel. We will discuss this effect in more detail later. In this case, it is necessary to have a dynamic range for power control of the order of 80 dB, which causes challenges for system implementation.
- Power control for forward (down)link (from base stations to mobiles): The overall power for all users is bounded by the transmitter amplifiers and the concerns to introduce less interference to other cells. The computational capability in a base station can be very strong. However, to perform optimization, the downlink channel conditions should be fed back from users, which will cause more communication overhead.

According to what is measured to determine power-control command, power-control techniques can be classified into the following categories:

- Power-strength-based: In strength-based schemes, the strength of a signal arriving at the base station from a mobile is measured to determine whether it is higher or lower than the desired strength. The command to lower or raise the transmit power is made accordingly.
- SINR-based: In SINR-based schemes, the measured quantity is the SINR, in which interference consists of channel noise and multiuser interference. Strength-based power control is easier to implement, but SINR-based power control reflects better system performance such as QoS and capacity. A serious problem associated with SINR-based power control is the potential to get positive feedback. This causes infeasibility

to endanger the stability of the system. Positive feedback arises in a situation in which one mobile under instructions from the base station has to raise its transmit power in order to deliver a desirable SINR to the base station, but the increase in its power also results in an increase in interference to other mobiles so that these other mobiles are then forced to also increase their power so as to maintain their link qualities. Therefore the interference also increases, and this results in positive feedback.

- BER-based: In BER-based power control, the BER is defined as an average number of erroneous bits compared with the original sequence of bits. If the signal and interference power are constant, the BER will be a function of the SINR. However, in reality, the SINR is time variant and thus the average SINR will not correspond to the average BER. In this case the BER is a better quality measure. Because channel coding is implemented in every practical system, power control can be based on the average number of erroneous frames as well. Other QoS measures, such as frame error rate (FER), can also be applied to represent users' conditions. However, to measure the accurate BER, the system needs some delay to accumulate a sufficient number of samples.

Depending on whether feedbacks exist, power-control techniques can be classified as follows:

- Closed-loop power control: Closed-loop power control uses feedback to control the transmitted power to effectively combat the changing channel conditions. For a wideband communication system, closed-loop power control is updated in less than 1 ms, which costs a lot of transmission bandwidth.
- Open-loop power control: In mobile communications systems using multiple low Earth orbital satellites, the fading occurs too rapidly for the closed-loop power control to track, because of the large round-trip propagation delay. In this case, the solution is open-loop power control, in which the mobile user estimates the channel state on the forward (down)link, and this estimate is used as a measure of the channel state on the reverse (up)link. These techniques can compensate for path loss and large-scale variations, such as shadowing, but it is impossible to compensate for multipath fading because reverse (up) and forward (down) links are not totally correlated.
- Combined closed-loop and open-loop power control: When the mobile is powered up, the mobile should estimate the transmitted power so that its power will not interfere with other mobiles. This estimation is usually done by open-loop power control. After the appropriate power is obtained, close-loop power control is applied so that the channel changing can be tracked.

Most power-control schemes require exact knowledge or perfect estimates of some instantaneous quantities such as SINR and BER. In some wireless networks, these quantities are hard to estimate perfectly. Consequently these power-control schemes may no longer work when these instantaneous variables are replaced with random estimates. So it is necessary to study the *statistical* power-control scheme based on statistical models.

There are other criteria to classify the power-control algorithms such as power control, in which the transmit power level is controlled in a continuous power domain and power control in which the transmit power level is controlled in the discrete power domain.

In addition, according to network infrastructures, there exist two main classifications:

- Centralized power control: A centralized controller has all the information about the established connections and channel gains and controls all the power levels in the network or part of the network. Centralized power control requires extensive control signaling and measurement overhead in the network and can be applied only in some special cases such as a single-cell case in practice. Nevertheless the study of centralized power control can be used to give bounds on the performance of the distributed algorithms.

  Centralized power-control problems have been investigated in [114, 351]. In these systems a centralized controller that has knowledge about all link gains in the system would allocate users' transmit power directly so that the SINRs for all links are above their targeted value.

- Distributed power control: A decentralized controller controls only the power of one single transmitter, and the algorithm depends on only local information, such as the measured SINR or the channel gain of the specific user. These algorithms can be implemented in practice. But in real systems there are a number of undesired effects. For example, measuring and control signaling take time, which results in a time delay in the system; the possible output power of the transmitters is constrained because of physical limits and quantization; the signals needed for control may not be available and have to be estimated; QoS is a subjective measure, and relevant objective measures have to be employed.

  Distributed power-control schemes operate in a decentralized sense. Each transmitter updates its own power based on the measured interference at the receiver of the link. It has been shown in [79, 92] that the distributed algorithms converge to the optimal solution for the network, in which the transmitter power is minimized for each transmitter and the link quality is satisfied.

In the next two sections, the centralized and distributed power-control schemes are described in detail.

3.4   Centralized Power Control

For centralized power control, the channel gains among users are known to the system, so that the power-control optimization can be conducted centrally. The advantages of the centralized approach are that it is noniterative and synchronous. The disadvantages are that the computational complexity and channel-estimation overheads increase rapidly when the number of users is large, especially for the wireless networks with distributed topology such as multicell cases.

In this section, we discuss the scheme for centralized power control from an eigenvalue point of view first. Then we present a SINR balancing scheme as an example [351]. Finally, we discuss how to modify the eigenvalue for differential services.

First, we discuss the case in which the interferences are much greater than the thermal-noise term and all users have the same targeted SINR γ_0. In matrix form, if we omit the thermal-noise term, (3.1) and (3.2) can be written as follows:

$$\frac{1}{\gamma_0}\mathbf{P} \geq \mathbf{FP}, \tag{3.4}$$

where $\mathbf{P} = [P_1, P_2, \ldots, P_K]^T$ is the power vector (K is total number of cochannel users), and $\mathbf{F}$ is a nonnegative matrix defined as

$$F_{ij} = \begin{cases} 0, & \text{if } j = i \\ \frac{G_{ji}}{G_{ii}} > 0, & \text{if } j \neq i \end{cases}. \tag{3.5}$$

The objective of a power-control scheme is to maintain the link quality by keeping the SINR above the threshold γ_0, that is, to adjust the power vector $\mathbf{P}$ so that (3.4) is satisfied. According to the Perron–Frobenius theorem [82], the maximum value of γ_0 for which there exists a positive $\mathbf{P}$ so that (3.4) is satisfied is $\frac{1}{\rho(\mathbf{F})}$, where $\rho(\mathbf{F})$ is the spectral radius (the maximal eigenvalue) of matrix $\mathbf{F}$. The power vector that satisfies (3.4) is the eigenvector corresponding to $\rho(\mathbf{F})$, which is positive.

In another problem, the power-control problem is defined as determining the power allocation that achieves the maximal achievable γ_0 as

$$\max_{\mathbf{P} \geq 0} \min_i \left\{ \frac{G_{ii} P_i}{\sum_{j \neq i} G_{ji} P_j} \right\}. \tag{3.6}$$

We can prove that the solution to the preceding max–min problem is equivalent to the power-control problem in (3.4) [351].

In the following, we first show the properties of an irreducible nonnegative matrix; then we prove that $\mathbf{F}$ defined in (3.5) is an irreducible nonnegative matrix. Finally, we illustrate how to achieve the unique optimal SNR by adjusting the power. The proof of the following theorem can be found in [82].

THEOREM 2 *Let* $\mathbf{F}$ *be a* $K \times K$ *irreducible nonnegative matrix with eigenvalues* $\{\lambda_i\}_{i=1}^K$. *Then:*

1. $\mathbf{F}$ *has a positive real eigenvalue* λ^* *with* $\lambda^* = \max\{|\lambda_i|\}_{i=1}^K$.
2. λ^* *above has an associated eigenvector* $\mathbf{P}^*$ *with strictly positive entries.*
3. λ^* *has algebraic multiplicity equal to 1.*
4. *All eigenvalues* λ *of* $\mathbf{F}$ *other than* λ^* *satisfy* $|\lambda| < |\lambda^*|$ *if and only if there is a positive integer* k *with all entries of* $\mathbf{A}^k$ *strictly positive.*
5. *The minimum real* λ, *so that the inequality*

$$\lambda\mathbf{P} \geq \mathbf{FP} \tag{3.7}$$

has solutions for $\mathbf{P} \geq \mathbf{0}$, *is* $\lambda = \lambda^*$.

6. *The maximum real* λ, *so that the inequality*

$$\lambda\mathbf{P} \leq \mathbf{FP} \tag{3.8}$$

has solutions for $\mathbf{P} \geq \mathbf{0}$, *is* $\lambda = \lambda^*$.

THEOREM 3 **F** *is an irreducible nonnegative matrix.*

Proof: The matrix **F** by definition has all the elements on its main diagonal equal to 0 and all other elements greater than 0. A $K \times K$ matrix **B** is reducible if there exists a permutation matrix **Q** so that

$$\mathbf{QBQ}^T = \begin{bmatrix} \mathbf{C} & \mathbf{D} \\ \mathbf{O} & \mathbf{E} \end{bmatrix} \tag{3.9}$$

with $\mathbf{C}_{r \times r}$ and $\mathbf{E}_{(K-r) \times (K-r)}$ for $1 \leq r \leq K - 1$. **O** is a matrix with all its entries equal to zero. Matrix **F** would be reducible only if it had at least one row with more than one 0 element. So **F** is an irreducible nonnegative matrix. **QED**

THEOREM 4 *There exists a unique* γ^* *that may be achieved by all the K mobiles using a given channel and is given by*

$$\gamma^* = \frac{1}{\lambda^*}. \tag{3.10}$$

where λ^* *is the largest real eigenvalue of the matrix* **F**. *The power vector* $\mathbf{P}^*$ *that achieves* γ^* *is the eigenvector corresponding to* λ^*.

Proof: Let us define the quantities

$$\gamma_{\min} = \min_{1 \leq i \leq K} \{\gamma_i\}, \tag{3.11}$$

$$\gamma_{\max} = \max_{1 \leq i \leq K} \{\gamma_i\}. \tag{3.12}$$

We have

$$\gamma_i \geq \gamma_{\min}, i = 1, \ldots, K. \tag{3.13}$$

The preceding equation can be written as

$$\frac{1}{\gamma_{\min}} P_i \geq \sum_{j=1}^{K} F_{ij} P_j, \forall i. \tag{3.14}$$

The preceding can be expressed in matrix form as

$$\lambda\mathbf{P} \geq \mathbf{FP}, \tag{3.15}$$

where

$$\lambda = \frac{1}{\gamma_{\min}}. \tag{3.16}$$

From Theorems 2 and 3, it follows that the minimum real λ, so that the preceding set of inequalities will hold true, is λ^*, which is the largest eigenvalue of **F** that is positive and has **P*** as the corresponding eigenvector.

We also have

$$\gamma_i \leq \gamma_{\max}, \forall i. \tag{3.17}$$

Following the same procedure as previously given, we have

$$\lambda \mathbf{P} \leq \mathbf{FP}, \tag{3.18}$$

where

$$\lambda = \frac{1}{\gamma_{\max}}. \tag{3.19}$$

The maximum λ for the preceding set of inequalities to hold true is λ^*. So we have a λ^* that is achievable by all mobiles and $\lambda^* = \gamma_{\min} = \gamma_{\max}$.   **QED**

Now we consider the thermal noise at the receivers, and different users have different targeted SINRs. The requirement for acceptable link quality is again

$$\Gamma_i \geq \gamma_i, \quad 1 \leq i \leq K,$$

or, in matrix form,

$$[\mathbf{I} - \mathbf{DF}]\mathbf{P} \geq \mathbf{u}, \tag{3.20}$$

where I is a $K \times K$ identity matrix,

$$\mathbf{D} = \text{diag}\{\gamma_1, \dots, \gamma_K\}, \tag{3.21}$$

and **u** is an elementwise positive vector whose elements are defined as

$$u_i = \gamma_i N_i / G_{ii},$$

where N_i is the thermal noise value for user i. The SINR thresholds (γ_i, $i = 1, \dots, K$) are achievable if there exists at least one solution vector **P** that satisfies (3.20).

Here we give an example of a power-control problem. We want to minimize the overall transmitted power under the constraint that each user have the SINR of no less than its targeted SINR, i.e.,

$$\text{minimize} \quad \sum P_i, \tag{3.22}$$

$$\text{subject to} \quad [\mathbf{I} - \mathbf{DF}]\mathbf{P} \geq \mathbf{u}.$$

If the spectral radius of **DF** is less than unity, i.e., $\rho(\mathbf{DF}) < 1$, $[\mathbf{I} - \mathbf{DF}]$ is invertible and positive [82]. In this case the network is *feasible* and the optimal solution to the power-control problem is given by

$$\mathbf{P}^* = [\mathbf{I} - \mathbf{DF}]^{-1}\mathbf{u}. \tag{3.23}$$

Table 3.1 Stepwise removal algorithm

---

1. Determine the maximal achievable SINR γ^* corresponding to $\mathbf{F}$.
   If $\gamma^* \geq \gamma_0$, use $\mathbf{P}^* = [\mathbf{I} - \mathbf{DF}]^{-1}\mathbf{u}$, else set $K' = K$.
2. Remove the cell k for which the maximum of the row and column
   sums $r_k = \sum_{j=1}^{K'} G_{kj}, r_k^T = \sum_{j=1}^{K'} G_{jk}$
   is maximized and form the $(K' - 1) \times (K' - 1)$ matrix $\mathbf{G}'$.
   Determine γ^* corresponding to $\mathbf{G}'$.
   If $\gamma^* \geq \gamma_0$, use the eigenvector P^*,
   else select $K' = K' - 1$ and repeat step 2.

It is interesting to note that this solution is Pareto optimal. That is, for each feasible $\mathbf{P} \geq 0$,

$$\mathbf{P} \geq \mathbf{P}^*.$$

In the subsequent discussion, we explain the scheme of SNR balancing as an example of centralized power control. The goal for this SNR balancing power-control scheme is to maintain the targeted SINR with as many users as possible while the overall power is minimized. When the system is less crowded and the targeted SINR for each user is relatively small, the CCIs are not severe. Under this condition, all users' power can be obtained from (3.23). If the network becomes crowded and the targeted SINR is relatively large, the CCIs are so large that not all users' targeted SINRs can be achieved. Under this condition, we need to consider how to reduce the interferences. One possible solution is to remove the users from sharing the channel. The following theorem [351] tells that some users' power should be zero in order for the other users to achieve the targeted SINR.

THEOREM 5 *At least one optimum power vector P^* has the form*

$$P_i \begin{cases} = 0, & i \notin \mathbf{S} \\ \neq 0, & i \in \mathbf{S} \end{cases}, \tag{3.24}$$

where $\mathbf{S} = \{i : \Gamma_i \geq \gamma_0\}$.

A stepwise removal algorithm is shown in Table 3.1. The basic idea is to remove one user at a time until the targeted SINR level γ_0 is achieved for the remaining users. In the first step, the algorithm checks if the requiring link qualities are satisfied for the current number of users. If yes, the algorithm returns the power allocation $\mathbf{P}^*$. Otherwise, in the second step, the algorithm tries to remove the user that might cause the highest interference to others. The steps continue until the remaining users can achieve the desired link qualities.

Until now, users' requirements for targeted SINRs were fixed and predefined. Because users' channel conditions are changing, it is possible that we can change the users' targeted SINRs so that the system performance can be improved. In the following discussion, we analyze how the system behaves if we change these targeted SINRs.

It has been shown by the following theorem that there exists the derivative of the spectral radius [203]:

THEOREM 6 *Let λ be a simple eigenvalue of DF, with right and left eigenvectors $\mathbf{x}$ and $\mathbf{y}$, respectively. Let $\tilde{\mathbf{F}} = \mathbf{DF} + \mathbf{E}$; then there exists a unique $\tilde{\lambda}$, eigenvalue of $\tilde{\mathbf{F}}$ so that*

$$\tilde{\lambda} = \lambda + \frac{\mathbf{y}^H \mathbf{E} \mathbf{x}}{\mathbf{y}^H \mathbf{x}} + O(\|\mathbf{E}\|^2). \tag{3.25}$$

In power control, we try to reduce the maximum eigenvalue, because the maximum eigenvalue is the key factor that affects the overall transmit power [114]. We assume $\mathbf{x}$ and $\mathbf{y}$ are the eigenvectors of the largest eigenvalue. We define $\mathbf{E} = \Delta\Gamma_i \mathbf{F}_i$, where

$$(\mathbf{F}_i)_{jk} = \begin{cases} 0, & j \neq i \\ (\mathbf{F})_{jk}, & j = i \end{cases}. \tag{3.26}$$

Then we can have the gradient to reduce the spectral radius as

$$g_i^\rho = \frac{\partial \rho(\mathbf{DF})}{\partial \Gamma_i} = \frac{\mathbf{y}^H \mathbf{F}_i \mathbf{x}}{\mathbf{y}^H \mathbf{x}}. \tag{3.27}$$

By changing the targeted SINRs according to the preceding gradient, we can improve the feasibility in a most efficient way. Consequently, the overall power is reduced. However, the QoS may be reduced. To compensate for the QoS loss, in [114], the average QoS of different users is maintained by projecting the preceding gradient to a plane where the average QoS is a constant. The preceding analysis of the system behavior under different QoS requirements opens a way to explore the time and multiuser diversity by adapting users' SINR requirements according to their channel conditions [118].

3.5　　Distributed Iterative Power Control

In wireless communication systems, mobile users adapt to a time-varying radio channel by regulating transmitter power. Power control is intended to provide each user an acceptable connection by eliminating unnecessary interference. Because centralized power control needs excessive channel-estimation overhead for channel conditions between cochannel users, it is difficult to be implemented in a distributed topology or when the number of users is large. In [79, 92], many distributed power-control schemes are provided and some analytical insights are developed. In this section, we discuss the framework in [347] to unify and extend convergence results for cellular radio systems employing distributed power-control methods. For a variety of systems, this framework shows that interference constraints derived from the users' SINR requirements share certain simple properties. These properties imply that an iterative power-control algorithm converges not only in synchronous cases but also in asynchronously situations, even if the users perform power adjustments with outdated or incorrect interference measurements.

In this framework, a broad class of power-controlled systems is discussed. The users' SINR requirements can be described by a vector inequality of interference constraints of the form

$$\mathbf{P} \geq \mathbf{I}(\mathbf{P}),\tag{3.28}$$

where $\mathbf{P} = [P_1 \ldots P_K]'$, P_i is the ith user's transmitted power, K is the total number of users, $\mathbf{I}(\mathbf{P}) = [I_1(\mathbf{P}) \ldots I_K(\mathbf{P})]'$, and $I_i(\mathbf{P})$ is the effective interference of other users that user i must overcome. The system is feasible if (3.28) is satisfied.

The definitions of interference function $I(\mathbf{P})$ can be different for different network requirements. One example is the *fixed assignment*. Under the fixed assignment, each user has a fixed targeted SINR level γ_i. The interference function can be expressed as

$$I_i(\mathbf{P}) = \frac{\gamma_i(\sum_{j \neq i} G_{ji} P_j + N_0)}{G_{ii}}.\tag{3.29}$$

The most important feature of this framework is the definition of *standard function*:

DEFINITION 3 *Interference function* $I(\mathbf{P})$ *is* standard *if, for all* $\mathbf{P} \geq 0$, *the following properties are satisfied:*

- *Positivity:* $I(\mathbf{P}) > 0$.
- *Monotonicity: If* $\mathbf{P} \geq \mathbf{P}'$, *then* $I(\mathbf{P}) \geq I(\mathbf{P}')$.
- *Scalability: For all* $\alpha > 1$, $\alpha I(\mathbf{P}) > I(\alpha \mathbf{P})$.

When $I(\mathbf{P})$ is a standard interference function, the iteration

$$\mathbf{P}(t + 1) = I(\mathbf{P}(t))\tag{3.30}$$

is called the *standard power-control algorithm*. Next we show the convergence of this algorithm under synchronous and asynchronous cases.

In the synchronous case, each user performs a power-control update at each time slot. Starting from any initial feasible power-control vector $\mathbf{P}$, n iterations of the standard power-control algorithm produce the power vector $I^n(\mathbf{P})$. The following theorems [347] provide the proof that the standard power-control algorithm converges to a unique optimal solution.

THEOREM 7 *If the standard power-control algorithm has a fixed convergence point, then that fixed point is unique.*

THEOREM 8 *If* $\mathbf{P}$ *is a feasible power vector, then* $I^n(\mathbf{P})$ *is monotone-decreasing sequence of feasible power vectors that converges to the unique fixed point* $\mathbf{P}^*$.

THEOREM 9 *If* $I(\mathbf{P})$ *is feasible, then starting from* $\mathbf{0}$, *the all-zero vector, the standard power-control algorithm produces a monotone-increasing sequence of* $I^n(\mathbf{P})$ *that converges to the fixed point* $\mathbf{P}^*$, *where n is the index for the iteration.*

THEOREM 10 *If* $I(\mathbf{P})$ *is feasible, then for any initial power vector* $\mathbf{P}$, *the standard power-control algorithm converges to the unique fixed point* $\mathbf{P}^*$.

In the asynchronous case, some users are allowed to perform power adjustments faster and execute more iterations than others. In addition, the asynchronous iteration allows users to perform these updates by using outdated information on the interference caused by other users.

THEOREM 11 *If there is a sequence of nonempty set* $\{X(n)\}$ *with* $X(n + 1) \subset X(n)$ *for all n satisfying the following two conditions,*

1. *synchronous convergence condition: for all n and* $x \in X(n)$, $f(x) \in X(n = 1)$. *If* $\{y^n\}$ *is a sequence so that* $y^n \in X(n)$ *for all n, then every limit point of* $\{y^n\}$ *is a fixed point of f;*
2. *box condition: For every n, there exists sets* $X_i(n) \in X$, *so that* $X(n) = X_1(n) \times X_2(n) \times \ldots \times X_N(n)$;

and the initial solution estimate $x(0)$ *belongs to the set* $X(0)$, *then every limit point of* $\{x(t)\}$ *is a fixed point of f.*

THEOREM 12 *If* $I(\mathbf{P})$ *is feasible, then from any initial power vector* $\mathbf{P}$, *the asynchronous standard power-control algorithm converges to* $\mathbf{P}^*$.

From the preceding framework, the power-control problem can be reduced to finding a power vector $\mathbf{P}$ satisfying $\mathbf{P} \geq I(\mathbf{P})$, which describes the fact that the transmitted power must overcome the interference. The common properties of the distributed power-control schemes can be categorized as *standard functions*. For different kinds of situations with a variety of constraints, the power-control schemes will converge to the unique optimum if the formulated problems satisfy the definition of the standard function. So this convergence analysis is the importance and contribution of this framework. Many other iterative approaches for other purposes other than power control can also be classified as the *standard function*, so that the convergence can be proved accordingly.

3.6   Statistical Power Control

For wireless communication systems, the distributed iterative power-control algorithms have been proposed to maintain reliable communications between mobiles and base stations. To derive convergence results, these algorithms require perfect measurements of SINR, interference, or BER. However, these quantities are often difficult to measure, and convergence results neglect the effect of stochastic measurements. This reason highlights the need for the study of stochastic power-control [314] schemes that make use of available measurements, evolve stochastically, and converge in a stochastic sense.

In stochastic power control, because the measurements are random, the convergence is quantified in terms of the mean square error (MSE) of the power vector from the optimal power vector that is the solution of a feasible power-control problem. In this section, we describe a stochastic power-control scheme.

For power control, the coefficient of $\mathbf{I}(\mathbf{P})$ must be estimated perfectly. The *basic idea* for stochastic power control is to replace some coefficient of $\mathbf{I}(\mathbf{P})$ with some stochastic measure. For example, the total received power can be estimated by an unbiased estimator. However, because the estimator may have a large variance compared with the real value and cause the power-control scheme to diverge, the estimated results cannot be applied to the distributed iterative power-control scheme directly.

In [314], a power-update approach was proposed that considers the previous power and current power updates as

$$\mathbf{P}(n+1) = (1 - a_n)\mathbf{P}(n) + a_n\mathbf{I}(\mathbf{P}(n)), \tag{3.31}$$

where a_n is a forgetting factor and $\mathbf{I}(\mathbf{P}(n))$ uses the estimated value of received power. Here a_n can select a fixed value ε. When $\varepsilon = 1$, it is the same as the power-control scheme; a_n can also be selected adaptively so that the convergence rate can be improved and the variance of final results can be minimized. There is a trade-off between the value of the MSE and convergence speed for different values of a_n.

The convergence of the preceding power-control scheme is specified in terms of the MSE at iteration n:

$$MSE(n) = E[\|\mathbf{P}(n) - \mathbf{P}^*\|^2], \tag{3.32}$$

where $\mathbf{P}^*$ is the optimal power-control vector. The following results are proved in [314]:

1. If $a_n = \epsilon$ and if ϵ is chosen sufficiently small, then finite lower and upper bounds on the MSE as the number of iteration grows can be calculated. In the limiting case as $\epsilon \to 0$, both lower and upper bounds on the limiting MSE as well as the limiting MSE itself go to zero.
2. If $a_n = \epsilon$ but ϵ is chosen too large, then the MSE may diverge even if the power-control algorithm would converge.
3. If $a_n = a/n$, then the algorithm converges to the optimal power vector $\mathbf{P}^*$ in the sense that $\lim_{n\to\infty} MSE(n) = 0$, irrespective of other system parameters. Here a is a constant.

3.7   DS-CDMA Power Control

The necessity for power control in FDMA-/TDMA-based cellular networks stems from the requirement for CCI management. This type of interference is caused by frequency reuse due to limited available frequency spectrum. By a proper power adjustment, the harmful effects of CCI can be reduced. This power control allows a more "dense" reuse of resources and thus higher system capacity over whole networks.

In DS-CDMA systems, power control allows users to share resources of the system equally among themselves. In the event that power control is not used, all mobiles transmit signals toward the base station with the same power, without taking into account the fading and the distance from the base station. Mobiles that are closer to the base station will cause significant interference to the mobiles that are farther from the base station because of nonzero cross correlation between signature sequences assigned to users. This effect is known as the near/far effect.

In [265], a framework for power control and resource management for DS-CDMA systems with different services is proposed. In particular, the minimal total transmitted power and the maximum total rate conditions are considered. The framework shares the same ideas with the centralized power-control scheme. The difference in the received SINR for user i can be written as

$$\Gamma_i = \frac{W}{r_i} \times \frac{P_i G_i}{\sum_{j \neq i}^{K} P_j G_j + N_0},\tag{3.33}$$

where G_i is the channel gain and r_i is the rate for user i, respectively. Here W/r_i is called the processing gain. By using a similar approach as in the previous sections, DS-CDMA power control can be implemented to combat slow fading and even fast fading.

IS95 CDMA (CDMAone) is the first CDMA-based digital cellular radio system standard pioneered by Qualcomm. The CDMA network provides for mobile voice communication as well as many new advanced services, like mobile fax and text messaging. IS95 is a promising 2G standard, and all 3G standards use CDMA. In the following discussion, we explain power control in the IS95 standard. In the IS95, there are open-loop and close-loop power controls for the uplink and close-loop power controls for the downlink case. Finally, we discuss power control in some newly developed 3G standards.

Open-loop power control is used during mobile's access attempts. In Figure 3.4, we show the block diagram for reverse (up)link open-loop power control. The mobile's transmit power is determined by measuring the received signal strength of the base station and estimating the forward (down)link path loss. Assuming a similar path loss for the reverse (up)link, the mobile uses this information to determine its transmitter

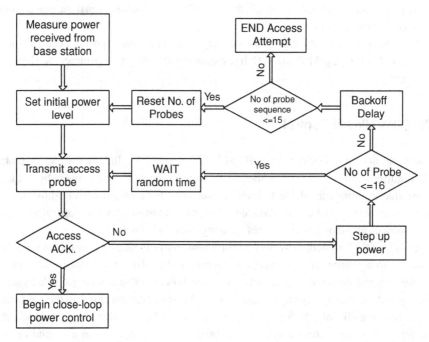

Figure 3.4 IS-95 open-loop power control for reverse (up)link.

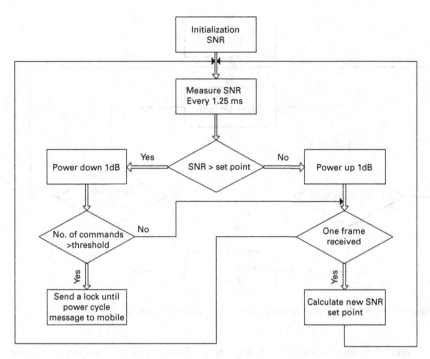

Figure 3.5 IS-95 closed-loop power control for reverse (up) link.

power. The first time a handset transmits, it will do so on its access channel as a reply to a message on the paging channel or to place an outgoing call. Because of the near/far problem and the risk of interfering with other mobiles in the cell, the handset initially transmits at a low power. It then makes successive attempts (or probes) to be heard, gradually increasing its power each time until the base station detects the message and acknowledges it. If not heard, the mobile waits a random amount of time before transmitting next its "access probe."

For reverse (up)link, close-loop power control is used when the user is admitted to the system and to combat the near/far effect when the user's channels fluctuate. A block diagram of reverse (up) power control is shown in Figure 3.5. Because of the separation of forward (down) and reverse (up) link by 45 MHz, reverse channel characteristics are not strongly correlated to forward (down) channel characteristics. Moreover, tight power control is needed. The power level is updated at 800 bps (1.25 ms) to attempt to maintain the SINR at the base station. The power-control subchannel uses puncturing techniques to transmit "1" or "0" to the mobile to raise or lower its power level. Closed-loop power control can be further subdivided into inner and outer loops. In the outer loop, the base station starts with a targeted FER and a targeted bit energy-to-noise spectral-density ratio (Eb/N0) required for achieving the targeted FER. The base station then continually measures the FER and adjusts the threshold Eb/N0 accordingly.

For the forward (down)link of IS-95, orthogonal Walsh codes are used to separate the mobiles' signals. Consequently the CCI is a smaller problem than on the reverse (up)link.

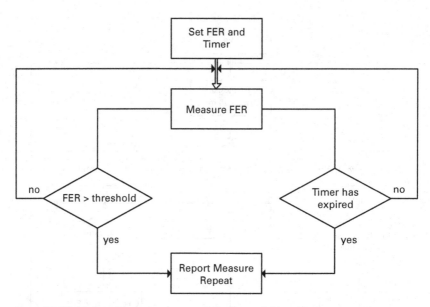

Figure 3.6 IS-95 closed-loop power control for forward (down)link

In Figure 3.6, we show the block diagram for forward (down)link power control. The power adjustment is based on FERs at the mobile and updated in a slow quality loop.

Finally, we briefly describe the power-control scheme for two 3G standards: CDMA2000 and WCDMA. The differences are as follows:

- CDMA2000: Unlike IS-95, in which fast closed-loop power control was applied only to the reverse (up)link, CDMA2000 channels can be power controlled at up to 800 Hz in both the reverse (up) and the forward (down)links. The reverse (up)link power-control command bits are punctured into the forward fundamental channel (F-FCH) or the forward dedicated control channel (F-DCCH), depending on the service configuration. The forward (down)link power-control command bits are punctured in the last quarter of the reverse-pilot channel (R-PICH) power-control slot.

  In the reverse (up)link, during gated transmission, the power-control rate is reduced to 400 or 200 Hz on both links. The reverse (up)link power-control subchannel may also be divided into two independent power-control streams, either both at 400 bps, or one at 200 bps and the other at 600 bps. This mechanism allows for independent power control.

  In addition to closed-loop power control, the power on the reverse (up)link of CDMA2000 is also controlled through an open-loop power-control mechanism. This mechanism inverses the slow fading effect that is due to path loss and shadowing. It also acts as a safety fuse when fast power control fails. When the forward (down)link is lost, closed-loop reverse (up)link power control is "freewheeling" and the terminal disruptively interferes with neighboring terminals. In such a case, the open loop reduces the terminal output power and limits the impact to the system. Finally, the outer-loop power drives closed-loop power control to the desired set point based on error statistics that it collects from the forward (down) link or reverse (up)link. Because of the expanded data-rate range and various QoS requirements, different users will have different outer-loop thresholds; thus different users will receive different power levels at the base station. In the reverse (up)link, CDMA2000 defines some nominal gain offsets based on various channel frame formats and coding schemes. The remaining differences will be corrected by the outer loop itself.

- WCDMA: Open-loop power control is the ability of the transmitter to set its output power to a specific value. It is used for setting initial uplink and downlink transmission powers when a user is accessing the network. The open-loop power-control tolerance is ±9 dB (normal conditions) or ±12 dB (extreme conditions). Inner-loop power control (also called fast closed-loop power control) in the uplink is the ability of the transmitter to adjust its output power in accordance with one or more transmit power-control (TPC) commands received in the downlink, in order to keep the received uplink SINR at a given SINR target. The transmitter is capable of changing the output power with step sizes of 1, 2, and 3 dB. The inner-loop power-control frequency is 1500 Hz. The forward closed-loop power control is faster than IS-95. On the reverse (up)link, power-control bits are transmitted in every time slot. Therefore, mobile power control updates at the rate of 1500 adjustments per second to maximize cell capacity.

Outer-loop power control is used to maintain the quality of communication at the level of bearer service quality requirement while using as low a power as possible. Uplink outer-loop power control is responsible for setting a targeted SINR in the base station for each individual uplink inner-loop power control. This targeted SINR is updated for each user according to the estimated uplink quality (block error ratio, BER) for each radio resource control connection. Downlink outer-loop power control is the ability of the receiver to converge to the required link quality set by the network in downlink.

3.8   Other Works and Summary

Transmitted power is an important resource for wireless network design. Efficient power-control schemes can achieve links' QoSs, reduce unnecessary CCI, and be implemented with small costs for overhead, signaling, infrastructure, etc. According to the different network scenarios, the power-control schemes can be classified as uplink and downlink, open loop and close loop, etc.

In the power-control research literature, the schemes can be classified into centralized and distributed schemes. For centralized power control, the underlying reasons for the behavior of power-controlled systems are explored. The implementation is limited by the channel-estimation overhead. So it can work only in a system with a small number of users and a centralized topology. On the other hand, the distributed power-control schemes provide implementable methods in practice. The disadvantages are low convergence speed and possible infeasibility. However, in most practical wireless networks, close-loop power control is applied in a distributed manner.

CDMA is an interference-limited system because all mobiles transmit at the same frequency and in the same time interval. The reverse (up)link near/far effect greatly reduces the system capacity. Power control is one very important task for the next generation of all CDMA wireless networks. Power-control frequency is greatly increased to combat not only propagation loss, shadow fading and slow fading, but for fast fading as well.

There are other works for power control. For example, in [151], to increase the convergence speed of distributed power control, a second-order scheme is proposed to use both current and previous iterations for power update. In [297], the quantized feedback for closed-loop power control is analyzed and the loop filter is designed to minimize power-control error. In [154], the power-control problem is formulated as a convex optimization problem and can be solved efficiently. In [179], a Kalman filter method for power control is proposed to predict the interference power in the future.

In [250] and [248], the authors consider a system with beam-forming capabilities in the receiver and power control. An iterative algorithm is proposed to jointly update the transmission power and the beam-former weights so that it converges to the jointly optimal beam-forming and transmission power vector. In [183], a downlink beam-forming technique is proposed that converts the downlink beam-forming problem into a virtual uplink one and takes into account the data-rate information of all users for the CDMA system.

In [354], it was observed that, for a fixed assignment of mobile users to base stations, the feasibilities of uplink and downlink are equivalent. The preceding works consider the uplink power-control problem for a fixed assignment of mobile users to base stations. Power allocation and base-station assignment can be integrated to attain a higher capacity and achieve a smaller allotment of transmitter power [125, 164]. In a network with power-control capability, as the mobiles move or new calls arrive, the constant reassignment of mobiles to base stations helps to find a feasible power allocation and provides a framework for the handoff.

4 Rate Adaptation

4.1 Introduction

Rate adaptation is one of the most important resource-allocation issues, because the system can adapt the users' rates so that the limited radio resources can be efficiently utilized. Compared with power control, rate adaptation gives a new dimension of freedom to change the information transmission rate over time, i.e., power control maintains the desired link quality, whereas rate adaptation adjusts this link quality. In this chapter, we give an overview of the rate-adaptation system: where and how the rates can be changed; what the challenges and constraints are; and how rate adaptation can be combined with other techniques.

According to the different ISO (International Organization for Standardization) layers, rate adaptation can be classified into three different types: source rate adaptation in the application layer, rate control for data communication in the network/MAC layer, and channel protection adaptation in the physical layers. They are briefly summarized here:

- **Source Rate Adaptation**
  This type of adaptation adjusts the quality of transmitting information. For example, the voice encoder can change the information rate according to the talking period and the silence period, as it is useless to have a high data rate for the silence period. For video transmission, the data rate is very bursty over time, because of the different video scenarios and different frames such as I, B, P frames. Because the capacity to deliver the information is limited by the communication systems, the design of the wireless network protocol shall carefully consider the source rate adaptation so that the received information has high quality using the limited system resources.
- **Rate Adaptation for Network/MAC Layer**
  The network/MAC layer utilizes buffers to accommodate the rate differences from those of source coders and those that channels can provide. Rate control is critical to optimize the buffer behaviors and maintain the QoS. Moreover, the multiple-access nature of the wireless channels requires rate control for distributed users.
- **Rate Adaptation for PHY Layer**
  The wireless channel gains and phases fluctuate over time, which causes fading in the communication links. To combat the fading effect, channel protections, such as the forward error correction (FEC) technique, are widely employed in wireless networks.

Because users need different error protections in different conditions, adaptive channel protection is essential to deliver information to the destiny. Other techniques such as adaptive modulation and adaptive processing gains are also discussed.

Overall, we will explain the degrees of freedom for rate adaptation over different ISO layers. In the specific wireless networks, some types of rate adaptation can be applied or combined together. In addition, rate control can be combined with power control to further improve the system performance. Some examples are shown for the combination of all these techniques. Moreover, rate adaptation under the multiuser scenario will also be discussed.

This chapter is organized as follows: In Section 4.2, different adaptive source encoders such as voice and video are discussed. In Section 4.3, rate control for data communication is reviewed. In Section 4.4, different adaptive channel protection techniques are explained. In Section 4.5, power control and rate adaptation are combined together.

4.2   Source Rate Adaptation

In this section, we discuss how to control the source rates for different types of services such as voice and video. Because the structures of coders for different services are different, the design concerns for source adaptation are different. We concentrate on how rate adaptation can be implemented, what the design concerns are, what the parameters that can be controlled are, and what the effects of these parameters are on the QoS.

4.2.1   Adaptive Voice Encoder

In the 2G of wireless cellular networks, voice payload is the primary service. The operator meets the increasing need for services by combining digital technology and special encoding techniques for voice. These voice encoders take advantage of predictable elements in human speech. Several low-data-rate encoders, such as linear predictive encoder, regular pulse-excited encoder, and codebook-excited encoder [242], are proposed for GSM, IS54, IS-95, CDMA2000, UMTS, and others.

The motivation to have rate adaptation for a voice encoder is based on two factors. First, the level of speech activity changes during a conversation. For example, it has been observed that silence periods account for approximately 65% of all time of a two-way communication [13]. There is no need to have a high data transmission rate for the silence time. By adapting the voice encoder rate, a lot of bandwidth can be saved. Second, when the system is overcrowded, instead of dropping users from the networks, the system can control the voice encoder to gracefully reduce the quality of reconstructed voice in a controllable and predictable way. By doing this, the system can avoid the annoying line dropping for the users and accommodate more users.

To show how the voice encoder adapts its output rate, we analyze a uniform scalar quantizer encoder on a uniform source with a random index assignment as an example. Suppose the input vector $\mathbf{x} \in \Re^n$, where n is the number of samples for each quantization.

The output of the source encoder has m bits that determine the source encoding rate. The source encoder quantizer distortion can be written as

$$D_s = \sum_{i=1}^{2^m} \int_{S_i} \|\mathbf{x} - \mathbf{y}_i\|^2 P_b(\mathbf{x})d\mathbf{x}, \qquad (4.1)$$

where $\{S_i\}_{i=1}^{2^m}$ is the partition of $\Re^n$ into disjoint regions, $\mathbf{y}_i$ is the quantized vector with index i, and $P_b(\mathbf{x})$ is the probability density function of $\mathbf{x}$. From [83], the source-induced distortion of such a quantizer is given by $D_s(m) = (2^{-2m}/12)$. To adapt the rate, the source encoder can change the number of output bits m. The distortion D_s is also modified according to (4.1).

In practice, the real-time adaptive voice source encoder has the key property that the output rate can be externally controlled. This can be implemented by using either variable-rate or embedded encoders. In the first case, the coder generates one bit stream for each of the possible encoding rates. Only one of these will be selected and transmitted based on the rate assignment. Using the embedded encoders presents the advantage that only one bit stream is generated, making the rate adaptation simply by dropping as many bits as necessary from the end of the bit stream, where less important data are located closer to the end. One example of an embedded coder is shown in Figure 4.1. The maximum output rate is eight frames per transmission. The priority is ordered from high to low. By dropping the frames ("bit-dropping mechanism") from the lowest priority, the reconstructed voice quality is reduced gradually. Some common voice coder subjective tests are the dynamic rhyme test (DRT), the diagnostic acceptability measure (DAM), and the mean opinion score (MOS). The DRT is an American National Standards Institute (ANSI) (S32.2-1989) speech intelligibility test that provides a score of communication systems analyzing the listener's ability to distinguish between 96 pairs of monosyllable rhyming words. The DAM and MOS are speech acceptability (quality) measures.

Although the bit-dropping mechanism is exclusive to the embedded stream, this term is used loosely to represent a reduction in the source rate, regardless of the particular source encoder implementation. The source coder is assumed to have a maximal output rate of η_{max} bps. The source rate controller has the output rate of R bps ($R \le \eta_{max}$), where R is the variable source coding rate.

Define $D_s(R)$ as the distortion-rate function of the user's source encoder transmitting at rate R. In most well-designed encoders, D_s is a convex and decreasing function.

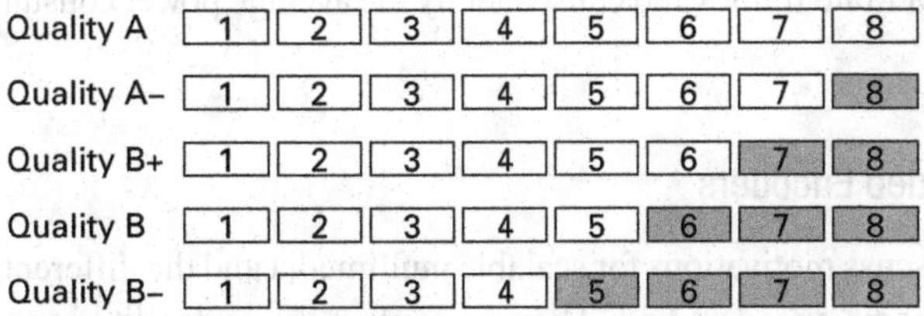

Quality A [1][2][3][4][5][6][7][8]

Quality A− [1][2][3][4][5][6][7][**8**]

Quality B+ [1][2][3][4][5][6][**7**][**8**]

Quality B [1][2][3][4][5][**6**][**7**][**8**]

Quality B− [1][2][3][4][**5**][**6**][**7**][**8**]

Figure 4.1 Example of embedded encoder.

Minimum distortion occurs at a maximum source rate η_{max}. Furthermore, the source encoder distortion-rate function [83, 152] can be approximated by

$$D_s = \delta 2^{2k(\eta_{max}-R)}, \tag{4.2}$$

where δ is the minimal distortion and k is a parameter depending on the encoder. This is a very general form that applies to the case of a Gaussian source with a MSE distortion or when the high-rate approximation holds. In the case of realistic encoders, we find that (4.2) constitutes a tight upper bound on the real distortion-rate curve. So the source coder output rate is one of the most important parameters for resource allocation for the voice encoder. Some practical speech encoder examples for rate adaptation are listed as follows:

1. GSM adaptive multirate (AMR) speech encoder uses eight source codecs with bit rates of 12.2, 10.2, 7.95, 7.40, 6.70, 5.90, 5.15, and 4.75 kbps.
2. The other example in practice is the Qualcomm code excited linear prediction (QCELP) variable-rate speech encoder. There are four data rates: full rate (9.6 kbps), half rate (4.8 kbps), quarter rate (2.4 kbps), and eighth rate (1.2 kbps).
3. Enhanced variable-rate codec (EVRC) is a speech codec used by CDMA networks. It was developed in 1995 to replace the QCELP codec. EVRC uses a relaxed code excited linear prediction technology, which improves speech quality with lower bit rates. EVRC supports three source rates of 9.6 kbps (full rate), 4.8 kbps (half rate), and 1.2 kbps (eighth rate). The lowest rate is not meant for speech signals, but for background noise.
4. EVRC was replaced with the selectable-mode vocoder (SMV) that retained the speech quality and at the same time improved network capacity.
5. Recently, however, SMV itself was replaced with the new CDMA2000 4GV codecs. 4GV is the next-generation 3GPP2 standards-based EVRC-B codec. 4GV is designed to allow service providers to dynamically prioritize voice capacity on their network as required.

There are many advantages of the variable-rate voice encoder. First, the system can reduce unnecessary data transmission during silence periods. A lot of capacity can be saved. Second, the system is allowed to have "soft capacity," which means that in the situation in which the number of users is greater than the normal system capacity, the average data rate for each user can be decreased slightly by forcing a small percentage of active speech frames to be encoded at a lower rate rather than a higher rate. Third, lower power is used during lower rate frames and consequently the average power consumption is reduced.

4.2.2   Scalability of Image/Video Encoders

In this subsection, we discuss motivations for scalable multimedia and the different types of scalability for multimedia encoders first. Then we explain the scalability in some of the popular image and video standards.

The scalability refers to the capability of recovering physically meaningful image or video information by decoding only partially compressed bit streams. There are many reasons for implementing scalable multimedia. First, the capabilities of different types of links can be significantly different. For example, for 2G wireless cellular phones, the data rate is several kilobits per second; for 3G wireless cellular phones, the data rate is up to several megabits per second; for WLANs, the data rate is up to 11 or 54 Mbps. Moreover, the capacities of links are also influenced by the fading channels and the number of users. So the transmissions of multimedia will be adapted to the variations of channel changes. Second, the QoS can be much different for different types of applications. For example, HDTV demands high-quality video transmissions and consequently high data rates, whereas the videophone may not need such a high quality of videos and may require a much lower data rate. In addition, the contents of videos and images affect the required overall data transmission volumes, which is another reason that scalability is important to multimedia transmission over wireless.

There are many possible applications for scalable multimedia. For example, the scalable multimedia transmission can provide important parts such as the base layer and less important parts such as the enhanced layers. Then the unequal error protection (UEP) can provide better error protection for the important parts while providing less error protection for the less important parts. With fixed and limited radio resources, this unequal error protection can provide a large performance gain for wireless transmission in response to channel error conditions. Another example is that the scalable transmission can offer the ability to adapt to the computing power at the receiving terminals.

Depending on the ability of scalability, the multimedia encoder can be classified as having coarse granularity, fine granularity, or an embedded encoder. For coarse granularity, there are only a few (two or three) layers. Each layer must be completely transmitted to enhance the recovered multimedia. For fine granularity, the multimedia transmission is divided into many small packets. The more packets transmitted, the better the reconstructed quality. For the extreme case, the bit stream can be truncated at any point. We call this an embedded encoder.

The scalability can be accomplished in various ways. We list the following types of scalability as examples [301].

- Spatial Scalability
  Spatial scalability is the ability to represent the same image or video with different resolutions or sizes. To produce the layered bit streams, a lower-resolution image is coded first to produce the first layer. Then the difference between the original image and the interpolated lower-resolution image is coded to produce the second layer. The spatial scalability is very useful when the receivers' displays have different sizes or when the progressive transmission is desired to give the user coarse quality first and then transmit the refinement to improve.
- Data Partitioning
  Data partitioning divides the bit stream of single-layer nonscalable multimedia data into two layers. The first layer comprises the critical information and is called the base

layer, which provides the basic multimedia service. The second layer provides improvement for the reconstructed multimedia and is called the enhancement layer, which can be transmitted if there are available transmission resources left. The enhancement layer can also have multiple sublayers to progressively enhance the reconstructed multimedia qualities.

- Temporal Scalability

  Temporal scalability for video is the ability of varying of frame rates. For example, for videoconferencing, the required frames per second may be low, whereas for the high-quality video broadcast, it is necessary for a large number of frames per second, because of the dramatic scene changes. Typically, a layered video codec with temporal scalability is encoded in such a way that the base layer includes all the even frames and the enhancement layer includes all the odd frames.

- SNR Scalability

  The image and video qualities are usually measured by the peak signal-to-noise ratio (PSNR). The scalability for PSNR is typically accomplished by quantizing with different quatization steps. The encoder first uses a coarse quantizer for the first layer. Then the subsequent layer contains the quantized difference between the original image and the reconstructed image from all the previous layers. In the receiver side, the image or video is reconstructed layer by layer as well. When SNR scalability is used, UEP can also be employed.

- Frequency Scalability

  Frequency scalability is applied when the multimedia frames carry different frequency components. The low-frequency components can provide a blurred version for the reconstructed multimedia. The higher-frequency components provide sharper versions. When the high-frequency components are dropped, frequency scalability can be obtained. This is especially useful for subband/wavelet or block-based decompositions.

- Hybrid Scalability

  Scalable encoders can combine different scalabilities to form a hybrid scalability for certain applications to fit different practical situations. Some examples are shown in [301], such as spatial and temporal hybrid scalability, SNR and spatial hybrid scalability, SNR and temporal hybrid scalability, and SNR, spatial, and temporal hybrid scalability.

Finally, we discuss the scalability in some of the existing multimedia standards. There are mainly three types of devices for digital video communications: digital television, video phone, and personal computer. Multimedia standards are developed to enable industry to provide multimedia services to customers with these devices. In the following discussion, we list some of the most important standards.

- JPEG and JPEG2000

  JPEG is the image compression standard developed by the Joint Photographic Experts Group. It works best on natural images (scenes). It uses a discrete cosine transform algorithm. JPEG2000 works best for computer-generated imagery and utilizes a discrete wavelet transform algorithm. Both standards have architectures with forward transform, quatiziation, and entropy coding. In addition, JPEG2000 has a layered structure and has many different functions. For a similar quality of PSNR, JPEG2000 compresses almost twice as much as JPEG. Both standards have a wide range of applications such as Internet, digital photography, medical imaging, wireless imaging, document imaging, digital cinema, image archives and databases, surveillance printing, and scanning facsimile.

- MPEG1, MPEG2, and MPEG4

These video standards were developed by the ISO, a network of National Standards Institutes from 148 countries working in partnership with international organizations, governments, industry, business, and consumer representatives.

MPEG1 was designed to produce VHS quality video at rate of 1.2 Mbps, which is suitable for CD-ROM. MPEG2 was designed to produce TV-quality pictures at data rates of 4–8 Mbps and high quality at 10–15 Mbps. It has a wide range of applications such as HDTV, DVD, and communications. The MPEG4 standard is designed to satisfy the requirements of the new generation of highly interactive multimedia applications such as interactivity with individual objects, scalability of contents, and a high degree of error resilience.

- H261, H263, and H264

These video standards have their roots in the ITU, where governments and the private sector coordinate global telecom networks and services. The main target for these video standards is for videoconferencing over existing, Public Switched Telephone Network (PSTN), Integrated Services Digital Network (ISDN), and telecommunication networks.

H.261 came out in 1990, intended for videoconferencing over PSTN synchronous circuits. H.261 is designed to run at multiples of 64-kbps data rates from 1x to 30x. These correspond to ISDN and E1/T1 circuits. H.261 is not designed for a packet-based network and does not work well over frame relay or transmission control protocol/Internet protocol (TCP/IP). When the next-generation standard ITU-T SG15 H.263 came out in February 1995 it then made H.261 obsolete. H.263 performs far better on packet networks like the Internet and can support more data rates and image sizes. Designed for low-bit-rate communications, it has replaced H.261 for videoconferencing in most applications and also dominates Internet videostreaming today.

H.264 is a new video-coding algorithm developed jointly by the ITU and the MPEG. H.264 provides a far more efficient mechanism for compressing and decompressing motion videos. This mechanism or algorithm requires significantly less bandwidth to transmit a motion image than has previously been possible. For videoconferencing, H.264 requires only 50% of the previously required bandwidth to provide the same quality of image. This also means that, if current bandwidth is maintained, substantially higher video quality will be achieved.

Overall, in the preceding subsection, the scalability of the image/video encoder was discussed, which provides one dimension of freedom to control the reconstructed qualities of image and video. Then the current image/video standards were briefly reviewed.

By utilizing the scalability, the encoder can adapt its behavior according to the link conditions, demand of QoS, receiver complexity, and others. Consequently the resource allocation in the wireless networks can be improved.

4.3   Rate Control for Network/MAC Layer

In this section, we discuss the rate control for the network/MAC layer. For the single-user case, we discuss the basic idea of queuing theory. For the multiple-user case, we briefly review the current packet multiple-access protocols discussed in the next chapter.

4.3.1   The Basics of Queuing Theory

For data transmission over wireless networks, one of the most important performance measures is the average delay required for delivering a packet from the source to the destination. The delay is influenced by the packet arrival rate, service rate, and others. Moreover, the maximal buffer size is also an important issue for practical implementation. In this subsection, we briefly review the related queuing theory, which is a primary methodological framework for analyzing network delay. Then we give some examples of the formulations of rate control for data transmissions. In addition, we also review the recently developed concepts of effective bandwidth and effective capacity.

Before we proceed further, let's understand the different components of delay in a messaging system. The total delay experienced by messages can be classified into the following categories:

- Processing Delay
  This is the delay between the times of receipt of a packet for transmission to the point of putting it into the transmission queue. On the receive end, it is the delay between the time of reception of a packet in the receive queue to the point of actual processing of the message. This delay depends on the CPU speed and CPU load in the system.
- Queuing Delay
  This is the delay between the point of entry of a packet in the transmit queue to the actual point of transmission of the message. This delay depends on the load on the communication link and on how many packets are currently buffered.
- Transmission Delay
  This is the delay between the transmissions of the first bit of the packet to the transmission of the last bit. This delay depends on the speed of the communication link.
- Propagation Delay
  This is the delay between the points of transmission of the last bit of the packet to the point of reception of the last bit of the packet at the other end. This delay depends on the physical characteristics of the communication link.
- Retransmission Delay
  This is the delay that results when a packet is lost and has to be retransmitted. This delay depends on the error rate on the link and the protocol used for retransmissions.

We will be dealing primarily with queuing delay. We begin our analysis of queuing systems by understanding Little's theorem:

THEOREM 13 *The average number of customers N can be determined from*

$$N = \lambda T, \tag{4.3}$$

where λ is the average customer arrival rate and T is the average service time for a customer.

Proof of this theorem can be obtained from any standard textbook on queuing theory [28]. Here we focus on an intuitive understanding of the result. Consider the example of a network in which the packet arrival rate (λ) doubles but the packets still spend the same amount of time in the network (T). This will double the number of packets in the network (N). By the same logic, if the packet arrival rate remains the same but the packet service time doubles, this will also double the total number of packets in the network.

With Little's theorem, we have developed some basic understanding of a queuing system. To further our understanding we classify the characteristics of a queuing system that have an impact on its performance. For example, queuing requirements will depend on the following factors:

- Arrival Process
  The probability density distribution determines customer arrivals in the system. In a messaging system, this refers to the message arrival probability distribution.
- Service Process
  The probability density distribution determines customer service times in the system. In a messaging system, this refers to the message transmission time distribution. Because message transmission is directly proportional to the length of the message, this parameter indirectly refers to message length distribution.
- Number of Servers
  In a messaging system, this refers to the number of links between the source and destination nodes.

Based on the preceding characteristics, queuing systems can be classified by the following convention: A/S/n, where A is the arrival process, S is the service process, and n is the number of servers. A and S can be any of the following:

- M (Markov): exponential probability density,
- D (Deterministic): all customers have the same value,
- G (General): any arbitrary probability distribution.

The following are examples of queuing systems that can be defined with this convention:

- M/M/1:
  This is the simplest queuing system to analyze. Here the arrival and service times are negative exponentially distributed (Poisson process). The system consists of only one server. This queuing system can be applied to a wide variety of problems as any

system with a very large number of independent customers can be approximated as a Poisson process. Using a Poisson process for service time, however, is not applicable in many applications and is only a crude approximation.

- M/D/n:
  Here the arrival process is Poisson and the service time distribution is deterministic. The system has n servers (e.g., a ticket booking counter with n cashiers). Here the service time can be assumed to be the same for all customers.

- G/G/n:
  This is the most general queuing system in which the arrival and service time processes are both arbitrary. The system has n servers. No analytical solution is known for this queuing system.

One of the examples for queuing theorem is the service in a supermarket. The customers come to the checkouts randomly with some rates. There are some employees at the checkouts to help the customers. Depending on the number of checkouts and employees as well as the service speeds, the queue lengths and average waiting times can be different. Queuing theorem can mathematically describe these behaviors.

With the knowledge of basic queuing theory, we now give some possible formulated problem to illustrate the design concerns of wireless communication networks.

The first concern is delay. For different applications, the delay constraints are different. For example, for the voice packet, the delay requirement is very strict, because the delayed packets can significantly reduce voice quality, whereas for services like e-mail, the delay can be arbitrarily long. So the problem formulation for such a kind of problem is usually constrained by the maximal delays for the specific types of applications. The optimization is performed by adapting the service rate from the link or the incoming packet rate from the source.

The second possible concern is maximal buffer size. Because of the practical implementation, the buffer size of the communication system is limited. So the problem formulation for this kind of problem has the constraint of the maximal buffer size. The resource allocation scheme tries to prevent buffer overflow. If the buffer does overflow, some packets will be selectively dropped.

One effective way to analyze the delay is the theory of effective bandwidth. The effective bandwidth is widely used in the wired network. Recently some works have applied these ideas to wireless channels and a link-layered channel model is defined as effective capacity [338]. We will utilize this model in some applications in Part III of this book.

4.4   Adaptation for PHY Layer

Channel protection adaptation such as adaptive channel coding and adaptive modulation schemes can be viewed as a low-complexity alternative for mitigating the channel-quality fluctuations of wireless channels. These schemes match the channel protection ability for transmission errors, according to the channel conditions. In this section, we discuss the rate adaptation for channel protection according to the channel conditions. We discuss the adaptive channel coding first. Then we discuss adaptive modulation. Adaptive coded modulation is a combination of the channel coding and adaptive modulation. Finally, we discuss the adaptive processing gain for CDMA networks.

4.4.1 Adaptive Channel Coding

The error-correcting code can be classified into three categories: blocking code, convolutional code, and turbo code. In this subsection, we briefly review how to adapt these codes.

Block codes are referred to as (n, k) codes. A block of k information bits is coded to become a block of n bits. $n - k$ bits are used for the channel protection. For a fixed block size n, if $n - k$ is large, the code usually has a large Hamming distance [235] and consequently has a high capability of detecting and correcting errors. If the transmitter knows the exact or part of channel information, it can select the different block codes to cope with different channel conditions, i.e., if the channel is bad, the information per block will be reduced; otherwise, more information can be transmitted within each block. The most popular block coding schemes are the Bose–Chaudhuri–Hochquenghem (BCH) code and the Reed–Solomon (RS) Code.

The BCH code is a multilevel, cyclic, error-correcting, variable-length digital code used to correct errors up to approximately 25% of the total number of digits. BCH codes are not limited to binary codes, but may be used with multilevel phase-shift keying (PSK) whenever the number of levels is a prime number or a power of a prime number. The RS code is an important subclass of nonbinary BCH error-correcting codes in which the encoder operates on multiple bits rather than on individual bits. The RS code is therefore ideally suited to use with m-ary modulation and copes well with bursts of errors. The RS code is specified as $RS(n,k)$ with s-bit symbols. This means that the encoder takes k data symbols of s bits each and adds parity symbols to make an n symbol codeword. There are $n - k$ parity symbols of s bits each. A RS decoder can correct up to t symbols that contain errors in a codeword, where $2t = n - k$. So, by adapting t, we can have different levels of channel protections.

A convolutional code extends the concept of a block code to allow memory from block to block. Each encoded symbol is therefore a linear combination of information symbols in the current block and a selected number of preceding blocks. Therefore, for example, if the final output is a 1 followed by a 0, then these two digits can have arrived at only a certain sequence of 0s and 1s preceding them. The longer the sequence, the easier it becomes for the receiver to detect where the received sequence deviates from a possible sequence and to correct one or more errors. Decoding of convolutional codes is based on the principle of the Viterbi decoding algorithm or sequential decoding.

Convolutional codes are commonly specified by three parameters; (n, k, m), where n is the number of output bits, k is the number of input bits, and m is the number of memory registers. The quantity k/n, the code rate, is a measure of the efficiency of the code. Commonly k and n parameters range from 1 to 8, m from 2 to 10, and the code rate

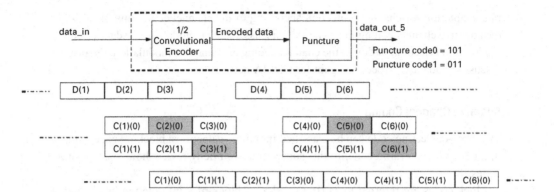

Figure 4.2 Illustration of puncture convolutional code.

from 1/8 to 7/8. To adapt the code rate for different channel protections, the transmitter can either use different encoders or punctured convolutional code.

A punctured convolutional code is one in which some bits are discarded at the transmitter to reduce the amount of data to be transmitted. This also reduces the amount of protection provided to the transmitted data by effectively reducing the coding rate. One illustrative example is shown in Figure 4.2. A rate-compatible puncture convolutional (RCPC) code [101] is a special kind of punctured convolutional code, where the rate compatibility restriction on the puncturing tables ensures that all code bits of high-rate codes are used by the lower-rate codes. This is very useful when the receivers detect error corruption and demand progressive error protections. This means the transmitter can send some extra bits to combine with the previously received code to form a new code with a lower data rate and higher error-correction ability. Moreover, only one kind of transceiver is needed for the RCPC codes with different rates.

With convolutional channel coding, the union bound [235] for the BER using channel coding can be expressed as

$$P_c(\Gamma) = \sum_{d=d_{\text{free}}}^{\infty} a_d P_d(\Gamma), \qquad (4.4)$$

where d_{free} is the free distance of the convolutional code, a_d is the total number of error events of weight d, and $P_d(\Gamma)$ is the probability that an incorrect path at distance d from the correct path is being chosen by the Viterbi decoder. When the hard decision is applied, $P_d(\Gamma)$ can be given by

$$P_d(\Gamma) = \begin{cases} \sum_{k=(d+1)/2}^{d} \binom{d}{k} (P_b)^k (1 - P_b)^{d-k}, & d \text{ is odd} \\ \frac{1}{2} \binom{d}{d/2} (P_b)^{d/2}(1 - P_b)^{d/2} \\ + \sum_{k=d/2+1}^{d} \binom{d}{k} (P_b)^k (1 - P_b)^{d-k}, & d \text{ is even} \end{cases}, \qquad (4.5)$$

where P_b is the uncoded BER depending on the modulations.

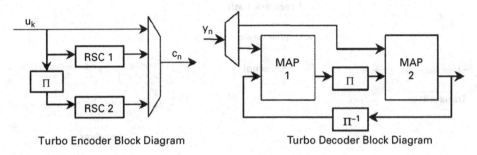

Turbo Encoder Block Diagram　　　　Turbo Decoder Block Diagram

Figure 4.3 Block diagram of turbo code.

The recently developed turbo code scheme [25] is capable of asymptotically achieving Shannonian performance limits at the cost of high complexity and latency. The basic idea of turbo codes is employing two convolutional codes in parallel with some kind of interleaving in between. The system is illustrated in Figure 4.3. The transmitter can adapt the coding rate by puncturing the output data like the punctured code previously mentioned.

A soft-in–soft-out (SISO) decoder receives as input a "soft" (i.e., real) value of the signal. The decoder then outputs for each data bit an estimate expressing the probability that the transmitted data bit was equal to one. In the case of turbo codes, there are two decoders for outputs from both encoders. Both decoders provide estimates of the same set of data bits, albeit in a different order. If all intermediate values in the decoding process are soft values, the decoders can gain greatly from exchanging information, after appropriate reordering of values. Information exchange can be iterated a number of times to enhance performance. At each round, decoders reevaluate their estimates, using information from the other decoder, and only in the final stage will hard decisions be made, i.e., each bit is assigned the value 1 or 0. Such decoders, although more difficult to implement, are essential in the design of turbo codes.

Low-density parity-check (LDPC) codes were invented by Gallager in 1960. They were not recognized for many years because they were thought to be impractical. Recently they were considered to have an iterative decoding structure. The inherent parallelism in decoding LDPC codes suggests their use in high-data-rate systems. Their performance is similar to turbo codes but they may have some implementation advantages. The adaptive rate can be obtained by puncturing lower rate codes. Some other works can be found in [193, 237, 254, 255].

4.4.2　Adaptive Modulation

Adaptive modulation is a promising technique to increase the data rate that can be reliably transmitted over fading channels. For this reason some form of adaptive modulation is being proposed or implemented in many next-generation wireless systems. The basic premise of adaptive modulation is a real-time balancing of the link budget in flat fading through adaptive variations of the transmitted power level, symbol

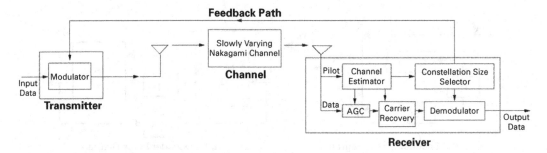

Figure 4.4 Structure of adaptive modulation.

transmission rate, constellation size, BER, or any combination of these parameters. Thus, without wasting power or sacrificing BER, the adaptive modulation schemes provide a higher average link spectral efficiency (bps/Hz) by taking advantage of fading through adaptation.

The basic structure of adaptive modulation scheme can be seen in Figure 4.4. The input data are modulated and transmitted to the channel. The receiver has channel estimation, automatic gain control, and carrier recovery for the demolulator to recover the input data. The best constellation size is selected according to the recovered data quality and is then sent back via the feedback path. If the link quality is overly good, a higher rate per symbol may be selected for the future transmission. If the link quality is bad, a lower rate per symbol will be chosen. The available adaptive modulation methods are adaptive MQAM, adaptive MPSK, adaptive PAM, adaptive orthogonal signal coherent detection, and others.

The performance approximations for the adaptive modulations are extensively studied in [49]. The BER approximation over an additive white Gaussian noise (AWGN) channel is given as a function of SNR Γ and rate T as

$$\text{BER}(\Gamma) \approx c_1 \exp\left(\frac{-c_2\Gamma}{2^{c_3 T} - c_4}\right), \tag{4.6}$$

where c_1–c_4 are constants that depend on the modulation methods. This approximation allows a continuous rate instead of a discrete rate. The approximation serves as a relaxation for adaptive modulation analysis using continuous programming methods.

In [312], the BER performances of BPSK, QPSK, 16 MQAM, and 64 MQAM modulation over Rayleigh fading are given by the following as functions of received symbol SNR:

$$P_b^{\text{BPSK}} = 0.5\left(1 - \sqrt{\frac{\Gamma}{1+\Gamma}}\right), \tag{4.7}$$

$$P_b^{\text{QPSK}} = 0.5\left(1 - \sqrt{\frac{\Gamma}{2+\Gamma}}\right), \tag{4.8}$$

$$P_b^{16\text{QAM}} = 0.5\left[\left(1 - \sqrt{\frac{\Gamma}{10+\Gamma}}\right) + \left(1 - \sqrt{\frac{9\Gamma}{10+9\Gamma}}\right)\right], \tag{4.9}$$

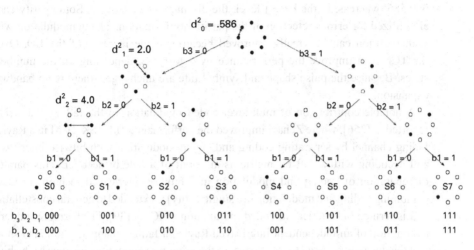

Figure 4.5 Partition of 8 PSK constellation to ever-increasing Euclidean distance subsets.

$$P_b^{64QAM} = \frac{1}{24}\left(14 - 7\sqrt{\frac{\Gamma}{42+\Gamma}} - 6\sqrt{\frac{9\Gamma}{42+9\Gamma}} + \sqrt{\frac{25\Gamma}{42+25\Gamma}}\right.$$

$$\left. -2\sqrt{\frac{81\Gamma}{42+81\Gamma}} - \sqrt{\frac{121\Gamma}{42+121\Gamma}} + \sqrt{\frac{169\Gamma}{42+169\Gamma}}\right). \quad (4.10)$$

These four modulation methods are widely used. The channel coded performance can be obtained by using (4.5).

4.4.3   Adaptive Coded Modulation

The modulation and coding schemes are usually concatenated directly. The purpose of modulation is transmitting as much information as possible given a limited bandwidth, while the purpose of channel coding is for error protection. Some joint schemes can provide additional gain. In this subsection, we discuss two examples.

Trellis code modulation (TCM) [315] was developed by Ungerboeck to show that, for bandlimited channels, substantial coding gains can be achieved by convolutional coding of signal levels (rather than coding of binary source levels). In TCM, modulation and coding are joined to increase the Euclidean distance (ED) between signal sequences. TCM applies the parity check on a per-symbol basis instead of the older technique of applying it to the bit stream and then modulating the bits. The key idea is called "Mapping by Set Partitions." This idea is to group the symbols in a treelike fashion and then separate them into two limbs of equal size. At each limb of the tree, the symbols are further apart. Although the tree is hard to visualize in multidimensions, a simple one-dimensional example illustrates the basic procedure, as shown in Figure 4.5. The figure shows how the 8 points of 8 PSK are successively partitioned into different groups such that the ED is increasing at each level. At the top level, the ED is 0.586, whereas, in the lowest level, the distance increases to 2. Some parity checks are utilized for error protection of the partition information. The demodulation within each partition can be greatly improved because of the increase of the ED. Overall the TCM can improve the performance by 3 to 6 dB, depending on the number of states. Because the pulse shape and symbol rate are unchanged, there is no bandwidth expansion.

A notable contribution of multilevel coded modulation in the fading channel was reported in [356], where Zehavi improved the performance of 8 PSK TCM in a Rayleigh fading channel by separating coding and signal modulation with bitwise interleaving, and decoding with a soft bit metric as an input of a Viterbi decoder. This paradigm of separation of coding and modulation in a bitwise interleaved fashion (instead of combining coding and modulation in a single entity) is extended to general constellations, and is termed bit-interleaved coded modulation (BICM) [39]. Furthermore, under the assumption of an independent interleaved Rayleigh fading channel, Caire also showed that the diversity gain of BICM is equal to the minimum Hamming distance of the binary code. Therefore, for the same coding complexity, BICM offers higher diversity gains over conventional TCM whose Hamming distance is limited to the number of distinct symbols rather than bits. This superior code diversity property of BICM in combination with the design flexibility that is due to the separation of coding and modulation has motivated many applications. Adaptive BICM was introduced in [225], where the authors proposed a time-domain adaptation of constellation size under the frequency nonselective fading channel.

4.4.4   Adaptive Processing Gain for CDMA

To deliver flexible, variable-data-rate services with high spectrum efficiency is the major objective of the design of future wireless systems. Current CDMA systems such as IS-95, CDMA2000, and UMTS standards can realize this objective. For voice communications, because of the strict delay QoS constraint, the problem is how to adapt the transmission power to maintain the targeted level SNR so as to maintain the QoS, whereas for other services, such as data and multimedia, users may require a variable bit rate. Thus, to optimally distribute the limited system resources among users for different types of services, CDMA systems need to have the ability for variable-rate services. In this section, we discuss the variable bit rates for CDMA system and one example of UMTS.

There are three approaches for variable bit rates in CDMA systems:

1. fixed processing gain, variable chip rate,
2. fixed chip rate and processing gain, multiple signatures,
3. fixed chip rate, variable processing gain.

In the first method, the chip rate is varied along with the symbol duration to keep the processing gain fixed. As the duration of a symbol is decreased, the bit rate of the user is increased while its immunity to multiaccess interference remains unchanged. However, the transmission bandwidth of the signal is increased, which causes unequal bandwidth expansion and complicates the frequency planning. Therefore the methods have not

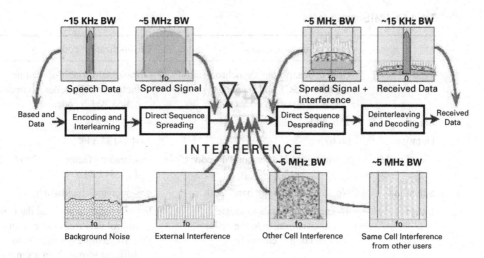

Figure 4.6 Interference and spreading diagram for CDMA.

been implemented. In the second approach, the symbol duration and the processing gain are kept fixed, and the user's bit rate is increased by assigning with multiple signatures. This is known as multicode CDMA. This method suffers self-interferences caused by transmitting several nonorthogonal signatures from the same user. This method was proposed and implemented by Bell Labs. In the third approach, the symbol duration is varied over a fixed-rate chip sequence. As the symbol duration is shortened the bit rate is increased, but at the same time the processing gain is decreased. Consequently, the immunity to multiaccess interference is decreased as the bit rate is increased. This method is implemented in UMTSs.

In UMTS the orthogonal variable-spreading factor (OVSF) is used for variable bit rates. TDD WCDMA uses spreading factors 4–512 to spread the baseband data over the 5-MHz band. The spreading factor in decibels indicates the process gain (spreading factor 128 = 21 dB process gain). The interference margin is calculated by

$$\text{interference margin} = \text{process gain} - (\text{required SNR} + \text{system losses}).$$

The required SNR is typically about 5 dB. System losses are defined as losses in the receiver path. System losses are typically 4–6 dB. The interference and spreading block diagram is shown in Figure 4.6. In Table 4.1, different types of codes are listed for variable bit rates of a UMTS.

4.5 Source–Channel Coding with Power Control

4.5.1 Joint Source–Channel Coding

Shannon's separation theorem states that source coding (compression) and channel coding (error protection) can be performed separately and sequentially while maintaining

Table 4.1 UMTS spread codes

	Synchronization codes	Channelization codes
Type	Gold codes, primary synchronization codes (PSC), and secondary synchronization codes (SSC)	Orthogonal variable-spreading factor (OVSF) codes, sometimes called Walsh codes
Length	256 chips	4–512 chips
Duration	66.67 s	1.04 s–133.34 s
No. of codes	1 primary code/16 secondary codes	= spreading factor 4 ... 256 UL, 4 ... 512 DL
Spreading	No, does not change bandwidth	Yes, increases bandwidth
Usage	To enable terminals to locate and synchronize to the cells' main control channels	UL: To separate physical data and control data from same terminal DL: To separate connection to different terminals in a same cell

optimality. However, this is true only in the case of asymptotically long block lengths of data. In many practical applications, the conditions of Shannon's separation theorem neither hold nor can be used as a tight approximation. For example, in real-time communications, like videoconferencing, any delay greater than 100 ms is not tolerable. Joint source–channel coding jointly optimizes the source–channel coders when the assumptions are invalidated, and thus achieves performance gains. So considerable interest has developed in various schemes of joint source–channel coding.

One easy example is a channel-capacity-limited video communication system. Both the source and channel coders need bits. Spending more bits on the source implies not enough channel protection, which leads to channel errors and received video quality is poor. Spending more bits on the channel implies enough protection and no transmission errors, but the source encoder is overcompressed and received video quality is again poor. There is a trade-off. A balanced point exists where the channel capacity is optimally allocated between source and channel to achieve the best-received video quality. In Figure 4.7, we show an example of trade-off for the source rate and the channel rate, where the total transmission bandwidth is fixed. We can see that the received PSNR will increase with the increasing of the source rate. This is because the source-introduced errors are reduced. But when the source rate is higher than a certain threshold, the reconstructed images quality drops quickly. This is because the channel-introduced errors begin to dominate the performance.

The joint source–channel-coding research pays a lot of attention to developing a low-complexity mechanism for the determination of rate allocation for source–channel coding of progressive sources. The system diagram is shown in Figure 4.8, in which the source and channel coders can be any family of channel codes and any progressive source coder. In addition, a rate-compatibility condition on the error-control code can be applied. The rate-control unit solves the optimization problem to devise an

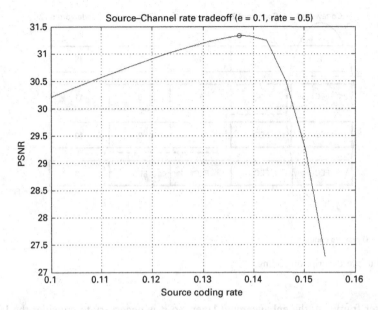

Figure 4.7 Example of a trade-off for source rate and channel rate.

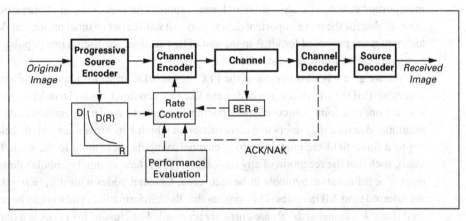

Figure 4.8 Block diagram for a typical joint source–channel coding.

efficient and fast solution. The reconstructed distortion can be written in the following form:

$$D = D_s + D_c \qquad (4.11)$$

where D_s is the source-coder-induced error and D_c is the channel-induced error. The problem is how to change the parameters to balance D_s and D_c under the limited resources to achieve the minimal recovered distortion D.

For multimedia data, different parts of the data have different importance. For example, the base layer of MPEG4 is more important and the beginning frames are more important

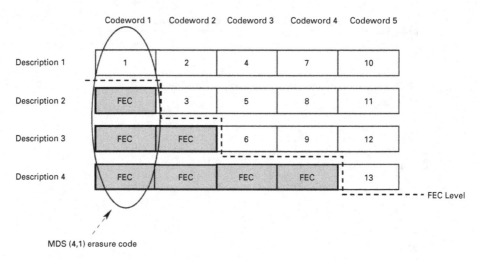

Figure 4.9 Multiple-description coding.

than the later frames in the enhancement layer. So it is necessary to consider the UEP for multimedia data. UEP can be implemented by adapting the channel-coding rate, modulation, or the processing gain, according to the importance of the data. Because the unimportant data can be assigned with fewer resources for transmission, more resources are available for the more important data to transmit with better channel protection. Many techniques are proposed for UEP. In the rest of this section, we discuss one popular UEP technique called multiple-description coding (MDC).

First, we give a brief overview of the FEC-based MDC in which maximum-distance-separable (MDS) (n, k) erasure codes are used to construct multiple independent bit streams under a joint source–channel coding framework. An (n, k) erasure code with minimum distance d_{min} refers to a construction in which k information symbols belonging to a finite field are encoded into n channel symbols (belonging to the same finite field), such that the reception of any $(n - d_{min} + 1)$ of the n channel symbols allows the original k information symbols to be recovered. Channel codes with $d_{min} = n - k + 1$ are referred to as MDS codes. This implies that the k information symbols can be recovered if any k channel symbols are correctly received. The class of RS codes is a popular class of codes possessing this property.

Based on this property, an unequal loss protection framework is shown in Figure 4.9, in which unequal amounts of parity symbols, using different-rate RS codes, are applied to embedded descriptions. This framework can provide a graceful degradation of delivered image quality as packet losses increase. In Figure 4.9, we illustrate the general mechanism for converting an embedded bit stream from a source encoder into multiple descriptions in which contiguous information symbols are spread across the multiple descriptions. The information symbols are protected against channel errors by systematic $(n = 4, k)$ MDS codes, with the level of protection depending on the relative importance of the information symbols. For example, in Figure 4.9, we show the construction of a $(4, 1)$ systematic MDS code (codeword 1) in which erasure of any three descriptions still allows us to reconstruct information symbol 1. For codeword 1, as long as one description arrives correctly, it can be recovered. For codeword 2, two descriptions need to be decoded correctly. For codewords 3 and 4, 3 out of 4 descriptions will be successfully received. For codeword 5, no error protection is added.

4.5.2   Degrees of Freedom

In this subsection, we first illustrate a single-user communication case, in which the user can adapt its rate, power, and BER to maximize its spectrum efficiency. The goal is to determine the impact on spectral efficiency of adapting various modulation parameters under different constellation restrictions and BER constraints, for a large class of modulation techniques and fading distributions. The overall rate can be written as

$$R = W \sum_{i=1}^{M-1} k_i \int_{\gamma_i}^{\gamma_{i+1}} P_r(\gamma) d\gamma \text{ bps}, \qquad (4.12)$$

where W is the bandwidth, R/W is the spectrum efficiency, k_i is the rate for modulation (for example, $k_i = 1$, for BPSK, $k_i = 2$, for QPSK, etc.), M is the number of modulations, γ_i is the threshold for modulation changing, and $P_r(\gamma)$ is the distribution for SNR γ. Suppose $P(\gamma)$ is the power for SNR γ. The average power can be written as

$$P_{\text{ave}} = \int_0^{\infty} P(\gamma) P_r(\gamma) d\gamma. \qquad (4.13)$$

The average BER, for MQAM, MPSK, and general form can be written as

$$\overline{\text{BER}} = \frac{\sum_{i=1}^{M-1} k_i \int_{\gamma_i}^{\gamma_{i+1}} \text{BER}(\gamma) P_r(\gamma) d\gamma}{\sum_{i=1}^{M-1} k_i \int_{\gamma_i}^{\gamma_{i+1}} P_r(\gamma) d\gamma}. \qquad (4.14)$$

Next, we study how to formulate an optimization problem across layers. The optimization can have the degrees of freedom such as rate, power, and BER. There are also trade-offs and constraints. For example, when the desired rate is high or the targeted BER is low, the power needs to be high. When the power is fixed, the higher the rate, the larger the BER. The problem can be formulated to optimize a certain performance metric, under some practical constraints. One common example to optimize the spectrum efficiency can be written as

$$\max R, \qquad (4.15)$$

$$\text{subject to} \begin{cases} \text{power constraint, such as } P_{\text{ave}} < \bar{P} \\ \text{BER constraint, such as } \overline{\text{BER}} < \text{BER}_0 \end{cases},$$

where BER_0 is the targeted BER. The parameters for adaptation can be power, rate, and even the desired BER. This problem can be solved using the Lagrangian method, which we will discuss in detail in Part II of this book. Because all the degrees of freedom are correlated with each other, using just one or two degrees of freedom in adaptation yields close to the maximum possible spectral efficiency obtained by utilizing all degrees of freedom. Therefore the parameters to adapt are chosen based on implementation considerations.

Next, we consider K cochannel users that may exist in a distinct cell. Assuming coherent detection is possible so that it is sufficient to model this multiuser system by an equivalent baseband model. We assume that the link gain is stable within each transmission frame. For the uplink case, the signal at the ith base station output is given by

$$x_i(t) = \sum_{k=1}^{K} \sqrt{G_{ki} P_k} s_k(t) + n_i(t),$$ (4.16)

where G_{ki} is the path loss from the kth mobile to the ith base station, P_k is the kth link's transmitting power, $s_k(t)$ is the message symbol, and $n_i(t)$ is the thermal noise. Then we can express the sampled received signal as

$$x_i(n) = \sum_{k=1}^{K} \sqrt{P_k G_{ki}} s_k(n) + n_i(n),$$ (4.17)

where $n_i(n)$ is the sampled thermal noise. The ith base station's output SINR is

$$\Gamma_i = \frac{P_i G_{ii}}{\sum_{k \neq i} P_k G_{ki} + N_i}, \quad i = 1, \ldots, K,$$ (4.18)

where $N_i = E\|n_i\|^2$. Write the preceding equations in matrix form. We have

$$(\mathbf{I} - \mathbf{DF})\mathbf{P} = \mathbf{u},$$ (4.19)

where $\mathbf{P} = [P_1, \ldots, P_K]^T, \mathbf{u} = [u_1, \ldots, u_K]^T, u_i = \Gamma_i N_i / G_{ii}, \mathbf{D} = \text{diag}\{\Gamma_1, \ldots, \Gamma_K\},$ and

$$[\mathbf{F}]_{ij} = \begin{cases} 0, & \text{if } j = i \\ \frac{G_{ji}}{G_{ii}}, & \text{if } j \neq i \end{cases}.$$ (4.20)

To have a feasible solution in (4.19), the spectral radius $\rho(\mathbf{DF})$, i.e., the maximum eigenvalue of $\mathbf{DF}$, must be less than 1 [85]. We will see that the overall transmit power is strongly affected by the value of $\rho(\mathbf{DF})$.

Adaptive modulation provides the system with the ability to match the effective bit rate (throughput) according to the interference and channel conditions. MQAM is the modulation method that has high spectrum efficiency. For a given system and a targeted BER, the throughput can be written as a function of the received SINR. Let T_i denote the ith user's throughput, which is the number of bits sent within each transmitted symbol. The throughput is assumed continuous, and the ith user's BER using different MQAM modulation with different throughput can be approximated as [49, 241]

$$\text{BER}_i \approx c_1 e^{-c_2 \frac{\Gamma_i}{2^{T_i} - 1}},$$ (4.21)

where $c_1 \approx 0.2$ and $c_2 \approx 1.5$ for MQAM. Rearrange (4.21). For a specific BER, the ith link's throughput is given by

$$T_i = \log_2\left(1 + c_3^i \Gamma_i\right),\tag{4.22}$$

where $c_3^i = -\frac{c_2^i}{\ln(\text{BER}_i/c_1^i)}$. Without loss of generality, we assume each user has the unit bandwidth. Define the overall network throughput as $T = \sum_{i=1}^K T_i$.

From the system designer point of view, one common goal is to maximize the overall network throughput under the constraints that the power allocation is feasible. The problem we will address becomes

$$\max_{T_i, P_i} T = \sum_{i=1}^K T_i \tag{4.23}$$

subject to power feasibility: $\rho(DF) < 1$.

The problem can be solved in the following two fashions:

- centralized algorithms, in which the channel gain information is collected and optimization is conducted in a central node. The optimization technique for such a problem can be the Lagrangian method or some convex/nonlinear optimization technique. However, the overhead for signaling can cause a considerable communication burden.
- distributed iterative algorithms, in which the distributed node update its power or rate according to its local information. For a fixed rate requirement, the problem is pure distributed power-control problem and has been discussed in the previous chapter. If the rates can be adaptive, coordination among distributed nodes to change the rates should be carefully designed, because the system might not be feasible for sufficiently large rates. Moreover, the feasible range for the problem in (4.23) is not convex. The rate adaptation and power control can fall into some local optima with low system performances.

The preceding two types of approaches can achieve the potential gains of rate adaptation and power control for the optimization in multiuser communication networks such as the problem in (4.23). The next question is that of where the gain comes from, power control or rate adaptation. In general cases of wireless networks, some observations can be observed as follows:

1. Using rate adaptation, even without any power control, provides a significant throughput advantage over using pure power control.
2. Combining rate adaptation and a suitable power-control scheme leads to a slightly higher throughput compared with rate adaptation only.
3. Most of the throughput gains may be realized with a modest transmission power range.

From the preceding observations, using rate adaptation without power control appears to be the most promising case. It provides a throughput that is close to optimal and certainly higher than that provided by power control. However, specifically for cellular networks, there are some different observations, as follows:

1. Adaptive modulation was shown to provide performance gain throughout the sector of wireless cellular networks.
2. Combining adaptive coding, based on BICM [39], with adaptive modulation provides an extra gain in addition to the gain that was obtained with adaptive modulation without coding (the preceding case 1). Users near the edge of the sector experience a higher performance gain than those near the base station.
3. Combining power control with adaptive modulation provides a performance gain similar to, but less than, the previous case.
4. Minimal performance gain is obtained when power control is combined with adaptive modulation and adaptive coding.

With regard to implementation, both rate adaptation and power control impose practical challenges. For rate adaptation, different modulation and coding schemes have to be implemented. Moreover, there shall be a reliable channel to feed back the selected rate without any delay. For power control, the link quality shall also be monitored and fed back. In the most of current standards, the power control performs much more frequently than the rate adaptation.

5 Multiple Access and Spectrum Access

5.1 Introduction

The available wireless radio resources are very limited, while there are an increasing number of mobile users. It is necessary to share a communication channel or physical communication medium among multiple users. The multiple-access scheme is a general strategy to allocate the limited resources, such as bandwidth and time, to guarantee the basic QoS, improve the system performances, and reduce the cost for the network infrastructures. Whereas multiple access considers the problem of allocating limited radio resources to multiple users, spectrum access decides whether an individual user can access a certain spectrum.

The basic idea of the multiple-access scheme is to combine several signals for transmission on a certain shared medium (e.g., a wireless channel). The signals are combined at the transmitter by a multiplexor (a "mux") and split up at the receiver by a de-multiplexor. Based on how to divide the limited radio resources to multiple users, the multiple-access schemes can be classified as time-division multiple access (TDMA), frequency-division multiple access (FDMA), code-division multiple access (CDMA), space-division multiple access (SDMA), and others.

The multiple-access schemes need to be dynamically coordinated for a number of reasons: The users' data flows might not have data to transmit, the channel conditions are different for different users, and the QoS such as the delay constraints are different for different types of payloads. Based on how to coordinate access for the radio resources, multiple-access schemes can be classified into two types: scheduling and random access. In scheduling, there is a centralized control, the base station, that controls which user can transmit by using specific resources such as the bandwidth at different times. In random access, there is no such centralized control. Users access and utilize the resources in a distributed way. If conflicts of resource usage occur, certain mechanisms are employed to avoid conflicting again in the future. These two types of schemes are employed in different scenarios depending on the networks' situations. For example, in cellular networks, it is possible for centralized control from the base station where scheduling can be employed. On the other hand, in the WLAN, mobile users distributively share the limited bandwidth. As a result, random-access schemes are widely deployed.

For spectrum access, we first discuss channel allocation and opportunistic spectrum access. For the networks with infrastructures such as cellular networks, the goals are to reduce the interference and adaptively accommodate more mobile users. We also discuss the scenarios without infrastructure, such as WiFi ad hoc model, Bluetooth channel allocation, and cognitive radio. Second, we consider handoff for the mobility case in which the mobiles move from one service area to another. Different types of handoff schemes are discussed, and a 3G system example is illustrated. Finally, we investigate admission control for the new call.

The organization of the rest of this chapter is as follows: In Section 5.2, we briefly review the current duplexing methods and multiple-access methods that are used in the current wireless communication systems. In Section 5.3, the scheduling schemes are discussed in detail. A cross-layer model is introduced, then a framework and the design trade-offs are discussed, basic approaches are briefly reviewed next, and finally, the cross-layer approaches in the wireless networking and resource-allocation literature are explained. In Section 5.4, several popular random-access schemes are illustrated. WLAN random-access protocol is discussed. An example of 3G system multiple-access schemes are briefly explained in Section 5.5. Then we study the spectrum-access schemes in Sections 5.6, 5.7, and 5.8 for channel allocation, handoff, and admission control, respectively. The chapter is summarized in Section 5.9.

5.2   Multiple-Access Methods

In this subsection [247], we first discuss duplexing. Then we explain the basic methods for multiple access. Finally, we show the different methods used in current wireless communication systems.

In wireless communication, it is desirable for each user to transmit and receive simultaneously, which is called duplexing. There are two techniques for duplexing. Frequency-division duplexing (FDD) provides two distinct frequency bands for transmitting and receiving. Time-division duplexing (TDD) uses different time slots for forward and reverse links. There are several pros and cons between FDD and TDD. For FDD, the radio frequency must be carefully designed to reduce the RF cost and handle the different power of transmitted and received radio signals. For TDD, there are transmission delays and the system is sensitive to propagation delays. So TDD is often employed in cordless phone and fixed wireless networks. In Figure 5.1, an illustrative example is shown for the FDD case and the TDD case.

For multiple users' communication, multiple-access schemes are developed to share simultaneously the limited bandwidth of a radio spectrum. FDMA, TDMA, frequency-hopped multiple access (FHMA), CDMA, SDMA, and orthogonal frequency-division multiple access (OFDMA) are major access techniques [247]. These multiple-access techniques have been widely used in current wireless communication systems such as GSM, IS-95, CT2, and DECT. The details of the preceding multiple-access schemes are explained as follows:

- In a FDMA system, an individual user is assigned to a unique frequency band or channel. No other user shares the same frequency band. The illustrative example is shown in Figure 5.2. In a practical cellular network implementation, the bandwidth of

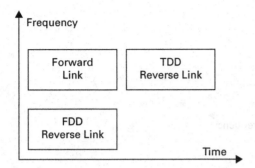

Figure 5.1 Illustrative example of TDD and FDD systems.

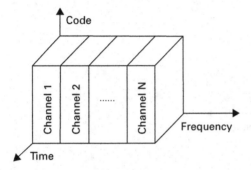

Figure 5.2 Illustration of FDMA systems.

a FDMA channel is relatively small (about 30 kHz) to support one call. The symbol length is long so that the intersymbol-interference problem is not severe. Consequently, the equalizer in the receiver can be simple. Moreover, fewer bits are needed for synchronization and frame overhead for the FDMA system. So the complexity of the FDMA system is lower than that of the TDMA system. FDMA systems might waste resources because the idle channel cannot be utilized by the system to transmit other users' information. A complex filter is necessary for mobile units and base stations to minimize the adjacent channel interferences. Because many channels share the same antenna in the base station, the power amplifier needs to be able to transmit in a high dynamic range. This can cause the problem of nonlinearity that can cause intermodulation frequencies to interfere with the other users. FDMA systems are implemented in first-generation wireless networks such as the Advanced Mobile Phone System (AMPS).

• In a TDMA system, the radio spectrum is divided into time slots, where each user occupies all spectrums at a specific time. Users take turns for transmission, so that an analog transmission cannot be employed and digital transmission is necessary. An illustrative example is shown in Figure 5.3. In each TDMA frame, some more overheads such as synchronization, guard time, and address are employed for timing and identifying. Because of the bursty transmission of TDMA systems, the battery life can be saved when the mobile user is not transmitting. The handoff process for

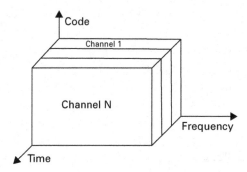

Figure 5.3 Illustration of TDMA systems.

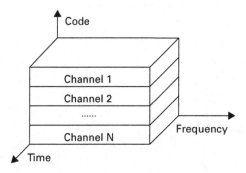

Figure 5.4 Illustration of CDMA systems.

TDMA systems is much simpler because the mobile can listen for other base stations during the idle frame. Because the symbol rate is much higher than that of FDMA, the intersymbol interference needs to be removed by more complex equalizers. Because the users are assigned with time slots only when they have packets to transmit, TDMA systems have higher spectrum efficiency than FDMA systems. This higher spectrum efficiency can increase the number of admitted users in the networks. The scheduling schemes discussed in the next subsection study how to decide when to transmit whose information.

- Spread-spectrum multiple access (SSMA) applies signals with bandwidths much higher than those of FDMA systems. A pseudo-noise (PN) code is assigned to each user so that its signal converts to a wideband noiselike signal. Different users' PN codes are orthogonal to each other, so that the receiver can recover the transmitted signal and suppress the interfering signals by multiplying the specific PN code. An illustrative example is shown in Figure 5.4. There are two types of SSMA: FHMA and direct-sequence code-division multiple access (DS-CDMA).

  1. In a FHMA system, at different times, each user transmits narrowband data over different frequencies according to the order of a certain PN code. The difference between FHMA and FDMA is that the frequency-hopped signal changes channels at rapid intervals. If the changing rate is greater than the symbol rate, the system is called a fast-hopping system. Otherwise, the system is called a slow-hopping

system. In practice, the FHMA system often employs modulations with a constant envelope and noncoherent receiver so that there is no problem for nonlinearity of the amplifier. There are two major advantages of the FHMA system. First, it has high security because it is hard to intercept the hopping signals. This is the reason for its wide applications in military applications. Second, FHMA is immune to fading because it rarely falls into deep fading for a long period of time. Hence, error-control coding and interleaving techniques can effectively combat the fading effects and improve the link qualities.

2. In a CDMA system, the narrowband signal is multiplied by a spread signal such as a PN code. The resulting signal has a much larger bandwidth. The bandwidth efficiency is not high for the signal user. But for a multiple-access system, CDMA is more efficient for spectrum usage, but is limited by the interference from the other users. In the receiver, the received power shall be carefully controlled because the user with higher received power will dominate and impair the decoding of the user with lower received power. This is called the near–far effect. Power control is used in CDMA systems to combat this effect. By careful power control, CDMA systems can have a soft capacity limit, which means there is no absolute limit on the maximal number of users in the system, as more users just imply more interferences. Multipath fading or frequency-selective fading can be overcome because of the inherent frequency diversity. The chip rate for CDMA systems is very high and a Rake receiver can be used to capture and collect the energy of different delayed signals from different multipaths. Soft handoff can be employed to select the best version of the signal at any time without switching the frequencies. In 3G wireless networks, most of the standards are based on CDMA schemes.

Multiuser detection (MUD) [318] deals with the demodulation of mutually interfering digital streams of information. Cellular telephony, satellite communication, high-speed data transmission lines, digital radio/television broadcasting, fixed wireless local loops, and multitrack magnetic recording are some of the communication systems subject to multiaccess interference. The superposition of transmitted signals may originate from nonideal characteristics of the transmission medium or it may be an integral part of the multiplexing method as in the case of CDMA. MUD (also known as cochannel interference suppression, multiuser demodulation, and interference cancellation) exploits the considerable structure of the multiuser interference in order to increase the efficiency with which channel resources are employed. Some basic approaches can be classified as follows:

(a) Optimal: maximum-likelihood sequence estimation.

(b) Suboptimal:
- Linear: Decorrelator, minimum MSE.
- Nonlinear: Multistage, decision feedback, successive interference cancellation.

The advantages of MUD are as follows:

(a) significant capacity improvement (2.8× upper link),

(b) more efficient uplink spectrum utilization,

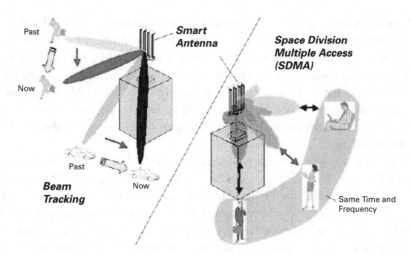

Figure 5.5 Illustration of SDMA systems.

(c) reduced MAI,

(d) More efficient power utilization because the near–far effect is reduced.

There are still some design challenges:

(a) processing complexity and cost,

(b) sensitivity and robustness,

(c) interference to neighboring cell,

(d) difficult to implement in downlink.

Some real implementations have been proposed from Datang telecommunication TD-SCDMA in China.

- In addition to the major multiple-access schemes, there are some possible hybrid schemes such as FDMA/CDMA, DS/FHMA, TD-CDMA, and TD-FHMA. For example, in a 3G standard, TD-SCDMA is proposed. In a GSM system, TD-FHMA is employed when a mobile can hop to new frequency to avoid deep fading.

- By using the antenna signal processing technique, SDMA separates users' signals in different directions of arrival (DOAs). With SDMA, multiple users with different DOAs are able to communicate at the same time using the same channel. In addition, the antenna can collect transmitting power from multipath components, combine them in an optimal manner, suppress interference from other users, and improve the received SINR. Consequently, the performance can be improved. One illustrative example is shown in Figure 5.5.

- Frequency-division multiplexing (FDM) is a technology that transmits multiple signals simultaneously over a single transmission path, such as a cable or wireless system. Each signal travels within its own unique frequency range (carrier), which is modulated by the data (text, voice, video, and others).

  The orthogonal FDM (OFDM) technique distributes the data over a large number of carriers that are spaced apart at precise frequencies. This spacing provides the "orthogonality" in this technique, which prevents the demodulators from seeing

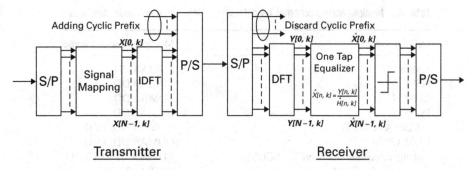

Figure 5.6 Illustration of OFDM Systems.

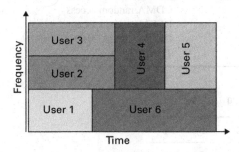

Figure 5.7 Illustration of OFDMA systems.

frequencies other than their own. The benefits of OFDM are high spectral efficiency, resiliency to RF interference, and lower multipath distortion. These benefits are useful because in a typical terrestrial broadcasting scenario there are multipath channels (i.e., the transmitted signal arrives at the receiver using various paths of different lengths). Because multiple versions of the signal interfere with each other [intersymbol interference (ISI)] it becomes very hard to extract the original information. OFDM has very long symbol duration, so it is immune to ISI. The OFDM system is illustrated in Figure 5.6.

In a multiuser scenario, the available techniques are OFDM-TDMA, OFDM-CDMA, and OFDMA. In OFDMA systems, each user occupies a subset of subcarriers and each carrier is assigned exclusively to only one user at any time, so that there are no intracell interferences. One example of OFDMA systems is shown in Figure 5.7. By allocating different users' transmissions to different time–frequency slots, the multiuser diversity, time diversity, and frequency diversity can be exploited to improve system performances. OFDMA technology has been endorsed in leading standards such as ETSI (European Telecommunications Standards Institute) DVB-RCT (digital video broadcasting-return channel terrestrial) and IEEE 802.16a,d and 16e.

- In random-access protocols, the channels are used by users attempting to access a single channel in an uncoordinated manner. Consequently, the transmissions are due to collisions by the multiple users. Many packet radio (PR) access techniques are developed to handle the collisions. PR is very easy to implement, but has low spectral

Table 5.1 Multiple-access schemes for different wireless communication standards

Cellular system	Multiple-access scheme
AMPS	FDMA/FDD
GSM	TDMA/FDD
IS-54	TDMA/FDD
IS-95	CDMA/FDD
WCDMA (UMTS)	CDMA/FDD, TDD
CDMA2000	CDMA/FDD, TDD
Datang Telecommunication TD-SCDMA	TDMA + CDMA/TDD
Flarion Flash OFDM	OFDMA/FDD
CT2 (cordless telephone)	FDMA/TDD
Digital European cordless telephone (DECT)	FDMA/TDD
IEEE 802.11b	CDMA/random access
IEEE 802.11a/g	OFDM/random access
Ultrawide Band	OFDM,CDMA/TDD
IEEE 802.16	TDMA/TDD

Table 5.2 Example of opportunistic scheduling

User/time	1	2	3	4	5	6
User 1	1	3	2	3	2	2
User 2	2	2	3	2	1	3
User 3	3	1	1	1	3	1

efficiency and may have delays. Some of the available PR access techniques are Aloha, carrier-sense multiple access (CSMA), carrier-sense multiple access With collision-detection (CSMA/CD), data-sense multiple access (DAMA), and packet reservation multiple access (PRMA). In Section 5.4, more details are given.

Overall, in Table 5.1, different multiple-access techniques are shown for various wireless communication networks.

5.3   Scheduling

Scheduling is a plan for performing work or achieving an objective, specifying the order, and allotting time for each part. For wireless transmissions, a scheduling policy is a rule that specifies which user is scheduled at each time slot for transmission. The basic goal for scheduling is to select the users to occupy the limited bandwidth for different times according to users' transmission need, the channel conditions, fairness, and other constraints, so as to maximize some performance matrices such as the network throughput and minimal delay.

To understand the scheduling, a simple example is given in Table 5.2. Suppose there are three users in the system and at each time only one user occupies all the bandwidth.

For different times, the users experience different channel conditions that result in different rates as shown in the table. If a fixed scheduling scheme is selected such as to schedule users according to the order $\{1, 2, 3, 1, 2, 3\}$, the resulting rates for all users are $\{4, 3, 2\}$. However, if we take advantage of the channel fluctuations and apply opportunistic scheduling like $\{3, 1, 2, 1, 3, 2\}$, the resulting rates are $\{6, 6, 6\}$, which are much better than those of the fixed scheduling.

In addition to the advantages of multiuser diversity and channel diversity, there are other design issues for the scheduling schemes. First, fairness shall be considered because the users in the same class may pay the same for their services. To define fairness, we can have different matrices such as short-term fairness, long-term fairness, max–min fairness, and proportional fairness. In addition, the delay issue for each individual user data flow shall be considered, because some services, such as voice, cannot suffer arbitrary delay of packets.

One major application of the scheduling is for downlink multiuser transmission, in which each user's data flow has its own buffer. The scheduler selects which user's buffer to use for transmission. If there exists a reliable feedback channel, uplink scheduling can also be employed. In addition, for ad hoc networks, scheduling can reduce the severe cochannel interferences that cannot be effectively removed by optimal power control.

The organization of this section is as follows: First, the cross-layer system model is given to explain the difference between the wireless scheduler and the wired scheduler. Second, a framework of opportunistic scheduling framework is studied and the design trade-offs are explained. Third, some basic scheduling principles are explained and some popular scheduling schemes are studied. Finally, some advanced cross-layer schemes are cited and introduced.

5.3.1   Cross-Layer System Model

To understand how to optimize the scheduling scheme, first the basic system models are studied in the PHY layer, MAC layer, and application layer for the wireless networks to demonstrate the differences in those in the wired networks. Then, reasons are given as to why wired network scheduling methods are not applicable to wireless networks.

For the PHY layer, the distinguishing characteristics of the channel models are as follows:

- Link quality for each link is varying over time. Moreover, the scheduler switches transmissions for different users. So it is possible for opportunity scheduling to take advantage of dynamic optimization.
- Channel errors are location dependent and bursty in nature.
- There is contention for the channel among multiple mobile users.
- Mobile users do not have global channel information.
- The scheduling schemes must take into consideration both uplink and downlink traffic.
- Mobile users are constrained with battery power and processing power.

For the MAC layer model in wireless networks, there are three major concerns:

- Scheduling and random access are two approaches for multiple users to share the same channel for transmission. The major difference is that there is centralized control for scheduling to coordinate which user shall occupy the channel at a certain time. On the other hand, for random-access networks, the distributed topology makes central scheduling impossible. Under this condition, the different users try to access the channel and avoid the potential collisions. Even though random access can be implemented by use of only local information, the throughput is severely less than the scheduling scheme, especially when the network is crowded. So, in cellular networks, the scheduling is employed for major data communications, whereas random access is used for some control channels. But in ad hoc networks, such as WLANs, random access is widely employed.
- Power control is employed in the scheduling scheme so as to maintain the basic link quality. Detailed descriptions for power control are discussed in the corresponding chapter.
- Rate adaptation is utilized for the scheduling scheme to fully take advantage of the channel fluctuations. If the channel is good, more bits are transmitted per symbol and there can be fewer protection bits.

For the application model, the scheduler typically seeks to satisfy the application's QoS and provide fair services to the users with the same priorities.

- The scheduler needs to select the user for transmission when there are data for transmission, the user's channel condition is good, and fairness among different users is maintained.
- For delay-sensitive applications such as voice and multimedia, the delayed packet over a deadline is useless and shall be dropped. The reconstructed signal has less QoS if too many packets are delayed.
- For error-sensitive applications such as data, the QoS can be achieved only if all the packets are transmitted errorless. To achieve the goal, more protection bits are necessary and there might be multiple layers of channel protections such as a convolutional code plus a cyclic redundancy check (CRC).
- For different applications, the data rates can be greatly different. For example, for voice payload, the data rates are from 4 to 12 kbps; for video phones, the data rates can be hundreds of kbps. Considering the different channel protections, the final transmission rate over channels can also vary.
- Because different users in the same priorities may pay the same for their QoS, it is necessary for the scheduling schemes to maintain fairness by scheduling the transmission slots fairly for different users. There are several available fairness definitions such as short-term fairness, long-term fairness, max–min fairness, and proportional fairness. The details are discussed in the rest of this chapter.

In a wired network, a "fluid model" scheduler provides fair queueing. The fair fluid model can be defined as follows [31]: For the set $B(t)$ in which the users have nonempty queues at time t, the allocated rate is

$$R_i(t) = \frac{r_i C(t)}{\sum_{j \in B(t)} r_j},$$

(5.1)

where r_i is the priority weight and $C(t)$ is the overall channel capacity viewed by all users at time t. For a period of time $(t, t + \tau)$, the queueing is fair if

$$\left| \frac{W_i(t, t + \tau)}{r_i} - \frac{W_j(t, t + \tau)}{r_j} \right| = 0, \forall i, j \in B(t, t + \tau), \tag{5.2}$$

where $W_i(t, t + \tau)$ is the overall rate assigned during time period $(t, t + \tau)$. We assume there is no change for nonempty queues and the channel capacity is fixed.

However, in wireless networks, such a fluid model cannot be employed. The basic reason is that the channel is changing over time and with the same bandwidth, different users can experience different channel capacities. So in the next subsection, we describe a wireless scheduling framework and discuss the various design trade-offs for the wireless schedulers.

5.3.2   Opportunistic Scheduling Framework and Trade-Offs

In this subsection, we discuss an opportunistic scheduling framework [185] to exploit the time-varying nature of the radio environment to increase the overall performance of the system under certain QoS and fairness requirements of users. Different categories of scheduling problems are identified and discussed in detail.

Suppose there are K users in the system. A time-slotted system is considered in which time is the resource to be shared among all users. At any given time, only one user can occupy a given channel within a cell. The wireless channel conditions are time varying, and the users experience different performances if they occupy the same channel. To quantify the user's performance, we define the following utility function.

Define the ith user's utility function at time slot n as $U_i(n)$, which represents the level of performance that will be experienced by user i if it is scheduled to transmit at time n. The value of $U_i(n)$ measures the "worth" of time slot n to the user i. For example, it can be a function of the user's channel condition. The better the channel condition, the larger the value of $U_i(n)$. Here we assume $U_i(n)$ is nonnegative $\forall, i, n$. The example of $U_i(n)$ can be rate, rate per joule (i.e, rate over power), or minus power.

To fully utilize the scarce radio resource and provide the highest-data-rate services, the opportunistic scheduling shall select the user with the best link quality. However, such a scheme can be unfairly biased, especially when the users are widely dispersed. Some users far away from the base station may rarely be able to access the channel, whereas some users close to the base station have access to the channel so often that the QoS is unnecessarily high or the desired high rates cannot be achieved by the practical implementation such as power and maximal rate. So it is necessary to define the constraints for the scheduling methods. Possible constraints can be the basic QoS constraint, the fairness constraint, practical constraint, and others.

Define $\mathbf{U}(n) = [U_1(n) \ldots U_K(n)]'$ as the performance vector at time n. Let $\mathbf{U} = \{\mathbf{U}(n), n = 1, 2 \ldots\}$ be the sequence of performance vectors. Define the constraint as $P(\mathbf{U}) \geq 0$ and policy as Q. The policy is feasible if it satisfies the constraints for all users. The system optimization goal is to maximize the system performance by exploiting time-varying channel conditions and, at the same time, maintaining certain user-oriented constraints such as

$$\arg \max_{Q} G(\mathbf{U}), \qquad (5.3)$$

$$\text{s.t. } P(\mathbf{U}) \geq 0,$$

where G is a function for the system optimization. The example of $G(\mathbf{U})$ can be $U_i(n)$, $\min_i U_i(n)$, $E[U_i(n)]$, etc.

The scheduling scheme can be employed for both uplink and downlink. For the uplink case, the selected user is notified to occupy the next assigned time slot through a reliable downlink control channel. For the downlink case, the base station selects the user according to the feedback downlink channel conditions from the mobiles. In general, downlink scheduling is more important because the data traffic transmission is highly asymmetric. For example, the download data from the Internet are much more than the uplink data such as e-mails. Uplink scheduling has another practical challenge for time synchronization if the duration of a time slot is short.

In a practical implementation, there are some design trade-offs, as follows:

1. First, there is a trade-off between fairness and system performance. For example, allowing only users with good channel conditions to transmit may result in high throughput, but meanwhile will sacrifice the transmissions of other users. In the literature, long-term fairness has been considered, such as proportional fairness [160], time-average fairness [118], and short-term fairness [119, 124] with delay constraints.
2. Second, there is a trade-off between centralized design and distributed design. To optimize the performance, the scheduler needs channel information so that the global optimum can be achieved. However, estimating the channel information requires many communication overheads and signaling, which reduce the spectrum efficiency of the system. For the distributed scheme, there is no need for these overheads and signaling. However, the performance is inevitably lower than that of the centralized system.
3. Third, there is a trade-off between separation and compensation. Separation denotes the degree to which the service of one user is unaffected by the behaviors and channel conditions of other flows. Compensation is provided for the user who is unable to obtain successful transmission now by allocating the time slots in the future. Because the time resource is limited for all users, individual behaviors and channel conditions are not sufficient to independently contribute to the overall system performances.

5.3.3   Basic Scheduling Approaches

In this subsection, we discuss some basic approaches for wireless scheduling methods. The fundamental ideas are illustrated first. Then some popular scheduling schemes are explained and compared. Finally, the desired properties of good wireless scheduling schemes are described.

First, we classify the existing schedules into four classes based on the underlying principles as follows:

- Max–min fairness class:
  In this type of scheduling, the philosophy is to allocate the resources, such as channels and time slots, to the user with the minimal received QoS. Some schemes with hierarchy QoS are also proposed with the similar ideas. The advantage of this type of scheduling scheme is the delivery of extreme fairness among the users with the same priority. However, this type of scheme does not consider the channel variances for different users and over different times, so the diversity, such as multiuser, time, and frequency, cannot be exploited. Consequently the achieved performance is low. This type of scheme is one extreme side of the spectrum for the trade-off between fairness and performance.

  In [307], scheduling policies for max–min fair allocation of bandwidth are proposed in wireless ad hoc networks. The basic idea behind max–min fairness is to first allocate equal bandwidth to all contending users. If a user cannot utilize its bandwidth, because of constraint elsewhere, then the residual bandwidth is distributed among others. Thus no user is penalized excessively, and a certain minimum QoS is guaranteed to all users. User satisfaction is often a concave function, i.e., satisfaction increases rapidly with increase in bandwidth in the low-bandwidth region, and increases slowly with increase in bandwidth in the high-bandwidth range. Thus total user satisfaction is often improved if all users obtain an equitable QoS, rather than some users having much better QoS at the expense of others. More important, this can be attained without assuming specific knowledge about the individual user-satisfaction functions. The fairness objective is to distribute bandwidth as evenly as possible, without unduly reducing the throughput. A bandwidth allocation is said to be max–min fair if it is not possible to increase the allocation of any user without hurting another user with a lower service rate. A fair scheduling is proposed, in which the dynamic weights are assigned to the flows such that the weights depend on the congestion in the neighborhood. The flows that constitute maximum weighted matching are scheduled. Some other works for fairness in ad hoc networks can also be found in [144].

- Maximal performance class:
  This type of scheduling scheme allocates the resources to the user who can achieve the best performance. By doing this, the overall system performance is optimized for each time slot so that the performance is also optimized over time. The advantage of this type of scheme is to have a high performance, because the scheduler opportunistically selects the user with the best channel in the current time slot. However, no fairness is considered in this type of scheme. Because users in the same priority pay the same for their services, their QoS shall be delivered fairly. This type of scheme is on the other extreme for the trade-off between fairness and performance, compared with the max–min fair class.

- Mixed strategy class:
  Because there are different types of mobile users, the requested services can have different natures and requirements. For example, for a voice packet, the data rate is relatively low but the delay constraint is strict. For data transmissions, the data rate can be arbitrary and some packets such as e-mail can suffer long delay. For multimedia transmissions, the data rate might be very high and the delay constraint is between

voice packets and data packets. To accommodate all these types of users, some mixed strategy scheduling schemes are proposed.

In [275], the authors address the following two problems: The first problem is that of how multiple real-time data users can be supported simultaneously with good QoS for all users. The QoS here is that the packet delays do not exceed given thresholds with high probability. The second problem is that of how a mixture of real-time and non-real-time users can be supported simultaneously. Here the real-time users receive their desired QoS and the non-real-time users receive the maximum possible throughput without compromising the QoS requirement of real-time users. The conclusion is drawn that intelligent scheduling algorithms in conjunction with token-based rate control provide an efficient framework for supporting a mixture of real-time and non-real-time data applications in a single carrier.

- Proportional fair class (trade-off between fairness and performance):
To balance fairness and performance, different criteria have been proposed. Among them, the most popular one is proportional fairness. The basic concept is to weight the scheduling priorities by both the channel conditions and the transmission history. If the channel conditions are good, the scheduler takes the opportunities to transmit more data. On the other hand, if one user has less transmission for a period of time, this user deserves to be scheduled with more transmission times. To illustrate the idea, we define the maximal transmission rate for the ith user at time T as $C_i(T)$ and the transmission history for this user is $R_i(t)$. The proportional fairness scheduler can be expressed as follows:

$$\arg\max_i \frac{C_i(T)}{\sum_{t=0}^{T-1} R_i(t) + \epsilon}, \tag{5.4}$$

where ϵ is a small positive constant to prevent the error of "divided by zero." If the user is selected, $R_i(T) = C_i(T)$; otherwise $R_i(T) = 0$. Because of the ergotic characteristics, the preceding scheduler provides the estimated data rate proportional to the mean of channel conditions. So the preceding proportional fair scheduler provides a trade-off between fairness and performance.

The performance of the preceding channel-aware scheduling algorithms has been exploited at the packet level for a static user population, often assuming infinite backlogs. In [33], the performance at the flow level is focused in a dynamic setting with random finite-size service demands. In certain cases, the user-level performance may be evaluated by means of a multiclass processor-sharing model in which the total service rate varies with the total number of users. The latter model provides explicit formulas for the distribution of the number of active users of the various classes, the mean response times, the blocking probabilities, and the mean throughput. In addition, in the presence of channel variations, greedy strategies that maximize throughput in a static scenario, may result in a suboptimal throughput performance for a dynamic user configuration and cause potential instability effects.

Emerging spread-spectrum high-speed data networks utilize multiple channels via orthogonal codes or frequency-hopping patterns such that multiple users can transmit concurrently. The preceding scheduling schemes select only one user for transmission per time. In [190], a framework is developed for opportunistic scheduling over multiple wireless channels. With a realistic channel model, any subset of users can be selected for data transmission at any time, albeit with different throughput and system resource requirements. The selection of the best users and rates from a complex general optimization problem is transformed into a decoupled and tractable formulation: a multiuser scheduling problem that maximizes total system throughput and a control-update problem that ensures long-term deterministic or probabilistic fairness constraints.

Several popular practical wireless scheduling schemes are explained and compared as follows.

- One of the first papers that addresses the wireless scheduling is [29], in which a channel-state-dependent packet scheduling (CSDPS) scheme is proposed. The basic idea is that a link status monitor monitors the link status for all mobiles. If any link is in a certain state with a bad channel response, the scheduler stops scheduling time slots to this link until time out. CSDPS improves the scheduling performance by taking into consideration the location-dependent and time-dependent channel states. Some improvements such as CSDPS class-based queueing (CSDPS CBQ) are proposed to maintain the fairness. To reduce the overhead for detecting the states with bad channel responses, a ready-to-send (RTS) and clear-to-send (CTS) pair can be employed to transmit a shorter packet between mobiles and base station before data packet transmission.
- Wireless fluid fair queueing (WFFQ) [191] is based on the fluid model and the wireless version of a fluid fair queueing model in wired networks. Idealized wireless fair queueing (IWFQ) is based on a weighted fair queueing scheme in wired networks. The difference is that the scheduler picks only the users with the good channel state.
- Channel-condition independent packet fair queueing (CIF-Q [221]) is proposed and fairness objectives are addressed, including the following:
  1. Delay and throughput guarantees: Delay bound and throughput for error-free flows shall be guaranteed, and not affected by other flows in error state.
  2. Long-term fairness: After a flow exits from a link error, as long as it has enough service demand, it shall be compensated, over a sufficiently long period, for all of its lost service when it is in error.
  3. Short-term fairness: The difference between the normalized services received by any two error-free flows that are continuously backlogged and are in the same state (i.e., leading, lagging, or satisfied) during a time interval shall be bounded.
  4. Graceful degradation for leading flows: During any interval when it is error free, a leading backlogged flow shall be guaranteed to receive at least a minimum fraction of the service it will receive in an error-free system.
- Server-based fair approach (SBFA) is proposed in [246], in which a part of the bandwidth is allocated to some compensation servers to compensate the transmission to the deferred users because of the bad channels, so that the long-term fairness can be achieved.

Table 5.3 Comparison of the algorithm properties

	Delay bound	Long-term throughput guarantee	Mechanism for wireless link variability	Short-term fairness	Preallocated resource for compensation	Integrated with MAC protocol
CSDPS			x			
CSDPS+CBQ		x	x			
IWFQ	x	x	x			
CIF-Q	x	x	x	x		
SBFA		x	x		x	
I-CSDPS	x	x	x			
PRADOS	x					x

- Improved CSDPS (I-CSDPS) is proposed in [86] to use a deficit round-robin scheduler combined with an explicit compensation counter. In each service round, the number of packets served in each queue is determined by two parameters: deficit counter (DC) and quantum size (QS), where DC records the total credit received and QS is the credit size (the number of bits or bytes). I-CSDPS has the following advantages: The compensation rate can be dynamically changed according to system load, and it is able to handle variable-sized packets. The disadvantage is that it does not impose any restriction on the flow that receives excessive service.
- Prioritized regulated allocation delay-oriented scheduling (PRADOS) is proposed for traffic scheduling in wireless ATMs in [232].

To compare scheduling schemes, we summarize the properties in Table 5.3 [41]. Here delay bound is defined only for flows with error-free wireless links. The long-term throughput guarantee implies that, as long as a flow has enough service demand and the link errors it experiences are sporadic, its throughput over a sufficiently long period can be maintained above a certain value.[1]

For a wireless packet scheduling scheme, the following objectives shall be achieved [41]:

1. Providing long-term fairness and throughput guarantees for flows with error-free links or sporadic link errors.
2. Achieving high wireless channel utilization.
3. Minimizing packet loss.
4. Providing delay (jitter, if possible) bound for flows with error-free links or sporadic link errors.
5. Supporting multiple classes of traffic with QoS differentiation.
6. Achieving low power consumption in mobile hosts.
7. Achieving medium algorithm complexity.

---

[1] For comparison, short-term fairness is defined in [221].

5.3.4   Cross-Layer Scheduling Approaches

In this subsection, we discuss some advanced scheduling approaches that consider some cross-layer optimizations. The basic principles are illustrated.

- Energy-efficient scheduling:
  The problem of minimizing the energy required for transmitting packets over a wireless network is studied based on the following observation: In many channel-coding schemes, lowering transmission power and increasing the duration of transmission leads to a significant reduction in transmission energy. In particular, it is observed that, for a given channel-coding scheme, if $w(\tau)$ is the energy expended for transmitting a packet over one unit of time, then $w(\tau)$ is a nonnegative, monotonically decreasing, and strictly convex function of τ. Based on this observation, for a variety of scenarios, the scheduling problem for off-line energy-efficient transmission reduces to a convex optimization problem. For the downlink channel, with a single transmitter and multiple receivers, an iterative algorithm, called MoveRight, is devised to yield the optimal off-line schedule. The MoveRight algorithm also optimally solves the downlink problem with additional constraints imposed by packet deadlines and finite transmission buffers. For the uplink (or multiaccess) problem, MoveRight optimally determines the off-line time-sharing schedule. A very efficient on-line algorithm, called MoveRightExpress, that uses a surprisingly small look-ahead buffer is proposed and is shown to perform competitively with the optimal off-line schedule in terms of energy efficiency and delay.

- Joint scheduling and power control:
  In [64], a cross-layer design framework is introduced to the multiple-access problem in contention-based wireless ad hoc networks. The motivation behind the coupling of scheduling and power control is twofold: (1) accounting for the wireless link volatility in the design of higher layer protocols, e.g., link scheduling, and (2) introducing a more realistic and accurate characterization of nonconflicting transmissions such as those satisfying a prespecified SINR constraint. The multiple-access problem is solved via two alternating phases, namely scheduling and power control. The scheduling algorithm is essential to coordinate the transmissions of independent users in order to eliminate strong levels of interference (e.g., self-interference) that cannot be overcome by power control. On the other hand, power is controlled in a distributed fashion to determine the admissible power vector, if one exists, that can be used by the scheduled users to satisfy their single-hop transmission requirements. This is done for two types of networks, namely TDMA and TDMA/CDMA wireless ad hoc networks.

- Space diversity:
  Multiuser diversity is a form of diversity inherent in a wireless network, provided by independent time-varying channels across the different users. The diversity benefit is exploited by tracking the channel fluctuations of the users and scheduling transmissions to users when their instantaneous channel quality is near the peak. The diversity gain increases with the dynamic range of the fluctuations and is thus limited

in environments with little scattering, slow fading, or both. In such environments, the use of multiple-transmission antennas to induce large and fast channel fluctuations is introduced in [320], so that multiuser diversity can still be exploited. The scheme can be interpreted as opportunistic beam-forming and true beam-forming gains can be achieved when there are sufficient users, even though very limited channel feedback is needed. Furthermore, in a cellular system, the scheme plays an additional role of opportunistic nulling of the interference created on users of adjacent cells.

- BER consideration scheduling:
  A MAC protocol called a wireless multimedia access control protocol with BER scheduling (WISPER [5]) for a CDMA-based systems is proposed. WISPER utilizes the novel idea of scheduling the transmission of multimedia packets according to their BER requirements. The scheduler assigns priorities to the packets and performs an iterative procedure to determine a good accommodation of the highest-priority packets in the slots of a frame so that packets with equal or similar BER requirements are transmitted in the same slots.

- Multimedia user-satisfaction factor scheduling:
  The goals are to define a new QoS measure for multimedia transmissions and to develop new scheduling schemes to improve system performances and maintain individual fairness among users. First, the QoS measure is quantified as a user-satisfaction factor (USF) for different delay constraints and different applications such as voice, video, and data. The proposed USF is a function of both the number of received bits and the delay-sensitivity profiles. Then, based on this QoS measure, four types of scheduling schemes are proposed, namely the max–min approach, the overall performance approach, the two-step approach, and the proportional approach. Simulation results show that these four scheduling schemes achieve better trade-offs between the system performance and individual fairness, compared with the weighted round-robin and the modified proportional fairness scheduling schemes.

- Some other scheduling works:
  Distributed fair scheduling [316] is proposed under the current WLAN standard. The distributed coordination function (DCF) is modified to support fair scheduling. Priority scheduling in multihop ad hoc networks [344] is proposed to combat the location-dependent contention and "hidden-terminal" problem.

5.4   Packet Radio Multiple-Access Protocols

For uplink communication systems or multipoint-to-point communication systems, too many communication overheads are needed to coordinate the transmissions of different users' data using limited radio bandwidth. Consequently, the distributed implementation is necessary. Under this requirement, it is desirable for a distributed user to decide when to occupy the channel and how to resolve the conflict if more than one user tries to occupy the channel at the same time. In addition, for multiple users, multiple-access protocols such as time-division multiplexing (TDM) and FDM are inefficient for bursty data network, because the delays for the transmission users might be unnecessarily long. It is better to share the medium so that the users needing it can use it. Multiple-access protocols, such as the PR access protocol, can dynamically allocate the channel to users with the packets to transmit and resolve the conflict of channel usages.

In PR multiple-access protocols, many users try to transmit bursts of data through a single channel in a distributed manner. The base station can detect the collisions of simultaneous transmission and broadcast ACK or NACK signals, indicating a successful transmission or an unsuccessful transmission, respectively. Then the collided users wait for a random time to retransmit. The PR multiple-access protocols are very easy to implement and they induce trivial overheads, but they have a relatively lower spectral efficiency than that of scheduling.

In this subsection, we discuss some popular classes of multiple-access protocols. Then we give two examples that are employed in WLANs and 3G networks.

5.4.1   Aloha

The first class of multiple-access protocols is Aloha, which was proposed in late 1960s by researchers at the University of Hawaii. The basic idea of Aloha is that each unbacklogged node simply transmits a newly arriving packet in the first slot after the packet arrival, thus risking occasional collisions but achieving very small delay if collisions are rare [28]. The scheme can be written as follows:

1. If there is a message, transmit it; otherwise, wait.
2. If transmission is successful, remove the transmitted message from buffer and go back to 1.
3. When a collision occurs, the collision nodes become backlogged and wait for a certain random number of time before retransmitting.

The performance of Aloha is briefly analyzed as follows: Assume that users try to send frames at random times (Poisson events). Let G be the average rate at which all users try to send their frames. The probability of trying to send k frames in two frame times is

$$P(k) = \frac{(2G)^k e^{-2G}}{k!}. \tag{5.5}$$

The probability is $P(0) = e^{-2G}$ when no other frames are sent. The throughput is the rate at which frames are sent multiplied by the probability that the transmission is successful, which is Ge^{-2G}. If frames can be transmitted only at the beginning of a slot, we call the modified scheme slotted Aloha, which doubles the performance of Aloha with throughput Ge^{-G}. The performance of unslotted Aloha and slotted Aloha is shown in Figure 5.8. For unslotted Aloha, the optimal throughput $1/(2e)$ appears when $G = 1/2$. For slotted Aloha, the optimum is $1/e$ when $G = 1$. When G is less than the optimal point, too many idle slots are generated; otherwise, too many collisions are generated. As we see from the figure, the Aloha protocols have poor utilization.

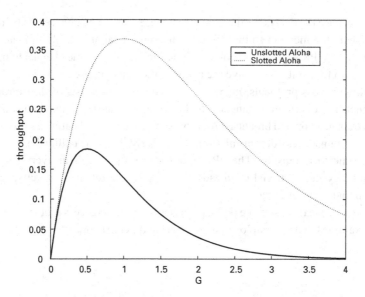

Figure 5.8 Performances of Aloha.

Beyond the throughput, the delay issue and buffer size issue shall also be considered for implementation. A detailed analysis of Aloha is beyond the scope of this book and can be studied in [28].

5.4.2   CSMA

One of the key properties of a LAN is that propagation delay between stations is small compared with frame transmission time. So if nodes can detect idle periods quickly, it is reasonable to end idle periods quicker and allow nodes to transmit after such detections. This type of scheme is called carrier-sense multiple access (CSMA), which can reduce the unnecessary idle time and consequently improve the system performance. This class of multiple-access algorithm is popular in wireless networks with low propagation delays. CSMA can be classified as follows:

- Nonpersistent CSMA: If collision happens, the mobile waits for a random time before retransmission of the packet. Because the transmission interval for a WLAN is much larger than the propagation delay, this scheme is popular for WLANs.
- 1-persistent CSMA
    1. If the medium is idle, then transmit.
    2. If the medium is not idle, then wait until it is and then transmit.
- p-persistent CSMA
    1. If the medium is idle, then transmit.
    2. If the medium is not idle, then wait until it is idle; once idle, then transmit with probability p, and wait for the next slot with probability $1-p$ and repeat.

The following are three variations for CSMA:

- The first one is CSMA/CA. In CSMA/CA, as soon as a node receives a packet to be sent, it checks to be sure the channel is clear (no other node is transmitting at the time). If the channel is clear, then the packet is sent. If the channel is not clear, the node waits for a randomly chosen period of time, and then checks again to see if the channel is clear. This period of time is called the backoff factor, and is counted down by a backoff counter. If the channel is clear when the backoff counter reaches zero, the node transmits the packet. If the channel is not clear when the backoff counter reaches zero, the backoff factor is set again, and the process is repeated.
- The second one is carrier-sense multiple access/collision detection (CSMA/CD), which is used by standard Ethernet networks. CSMA/CD enables devices to detect a collision. After detecting a collision, a device waits for a random delay time and then attempts to retransmit the message. If the device detects a collision again, it waits twice as long to try to retransmit the message. This is known as exponential backoff.
- The third one is data-sense multiple-access (DSMA). In this protocol, a mobile user decodes a busy–idle signal from a forward control channel before trying to transmit data in the reverse channel. This protocol is used in a cellular digital packet data (CDPD) cellular network.

For the users with data streaming or a deterministic packet rate, the FDMA, TDMA, or CDMA system can provide reliable QoS. On the other hand, when the packet arrival is bursty and the packet length is short, Aloha and CSMA schemes provide simple solutions. However, if the packet length is long, the payload is high, or there are some delay requirements, some reservation protocols can be utilized to get a trade-off between the traditional multiple-access protocols (such as TDMA) and the random-access protocols (such as Aloha).

- Reservation Aloha [305]: In this protocol, it is possible for high-priority users to reserve time slots for their packet transmission based on TDM.
- PRMA [88] is introduced as a way to integrate voice and data in microcellular wireless communication systems. In PRMA, voice and data subsystems are logically separated. The available bandwidth is dynamically partitioned between voice and data, in the sense that unused voice slots are assigned to data traffic. Furthermore, PRMA exploits the fact that the voice activity has a talk spurt–silence model. This model assumes that a voice has its maximum rate during talk spurts and carries no information during silence periods. As a result, in PRMA the voice terminals are granted access to the channel only during talk spurts and give up the use of channel during silence periods. During silence-to-talk spurt transitions, a call has to contend to access the channel using technics similar to those of slotted ALOHA. Once the call gains access to the channel successfully, it keeps that slot in subsequent frames for the rest of its talk spurt. This contention causes a certain bandwidth inefficiency because of collision and blank slots that occur during the contention period.

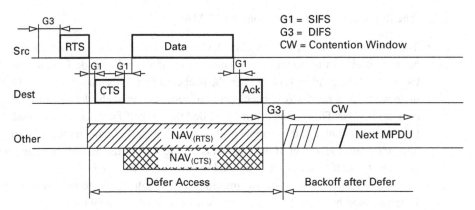

Figure 5.9 Timing diagram of RTS/CTS.

5.4.3   Wireless LAN (RTS/CTS)

The IEEE 802.11 MAC protocol supports two kinds of access methods: the DCF and the point coordination function (PCF). The DCF is the basic access mechanism using CSMA/CA and must be implemented in all stations. In contrast, the PCF is optional and is based on polling controlled by a point coordinator.

The required multiple-access method for IEEE 802.11 is the DCF that uses CSMA/CA. This requires each station to listen for other users. If the channel is idle, the station may transmit. However, if it is busy, each station waits until transmission stops, and then enters into a random backoff procedure. This prevents multiple stations from seizing the medium immediately after completion of the preceding transmission.

Packet reception in a DCF requires ACK from the receiver, as shown in Figure 5.9. The period between completion of packet transmission and start of the ACK frame is one short interframe space (SIFS). ACK frames have a higher priority than other traffic. Fast acknowledgement is one of the salient features of the 802.11 standard, because it requires ACKs to be handled at the MAC sublayer.

Transmissions other than ACKs must wait at least one DCF interframe space (DIFS) before transmitting data. If a transmitter senses a busy medium, it determines a random backoff period by setting an internal timer to an integer number of slot times. On expiration of a DIFS, the timer begins to decrement. If the timer reaches zero, the station may begin transmission. However, if the channel is seized by another station before the timer reaches zero, the timer setting is retained at the decremented value for subsequent transmission.

The method just described relies on the physical carrier sense. The underlying assumption is that every station can "hear" all other stations. This is not always the case. Referring to Figure 5.10, the access point (AP) is within range of Station-A (STA-A), but STA-B is out of range. STA-B will not be able to detect transmissions from STA-A, and the probability of collision is greatly increased. This is known as the hidden node.

To combat this problem, a second carrier-sense mechanism is available. Virtual carrier-sense enables a station to reserve the medium for a specified period of time through the

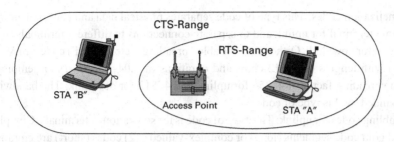

Figure 5.10 Illustration of hidden-node problem.

use of RTS/CTS frames. In the case previously described, STA-A sends a RTS frame to the AP. The RTS will not be heard by STA-B. The RTS frame contains a duration/ID field that specifies the period of time for which the medium is reserved for a subsequent transmission. The reservation information is stored in the network allocation vector (NAV) of all stations detecting the RTS frame.

On receipt of the RTS, the AP responds with a CTS frame, which also contains a duration/ID field specifying the period of time for which the medium is reserved. When STA-B does not detect the RTS, it will detect the CTS and update its NAV accordingly. Thus collision is avoided even though some nodes are hidden from other stations. The RTS/CTS procedure is invoked according to a user-specified parameter. It can be used always, never, or for packets that exceed an arbitrarily defined length.

As already mentioned, the DCF is the basic media-access control method for 802.11, and it is mandatory for all stations. The PCF is an optional extension to the DCF. The PCF provides a TDD capability to accommodate time-bounded, connection-oriented services such as cordless telephony.

5.5   3G System Multiple Access

The main IMT-2000 standardization effort was to create a new air interface that would increase frequency usage efficiency. The WCDMA–air interface is selected for paired frequency bands (FDD operation) and TD-CDMA (TDD operation) for unpaired spectra. 3G CDMA2000 standard is created to support IS-95 evolution. Both 3G standards select CDMA as the multiple-access protocol. In this section, we briefly describe the CDMA and random-access protocols in the UMTS (WCDMA) wireless systems.

There are four types of CDMA codes that are used in a UMTS system. Their properties and usages are summarized as follows:

- Synchronization codes: This type of code enables terminals to locate and synchronize to the cells' main control channels. Gold codes with lengths of 256 chips and durations of 66.67 μs are employed with 1 primary code and 16 secondary codes. The codes are classified as primary synchronization codes (PSCs) and secondary synchronization codes (SSCs). There is no bandwidth expansion for this type of code.

- Channelization codes: This type of code separates physical data and control data from the same terminal for uplink and to separate connections to different terminals in the same cell for downlink. Orthogonal variable-spreading factor (OVSF) codes (or Walsh codes) with lengths of 4–512 chips and durations of 1.04–133.34 μs are employed with a spreading factor of 4–256 for uplink and 4–512 for downlink. The bandwidth is expanded by this type of code.
- Scrambling codes for uplink: This type of code is for separation of terminals. Complex-valued gold code segments (long) or complex-valued S(2) codes (short) are employed with 38,400 chips or 256 chips with time durations of 10 ms or 66.67 μs. The number of codes is 16,777,216 and there is no expansion for bandwidth.
- Scrambling codes for downlink: This type of code is for separation of sectors. Complex-valued gold code segments are employed with 38,400 chips with a time duration of 10 ms. The number of codes is 512 primary/15 secondary for each primary code, and there is no expansion for bandwidth.

UMTS random-access channel (RACH) is an uplink common transport channel used to carry control information or short data packets from a mobile station. The RACH is always received from the entire cell. The RACH is characterized by a collision risk and by a transmission using open-loop power control. The physical RACH is designed based on a slotted Aloha approach in which a random-access burst can be transmitted in different access slots, spaced 1.25 ms apart. When a mobile station does not receive an acknowledgment for its access request within a predefined time-out period, it backs off and retransmits its access request.

5.6   Channel Allocation/Opportunistic Spectrum Access

Because the radio spectrum is limited, a given radio spectrum is to be divided into a set of disjointed channels that can be used simultaneously while minimizing interference in adjacent channels by allocating channels appropriately (especially for traffic channels). Frequency allocation should be carefully planned to avoid degradation caused by CCI. In Figure 5.11, a traditional channel-allocation scheme with adjacent cells using other frequency channels is illustrated.

With the control of a central processor, the channel-allocation schemes are coordination based (planning based). If the distributed schemes are necessary, some measurement-based schemes have been proposed. Coordination-based channel-allocation schemes can be divided in general into fixed channel-allocation schemes (FCA schemes), dynamic channel-allocation schemes (DCA schemes), hybrid channel-allocation schemes (HCA schemes, combining both FCA and DCA techniques), and allocation schemes with borrowing.

- FCA
  Channels can be used only in designated cells. Different groups of radio channels may be assigned to adjacent cells, but the same groups must be assigned to cells

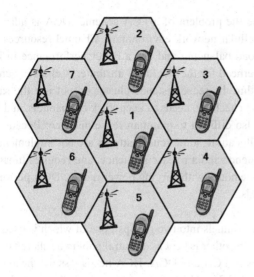

Figure 5.11 FCA.

separated by a certain distance (reuse distance) to reduce CCI. FCA is the easiest channel-allocation scheme, but provides the worst channel utilization. FCA is used in AMPS. The advantages of FCA are the fixed characteristics for transceivers (power, frequency), good performance under uniform and/or high traffic loads, no requirement for run-time coordination because cells independently decide their channel-allocation decisions. However, hot spots or localized congestion is a problem. For example, in a stadium, during game time there are not enough channels for the large number of calls, whereas at other times there are few call requests.

• DCA

In DCA schemes, all channels are kept in a central pool and are assigned dynamically to new calls as they arrive in the system. After each call is completed, the channel is returned to the central pool. It is fairly straightforward to select the most appropriate channel for any call based simply on current allocation and current traffic, with the aim of minimizing the interference. To avoid CCI, any channel that is used in one cell can be reassigned simultaneously to another cell in the system only if the distance between two cells is larger than the minimum reuse distance. For example, the central pool holds all channels. Cell "i" can use any channel. It may be reused by another cell "j" if both cells maintain reuse distance or locking distance.

DCA schemes can be centralized or distributed. The centralized DCA scheme involves a single controller selecting a channel for each cell. The distributed DCA scheme involves a number of controllers scattered across the network [mobile switching centers (MSCs)]. Centralized DCA schemes can theoretically provide the best performance. However, the enormous amount of computation and communication among base stations leads to excessive system latencies and renders centralized DCA schemes impractical. Nevertheless, centralized DCA schemes often provide a useful benchmark to compare practical decentralized DCA schemes.

A DCA scheme can overcome the problem of a FCA scheme. DCA is adaptive to traffic load changes and the cellular network environment. Channel resources are better utilized. DCA improves load balancing and has a better performance in low traffic. However, DCA has an overhead for the "locking" attribute: A channel can be assigned only if all the cells within the reuse distance cluster do not use the same channel. DCA requires exhaustive checking, needs to keep track of channels, and has channel-assignment delay. It is also difficult to maintain reuse distance: Because of the dynamic nature of channel allocation, some reuse patterns are not implemented. DCA also depends on a call sequence and a release sequence, and requires tunable-frequency base stations and mobile nodes with varying power ranges. DCA performs worse than FCA under heavy loads.

- HCA

  HCA divides the total number of channels into two groups , one of which is used for fixed allocation to the cells, while the other is kept as a central pool to be shared by all users. HCA mixes the advantages of FCA and DCA. HCA needs less overhead than DCA and has less complexity. But HCA still performs worse than FCA for a heavy traffic load. The optimum ratio of dynamical channels to fixed channels depends on the traffic load.

- Channel borrowing

  Channels may be shared between adjacent cells and are not reallocated but are reused by a different node. Channel borrowing can be performed between one cell with a base station of a neighbor cell. Base stations that are not parents are chosen based on proximity. Local management between adjacent cells is necessary. All locations may not have nonparent base stations, and low-quality connections might occur in base stations that are not parents. These types of schemes require fine-grain power control of nodes and base stations and might increase the probability of channel interference. Some existing channel-borrowing schemes are listed as follows:

  1. Simple borrowing (SB): A nominal channel set is assigned to a cell, as in the FCA case. After all nominal channels are used, an available channel from a neighboring cell is borrowed.

  2. Simple borrowing from the richest (SBR): Channels that are candidates for borrowing are available channels nominally assigned to one of the adjacent cells of the acceptor cell. If more than one adjacent cell has channels available for borrowing, a channel is borrowed from the cell with the greatest number of channels available for borrowing.

  3. Basic algorithm (BA): This is an improved version of the SBR strategy that takes channel locking into account when selecting a candidate channel for borrowing. This scheme tries to minimize the future call-blocking probability in the cell that is most affected by the channel borrowing.

  4. Basic algorithm with reassignment (BAR): This scheme provides for the transfer of a call from a borrowed channel to a nominal channel whenever a nominal channel becomes available.

  5. Borrow first available (BFA): Instead of trying to optimize when borrowing, this algorithm selects the first candidate channel it finds.

Besides the coordination-based approach, a distributed measurement-based method can alleviate the processing time of the central processor. Some available schemes are listed as follows: In [48], frequency assignment and portable access are separately performed. An adaptive threshold concept [328] has been proposed for time-slot selection. A directed-retry scheme is proposed in [61]. A load sharing concept is proposed in [155]. Time-slot selection algorithms are investigated in [329, 330].

For channel allocation, it is worth mentioning the graph-coloring theory. In graph theory, graph coloring is an assignment of "colors" (red, blue, and so on, but consecutive integers starting from 1 can be used without loss of generality), to certain objects in a graph. Such objects can be vertices, edges, faces, or a mixture of those. One of the most important theories is the four-color theorem, which asserts that every planar graph—and therefore every "map" on the plane or sphere—no matter how large or complex, is four-colorable. Graph coloring enjoys many practical applications as well as theoretical challenges, especially for dynamic and ad hoc scenarios.

The previously mentioned schemes are mostly used in the cellular networks. Next, we discuss channel allocations in some other wireless networks. As mentioned in the previous chapter, parts of standards for WiMAX employ OFDMA technology. Because each user has different distances and multipath/Doppler scenarios, the channels for different users fluctuate over time and frequency. If the channel-allocation scheme can fully take advantage of these dynamic natures and transmit users' information only when their channels are good, the high diversity of multiuser, frequency, and time can be achieved. We will discuss OFDMA channel allocation in later chapters in detail.

In IEEE 802.11b standards for WiFi, there are a total of 14 overlapping channels with a bandwidth of 22 MHz for each channel. The overall bandwidth is from 2.412 to 2.484 GHz. Depending on different countries, the available channels are different. If multiple users try to access simultaneously, different channels can be selected to avoid interference. The channel allocation should be allocated in a distributed way because WiFi networks are infrastructureless (see Figure 5.12).

For the Bluetooth standard in WPANs, frequency hopping is employed to avoid interference, as shown in Figure 5.13. Within each piconet, the channel allocation is performed by a master node to communicate with the slave nodes. The channel allocation has to consider the interference from the other applications such as WiFi.

Finally, we discuss opportunistic/dynamic spectrum access, which enables the opportunistical/dynamic management (allocation, deallocation, sharing) of radio resources (time slots, frequency carriers, codes) within a single-access system or between different radio-access systems. The RF spectrum is becoming scarce. Conventional fixed-spectrum allocation, however, results in low utilization of the allocated spectrum. Opportunistic/dynamic spectrum access can improve spectral efficiency, increase capacity, and improve ease of access to the spectrum, as smaller parties are enabled to enter the market as costs are reduced. So it is a natural migration of spectrum to those users that will use it most efficiently. In addition, the FCC has recently investigated the efficient spectrum usage for cognitive radio, which is a novel technology that improves spectrum utilization by allowing secondary networks (users) to borrow unused radio spectra from primary licensed networks (users) or to share the spectrum with the

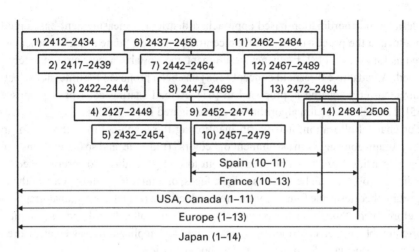

Figure 5.12 Channel allocation for 802.11b standards [247].

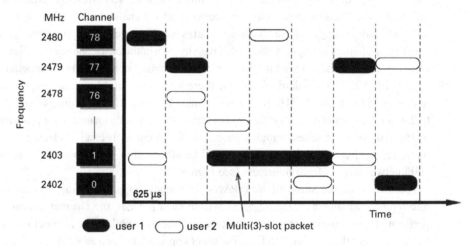

Figure 5.13 Frequency hopping for Bluetooth.

primary networks (users). As an intelligent wireless communication system, cognitive radio is aware of the RF environment, selects the communication parameters (such as carrier frequency, bandwidth, and transmission power) to optimize the spectrum usage, and adapts its transmission and reception accordingly. The process for spectrum access is the first to sense what spectrum is available, then to gain access to some of the available spectrum, next to use the available spectrum, and finally to release the used spectrum. Significant research is necessary to investigate how to access the spectrum, and we will investigate this in detail in the later chapters.

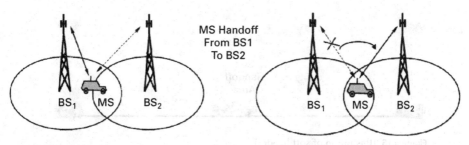

Figure 5.14 Illustration of handoff.

5.7   Handoff

In this section, we first study the basic concepts for handoff. Then we classify the different handoff schemes. The basic handoff models are illustrated. Finally, the handoff schemes in the current cellular standards are briefly studied.

In wireless networks, the term handoff/handover [357] refers to the process of transferring an ongoing call or data session from one channel connected to the core network to another. In satellite communications it is the process of transferring satellite control responsibility from one earth station to another without loss or interruption of service.

In wireless communications, there are two reasons why a handoff (handover) might be conducted: If the phone has moved out of range from one cell site [base station (BS)] and can get a better radio link from a stronger transmitter, or if one BS is full the connection can be transferred to another nearby BS. An example of handoff is shown in Figure 5.14. In urban mobile cellular radio systems, especially when the cell size becomes relatively small, the handoff procedure has a significant impact on system performance. Blocking probability of originating calls and the forced termination probability of ongoing calls are the primary criteria for indicating performance.

Next, we study the basic types of handoff schemes.

1. Hard handoff

   A hard handoff requires the mobile to break the connection with the old BS prior to making the connection with the new one. Hard handoff is also called "break-before-make." TDMA and FDMA systems use a hard handoff when the mobile is moving from one cell site to another. A hard handoff can increase the likelihood of a dropped call.

   CDMA phones use a hard handoff when moving from a CDMA system to an analog system because soft handoff is not possible in analog systems. A pilot beacon unit (PBU) at the analog cell site alerts the phone that it is reaching the edge of CDMA coverage. The phone switches from digital to analog mode as during the hard handoff. The CDMA hard handoff may be used when moving from a CDMA network to an analog one. Analog to CDMA handoff is not available because of the limitations of analog technology.

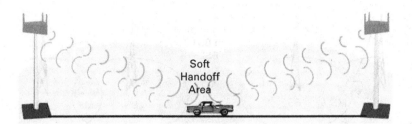

Figure 5.15 Illustration of soft handoff.

2. Soft Handoff

A soft handoff establishes a connection with the new BS prior to breaking the connection with the old one. This is possible because CDMA cells use the same frequency and because the mobile uses a Rake receiver. The CDMA mobile assists the network in the handoff. The mobile detects a new pilot as it travels to the next coverage area. The new BS then establishes a connection with the mobile. This new communication link is established while the mobile maintains the link with the old BS. Soft handoff is also called "make-before-break," as illustrated in Figure 5.15.

There are two variations of soft handoff involving handoff between sectors within a BS: softer and soft–softer. The softer handoff occurs between two sectors of the same BS. The BS decodes and combines the voice signal from each sector and forwards the combined voice frame to the BC. The soft–softer handoff is combination handoff involving multiple cells and multiple sectors within one of the cells.

An idle handoff occurs when the phone is in idle mode. The mobile will detect a pilot signal that is stronger than the current pilot. The mobile is always searching for the pilots from any neighboring BSs. When it finds a stronger signal, the mobile simply begins attending to the new pilot. An idle handoff occurs without any assistance from the BS.

3. Network-controlled, mobile-assisted, and mobile-controlled handoffs

In the network-controlled handoff, the network makes the handoff decisions based on the measurement of mobile links at the BSs. The handoff process includes data transmission, channel switching, and network switching. The information for all link qualities is needed for the network to properly conduct handoff. The network-controlled handoff is used in first-generation analog systems such as TACS (total-access communication system) and AMPS (advanced mobile phone system).

For mobile-assisted handoff, the mobiles make the measurement and the network makes the decision. For example, in a circuit-switched GSM, the BS controller obtains the measurement from the mobiles and manages the handoff. Because the measurement needs to feed back, the procedure duration can be 1 s.

For mobile-controlled handoff, each mobile measures the signal strength from the nearby BSs and interference levels in all channels, and then completely controls the handoff process. A handoff can be initiated if the link of the serving BS is lower than that of another BS by a certain threshold. The reaction time can be very short.

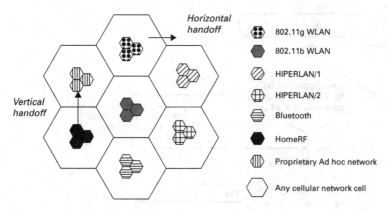

Figure 5.16 Illustration of vertical handoff.

4. Vertical handoff

Vertical handoff refers to a network node changing the type of connectivity it uses to access a supporting infrastructure, usually to support node mobility. For example, a suitably equipped laptop might be able to use both a high-speed WLAN and a cellular technology for Internet access. WLAN connections generally provide higher speeds, whereas cellular technologies generally provide more ubiquitous coverage. Thus the laptop user might want to use a WLAN connection whenever one is available and to "fail over" to a cellular connection when the WLAN is unavailable. Vertical handoff refers to the automatic failover from one technology to another in order to maintain communication. This is different from a "horizontal handoff" between different wireless APs that use the same technology in that a vertical handoff involves changing the data link-layer technology used to access the network.

The system illustration of vertical handoff is shown in Figure 5.16. Vertical handoff between WLAN and UMTS/CDMA2000 has attracted a great deal of attention in all the research areas of the wireless network because of the benefit of utilizing the higher bandwidth and lower cost of a WLAN as well as better mobility support and larger coverage of 3G cellular system.

802.21 is an IEEE emerging standard. The standard supports algorithms enabling seamless handover between networks of the same type as well as handover between different network types, also called media-independent handover (MIH). The standard provides information to allow handing over to and from cellular, GSM, GPRS, WiFi, Bluetooth, and 802.11 networks through different handover mechanisms.

The traffic models for handoff are used to model the users' mobility, so that the handoff schemes can be efficiently designed. We briefly introduce these traffic models as follows.

1. Hong and Rappaport's Traffic Model [135]: In this model, the user's location is uniformly distributed within a radius, and the movement of users does not change within each cell.

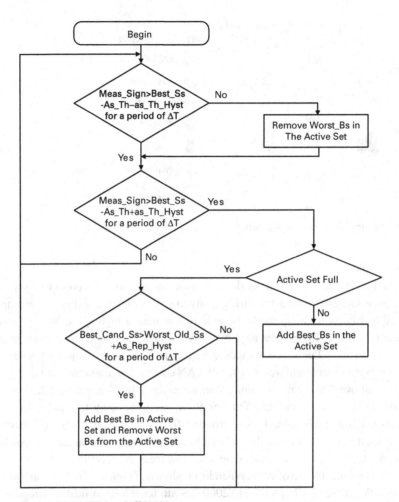

Figure 5.17 UMTS handoff [373].

2. Traffic Model of El-Dolil et al. [63]: It is an extension of Hong and Rappaport's traffic model to the highway microcellular radio networks, in which the highway is segmented into a microcell with small BSs radiating cigar-shaped mobile radio signals along the highway.

3. Steele and Nofal's Traffic Model [290]: This model is based on city street microcells, for pedestrians making calls while walking along streets.

4. Xie and Kuek's Traffic Model [340]: A uniform density of mobile users is assumed, and a user is equally likely to move in any direction with respect to the cell boundary.

5. Approximated Traffic Model of Zeng et al. [358]: Under the assumption of small blocking probability and termination probability, the average number of occupied channels can be approximated by Little's formula.

Finally, we discuss the handover process as described in TR 25.922 of the 3GPP specifications [373]. The detailed algorithm block diagram is shown in Figure 5.17.

- Measurement: Meas_Sign is the received signal code power (RSCP) over the received signal-strength indicator (RSSI). RSCP is the power carried by the decoded pilot channel, and RSSI is the total wideband received power within the channel bandwidth. Some filtering is necessary to remove the effects of fast fading. Best_Ss is the best measured cell present in the active set. Worst_Old_Ss is the worst measured cell present in the active set. Best_Cand_Set is the best measured cell present in the monitored set.
- Sets: We define the set as the list of cells or Node B's, the active set as the list of cells having a connection with the mobile station, and the monitored set as the list of neighboring cells whose pilot channel is continuously measured but not strong enough to be added to the active set.
- Thresholds: The main parameter in the soft-handover algorithm is the threshold for soft handover As_Th. Its value determines the amount of users that are in soft-handover mode and hence influences the system capacity and coverage. Roughly stated, it is the maximum difference in the SINR that two pilot signals can have so their cells can coexist in the active set. As_Th_Hyst is the hysteresis for the As_Th threshold, and As_Rep_Hyst is the replacement hysteresis.
- Algorithm:
  1. Adding: If Meas_Sign is greater than (Best_Ss As_Th + As_Th_Hyst) for a period of ΔT and the active set is not full, the best cell outside the active set is added to the active set.
  2. Removing: If Meas_Sign is below (Best_Ss As_Th As_Th_Hyst) for a period of ΔT remove worst cell in the active set.
  3. Replacing: If active set is full and Best_Cand_Ss is greater than (Worst_Old_Ss + As_Rep_Hyst) for a period of ΔT add the best cell outside the active set and remove the worst cell in the active set.

5.8   Admission Control

Admission control, or call admission control (CAC), can be defined as a process for managing the arriving traffic (at the call, session, or connection level) based on some predefined criteria. CAC is an algorithm that admits/rejects arriving users to optimize some objective function. Meanwhile, CAC guarantees the QoS of new users as well as of the active users. CAC is usually used to guarantee the signal quality (SIR, Eb/I0, or BER), guarantee the call-dropping probability, give priority to some classes, maximize revenue, conduct fair resource sharing, guarantee transmission rate, and guarantee packet-level QoS.

CAC is used in wireless networks to optimize the system performance and guarantee the QoS. Various approaches and techniques have been proposed for different models and environments. The admission control schemes are classified by different criteria, listed as follows:

- Admission criteria
  The admission criteria are typically based on the threshold-based mechanism. In this mechanism, a metric is defined to represent resource availability, number of users,

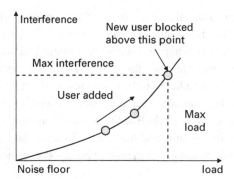

Figure 5.18 CDMA admission control.

SINR or interference, (Tx or Rx) power, or the overload probability. When a new user comes to the network, there is an increase for this metric. If the resulting new value is greater than a threshold, the user will be rejected. Otherwise, the user will be admitted. The threshold is determined so that the already admitted users can have the desired link quality.

For the systems with hard capacity such as TDMA or FDMA systems, the metric can represent the number of users or the channels, and the incoming user's traffic pattern is also known. For systems with soft capacity, like a CDMA system, the SINR can be utilized to adaptively judge if the system can admit more users. The illustration of this type of CAC is shown in Figure 5.18.

• Decision and information level
A CAC algorithm can operate in a distributed, semidistributed, or centralized manner, depending on whether the information can be obtained locally, semilocally, or globally. In distributed algorithms, the BS makes the decision alone for the admission control. In semidistributed schemes, the neighbor BSs exchange information and collaborate together. In the centralized manner, all information is collected and sent to a centralized point to make the admission decisions.

• Other issues
First, the threshold for admission control can be fixed and statistically obtained, or the threshold can be adaptively modified according the dynamics of the wireless channels and users' traffic patterns. Obviously the adaptive schemes can provide better performance but with a higher cost of signaling and complexity. Second, the admission control can also be prioritized because the different types of users have different hierarchies. Finally, the decision can be made by constrained optimization or other techniques, such as fuzzy logic.

In [224], some of the existing CAC approaches are investigated.

• Guard channel approach
Some portion of the wireless resources is reserved for a new admission call or handoff calls so that the call-failure probability and handoff call-dropping probability can be maintained below the target level. For a modified version, the fractional guard channel

approach, new calls are gradually blocked according to a probability related to the current status (i.e., the number of ongoing calls) of the network.

- Cooperative and noncooperative approaches

  For cooperative approaches, the neighboring cells exchange information about the network status so that a resource reservation can be made in advance accurately. The noncooperative approaches use prediction techniques, such as the ARMA (autoregressive moving average model) model, to project the amount of the resources required locally so that the resources can be reserved in advance without the need for information exchange among neighboring cells.

- Mobility-based approach

  Mobility information, such as position and direction of movement, of mobiles can be employed to enhance the accuracy of the resource reservation.

- Pricing-based approach

  Dynamic pricing is used to limit the call arrival rate so that the maximum utility and revenue of the system are achieved. The pricing itself is determined in a centralized one, whereas the admission control is conducted distributively.

Some other radio resource-allocation schemes, like channel allocation, BS assignment, scheduling or buffer management, power control, and bandwidth reservation can also be integrated with CAC. Different CAC approaches can be evaluated by different criteria, such as optimality, complexity, information exchange, stability, scalability, and speed.

CAC and handoff are related in the sense that handoff can be viewed as CAC in the new cell. However, there are significant differences. First, dropping a call is more annoying to customers than blocking a call. So the priority of handoff is higher than CAC. Moreover, a handoff mobile user will release a certain channel. If the neighboring cells can cooperate, the resource-allocation challenges for handoff are less than for CAC.

In future wireless networks, because of the heterogeneous environment, there exist several types of wireless access technologies, so CAC schemes must be able to handle vertical handoff and special modes of connection such as ad hoc on cellular. As a result, the CAC schemes need to accommodate multiple types of users and applications with different QoS requirements. With great demands for multimedia applications, system utilization and QoS performance can be improved by adjusting the CAC, depending on the state of the network and users' QoS requirements. Finally, both call-level QoS and packet-level QoS need to be considered to design CAC algorithms so that the call-packet-dropping and call-packet-blocking probabilities can be maintained at the target level.

5.9   Summary

In this chapter, we discussed how multiple users share the limited bandwidth. To compare the different schemes' advantages and disadvantages, we listed different schemes as follows [150]:

- CDMA: high-frequency efficiency in cellular systems, soft-handover capability, but requirement for power control.
  1. Multicarrier MC-CDMA: suitable applications for frequency-selective fading channel, but requirement for highly linear amplification.
  2. DS-CDMA: robustness to multipath fading by use of Rake receiver.
  3. TDD-CDMA: accurate power-control capability, asymmetric communication capability, but requirement for long overhead for synchronization.
  4. MC/DS-CDMA: combination of advantages of both MC-CDMA and DS-CDMA, high data transmission under severe fading channel.
- FDMA: easy implementation, small ISI, but low flexibility in channel allocation and small channel capacity in cellular system.
- OFDMA: flexibility of assignment subcarriers to users, high-data-rate transmission for downlink, but requirement for quasi-synchronization in uplink, large guard interval to compensate for synchronization.
- TDMA: compatibility with 2G systems, requirement for accurate channel equalization in high-rate communication, but low flexibility in channel allocation, large overhead for synchronization, easier handover than FDMA.
- SDMA: requirement of adaptive antenna-array techniques, capacity to share the same frequency and time slot for multiple users, but sensitivity to location of users, increase of hardware complexity to track signals, requirement for internal handover.

For multiple-access schemes such as FDMA, TDMA, and CDMA, the channel is assigned to the users regardless of whether there are data packets to transmit. To dynamically utilize the limited radio resources, scheduling and random access are proposed to allocate the channels to the users when necessary. For scheduling, there exists a coordinator that can control by whom and when the channel will be occupied. For random access, distributed users access the channel randomly. Then some collision detection and avoidance schemes are proposed to improve the throughput. Of course, other design issues such as fairness, QoS, and delay shall be considered. Overall, the scheduling and random-access schemes are still open issues, especially in a cross-layer design.

From an individual user's point of view, the spectrum-access techniques provide mobile users the available spectrum to be utilized for communication. When a new call comes, admission control policy decides if the call can be accepted by the system without hurting the current existing users' link quality. Then a channel is allocated to the user, and this channel is released by the user after the call ends. If the user moves to different service areas, handoff technologies are employed for users to transparently change the APs.

Part II

Optimization Techniques for Resource Allocation

6   Optimization Formulation and Analysis

6.1   Introduction

In wireless networks, the available radio resources such as bandwidth are very limited. On the other hand, the demands for wireless services are exponentially increasing. Not only is the number of users booming, but also more bandwidth is required for the new services such as video telephony, TV on demand, wireless Internet, and wireless gaming. How to accommodate all these requirements has become an emergent research issue in wireless networking. Resource allocation and its optimization are general methods to improve the network performances.

The design of wireless networking is usually conducted in two different styles. For physical-layer researchers, the bandwidth is very limited and optimization is critical to approach the optimality, such as the Shannon capacity. On the other hand, for higher-layer researchers, it is mostly impossible to have any analytical solution. Therefore the design criteria is often heuristic. There are trade-offs between these two types of approaches. One of our major goals is to present these trade-offs so that better implementations can be put into practice.

In this chapter, we discuss how to formulate the wireless networking problem as a resource-allocation optimization issue. Specifically, we study what the resources are, what the parameters are, what the practical constraints are, and what the optimized performances across the different layers are. In addition, we address how to perform resource allocation in multiuser scenarios. The trade-offs between the different optimization goals and different users' interests are also investigated. The goal is to provide readers with a new perspective from the optimization point of view for wireless networking and resource-allocation problems.

This chapter is organized in the following way: In Section 6.2, we discuss the basic formulation of the wireless network and resource-allocation problem. In Section 6.3, we discuss how to judge whether a solution is optimal. In Section 6.4, we explain the important concept of duality. To obtain the closed-form solution, some approximations of the constraint and optimization goal functions are illustrated in Section 6.5. A cross-layer multiuser resource optimization example is shown in Section 6.6 by formulating a constrained optimization problem, using an approximation.

6.2 Constrained Optimization

Many wireless resource-allocation problems can be formulated as constrained optimization problems, which can be optimized from the network point of view or from the individual point of view. The general formulation can be written as

$$\min_{\mathbf{x} \in \Omega} f(\mathbf{x}), \tag{6.1}$$

$$\text{s.t.} \begin{cases} g_i(\mathbf{x}) \leq 0, & \text{for } i = 1, \ldots, m \\ h_j(\mathbf{x}) = 0, & \text{for } j = 1, \ldots, l \end{cases},$$

where $\mathbf{x}$ is the parameter vector for optimizing the resource allocation, Ω is the feasible range for the parameter vector, and $f(\mathbf{x})$ is the optimization goal matrix, objective goal, or utility function that represents the performance or cost. Here $g_i(\mathbf{x})$ and $h_j(\mathbf{x})$ are the inequality and equality constraints for the parameter vector, respectively. The optimization process finds the solution $\bar{\mathbf{x}} \in \Omega$ that satisfies all inequality and equality constraints. For an optimal solution, $f(\bar{\mathbf{x}}) \leq f(\mathbf{x}), \forall \mathbf{x} \in \Omega$.

If the optimization goal, the inequality constraints, and the equality constraints are all linear functions of the parameter function $\mathbf{x}$, then the problem in (6.1) is called a *linear program*. One important characteristic of a linear program problem is that there is a global optimal point that is very easy to obtain by linear programming. On the other hand, one major drawback of a linear program is that most of the practical problems in wireless networking and resource allocation are nonlinear. Therefore it is hard to model these practical problems as linear programs. If either the optimization goal or the constraint functions are nonlinear, the problem in (6.1) is called a *nonlinear program*. In general, there are multiple local optima in a nonlinear program and to find the global optimum is not an easy task. Furthermore, if the feasible set Ω contains some integer sets, the problem in (6.1) is called an *integer program*. Most integer programs are nondeterministic-polynomial-hard (NP-hard) problems that cannot be solved by polynomial time.

One special kind of nonlinear program is the convex optimization problem in which the feasible set Ω is a convex set, and the optimization goal and the constraints are convex/concave/linear functions. A convex set is defined in Definition 4.

DEFINITION 4 *A set Ω is* convex *if for any $\mathbf{x}_1, \mathbf{x}_2 \in \Omega$ and any θ with $0 \leq \theta \leq 1$, we have*

$$\theta \mathbf{x}_1 + (1 - \theta)\mathbf{x}_2 \in \Omega.$$

A convex function f is defined in Definition 5.

DEFINITION 5 *A function f is* convex *over $\mathbf{x}$ if the feasible range Ω of parameter vector $\mathbf{x}$ is a convex set, and if for all $\mathbf{x}_1, \mathbf{x}_2 \in \Omega$ and $0 \leq \theta \leq 1$,*

$$f[\theta \mathbf{x}_1 + (1 - \theta)\mathbf{x}_2] \leq \theta f(\mathbf{x}_1) + (1 - \theta)f(\mathbf{x}_2).$$

A function is strictly convex if the strict inequality is held whenever $\mathbf{x}_1 \neq \mathbf{x}_2$ and $0 < \theta < 1$. A function is called concave is $-f$ is convex.

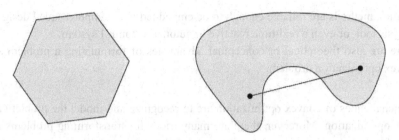

Figure 6.1 Convex-set examples: left, convex set, right, unconvex set.

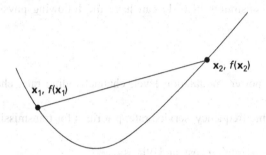

Figure 6.2 Convex-function example.

In Figure 6.1, we show two examples of a convex set and an unconvex set. In Figure 6.2, we show the example of a convex function. If the function is differentiable, and if either of the following two conditions hold, the function is a convex function:

$$\text{First-order condtion: } f(\mathbf{x}_2) \geq f(\mathbf{x}_1) + \nabla f(\mathbf{x}_1)^T(\mathbf{x}_2 - \mathbf{x}_1); \tag{6.2}$$

$$\text{Second-order condtion: } \nabla^2 f(\mathbf{x}) \succeq 0. \tag{6.3}$$

One important application of the convex function is Jensen's inequality. Suppose function f is convex and the parameter $\mathbf{x}$ has any arbitrary random distribution over Ω; then the following equality holds

$$f(E\mathbf{x}) \leq Ef(\mathbf{x}), \tag{6.4}$$

where E is expectation.

The advantages of convex optimization for wireless network and resource-allocation problem are shown as follows:

- There are a variety of applications such as automatic control systems, estimation and signal processing, communications and networks, electronic circuit design, data analysis and modeling, statistics, and finance.
- Computation time is usually quadrature. Problems can then be solved, very reliably and efficiently, using interior-point methods or other special methods for convex optimization.

- Solution methods are reliable enough to be embedded in a computer-aided design or analysis tool, or even a real-time reactive or automatic control system.
- There are also theoretical or conceptual advantages of formulating a problem as a convex optimization problem.

The challenges of convex optimization are to recognize and model the problem as a convex optimization. Moreover, there are many tricks for transforming problems into convex forms.

We have discussed the basics for constrained optimization problems. Next we will see how the problem can be formulated. In wireless networking and resource allocation, the parameters, functions, and constraints in (6.1) can have the following physical meanings:

- **Parameters**:
  1. Physical layer: transmitted power, modulation level, channel-coding rate, channel/code selection, and others.
  2. MAC layer: transmission time/frequency, service rate, priorities for transmission, and others.
  3. Network layer: route selection, routing cost, and others.
  4. Application layer: source-coding rate, buffer priority, packet arrival rate, and others.
- **Optimization goals**:
  1. Physical layer: minimal overall power, maximal throughput, maximal rate per joule, minimal BER, and others.
  2. MAC layer: maximal overall throughput, minimal buffer overflow probability, minimal delay, and others.
  3. Network layer: minimal cost, maximal profit, and others.
  4. Application layer: minimal distortion, minimal delay, and others.
- **Constraints**:
  1. Physical layer: maximal mobile transmitted power, available modulation constellation, available channel-coding rate, limited energy, and others.
  2. MAC layer: contentions, limited time/frequency slot, limited information about other mobiles, and others.
  3. Network layer: maximal hops, security concerns, and others.
  4. Application layer: the base-layer transmission, limited source rate, strict delay requirement, security, and others.

After formulating the constrained optimization problem over wireless networking and resource allocation, we need to find solutions. In this chapter, we explain how to get the closed-form solution. If the closed-form solution cannot be obtained, other methods such as programming are discussed in the other chapters in Part II of this book. One of the most important methods used to find a closed-form solution for constrained optimization is the Lagrangian method, which has the following steps:

1. Rewrite (6.1) as a Lagrangian multiplier function J as

$$J = f(\mathbf{x}) + \sum_{i=1}^{m} \lambda_i g_i(\mathbf{x}) + \sum_{j=1}^{l} \mu_j h_j(\mathbf{x}), \qquad (6.5)$$

where λ_i and μ_j are Lagrangian multipliers.

2. Differentiate J over $\mathbf{x}$ and set to zero as

$$\frac{\partial J}{\partial \mathbf{x}} = 0. \qquad (6.6)$$

3. From (6.6), solve λ_i and μ_j.
4. Replace λ_i and μ_j in the constraints to get optimal $\mathbf{x}$.

Notice that the difficulty in the Lagrangian method is step 3 and step 4, where the closed-form solution is obtained for the Lagrangian multipliers. Some approximations and mathematical tricks are necessary to obtain the closed-form solutions.

One simple example is shown as follows. Suppose we have N channels to transmit data. The total rate of these N channels is R_0. Different channels have different channel conditions. The problem is how to find the optimal power allocation to different channels so as to minimize the overall transmitted power under the constraint of the required throughput. The solution is the famous water-filling algorithm.

Suppose the power to the ith channel is P_i. The channel gain for this channel is G_i and the noise power is σ^2 for all channels. The capacity for the ith channel can be written as

$$R_i = W \log_2 \left(1 + \frac{P_i G_i}{\sigma^2} \right). \qquad (6.7)$$

The constrained optimization problem can be written as

$$\min_{P_i} \sum_{i=1}^{N} P_i, \qquad (6.8)$$

$$\text{s.t.} \sum_{i=1}^{N} R_i = R_0.$$

By using the Lagrangian method, we can have a closed-form solution for the water-filling problem. We write the Lagrangian multiplier as

$$J = \sum_{i=1}^{N} P_i + \lambda \left[P_{max} - \sum_{i=1}^{N} W \log_2 \left(1 + \frac{P_i G_i}{\sigma^2} \right) \right]. \qquad (6.9)$$

Setting $\frac{\partial J}{\partial P_i} = 0, \forall i$, we can have the optimal solution of the water filling as

$$P_i = \left(\lambda - \frac{\sigma^2}{G_i} \right)^+, \qquad (6.10)$$

where $y^+ = \max(y, 0)$ and λ is obtained by bisection search of the following expression:

$$\sum_{i=1}^{N} W \log_2 \left[1 + \frac{(\lambda - \frac{\sigma^2}{G_i})^+ G_i}{\sigma^2} \right] = R_0. \qquad (6.11)$$

6.3   Optimality

In this section, we discuss the optimality of the solution, i.e., how to determine whether the solution is optimal, and in what sense the solution is optimal. The organization of this section is as follows: The optimality for the unconstrained problem is discussed first. Then Fritz John conditions and Karush–Kuhn–Tucker (KKT) conditions are explained. Finally, second-order conditions are illustrated.

Before we discuss the unconstrained optimality, we define the global optimum and local optimum as follows:

DEFINITION 6 *Consider* min $f(\mathbf{x})$ *over* Ω *and let* $\bar{\mathbf{x}} \in \Omega$. *If* $f(\bar{\mathbf{x}}) \le f(\mathbf{x})$, $\forall \mathbf{x} \in \Omega$, *and then* $\bar{\mathbf{x}}$ *is called a global minimum. If* $f(\bar{\mathbf{x}})$ *is no more than the neighbor of* $\bar{\mathbf{x}}$, $\bar{\mathbf{x}}$ *is called a local minimum.*

Some necessary and sufficient conditions are as follows for the optimality:

- Necessary conditions:
  1. First-order necessary condition: If $f(\mathbf{x})$ is differentiable at $\bar{\mathbf{x}}$. If $\bar{\mathbf{x}}$ is a local minimum, then $\nabla f(\bar{\mathbf{x}}) = 0$.
  2. Second-order necessary condition: If $f(\mathbf{x})$ is twice differentiable at $\bar{\mathbf{x}}$. If $\bar{\mathbf{x}}$ is a local minimum, then $\nabla f(\bar{\mathbf{x}}) = 0$ and Hessian matrix $\mathbf{H}(\bar{\mathbf{x}})$ is positive semidefinite.
- Sufficient conditions:
  1. First-order necessary and sufficient condition: If $f(\mathbf{x})$ is pseudoconvex at $\bar{\mathbf{x}}$. Then, $\bar{\mathbf{x}}$ is a global minimum if and only if $\nabla f(\bar{\mathbf{x}}) = 0$.
  2. Second-order sufficient condition: If $f(\mathbf{x})$ is twice differentiable at $\bar{\mathbf{x}}$. If $\nabla f(\bar{\mathbf{x}}) = 0$ and Hessian matrix $\mathbf{H}(\bar{\mathbf{x}})$ is positive definite, then $\bar{\mathbf{x}}$ is a strict local minimum.

It is worth mentioning that some points on the curves satisfying the necessary conditions might not be the true optimum. For example, in Figure 6.3, we show $z = x^2 - y^2$. We can see that the necessary condition of the first-order differential equal to zero is satisfied for the saddle point at $(0, 0)$. However, the saddle point is not a local optimum.

Next we list a few theorems for the optimality of (6.1) as follows:

- Fritz John necessary conditions:
  Let $\bar{\mathbf{x}}$ be a feasible solution, and let $I = \{i : g_i(\bar{\mathbf{x}}) = 0\}$. Suppose $g_i, i \in I$ is continuous at $\bar{\mathbf{x}}$. f and $g_i, i \in I$ are differentiable at $\bar{\mathbf{x}}$, and $h_j, \forall j$ is continuous and differentiable at $\bar{\mathbf{x}}$. If $\bar{\mathbf{x}}$ is a local minimum, then there exist scalars u_0, u_i, for $i \in I$ and v_j for $j = 1, \ldots, l$ such that

$$u_0 \nabla f(\bar{\mathbf{x}}) + \sum_{i \in I} u_i \nabla g_i(\bar{\mathbf{x}}) + \sum_{j=1}^{l} v_j \nabla h_j(\bar{\mathbf{x}}) = \mathbf{0}, \qquad (6.12)$$

$$u_0, u_i \ge 0, \forall i \in I, (u_0, \mathbf{u}_I, \mathbf{v}) \ne (0, \mathbf{0}, \mathbf{0}),$$

  where $\mathbf{u}_I$ is the vector whose component is u_i and $\mathbf{v} = (v_1 \ldots v_l)^T$.
- Fritz John sufficient conditions:

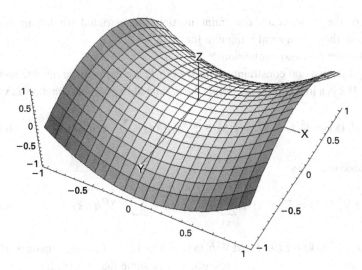

Figure 6.3 Example of a saddle point.

Define $S = \{\mathbf{x} : g_i(\mathbf{x}) \leq 0, \text{ for } i \in I, h_j(\mathbf{x}) = 0, \text{ for } j = 1, \ldots, l\}$. If h_j for $j = 1, \ldots, l$ are affine and $\nabla h_j(\bar{\mathbf{x}})$, $j = 1, \ldots, l$ are linearly independent, and if there exists an ε-neighborhood $N_\varepsilon(\bar{\mathbf{x}})$ of $\bar{\mathbf{x}}, \varepsilon > 0$, such that f is pseudo-convex on $S \cap N_\varepsilon(\bar{\mathbf{x}})$, then $\bar{\mathbf{x}}$ is a local minimum for (6.1).

- **KKT necessary conditions:**
Suppose f and $g_i, i \in I$, are differentiable at $\bar{\mathbf{x}}, g_i, i \in I$, is continuous at $\bar{\mathbf{x}}$, and $h_j, \forall j$ is continuous and differentiable at $\bar{\mathbf{x}}$. Suppose $\nabla g_i(\bar{\mathbf{x}}), \forall i \in I$ and $\nabla h_j(\bar{\mathbf{x}}), j = 1, \ldots, l$ are linearly independent. If $\bar{\mathbf{x}}$ is a local optimum, then unique scalars $u_i, i \in I$ and v_j for $j = 1, \ldots, l$ exist such that

$$\nabla f(\bar{\mathbf{x}}) + \sum_{i \in I} u_i \nabla g_i(\bar{\mathbf{x}}) + \sum_{j=1}^{l} v_j \nabla h_j(\bar{\mathbf{x}}) = \mathbf{0}, \tag{6.13}$$

$$u_i \geq 0, \forall i \in I.$$

- **KKT sufficient conditions:**
Suppose KKT conditions hold at $\bar{\mathbf{x}}$, i.e., there exist scalars $\bar{u}_i \geq 0$ for $i \in I$ and $\bar{v}_j$ for $j = 1, \ldots l$ such that

$$\nabla f(\bar{\mathbf{x}}) + \sum_{i \in I} \bar{u}_i \nabla g_i(\bar{\mathbf{x}}) + \sum_{j=1}^{l} \bar{v}_j \nabla h_j(\bar{\mathbf{x}}) = \mathbf{0}. \tag{6.14}$$

Let $J = \{j : \bar{v}_j > 0\}$ and $K = \{j : \bar{v}_j < 0\}$. Suppose f is pseudo-convex at $\bar{\mathbf{x}}$, g_i is quasi-convex at $\bar{\mathbf{x}}$ for $i \in I$, h_j is quasi-convex at $\bar{\mathbf{x}}$ for $j \in J$, and h_j is quasi-concave at $\bar{\mathbf{x}}$ for $j \in K$. Then $\bar{\mathbf{x}}$ is a global optimal solution to (6.1). If the generalized convexity assumptions on the objective and constraint functions are restricted to a domain $N_\varepsilon(\bar{\mathbf{x}})$ for some $\varepsilon > 0$, then $\bar{\mathbf{x}}$ is a local minimum for (6.1).

- KKT second-order necessary conditions:

  Suppose the objective and constraints defined in (6.1) are twice differentiable and Ω is not empty. If $\bar{\mathbf{x}}$ is a local optimum, define the restricted Lagrangian function $L(\mathbf{x})$ as

$$L(\mathbf{x}) = \phi(\mathbf{x}, \bar{\mathbf{u}}, \bar{\mathbf{v}}) = f(\mathbf{x}) + \sum_{i \in I} \bar{u}_i g_i(\mathbf{x}) + \sum_{j=1}^{l} \bar{v}_j h_j(\mathbf{x}). \tag{6.15}$$

Denote its Hessian at $\bar{\mathbf{x}}$ by

$$\nabla^2 L(\bar{\mathbf{x}}) = \nabla^2 f(\bar{\mathbf{x}}) + \sum_{i \in I} \bar{u}_i \nabla^2 g_i(\bar{\mathbf{x}}) + \sum_{j=1}^{l} \bar{v}_j \nabla^2 h_j(\bar{\mathbf{x}}), \tag{6.16}$$

where $\nabla^2 f(\bar{\mathbf{x}})$, $\nabla^2 g_i(\bar{\mathbf{x}})$ for $i \in I$ and $\nabla^2 h_j(\bar{\mathbf{x}})$ for $j = 1, \ldots, l$ are the Hessians of f, g_i for $i \in I$, and h_j for $j = 1, \ldots, l$ respectively. Assume that $\nabla g_i(\bar{\mathbf{x}})$ for $i \in I$ and $\nabla h_j(\bar{\mathbf{x}})$ for $j = 1, \ldots, l$ are linearly independent. Then $\bar{\mathbf{x}}$ is a KKT point and

$$\mathbf{d}^T \nabla^2 L(\bar{\mathbf{x}}) \mathbf{d} \geq 0 \tag{6.17}$$

for all $\mathbf{d} \in \{\mathbf{d} \neq \mathbf{0} : \nabla g_i(\bar{\mathbf{x}})^T \mathbf{d} \leq 0\}$ for all $i \in I$, $\nabla h_j(\bar{\mathbf{x}})^T \mathbf{d} = 0$ for all $j = 1, \ldots, l$.

- KKT second-order sufficient conditions:

  Suppose the objective and constraints defined in (6.1) are twice differentiable and Ω is not empty. Let $\bar{\mathbf{x}}$ be a KKT point for (6.1), with Lagrangian multiplier $\bar{u}$ and $\bar{v}$ associated with the inequality and equality constraints, respectively. Denote $I^+ = \{i \in I : \bar{u}_i > 0\}$ and $I^0 = \{i \in I : \bar{u}_i = 0\}$. Define the restricted Lagrangian function $L(\mathbf{x})$ and its Hessian. Define the cone

$$C = \{\mathbf{d} \neq \mathbf{0} : \begin{array}{ll} \nabla g_i(\bar{\mathbf{x}})^T \mathbf{d} = 0 & \text{for } i \in I^+, \\ \nabla g_i(\bar{\mathbf{x}})^T \mathbf{d} \leq 0 & \text{for } i \in I^0, \\ \nabla h_j(\bar{\mathbf{x}})^T \mathbf{d} = 0 & \text{for } j = 1, \ldots, l. \end{array} \tag{6.18}$$

Then if $\mathbf{d} \in \{\mathbf{d} > \mathbf{0}\}$ for all $\mathbf{d} \in C$, we have that $\bar{\mathbf{x}}$ is a strict local minimum for (6.1).

The optimality discussed in this section can be used for different scenarios. For example, it can prove the optimality of a certain solution; it can determine the termination criteria for the adaptive algorithm; it can be used for convergence analysis.

6.4   Duality

In this section, we define the duality concept for the constrained optimization. Under some convexity assumptions and constraints properties, the primal and dual problems have the same optimal objective values, so it is possible to solve the prime problem by considering the dual problem and to develop very efficient algorithms. We define the dual problem first, and then introduce the duality theorem. Some properties are discussed, and finally dual methods are illustrated with the examples.

Consider (6.1) as the prime problem. The Lagrangian dual problem of (6.1) can be defined as

$$\max \theta(\mathbf{u}, \mathbf{v}), \tag{6.19}$$

$$\text{s.t. } \mathbf{u} \geq \mathbf{0},$$

where $\theta(\mathbf{u}, \mathbf{v}) = \inf\{f(\mathbf{x}) + \sum_{i=1}^{m} u_i g_i(\mathbf{x}) + \sum_{j=1}^{l} v_j h_j(\mathbf{x}) : \mathbf{x} \in \Omega\}$.

For example, the prime problem for linear programming in standard form is

$$\min_{\mathbf{x}} \mathbf{c}^T \mathbf{x}, \tag{6.20}$$

$$\text{s.t. } \begin{cases} \mathbf{Ax} = \mathbf{b} \\ \mathbf{x} \geq \mathbf{0} \end{cases},$$

where $\mathbf{x}$ is the optimized vector with dimension $N \times 1$, $\mathbf{c}$ and $\mathbf{b}$ are constant vectors, and $\mathbf{A}$ is a constant matrix. The dual function is

$$\theta(\mathbf{u}, \mathbf{v}) = \inf_{\mathbf{x}} \left[\mathbf{c}^T \mathbf{x} - \sum_{i=1}^{N} \lambda_i x_i + \mathbf{v}^T (\mathbf{Ax} - \mathbf{b}) \right]$$

$$= -\mathbf{b}^T \mathbf{v} + \inf_{\mathbf{x}} (\mathbf{c} + \mathbf{A}^T \mathbf{v} - \lambda)^T \mathbf{x}, \tag{6.21}$$

where $\lambda = [\lambda_1 \ldots \lambda_N]^T$. Because a linear function $(\mathbf{c} + \mathbf{A}^T \mathbf{v} - \lambda)$ is unbounded below, $\theta(\mathbf{u}, \mathbf{v}) = -\infty$ only when $\mathbf{c} + \mathbf{A}^T \mathbf{v} - \lambda = 0$. We have the dual problem as follows

$$\max \ \theta(\mathbf{u}, \mathbf{v}) = \begin{cases} -\mathbf{b}^T \mathbf{v} \ \mathbf{c} + \mathbf{A}^T \mathbf{v} - \lambda = 0 \\ -\infty \quad \text{otherwise} \end{cases}, \tag{6.22}$$

$$\text{s.t. } \lambda \geq \mathbf{0}.$$

One of the major applications of the dual problem is the duality theorem. The following weak duality theorem states that the objective value of any feasible solution to the dual problem is a lower bound on the objective value of any feasible solution to the primal problem.

THEOREM 14 *Weak Duality Theorem: If* $\mathbf{x}$ *is a feasible solution to the prime problem in (6.1) and* $(\mathbf{u}, \mathbf{v})$ *are the feasible solution to the dual problem in (6.19), then* $f(\mathbf{x}) \geq \theta(\mathbf{u}, \mathbf{v})$. *The duality gap is defined as* $f(\mathbf{x}) - \theta(\mathbf{u}, \mathbf{v})$.

Under convexity assumptions and constraint qualification, the duality gap is zero under the following strong duality theorem.

THEOREM 15 *Strong Duality Theorem: Let* Ω *be convex set,* g_i *be convex, and* h_j *be affine. Suppose there exists an* $\mathbf{x} \in \Omega$ *that* $\mathbf{g}(\mathbf{x}) \leq \mathbf{0}$ *and* $\mathbf{h}(\mathbf{x}) = \mathbf{0}$, *and* $\mathbf{0}$ *is the interior point for* $\mathbf{h}$ *[Slater's (interiority) condition]. Then the optimal point of a prime problem is the same as the optimal point of a dual problem, i.e.,*

$$\inf\{(6.1)\} = \sup\{(6.19)\}. \tag{6.23}$$

The necessary and sufficient condition for the zero duality gap is the existence of a saddle point defined by Definition 7.

DEFINITION 7 $(\bar{\mathbf{x}}, \bar{\mathbf{u}}, \bar{\mathbf{v}})$ *with* $\bar{\mathbf{x}} \in \Omega$ *and* $\bar{\mathbf{u}} \geq \mathbf{0}$ *if and only if*

1. $\phi(\bar{\mathbf{x}}, \bar{\mathbf{u}}, \bar{\mathbf{v}}) = \min \left[f(\mathbf{x}) + \sum_{i \in I} \bar{u}_i g_i(\mathbf{x}) + \sum_{j=1}^{l} \bar{v}_j h_j(\mathbf{x}) \right],$
2. $\mathbf{g}(\bar{\mathbf{x}}) \leq \mathbf{0}, \mathbf{h}(\bar{\mathbf{x}}) = \mathbf{0},$
3. $\bar{\mathbf{u}}^T \mathbf{g}(\bar{\mathbf{x}}) = 0.$

If the primal and dual problems are equal, it is possible to solve the prime problem indirectly by solving the dual problem. For nonlinear nonconvex problems, the duality gap might not be zero. Sometimes, in practice, the dual problem will need less information to optimize, and consequently it is easier to solve. In addition, prime and dual problems can be solved iteratively to find the optimal solution. One solution is called cutting planes or outer-linearization methods as follows:

1. Initialization
2. Solve primal optimization.
3. Solve dual optimization.
4. Go back to step 1, until an optimal condition, such as the KKT condition, is satisfied.

6.5   Approximations

The difficulty in obtaining the analytical results from (6.1) is because of the nonlinear and noncovexity of the constraints and the optimization goal. This makes the Lagrangian multiplier function in (6.5) hard to differentiate and to obtain the optimal points. If some approximations for the constraints or the optimization goal functions can be obtained under some conditions, we can solve (6.1) by differentiating (6.5) and putting the results back into the constraints to get the optimal Lagrangian multiplier. There are several methods to achieve the preceding idea. We classify them as follows:

1. Parameterized approximation
   In this method, the nonlinear or nonconvex function is approximated by the parameterized function. The goal is to obtain the optimal parameters such that the approximation errors can be minimized.

   The most common approximation is *linear approximation*. Suppose the original function is $f(x)$; the linear approximation can be written as

$$\min_{a,b} \int_c^d \| f(x) - (ax + b) \|^2 dx, \tag{6.24}$$

   where $[c, d]$ is the interested region where the approximation needs to be accurate.

   Another type of approximation is *polynomial or Tyler expansion*. The original function can be expanded as

$$f(x) = c_0 + c_1(x - x_0) + c_2(x - x_0)^2 +, \ldots, \tag{6.25}$$

   where x_0 is the point where the series are expanded, and $c_0, c_1, c_2, \ldots,$ are constant.

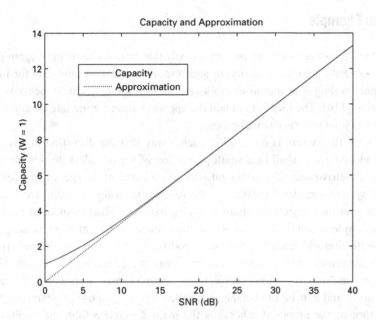

Figure 6.4 Capacity and its approximation.

In general, for any convex function $f'(x; \mathbf{a})$ of x with parameter vector $\mathbf{a}$, the approximation within region $[c, d]$ can be written as

$$\min_{\mathbf{a}} \int_c^d \| f(x) - f'(x; \mathbf{a}) \|^2 dx. \tag{6.26}$$

2. Omitting unimportant parts

The basic idea for this type of approximation is that even though the function itself is not convex, within a certain range, some components can be omitted. Consequently, the approximation is convex.

One good example is the channel capacity function:

$$C = W \log_2(1 + \text{SNR}), \tag{6.27}$$

where C is the capacity, W is the bandwidth, and SNR is the signal-to-noise ratio. As shown in Figure 6.4, the function in (6.27) is not convex. However, when the SNR is high, i.e., $\text{SNR} \gg 1$, we can omit one and have

$$C' = W \log_2(\text{SNR}). \tag{6.28}$$

C' is a convex function and shown in Figure 6.4 as well. As we can see, the approximation is very tight when the SNR is larger than 15 dB, which fits some of the communication situations. This approximation has been employed in some networking optimizations in [34].

6.6   Application Example

In this section, we give one example on how to apply the approximation and Lagrangian methods to solve the wireless networking and resource-allocation problems for joint source–channel coding over multiuser cross-layer design for subjectively better voice communication [110]. The idea is to obtain the approximation for the rate control over the distortion of joint source–channel coding.

For each user, the system is operated in such a way that the distortion that is due to channel-induced errors shall be a small proportion of the overall distortion, so that the system will behave according to the rate-distortion curve of the speech encoder. In doing so, the system considers the effects on the reconstructed signal qualities and takes into consideration the subjectively more annoying nature of channel-induced random errors. For example, when the channel is bad, there are more transmitted bits assigned to channel protection and fewer bits for source coding. This reduces the channel errors but increases source-coding distortions. For the reconstructed received voice, this kind of source-coder-induced distortion is subjectively better, behaves according to the rate-distortion curve, and can be predictably controlled by the proposed system. Such a dynamic nature of the proposed scheme is the main difference from the traditional joint source-channel-coding scheme that minimizes the end-to-end distortion without considering the subjective quality of reconstructed speech. In the rest of this section, we discuss the real-time source encoder and channel-coding modules of the proposed system in detail.

In the proposed system, the real-time source encoder has the key property that the output rate can be externally controlled. This can be implemented by use of either variable-rate or embedded encoders. In the first case, the coder generates one bit stream for each of the possible encoding rates. Only one of these will be selected and transmitted, based on the rate assignment. Using embedded encoders presents the advantage that only one bit stream is generated, making the rate adaptation simply by dropping as many bits as necessary from the end of the bit stream, where less important data are located. Although the "bit-dropping mechanism" is exclusive to the embedded stream, this term is used loosely to represent a reduction in the source rate, regardless of the particular source-encoder implementation. The source coder is assumed to have the maximal output rate η_{max} bits/s. The source rate controller has the output rate $r R$ bits/s ($r R \leq \eta_{max}$), where r is the variable channel-coding rate and R is the fixed channel-transmission rate. Then the data streams are encoded by the channel-coding module. BPSK modulation is employed with power control at the amplifier.

Define $D_s(r R)$ as the distortion-rate function of the user's source encoder transmitting at rate $r R$. In most well-designed encoders, D_s is a convex and decreasing function. The minimum distortion occurs at maximum source rate η_{max}. Furthermore, the source codec distortion-rate function [83, 152] is approximated by

$$D_s = \delta 2^{2k(\eta_{max} - r R)}, \tag{6.29}$$

where δ is the minimal distortion and k is a parameter depending on the encoder. This is a very general form that applies to the case of a Gaussian source with MSE distortion or when the high-rate approximation holds. In the case of realistic encoders, we find that (6.29) constitutes a tight upper bound on the real distortion-rate curve.

We use channel codecs with adjustable rates to jointly adjust rates in both source and channel codecs according to the needs for distortion controls and channel protections. In this example, a rate-compatible punctured convolutional (RCPC) code [101] is employed for channel coding, because of its wide range of channel-coding rates and simplicity. A family of RCPC codes is described by the mother code of rate $\frac{1}{M}$. The output of the coder is punctured periodically following puncture tables. The puncturing period Q determines the range of channel coding rates $r = \frac{Q}{Q+l}$, $l = 1, \ldots, (M-1)Q$, between $\frac{1}{M}$ and $\frac{Q}{Q+1}$ with different channel-error-protection abilities. Moreover, only one Viterbi receiver is needed for the RCPC codes with different rates, which reduces the system complexity.

For simplicity, all transmitted bits are assumed equally important for error-protection purposes. Unequal error protection can be employed in a similar way. Because channel-induced errors are perceptibly more annoying than source-encoding distortions, the *design goal* is that channel-induced errors will account for less than a small proportion of the overall end-to-end distortion. To meet the design goal, the received SINR Γ shall be no less than a targeted SINR. At a constant transmission rate, a reduction in the source-encoding rate allows for a decrease in channel-code rate, and as a result, increases the channel protection. Thus the targeted SINR is also a function of the source-encoding rate, or equivalently, a function of the channel-coding rate. In the rest of this subsection, we develop a model for this targeted SINR as a function of rate under the design goal.

To develop a relation between channel-coding rate and targeted SINR, we analyze a uniform scalar quantizer encoding a uniform source with a random index assignment as an example. To keep the analysis mathematically tractable, we consider only two sources of distortions: source-encoding (compression) and channel-induced distortions. We will study more complicated and practical encoders through the later simulations. Suppose the input vector $\mathbf{x} \in \Re^K$, where K is the number of samples for each quantization. The output of the source encoder has m bits that determine the source-encoding rate. The source-encoder quantizer distortion can be written as

$$D_s = \sum_{i=1}^{2^m} \int_{S_i} \|\mathbf{x} - \mathbf{y}_i\|^2 P_b(\mathbf{x}) d\mathbf{x}, \tag{6.30}$$

where $\{S_i\}_{i=1}^{2^m}$ is the partition of $\Re^K$ into disjoint regions, $\mathbf{y}_i$ is the quantized vector with index i, and $P_b(\mathbf{x})$ is the probability density function of $\mathbf{x}$. From [83], the source-encoding distortion of such a quantizer is given by $D_s(m) = \frac{2^{-2m}}{12}$. Define $q(j|i), i \neq j$, as the probability that the decoded vector index is j, where the transmitted vector index is i. Suppose channel errors happen randomly and independently with respect to the source-encoder index. The probability of having a decoding error is $P_e = \sum_{j=1}^{2^m} q(j|i)$, which is the same for all i. The overall distortion after the channel decoding at the receiver can be written as

$$D(m, P_e) = \sum_{i=1}^{2^m} \sum_{j=1}^{2^m} q(j|i) \int_{S_i} \|\mathbf{x} - \mathbf{y}_i\|^2 P_b(\mathbf{x}) d\mathbf{x}. \tag{6.31}$$

From (6.31), the expected MSE of the uniform scalar quantizer for a uniform source with a random index assignment is

$$D(m, P_e) = \frac{2^{-2m}}{12} + \frac{P_e}{6} \left(1 + 2^{-m}\right) \approx D_s(m) + \frac{P_e}{6} = D_s(m) + D_c, \tag{6.32}$$

in which the second term is the channel-induced distortion that is defined as D_c.

Define ζ as the proportion of channel-induced distortion over the overall distortion. To implement the design goal $\frac{D_c}{D} \leq \zeta$, we need to let the system achieve a certain targeted error probability P_e^t. Obviously $P_e = P_e^t$, when the system converges. From (6.32), we can write P_e^t as

$$P_e^t = \frac{6\zeta D_s(m)}{1 - \zeta} = \frac{\zeta 2^{-2m}}{2(1 - \zeta)}. \tag{6.33}$$

Next we briefly analyze the actual error probability P_e for the RCPC code. P_e is determined by the channel condition, coding structure, and SINR at the receiver. For BPSK modulation and Hamming distance d, the conditional pairwise error probability is given by [239]

$$P_e(d|\{\alpha_i\}) = Q\left(\sqrt{2\Gamma}\right), \tag{6.34}$$

where Γ is the SINR. Then the average error probability over fading channel statistics is given by

$$P_e(d) = \left(\frac{1 - \mu}{2}\right)^d \sum_{k=0}^{d-1} \binom{d - 1 + k}{k} \left(\frac{1 + \mu}{2}\right)^k, \tag{6.35}$$

where $\mu = \sqrt{\frac{\Gamma}{1+\Gamma}}$. When $\Gamma \gg 1$, the preceding equation can be simplified as

$$P_e(d) \approx \left(\frac{1}{4\Gamma}\right)^d \binom{2d - 1}{d}. \tag{6.36}$$

From [362], a tighter upper bound for any coded frame lengths larger than the constraint length is given by

$$P_e \approx \sum_d [(l_{m'} - 1)a_d + b_d] P_e(d), \tag{6.37}$$

where m' is the information frame size, $l_{m'}$ is the number of branches of the trellis that are in a frame, and a_d and b_d are some constants related to the coder.

The system will allocate resources such that the actual BER is the same as the targeted BER, i.e., $P_e = P_e^t$. Because P_e is a function of SINR and P_e^t is a function of r in (6.33), we obtain the relation between the SINR and the channel-coding rate. We plot the relation of Γ in decibels versus channel-coding rate in Figure 6.5 for RCPC code with memory 4,

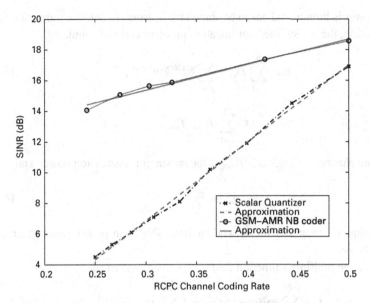

Figure 6.5 Required SINR vs. RCPC rate with $\zeta = 0.03$.

puncture period 8, where there are 20 information bits per frame, transmission rate is 24.4 kbps, $K = 1$, $m = 8$, and $\zeta = 0.03$. The curve shows almost a linear relation, because (6.33) is an exponential form of m and consequently an exponential form of channel-coding rate r, and (6.37) is a sum of polynomials in Γ. The targeted SINR as a function of channel-coding rate, when the design goal is achieved, can be approximated accurately by

$$\gamma \approx 2^{Ar+B}, \tag{6.38}$$

where γ is the required targeted SINR, A and B are fixed parameters of the error control coding and ζ. The received SINR shall be no less than this targeted SINR, i.e., $\Gamma \geq \gamma$.

Furthermore, Figure 6.5 shows the targeted SINR, in $\log_2$ scale, as a function of the RCPC channel-encoding rate, where the channel-induced distortion is less than 3% of that of the end-to-end distortion, i.e., $\zeta = 0.03$. The figure shows the different operation modes for the proposed system. To achieve the design goal, the required SINR is a function of the RCPC channel-coding rate. As we can see from the figure, the approximation in (6.38) is a good approximation for the qualitative behavior of the practical voice encoder as well. Notice that the curve of the GSM-AMR NB (narrowband) coder simulations differs from that of the analysis results. This quantitative difference is due to, in part, not including modeling and error concealment, distortion in the analysis results. Nevertheless, both results suggest that the linear approximation in log scale is a good choice for the relation in (6.38).

By using the approximation in (6.38), we can solve the resource-allocation problem in a closed form. The following is a simple example. Suppose there is a system in which the overall power is limited and has to be shared by N users. To achieve minimal overall system distortion, the constrained optimization problem can be formulated as

$$\min_{r_i} \sum_{i=1}^{K} D_s^i = \sum_{i=1}^{K} \delta 2^{2k(\eta_{\text{max}} - r_i R)}, \tag{6.39}$$

$$\text{s.t.} \sum_{i=1}^{K} P_i \le P_{\text{max}},$$

where r_i is the channel-coding rate, P_{max} is the maximal transmission power, and

$$P_i = \frac{\gamma_i \sigma^2}{G_i}, \tag{6.40}$$

with noise value σ^2, channel gain G_i, and required SNR r_i in (6.38). Here we suppose BPSK modulation.

The Lagrangian multiplier function for (6.39) can be written as

$$J = \sum_{i=1}^{K} \delta 2^{2k(\eta_{\text{max}} - r_i R)} + \lambda \left(\sum_{i=1}^{N} P_i - P_{\text{max}} \right). \tag{6.41}$$

By differentiating (6.41) and setting to zero, we have the expression of r_i as function of λ. By substituting it into the constraint in (6.39), we have the expression for λ. Finally, the closed-form solution can be obtained by the similar approach of water filling in (6.10). There are more constraints that can be added to (6.39), for example, the rate range and the positive power. Optimization techniques other than Lagrangian can also be employed in a similar way.

6.7   Summary

In this chapter, we study how to formulate the multiuser cross-layer wireless networking and resource-allocation problem as a constrained optimization problem. We discuss what the network optimization goals are, what the practical constraints are, and what the basic technologies are.

We give the basic mathematics to understand how to solve the constrained optimization problem. Some basic solutions and concepts of optimality are discussed. Specifically, we discuss the convex concept, Lagrangian method, optimality condition, and duality. Some examples are given to show the basic ideas.

To get the closed-form solution, the constraints and optimization goal need to be either linear or convex. Most of the practical constraints and optimization goals cannot satisfy the requirements, so some approximations need to be performed within some practical regions. Some methods are discussed to illustrate the ideas.

Finally, we have one practical example to explain how to use the approximation and then obtain the closed-form solutions. Unfortunately, the closed-form solution is hard to obtain in most of the practical systems. There are three major reasons:

1. The approximation cannot be employed, so the constrained optimization problem has many local optima. To solve the problem, adaptive methods such as the programming method are necessary to gradually find the better solution. The details will be discussed in the chapter on programming.

2. The only information that can be obtained is local information. Sometimes it is impractical to obtain the global information to optimize the constrained problem. To overcome this challenge, game theory can be employed to study the behavior of distributed optimization. Some methods are also proposed to improve the outcome of the game so as to approach the global optimization that we achieve by knowing all information and then applying the methods in this chapter or in the chapter on programming. The details will be discussed in the chapter on game theory.

3. Finally, some parameters are discrete in nature, such as the modulation level, channel-coding rate, source rate, and channel assignment. We will discuss optimization for this type of problem in detail in the chapter on integer optimization.

7 Mathematical Programming

7.1 Introduction

In mathematics, the term optimization refers to the study of problems that have the following forms:

- given: a function $f : A \to R$ from a certain set A to the real numbers;
- sought: an element x_0 in A such that $f(x_0) \leq f(x)$, $\forall x \in A$ ("minimization") or such that $f(x_0) \geq f(x) \forall x \in A$ ("maximization").

Typically, A is a certain subset of the Euclidean space $\mathbf{R}^n$, often specified by a set of constraints, equalities, or inequalities that the members of A have to satisfy. The elements of A are called feasible solutions. The function f is called an objective function, or cost function. A feasible solution that minimizes (or maximizes, if that is the goal) the objective function is called an optimal solution. The domain A of f is called the search space, and the elements of A are called candidate solutions or feasible solutions.

Such a formulation is sometimes called a mathematical program. Many real-world and theoretical problems may be modeled in this general framework. In this chapter, we discuss the following major subfields of the mathematical programming:

1. Linear programming (LP) studies the case in which the objective function f is linear and the set A is specified using only linear equalities and inequalities.
2. Convex programming studies the case in which the constraints and the optimization goals are all convex or linear.
3. Nonlinear programming (NLP) studies the general case in which the objective function or the constraints or both contain nonlinear parts.
4. Dynamic programming studies the case in which the optimization strategy is based on splitting the problem into smaller subproblems or considers the optimization problems over time.

In the chapter on integer/combinatorial optimization, we will explain the following programming methods in detail:

1. Integer optimization studies the programming problems in which some or all variables are constrained to take on integer values.
2. Combinatorial optimization is investigated with problems in which the set of feasible solutions is discrete or can be reduced to a discrete one.

Some other subfields can be easily studied by the same principles discussed in this and the next chapters.

1. Stochastic programming studies the case in which some of the constraints depend on random variables.
2. Infinite-dimensional optimization studies the case in which the set of feasible solutions is a subset of an infinite-dimensional space, such as a space of functions.

In general, mathematical programming problems can be solved using the following approaches: For twice-differentiable functions, unconstrained problems can be solved by finding the points where the gradient of the objective function is 0 (that is, the stationary points) and using the Hessian matrix to classify the type of each point. If the Hessian is positive definite, the point is a local minimum, if negative definite, a local maximum, and if indefinite, a certain kind of saddle point. If the objective function is convex over the region of interest, then any local minimum will also be a global minimum. There exist robust and fast numerical techniques for optimizing doubly differentiable convex functions. One can find the stationary points by starting with a guess for a stationary point, and then iterate toward it by using methods such as gradient descent, Newton's method, conjugate gradient, and line search.

Constrained problems can often be transformed into unconstrained problems with the help of Lagrangian multipliers. There are a few other popular methods: random-restart hill climbing, simulated annealing (SA), stochastic tunneling, genetic algorithms (GAs), evolution strategy, differential evolution, and particle-swarm optimization.

In this chapter, we study how to formulate the resource-allocation problem as a mathematical program problem. The purpose is not to study the programming itself, which can be found in detail in [20] and [34]. The organization of this chapter is as follows: First, we discuss LP. Then, convex optimization is illustrated. NLP and dynamic programming are examined next. Finally, a wireless networking and resource-allocation example is studied.

7.2   Linear Programming

Linear programming (LP) is the problem of maximizing/minimizing a linear function over a convex polyhedron. LP is extensively used in engineering. An example of an engineering application is improving network performances for different users using the same radio resources. Linear programming can be solved using the simplex method [56, 337] that runs along polytope edges of the visualization solid to find the best answer. Khachian [161] found an $O(x^5)$ polynomial-time algorithm. A much more efficient polynomial-time algorithm was found by Karmarkar [156]. This method goes through the middle of the solid (making it a so-called interior-point method), and then transforms and warps. Interior-point methods were known as early as the 1960s in the form of the barrier-function methods.

LP is a problem that can be expressed as follows *standard form*:

$$\min \mathbf{cx}, \tag{7.1}$$

$$\text{subject to } \begin{cases} \mathbf{Ax} = \mathbf{b} \\ \mathbf{x} \geq 0 \end{cases},$$

where $\mathbf{x}$ is the vector of variables to solve, $\mathbf{A}$ is a matrix of known coefficients, and $\mathbf{c}$ and $\mathbf{b}$ are vectors of known coefficients. The expression $\mathbf{cx}$ is called the objective function, and the equations $\mathbf{Ax} = \mathbf{b}$ are called the constraints. The matrix $\mathbf{A}$ is generally not square and usually $\mathbf{A}$ has more columns than rows, and $\mathbf{Ax} = \mathbf{b}$ is therefore quite likely to be underdetermined, leaving great latitude in the choice of $\mathbf{x}$ with which to minimize $\mathbf{cx}$.

Two families of solution techniques are in wide use today. Both visit a progressively improving series of trial solutions until a solution is reached that satisfies the conditions for an optimum. The *simplex methods*, introduced by Dantzig about 50 years ago, visit "basic" solutions computed by fixing enough of the variables at their bounds to reduce the constraints $\mathbf{Ax} = \mathbf{b}$ to a square system, which can be solved for unique values of the remaining variables. Basic solutions represent extreme boundary points of the feasible region defined by $\mathbf{Ax} = \mathbf{b}$, $\mathbf{x} \geq \mathbf{0}$, and the simplex method can be viewed as moving from one such point to another along the edges of the boundary. *Barrier* or *interior-point methods*,[1] by contrast, visit points within the interior of the feasible region. These methods derive from techniques for NLP that were developed and popularized in the 1960s by Fiacco and McCormick, but their application to LP dates back only to Karmarkar's innovative analysis in 1984.

First, the solution of a LP problem falls on the boundary of the feasible region by the following theorem:

THEOREM 16 *Extreme point (or simplex filter) theorem: If the maximum or minimum value of a linear function defined over a polygonal convex region exists, then it is found at the boundary of the region.*

This theorem implies that a finite number of extreme points indicate a finite number of solutions. Hence the search is reduced to a finite set of points. However, a finite set can still be too large for practical purposes. The simplex method provides an efficient systematic search guaranteed to converge in a finite number of steps.

LP problems must be converted into augmented form before being solved by the simplex algorithm. This form introduces nonnegative slack variables to replace nonequalities with equalities in the constraints. The problem can then be written in the following augmented form:

$$\begin{pmatrix} 1 & -\mathbf{c}^T & 0 \\ 0 & \mathbf{A} & \mathbf{I} \end{pmatrix} \begin{pmatrix} Z \\ \mathbf{x} \\ \mathbf{x}_s \end{pmatrix} = \begin{pmatrix} 0 \\ \mathbf{b} \end{pmatrix}, \tag{7.2}$$

---

[1] We will discuss barrier or interior-point methods in detail in the section on NLP.

where **x** is the variable vector from the standard form in (7.1), $\mathbf{x}_s$ is the introduced slack variable vector from the augmentation process, c contains the optimization coefficients, **A** and **b** describe the system of constraint equations, and Z is the variable to maximize.

The system is typically underdetermined, because the number of variables exceeds the number of equations. The difference between the number of variables and the number of equations gives us the degrees of freedom associated with the problem. Any solution, optimal or not, will therefore include a number of variables of arbitrary value. The simplex algorithm uses zero as this arbitrary value, and the number of variables with value zero equals the degrees of freedom.

Values of the nonzero value are called basic variables, and values of the zero value are called nonbasic variables in the simplex algorithm. The augmented form simplifies finding the initial basic feasible solution. The simplex method provides an efficient systematic search guaranteed to converge in a finite number of steps. The algorithm is as follows:

1. Begin the search at an extreme point (i.e., a basic feasible solution).
2. Determine if the movement to an adjacent extreme can improve on the optimization of the objective function. If not, the current solution is optimal. If, however, improvement is possible, then proceed to the next step.
3. Move to the adjacent extreme point that offers (or, perhaps, appears to offer) the most improvement in the objective function.
4. Continue steps 2 and 3 until the optimal solution is found or it can be shown that the problem is either unbounded or infeasible.

In 1972, Klee and Minty gave an example of a LP problem in which the polytope P is a distortion of an n-dimensional cube. They showed that the simplex method as formulated by Dantzig visits all $2n$ vertices before arriving at the optimal vertex. This shows that the worst-case complexity of the algorithm is exponential time. Similar examples have been found for other pivot rules. The problem is an open question of whether there is a pivot rule with polynomial-time worst-case complexity. Nevertheless, the simplex method is remarkably efficient in practice. Attempts to explain this efficiency employ the notion of average complexity or (recently) smoothed complexity.

The importance of LP derives in part from its many applications (see subsequent discussion) and in part from the existence of good general-purpose techniques for finding optimal solutions. These techniques take as input only a LP in the standard form, and determine a solution without reference to any information concerning the LP's origins or special structure. These techniques are fast and reliable over a substantial range of problem sizes and applications. Some examples that are not only for wireless resource allocations are given as follows:

1. Assignment problem
   Choose an assignment of people to jobs so as to minimize total cost. The ordinary model is that of a matching problem. Although the usual assignment problem is solvable in polynomial time (as a linear program), important extensions are the NP-complete problems. We discuss this problem in detail in the next chapter.

2. Blending problem

   A blend is a certain combination of materials to make another material. Given raw materials, their blends make intermediate materials, called stock, and/or final materials, called products. (There can be other blending activities that make products from stocks.) The raw materials have purchase costs, and the blending activities have costs of operation and maintenance. The products have either fixed demands or selling prices, and the problem is finding blends that minimize total net cost (or maximize profit). This arises in refinery operations, in the food industry, and in other process industries. The problem can sometimes be solved with linear programming, but there can be complicating relations that require nonlinear programming, such as the pooling problem.

3. Critical path

   A longest path in a network is studied, in which length units are time durations. The nodes represent tasks, and the arcs represent precedence constraints. The path is critical because the associated tasks determine the total completion time of the project. Moreover, at least one of their duration times must decrease in order to decrease the total completion time.

On the other hand, there are some limitations of LP. In practice, the objective goals and the constraints are rarely linear functions of the optimization parameters. Under nonlinear conditions, there might be many local optima and the simplex algorithm cannot find good solutions. In the next few sections, we discuss more generalized programming methods that deal with the more complicated problems.

7.3   Convex Optimization

A convex optimization problem can be defined as

$$\min f_0(x), \tag{7.3}$$

$$\text{subject to } \begin{cases} f_i(x) \le 0, i = 1, \ldots, m \\ \mathbf{a}_i^T \mathbf{x} = \mathbf{b}_i, i = 1, \ldots, p \end{cases},$$

where the objective function f_0 is convex, the inequality constraint functions $f_1 \ldots f_m$ are convex, and the equality constraint functions $g_i(\mathbf{x}) = \mathbf{a}_i^T \mathbf{x} - \mathbf{b}_i$ are affine.

A fundamental property of convex optimization problems is that any locally optimal point is also globally optimal. Moreover, by using the duality theory studied in the previous chapter, the optimality conditions can be easily identified. Some typical convex programming problems are as follows:

- LP

  Because linearity is a special kind of convexity, LP is a special kind of convex optimization. We discussed the details in the previous section.

- Quadratic optimization problems
  A problem is called a quadratic program if the objective function is quadratic and the constraint functions are affine as

$$\min \mathbf{x}^T \mathbf{P} \mathbf{x} + 2\mathbf{q}^T \mathbf{x} + \mathbf{r}, \tag{7.4}$$

$$\text{subject to } \begin{cases} \mathbf{G}\mathbf{x} \leq \mathbf{h} \\ \mathbf{A}\mathbf{x} = \mathbf{b} \end{cases},$$

where $\mathbf{P} = \mathbf{P}^T \geq 0$.

- Geometric programming

DEFINITION 8 *A function f is called monomial if*

$$f(x) = c x_1^{a_1} x_2^{a_2} \ldots x_n^{a_n}, \tag{7.5}$$

where $c \geq 0$ and $a_i \in \mathbf{R}$.

DEFINITION 9 *A function f is called posynomial if*

$$f(x) = \sum_{k=1}^{K} c_k x_1^{a_{1k}} x_2^{a_{2k}} \ldots x_n^{a_{nk}}, \tag{7.6}$$

where $c_k \geq 0$ and $a_{ik} \in \mathbf{R}$.

An optimization problem is called geometric programming if

$$\min f_0(x), \tag{7.7}$$

$$\text{subject to } \begin{cases} f_i(x) \leq 1, i = 1, \ldots, m \\ h_i(x) = 1, i = 1, \ldots, p \end{cases},$$

where $f_0, \ldots, f_m$ are posynomials and $h_1, \ldots, h_p$ are monomials.

- Semidefinite programming
  The field of semidefinite programming (SDP) or semidefinite optimization (SDO) deals with optimization problems over symmetric positive semidefinite matrix variables with linear cost functions and linear constraints. Popular special cases are LP and convex quadratic programming with convex quadratic constraints. A SDP problem is

$$\min \mathbf{c}^T \mathbf{x}, \tag{7.8}$$

$$\text{subject to } F(\mathbf{x}) \geq 0,$$

where

$$F(\mathbf{x}) = F_0 + \sum_{i=1}^{m} x_i F_i, \tag{7.9}$$

and $F(\mathbf{x})$ is positive semidefinite.

For the preceding classes of convex optimization, we can utilize a duality theory that is similar to LP. There are some good algorithms and robust, reliable software available. They can also be employed in a wide variety of new applications.

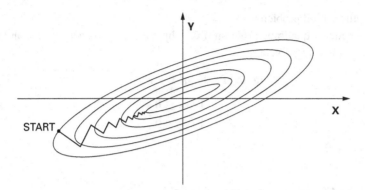

Figure 7.1 Gradient descendent example.

The other advantage of convex optimization is the fact that there are many simple methods to find the global optimum. We discuss the solutions of unconstrained optimization first, and then we generalize the solutions to constrained optimizations. For the unconstrained cases, some methods are as follows:

- Gradient method
  The gradient method finds the nearest local minimum of a function with the assumption that the gradient of the function can be computed. The method of steepest descent, also called the gradient descent method, starts at a point x_0 and, as many times as needed, moves from x_i to x_{i+1} by minimizing along the line extending in the direction of $-\nabla f(x)$, the local downhill gradient. The pseudo-code is given as follows:

Gradient Descent Method
- given a feasible starting point x_0,
- repeat
  1. Calculate the gradient $\nabla f(x)$.
  2. Line search: choose the step size t that optimize $f(x_i) - t \nabla f(x_i)$.
  3. Update: $x_{i+1} = x_i - t \nabla f(x_i)$.
- until the stopping criteria, such as the KKT condition (defined in the previous chapter) or the desired accuracy, are satisfied.

One of the examples of the gradient method is shown in Figure 7.1. As seen, the method of steepest descent is simple, easy to apply, and each iteration is fast. It is also very stable; if the minimum points exist, the method is guaranteed to locate them. But, even with all these positive characteristics, the method has one very important drawback: It generally has slow convergence. For badly scaled systems, i.e., if the eigenvalues of the Hessian matrix at the solution point are different by several orders of magnitude, the method can end up spending a long number of iterations before locating a minimum point. It starts out with a reasonable convergence, but the progress gets slower and slower.

- Newton method

  In mathematics, Newton's method is a well-known algorithm for finding roots of equations in one or more dimensions. It can also be used to find local maxima and local minima of functions by noticing that, if a real number x^* is where the first-order gradient of a function $f(x)$ is equal to zero, one can solve for x^* by applying Newton's method to $f'(x)$.

  DEFINITION 10 *Given the second-order deferential exists, the Newton step is defined as*

  $$\mathbf{v}_{nt} = -\nabla^2 f(\mathbf{x})^{-1} \nabla f(\mathbf{x}), \tag{7.10}$$

  which can have the following interpretations:

  $\mathbf{v}_{nt}$ is the minimizer of second-order approximation

  $$\hat{f}(\mathbf{x} + \mathbf{v}) = f(\mathbf{x}) + \nabla f(\mathbf{x})^T \mathbf{v} + \frac{1}{2} \mathbf{v}^T \nabla^2 f(\mathbf{x}) \mathbf{v}. \tag{7.11}$$

  $\mathbf{v}_{nt}$ is the steepest descent direction in the Hessian norm.
  $\mathbf{v}_{nt}$ is the solution of linearized optimality condition for the first-order deferential $\nabla f(\mathbf{x}^*) = 0$.
  The Newton step is independent of linear changes of the parameters.

  DEFINITION 11 *The Newton decrement is defined as*

  $$\lambda(\mathbf{x}) = [\nabla f(\mathbf{x})^T \nabla^2 f(\mathbf{x})^{-1} \nabla f(\mathbf{x})]^{\frac{1}{2}}. \tag{7.12}$$

  $\lambda^2/2$ *is an estimate of* $f(x) - f^*$, *based on the quadratic approximation of f at* $\mathbf{x}$ *[34]*.

  The geometric interpretation of Newton's method is that at each iteration one approximates $f(\mathbf{x})$ by a quadratic function around $\mathbf{x}$, and then takes a step toward the maximum/minimum of that quadratic function. The detailed steps are given as follows:

  Newton's method
  - Given a feasible starting point $\mathbf{x}_0$, tolerance $\epsilon > 0$.
  - Repeat
    1. Calculate the Newton step and decrement.
    2. Quit if $\frac{\lambda^2}{2} \leq \epsilon$.
    3. Line search: choose the step size t.
    4. Update: $\mathbf{x}_{i+1} = \mathbf{x}_i + t\mathbf{v}_{nt}$.

  Newton's method converges much faster toward a local maximum or minimum than the gradient descent. However, to use Newton's method one needs to know the Hessian of $f(\mathbf{x})$, which sometimes can be difficult to compute. There exist various quasi-Newton methods, in which an approximation for the Hessian is used instead. Another drawback of Newton's method is that finding the inverse of the Hessian can be an expensive operation.

- Other methods

  1. Conjugate gradients:

     This method is an attempt to mend this problem by "learning" from experience. It uses conjugate directions instead of the local gradient for going downhill. The method proceeds by generating vector sequences of iterates (i.e., successive approximations to the solution), residuals corresponding to the iterates, and search directions used in updating the iterates and residuals. Although the length of these sequences can become large, only a small number of vectors need to be kept in memory. In every iteration of the method, two inner products are performed to compute updated scalars that are defined to make the sequences satisfy certain orthogonality conditions. If the vicinity of the minimum has the shape of a long and narrow valley, the minimum is reached in far fewer steps than the case using the method of steepest descent.

  2. Secant method:

     For the Newton method, the Hessian matrix can be laborious both to calculate and to invert for systems with a large number of dimensions. The idea of the secant method is not to use the Hessian matrix directly, but rather to start the procedure with an approximation to the matrix, which, as one gets closer to the solution, gradually approaches the Hessian.

  3. Stochastic (or "on-line") gradient descent:

     The true gradient is approximated by the gradient of the cost function evaluated on only a single training example. The parameters are then adjusted by an amount proportional to this approximate gradient. Therefore the parameters of the model are updated after each training example. For large data sets, on-line gradient descent can be much faster than batch gradient descent.

  4. Broyden–Fletcher–Goldfarb–Shanno (BFGS) method:

     Quasi-Newton or variable metric methods can be used when the Hessian matrix is difficult or time consuming to evaluate. Instead of obtaining an estimate of the Hessian matrix at a single point, these methods gradually build up an approximate Hessian matrix by using gradient information from some or all of the previous iterates visited by the algorithm.

In the rest of this section, we discuss how to solve the constrained optimization. First, by using KKT condition, sometime we can prove that the constraints can be eliminated without affecting the final solutions. Moreover, by utilizing the dual problem, some constraints can be removed without loss of the optimality. If the constraints cannot be reduced, the following methods are the general approaches to solving the problems.

- Projected gradient method/Newton method

  If there is no unequal constraint, we can project the gradient or search direction according to the equality constraint. Suppose the optimization problem is

$$\min f(\mathbf{x}), \tag{7.13}$$
$$\text{subject to } \mathbf{Ax} = \mathbf{b}.$$

For the projected gradient method, the project gradient $\mathbf{v}_{pg}$ can be written as:

$$\begin{pmatrix} \mathbf{I} & \mathbf{A}^T \\ \mathbf{A} & \mathbf{0} \end{pmatrix} \begin{pmatrix} \mathbf{v}_{pg} \\ \mathbf{w} \end{pmatrix} = \begin{pmatrix} -\nabla f(\mathbf{x}) \\ \mathbf{0} \end{pmatrix}, \tag{7.14}$$

where $\mathbf{w}$ is an estimate of the Lagrangian multiplier.

For the projected Newton method, the projected Newton step $\mathbf{v}_{nt}$ can be expressed as

$$\begin{pmatrix} \nabla^2 f(\mathbf{x}) & \mathbf{A}^T \\ \mathbf{A} & \mathbf{0} \end{pmatrix} \begin{pmatrix} \mathbf{v}_{nt} \\ \mathbf{w} \end{pmatrix} = \begin{pmatrix} -\nabla f(\mathbf{x}) \\ \mathbf{0} \end{pmatrix}. \tag{7.15}$$

- Interior-point method/barrier method
  When the searching point is approaching the boundary of the feasibility, the approach is to add the penalty to the objective, so that the solution is always satisfying the unequal constraints. We will discuss these approaches in detail in the next section on NLP.
- Cutting-planes method
  The basic idea of this approach is to find a hyperplane so that the searching space for the optimal solution can be greatly reduced. We discuss this method more in the next chapter on integer programming.

7.4   Nonlinear Programming

NLP is a problem that can be put into the form

$$\min F(\mathbf{x}), \tag{7.16}$$

$$\text{subject to } \begin{cases} g_i(\mathbf{x}) = 0, & \text{for } i = 1, \ldots, m_1, \text{ where } m_1 >= 0 \\ h_j(\mathbf{x}) \geq 0, & \text{for } j = m_1 + 1, \ldots, m, \text{ where } m \geq m_1 \end{cases},$$

where F is one scalar-valued function of variable vectors $\mathbf{x}$. We seek to minimize F subject to one or more other such functions that serve to limit or define the values of the variable vector. F is called the "objective function," and the various other functions are called the "constraints." Because the objective function or the constraints can be nonlinear, the optimization in (7.16) is called NLP.

One of the greatest challenges in NLP is that some problems exhibit "local optima," i.e., the solutions merely satisfy the requirements on the derivatives of the functions but are not necessarily good. These situations are similar to multiple peaks. It is difficult for an algorithm that tries to move from point to point only by climbing uphill, because the peak it achieves might not be the highest. Algorithms that propose overcoming this difficulty are termed "global optimization." Next, we first discuss how to find a local optimum from an initialization. Specifically, we discuss the barrier method/interior-point method. Then we study the techniques on how to find the global optimum.

The idea of encoding the feasible set using a barrier and designing barrier methods was studied in the early 1960s by Fiacco-McCormick and others. These ideas were mainly developed for general nonlinear programming. Nesterov and Nemirovskii came up with a special class of such barriers that can be used to encode any convex set. They guarantee that the number of iterations of the algorithm is bounded by a polynomial in the dimension and accuracy of the solution.

In constrained optimization, a barrier function is a continuous function whose value on a point increases to infinity as the point approaches the boundary of the feasible area. It is used as a penalizing term for violations of constraints. The barrier function will also be convex and smooth. The two most common types of barrier functions are inverse barrier functions and logarithmic barrier functions, such as

$$I_{\text{inv}} = \begin{cases} \sum_{j=m_1}^{m} \frac{1}{h_j(\mathbf{x})} & \text{if } h_j \geq 0, \, j = m_1, \ldots, m \\ +\infty & \text{otherwise} \end{cases}, \tag{7.17}$$

and

$$I_{\text{log}} = \begin{cases} -\sum_{j=m_1}^{m} \log[h_j(\mathbf{x})] & \text{if } h_j \geq 0, \, j = m_1, \ldots, m \\ +\infty & \text{otherwise} \end{cases}, \tag{7.18}$$

respectively.

By adding the barrier function to the objective function $F(\mathbf{x})$, the problem in (7.16) becomes

$$\min t F(\mathbf{x}) + I(\mathbf{x}), \tag{7.19}$$

$$\text{subject to } g_i(\mathbf{x}) = 0, \quad \text{for } i = 1, \ldots, m_1.$$

In the extreme case, in which t is large enough, the barrier function I becomes an ideal barrier function, and the problem in (7.19) becomes the problem in (7.16).

The barrier method (path-following algorithm) tries to solve the problem in (7.16) by solving a sequence of the simplified problem in (7.19). In other words, the method computes the optimal $\mathbf{x}^*$ for a sequence of increasing values of t until the solution is close enough to the original problem. The details of the barrier method are as follows:

Barrier method
- Given a feasible initialization $\mathbf{x}_0$, tolerance $\epsilon > 0$, and $t > 0$, $\mu > 1$.
- Repeat.
  1. Calculate $\mathbf{x}^*$ in (7.19).
  2. $\mathbf{x} = \mathbf{x}^*$.
  3. If the tolerance is satisfied, return $\mathbf{x}$.
  4. $t = \mu t$.

There is a trade-off in the choice of μ. If μ is small, the complexity in solving the problem in (7.19) for each iteration is small, and the iterations closely follow the central path within the feasible range that can avoid possible local optimal. However, it needs more iterations. On the other hand, if μ is large, the barrier function converges fast to the ideal one, but the complexity for solving (7.19) increases and a certain possible local optimum might appear.

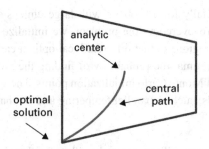

Figure 7.2 Interior-point method example.

Interior-point methods are a certain class of algorithms used to solve linear and nonlinear convex optimization problems. These algorithms have been inspired by the algorithms of Narendra Karmarkar for LP, developed in 1984. The basic elements of the method consist of a self-concordant barrier function used to encode the convex set. Mehrotra's predictor–corrector algorithm is a common implementation of an interior-point method. The primal–dual interior-point method proposed by Kojima, Mizuno, and Yoshise is also widely used. The interior-point method starts from the analytic center (which was proposed by Sonnevend and Megiddo), then follows the central path, and finally converges to an optimal solution. One example is shown in Figure 7.2.

To achieve the global optimum, the initial starting point for the algorithms plays an important role. One popularly used approach starts with some heuristics with good solutions, and then uses a certain algorithm to converge to a better solution. In most cases, some good solutions can be obtained. However, the global optimum is not guaranteed. To get the global optimum, there are some other methods [379] that are widely employed.

1. Monte Carlo methods
Monte Carlo methods are a class of computational algorithms for simulating the behavior of various physical and mathematical systems. They are distinguished from other simulation methods (such as molecular dynamics) by being stochastic, which are nondeterministic in a certain manner. The algorithm usually uses random numbers (or more often pseudo-random numbers) as opposed to deterministic algorithms.

Interestingly, the Monte Carlo method does not require truly random numbers to be useful. Much of the most useful techniques use deterministic, pseudo-random sequences, making it easy to test and rerun simulations. The only quality usually necessary to make good simulations is for the pseudo-random sequence to appear "random enough" in a certain sense. That is, they must either be uniformly distributed or follow another desired distribution when a large enough number of elements of the sequence are considered. Because of the repetition of algorithms and the large number of calculations involved, Monte Carlo is a method suited to calculation by a computer, utilizing many techniques of computer simulation.

For the optimization problems, especially for a problem with large dimensions, there might be a lot of local optima. To overcome the problem, we initialize the algorithm randomly within the feasible region, so that different local optima can be converged to. By comparing the local optima, the probability of finding the global optimum is increased with the number of Monte Carlo initialization points. There are some variations of Monte Carlo methods, such as parallel tempering and stochastic tunneling.

2. Simulated annealing

Simulated annealing (SA) is a generic probabilistic meta-algorithm for the global optimization problem. It was independently invented by S. Kirkpatrick, C. D. Gelatt, and M. P. Vecchi in 1983, and by V. Cerny in 1985.

The name and inspiration come from annealing in metallurgy, a technique involving heating and controlled cooling of a material to increase the size of its crystals and reduce their defects. The heat causes the atoms to become unstuck from their initial positions (a local minimum of the internal energy) and wander randomly through states of higher energy; the slow cooling gives them more chances of finding configurations with lower internal energy than the initial one.

By analogy with this physical process, each step of the SA algorithm replaces the current solution with a random "nearby" solution. This solution is chosen with a probability that depends on the difference between the corresponding function values and on a global parameter T (called the temperature). The temperature is gradually decreased during the process. The dependency is such that the current solution changes almost randomly when T is large, but increasingly "downhill" as T goes to zero. The allowance for "uphill" moves saves the method from becoming stuck at local minima.

At each step, the SA heuristic considers a certain neighbor of the current state and probabilistically decides between moving the system to a neighbor state or staying in the current state. The probabilities are chosen so that the system ultimately tends to move to states of lower energy. The probability is large when the temperature is high so that the algorithm will not be stuck in a certain local optima. On the other hand, the probability is low because the probability of local optima is low. When the temperature is zero, the algorithm reduces to the greedy algorithm. Typically this step is repeated until the system reaches a state that is good enough for the application, or until a given computation budget has been exhausted. It can be shown that, for any given finite problem, the probability that the SA algorithm terminates with the global optimal solution approaches 1 as the annealing schedule is extended.

3. Generic algorithm

A genetic algorithm (GA) is a search technique used in computer science to find approximate solutions for optimization. GAs are a particular class of evolutionary algorithms that use techniques inspired by evolutionary biology such as inheritance, mutation, natural selection, and recombination (or crossover).

GAs are typically implemented as a computer simulation in which a population of abstract representations (called chromosomes) of candidate solutions (called individuals) to an optimization problem evolves toward better solutions. Traditionally, solutions are represented in binary as strings of 0's and 1's, but different encodings are also possible. The evolution starts from a population of completely random individuals and happens in generations. In each generation, the fitness of the whole population is evaluated. Then multiple individuals are stochastically selected from the current population (based on their fitness) and modified (mutated or recombined) to form a new population, which becomes current in the next iteration of the algorithm. The details of the algorithm are shown as follows:

Genetic algorithm
- Choose the population of random initializations.
- Repeat.
  (a) Evaluate the individual performance of a certain proportion of the population.
  (b) Select pairs of best-ranking individuals to reproduce.
  (c) Apply crossover operator, which determines the probabilities that the two selected individuals will be actually combined together for the offspring.
  (d) Apply mutation operator, in which a small probability of mutation is added to the offspring.
- Until terminating condition.

7.5   Dynamic Programming

One of the major categories of programming methods is dynamic programming. Basically, the dynamic programming approach solves a high-complexity problem by combining the solutions of a series of low-complexity subproblems. Dynamic programming relies on a principle of optimality, which states that, in an optimal sequence of decisions or choices, each subsequence must also be optimal. To understand the basic problem formulation, we define the following concepts:

- State: A state is a configuration of a system and is identified by a label that indicates the properties corresponding to that state.
- Stage: A stage is a single step in the system undergoing a certain process and corresponds to the transition from one state to an adjacent state.
- Action: At each state there is a set of actions available from among which choices must be made.
- Policy: A policy is a set of actions, one for each of a number of states. An optimal policy is the best set of actions in accordance with a given objective.
- Return: A return is something that a system generates over one stage of a process. A return is usually something like a profit, a cost, a distance, a yield, or consumption of a product.
- Value of a state: The value of a state is a function of returns (sum) generated when the system starts in that state and a particular policy is followed. The value of a state under an optimal policy is the optimal value.

Table 7.1 Classifications of dynamic programming

Discrete	Continuous
Finite horizon	Infinite horizon
Deterministic	Stochastical
Constrained	Unconstrained
Perfect state information	Imperfect state information
Single objective	Multiple objective

Depending on the different characteristics of the preceding definitions, the dynamic programming problems can be classified as those in Table 7.1. Dynamic programming usually takes one of two approaches:

- Top-down approach:
  The problem is broken into subproblems, and these subproblems are solved and the solutions remembered, in case they need to be solved again. This is recursion and memorization combined together.
- Bottom-up approach:
  All subproblems that might be needed are solved in advance and then used to build up solutions to larger problems. This approach is slightly better in stack space and number of function calls, but it is sometimes not intuitive to figure out all the subproblems needed for solving the given problem.

In practice, to formulate the problem, the following steps are required.

1. Characterize the structure of an optimal solution.
2. Define the value of an optimal solution recursively.
3. Compute optimal solution values either top-down with caching or bottom-up in a table.
4. Construct an optimal solution from computed values.

In general, some typical applications are as follows.

- Problems that involve a sequence of decisions in time: production planning, stock control, investment decision making, and replacement policy.
- Problems in which a sequence of decisions is not directly related to time: sequential production process, optimal path problems, optimal search problems.
- Allocation problems: deciding a sequence of allocations of one or more limited resources.
- Combinatorial and graph-theoretic problems: scheduling, sequencing, and set partitioning.

Next, we give three of the most important dynamic programming examples that are widely employed in wireless communications.

- Shortest path for routing
  Problem definition: Given a connected graph $G = (V, E)$, a weight $d : E \to R^+$, and two vertices s and v in V, such that the sum of the weight of its constituent edges is

minimized. Dijkstra's algorithm is known to be one of the best algorithms to find a shortest path. The time required by Dijkstra's algorithm is $O(|V|^2)$. It can be reduced to $O(|V| \log |V|)$. The detailed pseudo-code for the algorithm is as follows:

1. Set iteration $i = 0$, initial point $u_0 = s$, route set $S_0 = u_0 = s$, cost function $L(u_0) = 0$, and $L(v) = \infty$ for $v \neq u_0$. If $|V| = 1$ then stop, otherwise go to step 2.
2. For each v in $V \backslash S_i$, replace $L(v)$ with min $L(v)$, $L(u_i) + d_v^{u_i}$. If $L(v)$ is replaced, put a label $(L(v), u_i)$ on v.
3. Find a vertex v that minimizes $\{L(v) : v \in V \backslash S_i\}$, say $u_i + 1$.
4. Let $S_{i+1} = S_i \cup \{u_i + 1\}$.
5. Replace i with $i + 1$. If $i = |V| - 1$ then stop, otherwise go to step 2.

- The Viterbi algorithm is a dynamic programming algorithm for finding the most likely sequence of hidden states, known as the Viterbi path, that result in a sequence of observed events, especially in the context of hidden Markov models. The forward algorithm is a closely related algorithm for computing the probability of a sequence of observed events. These algorithms form a subset of information theory.

  The Viterbi algorithm was originally conceived as an error-correction scheme for noisy digital communication links, finding universal application in decoding the convolutional codes used in both CDMA and GSM digital cellular, dial-up modems, satellite, deep-space communications, and 802.11 WLANs. The basic idea is to store a buffer of received signals and determine the most likely sequence of the original transmitted sequence. It is now also commonly used in speech recognition, keyword spotting, computational linguistics, and bioinformatics. For example, in speech-to-text speech recognition, the acoustic signal is treated as the observed sequence of events, and a string of text is considered to be the "hidden cause" of the acoustic signal. The Viterbi algorithm finds the most likely string of text given the acoustic signal.

  The algorithm is not general; it makes a number of assumptions. First, both the observed events and hidden events must be in a sequence. This sequence often corresponds to time. Second, these two sequences need to align, and an observed event needs to correspond to exactly one hidden event. Third, computing the most likely hidden sequence up to a certain point t must depend on only the observed event at point t and the most likely sequence at point $t - 1$. These assumptions are all satisfied in a first-order hidden Markov model.

  The terms "Viterbi path" and "Viterbi algorithm" are also employed to relate dynamic programming algorithms that discover the single most likely explanation for an observation. For example, in stochastic parsing a dynamic programming algorithm can be used to discover the single most likely context-free derivation (parse) of a string, which is sometimes called the "Viterbi parse."

  In Figure 7.3, we give an example of Viterbi decoding. There are two states A and B with the transition probability, as shown. In each state, two different observations x and y can be detected. The observed sequence is xxy. In the Viterbi algorithm, there is a decoding depth that represents the number of symbols that are jointly decoding. The deeper the decoding depth, the higher the error-correction capability, but also the higher the computation complexity and the latency. The question is this: What is the

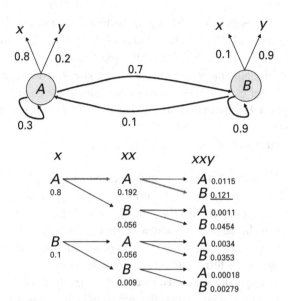

Figure 7.3 Viterbi algorithm example.

most likely sequence of states? From the example, we can see that the sequence AAB has the largest probability of $0.8 \times 0.3 \times 0.8 \times 0.7 \times 0.9 = 0.121$.

- Bellman equations occur in dynamic programming. A Bellman equation is also called an optimality equation or a dynamic programming equation. Suppose the action for time n is α_n, state is s_n, reward function is $R(s_n, \alpha_n)$, and the forgetting factor is β. The discounted time-average reward for an infinite time scale can be expressed as

$$V = \sum_{n=1}^{\infty} \beta^n R(s_n, \alpha_n). \tag{7.20}$$

Suppose we consider the stationary solution and the user will select the deterministic action as $a_n = \pi(s_n)$, where π is the policy.

The stochastic action is characterized by $P[s'|s, \pi(s)]$ which means the probability of the next state under the condition of the current state and policy. We can omit the time n and have the Bellman equation as

$$V^{\pi}(s) = R(s) + \beta \sum_{s'} P[s'|s, \pi(s)]V^{\pi}(s'), \tag{7.21}$$

whereas the equation for the optimal policy is referred to as the Bellman optimality equation:

$$V^*(s) = R(s) + \max_{\alpha} \beta \sum_{s'} P(s'|s, \alpha)V^*(s'). \tag{7.22}$$

The difference is that, rather than taking the action prescribed by a certain policy, we take the action that gives the best expected return.

7.6   A Wireless Resource-Allocation Example

In this section, we study a wireless resource allocation to utilize the linear/nonlinear/convex optimization as well as dynamical programming. With the increasing demand for wireless services, the ability to provide higher network throughput by a limited spectrum bandwidth is one of the important design considerations. Two detrimental effects to reduce the spectrum efficiency are cochannel interference and the time-varying nature of the channel. Power control and adaptive modulation are the important approaches to combat these effects. In power control, the transmitted power is continuously adjusted so that the receiver's SINR is maintained for a desired link quality. The main problem for power control is feasibility in the system with cochannel interferences. Under some channel conditions and receiver's SINR requirements, no matter how large the transmitted power is, the receiver's SINR cannot be high enough, i.e., there is no power-allocation solution. In adaptive modulation, each link's throughput is allocated according to the current channel conditions. The spectrum efficiency can be potentially increased. The main problem for adaptive modulation in multiaccess systems is fairness. Without the fairness constraint, the link with the best channel condition will occupy almost all the radio resources for transmission, but the others stop transmission. This situation is not fair for the services that users have paid for.

Much research has been done for both power control and adaptive modulation [49, 241, 250]. Most of the previous works do the optimization based on only current channel conditions and do not take into account the time diversity. In this example, we maximize the overall network throughput when each user's time-average throughput is a constant, which is determined by the user's service. We divide the problem heuristically into two subproblems. First, we find the throughput range that each user can select for different times to ensure fairness. Second, we find the best throughput allocation to different users at each time that generates the largest network overall throughput under the constraints that the power is feasible and throughput is within the ranges. From the simulation results, our scheme can increase the overall network throughput about 10% compared with that of previous works [250].

Consider K cochannel links that may exist in a distinct cell. Each link consists of a mobile and its assigned base station. Assume coherent detection is possible so that it is sufficient to model this multiuser system by an equivalent baseband model. We assume that link gain is stable within each transmission frame. For the uplink cases, the signal at the ith base station output is given by

$$x_i(t) = \sum_{k=1}^{K} \sqrt{G_{ki} P_k} s_k(t) + n_i(t), \tag{7.23}$$

where G_{ki} is the path loss from the kth mobile to the ith base station, P_k is the kth link's transmitting power, $s_k(t)$ is the message symbol, and $n_i(t)$ is the thermal noise. Then we can express the sampled received signal as

$$x_i(n) = \sum_{k=1}^{K} \sqrt{P_k G_{ki}} s_k(n) + n_i(n), \tag{7.24}$$

where $n_i(n)$ is the sampled thermal noise. The ith base station's output SINR is

$$\Gamma_i = \frac{P_i G_{ii}}{\sum_{k \neq i} P_k G_{ki} + N_i}, \quad i = 1, \ldots, K, \tag{7.25}$$

where $N_i = E\|n_i\|^2$. Write the preceding equations in matrix form. We have

$$(\mathbf{I} - \mathbf{DF})\mathbf{P} = \mathbf{u}, \tag{7.26}$$

where $\mathbf{P} = [P_1, \ldots, P_K]^T, \mathbf{D} = \text{diag}\{\Gamma_1, \ldots, \Gamma_K\}, \mathbf{u} = [u_1, \ldots, u_K]^T, u_i = \Gamma_i N_i / G_{ii}$, and

$$[\mathbf{F}]_{ij} = \begin{cases} 0, & \text{if } j = i \\ \frac{G_{ji}}{G_{ii}}, & \text{if } j \neq i \end{cases}. \tag{7.27}$$

To have a feasible solution in (7.26), the spectral radius $\rho(\mathbf{DF})$, i.e., the maximum eigenvalue of $\mathbf{DF}$, must be less than 1 [85]. We will see that the overall transmitted power is strongly affected by the value of $\rho(\mathbf{DF})$.

Adaptive modulation provides the system with the ability to match the effective bit rate (throughput) according to the interference and channel conditions. MQAM is the modulation method that has high spectrum efficiency. For a given system and a targeted BER, the throughput can be written as a function of the received SINR. Let T_i denote the ith user's throughput, which is the number of bits sent within each transmitted symbol. The throughput is assumed continuous, and the ith user's BER using different MQAM modulations with different throughputs can be approximated as [49, 241]

$$\text{BER}_i \approx c_1 e^{-c_2 \frac{\Gamma_i}{2^{T_i} - 1}}, \tag{7.28}$$

where $c_1 \approx 0.2$ and $c_2 \approx 1.5$ for MQAM. Rearrange (7.28). For a specific BER, the ith link's throughput is given by

$$T_i = \log_2 \left(1 + c_3^i \Gamma_i\right), \tag{7.29}$$

where $c_3^i = -\frac{c_2^i}{\ln(\text{BER}_i / c_1^i)}$. Without loss of generality, we assume each user has the unit bandwidth. Define the overall network throughput as $T = \sum_{i=1}^K T_i$.

In this example, we maximize the overall network throughput under the constraints that the power allocation is feasible and each user's time-average throughput is fixed. The problem we will address becomes

$$\max_{T_i, P_i} T = \sum_{i=1}^K T_i, \tag{7.30}$$

$$\text{subject to} \begin{cases} \text{feasibility: } \rho(\mathbf{DF}) < 1 \\ \text{Fairness: } \lim_{N \to \infty} \frac{\sum_{n=1}^N T_i(n)}{N} = \text{const. } \forall i \end{cases},$$

where const. is the user's time-average throughput and $T_i(n)$ is the throughput at time n.

It is very hard to get an analytic solution for the problem defined in (7.30). To simplify the problem, we heuristically divide the problem into two subproblems for different times and different users, respectively. First, to ensure fairness, we track the history of each

Example for Problem Partition

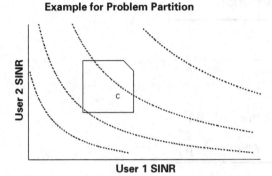

Figure 7.4 Two-users example.

user's throughput and determine the range of modulation level that each user can select at different times. Second, for the whole network, we determine the optimal throughput allocation to different users within the ranges at each time, which generates the highest overall network throughput.

A two-user example is shown in Figure 7.4. On each dotted curve, network throughput $T = T_1 + T_2 = $ const., where different lines have different constants. C is the range of users' throughput and where the system is feasible. The first subproblem changes C for different times to ensure fairness. For example, if user 1 experiences a bad channel response now and selects small throughput, user 1 will be more aggressive to transmit its data in the future. Consequently, range C will move to the right within the practical range, where user 1 can select higher throughput. The second subproblem is to find the point that generates the maximum network throughput within C at each time.

To implement a scheme to ensure fairness, we develop the following algorithm. Instead of having a fixed throughput range for each link, we can adaptively change the throughput range, which takes into account the links' throughput history. Suppose the ith link can select $T^{\mathrm{ave}} - T_i^{\mathrm{neg}}(n) \le T_i(n) \le T^{\mathrm{ave}} + T_i^{\mathrm{pos}}(n)$ at time n, where the average throughput $T^{\mathrm{ave}} = \sum T_i / K$. Each time $T_i^{\mathrm{neg}}(n)$ and $T_i^{\mathrm{neg}}(n)$ are modified by the current $T_i(n)$. When $T_i(n)$ is smaller than T^{ave}, $T_i^{\mathrm{neg}}(n+1)$ is decreased and $T_i^{\mathrm{pos}}(n+1)$ is increased so that the ith user has to be more aggressive to transmit its data in the near future. On the other hand, when $T_i(n)$ is larger than T^{ave}, $T_i^{\mathrm{neg}}(n+1)$ and $T_i^{\mathrm{pos}}(n+1)$ are modified in the opposite way. To track the history of $T_i(n)$, we define $T_i^{\mathrm{pos}}(n+1) = T_i^{\mathrm{pos}}(n) - [T_i(n) - T^{\mathrm{ave}}]\beta$, $T_i^{\mathrm{neg}}(n+1) = T_i^{\mathrm{neg}}(n) + [T_i(n) - T^{\mathrm{ave}}]\beta$, $0 < \beta < 1$, where β is a value that depends on how fast the channel changes. If the channel changes fast, β shall select a larger number to keep track of channel changing. Otherwise, β shall select a smaller number to have a smooth effect. Define $W = $ window size as the maximum throughput difference that each user can select from the average throughput T^{ave}. Our proposed algorithm is given by Table 7.2.

Next, we study how to find the maximum overall throughput at each time. First, the gradient of overall throughput with respect to the targeted SINR is given by

$$g_i^T = \frac{\partial T}{\partial \Gamma_i} = \frac{c_3}{1 + c_3 \Gamma_i}, \quad i = 1, \dots, K. \tag{7.31}$$

Table 7.2 Adaptive algorithm for throughput range

Initial:
$$T_i^{\text{pos}}(0) = W, \quad T_i^{\text{neg}}(0) = W,$$

Iteration:
Calculate $T^{\text{ave}}(n)$
$$T_i^{\text{pos}}(n+1) =$$
$$\min\left(\max\left\{T_i^{\text{pos}}(n) - \beta\left[T_i(n) - T^{\text{ave}}(n)\right], 0\right\}, W\right)$$
$$T_i^{\text{neg}}(n+1)$$
$$= \min\left(\max\left\{T_i^{\text{neg}}(n) + \beta\left[T_i(n) - T^{\text{ave}}(n)\right], 0\right\}, W\right)$$

Starting from any feasible point, we can enlarge each user's targeted SINR according to (7.31) to increase the overall network throughput, until we hit the boundary, i.e., $\rho(\mathbf{DF}) = 1 - \epsilon$, where ϵ is a small number. In this example, we use $\epsilon = 0.05$. Then we calculate $T^{\text{ave}} = \sum T_i / K$.

In the next step, we find the gradient $\partial\rho(\mathbf{DF})/\partial\Gamma_i$ and then project this gradient onto the plane where $T = $ const. Then we move along this modified gradient so that $\rho(\mathbf{DF})$ is reduced, when the overall throughput T is maintained as a constant. The overall transmitted power can be reduced consequently. The iteration stops when T_i reaches the boundary or a certain stable point.

The following theorem shows the existence of the derivative of the spectral radius [203].

THEOREM 17 *Let λ be a simple eigenvalue of **DF**, with right and left eigenvectors **x** and **y**, respectively. Let $\tilde{\mathbf{F}} = \mathbf{DF} + \mathbf{E}$; then there exists a unique $\tilde{\lambda}$, eigenvalue of $\tilde{\mathbf{F}}$ such that*

$$\tilde{\lambda} = \lambda + \frac{\mathbf{y}^H \mathbf{E}\mathbf{x}}{\mathbf{y}^H \mathbf{x}} + O(\|E\|^2). \tag{7.32}$$

In our application, we try to reduce only the maximum eigenvalue. It will be shown in the simulation that the maximum eigenvalue is the key factor that affects the overall transmitted power. We assume **x** and **y** are the eigenvectors of the largest eigenvalue. We define $\mathbf{E} = \Delta\Gamma_i \mathbf{F}_i$, where

$$(\mathbf{F}_i)_{jk} = \begin{cases} 0 & j \neq i \\ (\mathbf{F})_{jk} & j = i \end{cases}. \tag{7.33}$$

Then we can have the gradient to reduce the spectral radius as

$$g_i^\rho = \frac{\partial\rho(\mathbf{DF})}{\partial\Gamma_i} = \frac{\mathbf{y}^H \mathbf{F}_i \mathbf{x}}{\mathbf{y}^H \mathbf{x}}. \tag{7.34}$$

If we change each user's SINR according to the preceding gradient g_i^ρ, the overall throughput may be reduced. We need to modify this gradient such that $T = $ const., the plane that is tangent to the curve $T = $ const. at $[\Gamma_1, \ldots, \Gamma_K]$ can be expressed as $\sum_{i=1}^{K} k_i x_i = $ const., where $k_i = c_3/(1 + c_3\Gamma_i)$. We calculate the modified gradient vector $\mathbf{g}^M$ by projecting vector $\mathbf{g}^\rho$ onto that plane.

Table 7.3 Throughput maximization algorithm

Initial:
$T_1, \ldots, T_K = $ *any feasible throughput allocation.*

Iteration: Stop when T_i stable

1. Throughput Maximization
   do {
   $\mathbf{g}^T = \triangledown T$;
   $\Gamma_i = \Gamma_i + \mu.\mathbf{g}_i^T \ \forall \ i$;
   while $(\rho(\mathrm{DF}) < 1 - \epsilon)$
   calculate T^{ave}

2. $\rho(\mathrm{DF})$ Reduction:
   do {
   $\mathbf{g}^\rho = \triangledown \rho(\mathrm{DF})$;
   $\mathbf{g}^M = \mathrm{projection}(\mathbf{g}^\rho)$;
   $\Gamma_i = \Gamma_i - \mu'.\mathbf{g}_i^M \ \forall \ i$;
   if $[T_i > T^{\mathrm{ave}} + T_i^{\mathrm{pos}}(n)]$ $T_i = T^{\mathrm{ave}} + T_i^{\mathrm{pos}}(n)$
   if $[T_i < T^{\mathrm{ave}} - T_i^{\mathrm{neg}}(n)]$ $T_i = T^{\mathrm{ave}} - T_i^{\mathrm{neg}}(n)$
   while (T_i *not stable or not reaching boundary*)

Power Allocation Update: P = DFP + u.

SINR Range Update:
   Update $T_i^{\mathrm{pos}}(n)$, $T_i^{\mathrm{neg}}(n)$.

We repeat the preceding two steps until the algorithm is stable. Then the resulting throughput allocation (i.e., SINR allocation) is assigned to different users at that time. The transmitted power is then calculated, and each user's throughput history is updated.

Our proposed adaptive algorithm is given in Table 7.3, where μ and μ' are small constants and the power-allocation update can be implemented in the distributed manner [250].

To evaluate the performance of our algorithm, a network with 50 hexagonal cells is simulated. The base stations are placed at the center of the cells. Two adjacent base stations do not share the same channels. The radius of each cell is 1000 m. In each cell, one user is placed randomly with a uniform distribution. In the simulations, we consider slow Rayleigh fading.

In Figure 7.5, we show the track of convergence of our proposed algorithm. First, the algorithm starts from a feasible throughput vector. Then the overall throughput is increased by assigning a larger targeted SINR for each user. The maximum eigenvalue of **DF**, i.e., $\rho(\mathbf{DF})$ is also increased consequently. When $\rho(\mathbf{DF})$ is small, the increasing speed of overall power is low. However, when $\rho(\mathbf{DF})$ is larger than a certain point, the overall power is increased very quickly, until there is no feasible solution, i.e., no matter how large the transmitted power is, the receiver's SINR cannot reach the targeted value. To prevent this situation from happening, when $\rho(\mathbf{DF}) > 0.95$, we use the second step of our adaptive algorithm to reduce $\rho(\mathbf{DF})$, while keeping overall throughput unchanged. We can see from the curve that our algorithm reduces $\rho(\mathbf{DF})$ and overall transmitted power

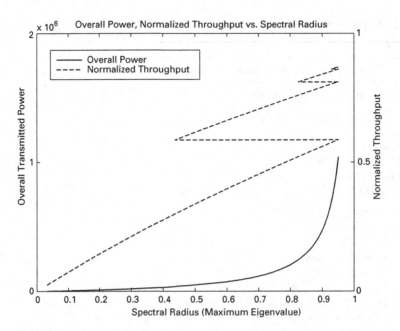

Figure 7.5 Convergence track.

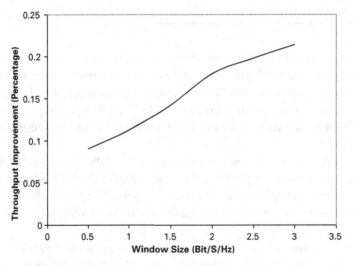

Figure 7.6 Throughput improvement vs. window size.

significantly. Then, at a certain point, the users' throughput reaches the boundary or is stable. The algorithm goes back to the first step to increase the overall throughput. The two steps of our algorithm are repeated alternatively until T_i, $\forall\, i$ are stable. The algorithm will stop within a few iterations. In addition, we find that the overall transmitted power follows the same track when we change $\rho(\mathbf{DF})$ with different throughput allocations. This proves that it is a good approach to minimize overall power by reducing $\rho(\mathbf{DF})$.

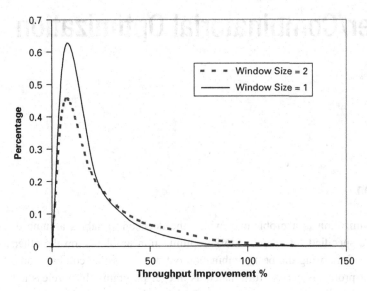

Figure 7.7 Histogram of throughput improvement.

In Figure 7.6, we show the effect of window size on the average throughput improvement (percentage), compared with the previous work [250] in which all users have the same fixed targeted SINR. The average throughput is improved when the window size goes larger. However, the system complexity and delay are increased as well. The improvement is saturated when the window size is large enough. This is because of the time-average constraint. If the user with good channel condition now is too greedy and requests too much throughput, it has to pay back by transmitting less in the future. In Figure 7.7, we show the histogram of throughput improvement. We can see that the overall throughput is improved about 10% in most of the cases. When the window size is increased, the distribution has a longer tail in the higher throughput improvement range.

In this example, we study how to use linear/nonlinear/convex/dynamic optimization for wireless resource-allocation problems. Each user can select a range of throughput. At each time, the links with bad channel conditions sacrifice their throughput, which reduces the unnecessary cochannel interference. The links with good channel conditions get more throughput, which increases the network throughput. We develop a heuristic algorithm to find the throughput allocation to generate the maximum network throughput at each time. For different times, we develop an adaptive algorithm to keep track of each user's throughput history so as to ensure fairness. The whole scheme can be interpreted that each user's throughput is "water filled" in different times, and for the whole system at any specific time, the overall network throughput is allocated to different links according to their channel conditions.

8 Integer/Combinatorial Optimization

8.1 Introduction

Discrete optimization is a problem in which the decision variables assume discrete values from a specified set. Combinatorial optimization problems, on the other hand, are problems of choosing the best combination out of all possible combinations. Most combinatorial problems can be formulated as integer programs. In wireless networking and resource allocation, integer/combinatorial optimization problems are investigated with the efficient allocation of limited resources to meet desired objectives when the values of some or all of the variables are restricted to be integral. Constraints on basic resources, such as modulation, channel allocation, and coding rate restrict the possible alternatives that are considered feasible. For example, in 3G cellular networks, discrete processing gains for different codes give users different bandwidths for transmission. In a WLAN, the available time slots are occupied by different users. Consequently the allocation of time is restricted to a discrete nature. In WiMAX or Flash-OFDM, the distinct time–frequency slot is also allocated to the admitted users. Moreover, for practical implementation, the coding rate and adaptive modulation can have only discrete values. Even for the power control, the minimal step for the current cellular system is 1 dB. To design future wireless networks, it is of importance to study these integer optimizations, especially from an industrial implementation point of view.

The versatility of the integer/combinatorial optimization model stems from the fact that, in many practical problems, activities and resources, such as channel, user, and time slot, are indivisible. Also, many problems have only a finite number of alternative choices (such as modulation) and consequently can appropriately be formulated as combinatorial optimization problems: The word combinatorial refers to the fact that only a finite number of alternative feasible solutions exist. Combinatorial optimization models are often referred to as integer programming models, in which programming refers to "planning" so that these are models used in planning in which some or all of the decisions can take on only a finite number of alternative possibilities.

Integer optimization is the process of finding one or more best (optimal) solutions in a well-defined discrete problem space. Such problems occur in almost all fields of management (e.g., finance, marketing, production, scheduling, inventory control, facility location and layout, database management), as well as in many engineering disciplines (e.g., optimal design of waterways or bridges, VLSI-circuitry design and testing, the layout of circuits to minimize the area dedicated to wires, design and analysis of data networks, solid-waste management, determination of ground states of spin glasses, determination of minimum-energy states for alloy construction, energy-resource-planning models, logistics of electrical power generation and transport, the scheduling of lines in flexible manufacturing facilities, and problems in crystallography). A survey of related applications of combinatorial optimization is given in [96]. In this chapter, we study how to use integer optimization for wireless networking and resource-allocation problems.

The major difficulty with these problems is that we do not have any optimality conditions to check if a given (feasible) solution is optimal or not. For example, in LP we do have an optimality condition: When a candidate solution is given, we'll check if there exists an "improving feasible direction" to move; if there isn't, then the solution is optimal. If we can find a direction in which to move that results in a better solution, then the solution is not optimal. There are no such global optimality conditions in discrete or combinatorial optimization problems. To guarantee a given feasible solution is optimal, the solution is "to compare" with every other feasible solution. To do this explicitly amounts to total enumeration of all possible alternatives that are computationally prohibitive because of the NP-completeness of integer programming problems. Therefore this comparison must be done implicitly, resulting in partial enumeration of all possible alternatives.

There are, at least, three different approaches for solving integer programming problems, although they are frequently combined into "hybrid" solution procedures in computational practice:

- relaxation and decomposition techniques,
- enumerative techniques,
- cutting-plane approaches based on polyhedral combinatorics.

This chapter is organized as follows: In Section 8.2, the general problem of integer optimization is illustrated. We also study how to formulate the wireless networking and resource-allocation problems as an integer optimization problem. In Section 8.3, we study a special case of the integer optimization problem, namely, the knapsack problem, and we explain its various applications. In Section 8.4, by using continuous relaxation and decomposition, we can solve the problem by using other methods such as linear/convex/nonlinear programming. In Section 8.5, we discuss the enumerative techniques to find optimal solutions. Specially, we focus on one of the most important enumerate methods, namely, branch-and-bound. In Section 8.6, we study the cutting-plane approaches based on polyhedral combinatorics. An example of wireless resource allocation is given in Section 8.7. Finally, we have the summary in Section 8.8.

8.2   General Problem

In this section, we first discuss the general problem formulation for integer optimization. Then we discuss the potential applications for wireless networking and resource allocation. We further illustrate the concerns for formulating the problem. Finally, an example of a minimal spanning tree is given for how to formulate the problem.

Most of the integer optimization research to date covers only the linear case. A survey of nonlinear integer programming approaches is given in [51]. The general problem formulation can be given by

$$\min_{\mathbf{x}, \mathbf{y}, \mathbf{z}} f(\mathbf{x}, \mathbf{y}, \mathbf{z}), \qquad (8.1)$$

$$\text{s.t.} \begin{cases} g_i(\mathbf{x}, \mathbf{y}, \mathbf{z}) \leq 0, & \text{for } i = 1, \ldots, m \\ h_j(\mathbf{x}, \mathbf{y}, \mathbf{z}) = 0, & \text{for } j = 1, \ldots, l, \\ \mathbf{x} \in \mathcal{R}, \mathbf{y} \in \{0, 1\}, & \text{and } \mathbf{z} \in \mathcal{I} \end{cases}$$

where function f is the objective function, function g_i is the equality constraint function, function h_j is the inequality constraint function, the component of vector $\mathbf{x}$ is a real value variable, the component of vector $\mathbf{y}$ is a variable of either 0 or 1, and the component of vector $\mathbf{z}$ is a integer value in a space $\mathcal{I}$. If $\mathbf{y} = \mathbf{0}$ and $\mathbf{z} = \mathbf{0}$, (8.1) becomes a nonlinear optimization case. If $\mathbf{z} = \mathbf{0}$, the problem in (8.1) is referred to as a pure 0–1 integer programming problem; if $\mathbf{y} = \mathbf{0}$, the problem in (8.1) is called a pure integer programming problem. Otherwise, the problem is a mixed integer programming problem.

For wireless networking and resource allocation, there are many potential applications of integer optimization. Next, we list some representative examples.

- **Network, routing, and graph problems**

  Many optimization problems can be represented by a network in which a network (or graph) is defined by nodes and by arcs connecting those nodes. Many practical problems arise around physical networks such as communication networks. In addition, there are many problems that can be modeled as networks even when there is no underlying physical network. For example, one can think of the assignment problem for which one wishes to assign a set of users to a certain set of jobs in a way that minimizes the cost of the assignment. Here one set of nodes represents the users to assign, another set of nodes represents the possible jobs, and there is an arc connecting a user to a job if that user is capable of performing that job.

  In addition, there are many graph-theoretic problems that examine the properties of the underlying graph or network. Such problems include the Chinese postman problem in which one wishes to find a path (a connected sequence of edges) through the graph that starts and ends at the same node, which covers every edge of the graph at least once and that has the shortest length possible. If one adds the restriction that each node must be visited exactly one time and drops the requirement that each edge be traversed, the problem becomes the notoriously difficult traveling salesman problem. Other graph problems include the vertex-coloring problem, whose object is to determine the minimum number of colors needed to color each vertex of the graph so that no pair of adjacent nodes (nodes connected by an edge) shares the same color; the edge-coloring problem, whose object is to find a minimum total weight collection of edges such that each node is incident to at least one edge; the maximum clique problem, whose object is to find the largest subgraph of the original graph such that every node is connected to every other node in the subgraph; and the minimum cut problem, whose object is to find a minimum weight collection of edges that (if removed) would disconnect a set of nodes s from a set of nodes t.

Although these combinatorial optimization problems on graphs might appear, at first glance, to be interesting mathematically but have little application to decision making in management or engineering, their domain of applicability is extraordinarily broad. The traveling salesman problem has applications in routing and scheduling, in large-scale circuitry design, and in strategic defense. The four-color problem (can a map be colored in four colors or fewer?) is a special case of the vertex-coloring problem. Both the clique problem and the minimum cut problem have important implications for the reliability of large systems.

- **Scheduling problem**

Space–time networks are often used in scheduling applications. Here one wishes to meet specific demands at different points in time. To model this problem, different nodes represent the same entity at different points in time. An example of the many scheduling problems that can be represented as a space–time network is the channel-assignment problem, which requires that one assign a specific favorable user to the channel to maximize the system performances or QoS. Each time, a channel must have one, and only one, user assigned to it, and a user can be assigned to a channel according to its instantaneous channel conditions as well as its transmission history or QoS.

- **Assignment problem**

The assignment problem is one of fundamental combinatorial optimization problems in the branch of optimization or operations research in mathematics. In its most general form, the problem is as follows:

There are a number of agents and a number of tasks. Any agent can be assigned to perform any task, incurring a certain cost that may vary depending on the assignment. It is required to perform all tasks by assigning exactly one agent to each task in such a way that the total cost of the assignment is minimized. If the numbers of agents and tasks are equal and the total cost of the assignment for all tasks is equal to the sum of the costs for each agent (or the sum of the costs for each task, which is the same thing in this case), then the problem is called a linear assignment problem. Commonly, when speaking of an assignment problem without any additional qualification, the linear assignment problem is meant. Other kinds are the quadratic assignment problem and the bottleneck assignment problem.

The assignment problem is a special case of another optimization problem known as the transportation problem, which is a special case of the maximal flow problem, which in turn is a special case of a linear program. Although it is possible to solve any of these problems by using the simplex algorithm, each problem has more efficient algorithms designed to take advantage of its special structure. Algorithms are known that solve the linear assignment problem within time bounded by a polynomial expression of the number of agents.

The restrictions on agents, tasks, and cost in the (linear) assignment problem can be relaxed, as shown in the subsequent example: Suppose that a taxi firm has three taxis (the agents) available, and three customers (the tasks) wishing to be picked up as soon as possible. For each taxi, the "cost" of picking up a particular customer will depend on the time taken for the taxi to reach the pickup point. The solution to the assignment problem will be whichever combination of taxis and customers results in the least total cost.

However, the assignment problem can be made rather more flexible than it first appears. In the preceding example, suppose that there are four taxis available, but still only three customers. Then a fourth task can be invented, perhaps called "sitting still doing nothing," with a cost of 0 for the taxi assigned to it. The assignment problem can then be solved in the usual way and still give the best solution to the problem.

Similar tricks can be played to allow more tasks than agents, tasks to which multiple agents must be assigned (for instance, a group of more customers than will fit in one taxi) or maximizing profit rather than minimizing cost.

The formal definition of the assignment problem (or linear assignment problem) can be written as follows: Each of n tasks can be performed by any of n agents. The cost of task i being accomplished by agent j is c_{ij}. Assign one agent to each task to minimize the total cost as

$$\min \sum_{i=1}^{n} \sum_{j=1}^{n} c_{ij} x_{ij}, \tag{8.2}$$

$$\text{s.t.} \begin{cases} \sum_{j=1}^{n} x_{ij} = 1, & \text{for all } i \\ \sum_{i=1}^{n} x_{ij} = 1, & \text{for all } j. \\ x_{ij} \in \{0, 1\} \end{cases}$$

Now we discuss the formulation considerations. The versatility of the integer programming formulation, as illustrated by the preceding examples, shall provide sufficient explanation for the high activity in the field of combinatorial optimization that is investigated with developing solution procedures for such problems. Because there are often different ways of mathematically representing the same problem, and because obtaining an optimal solution to a large integer programming problem in a reasonable amount of computer time may well depend on the way it is "formulated," much recent research has been directed toward the reformulation of integer programming problems. In this regard, it is sometimes advantageous to increase (rather than decrease) the number of integer variables, the number of constraints, or both. Moreover, a certain problem has its nature that can hardly be formulated as (8.1), but it has wide ranges of applications. In the following discussion, we give an example of a *minimum spanning tree* on how to formulate the problem as integer programming.

Given a connected, undirected graph, a spanning tree of that graph is a subgraph that is a tree and connects all the vertices together. A single graph can have many different spanning trees. We can also assign a weight to each edge, which is a number representing

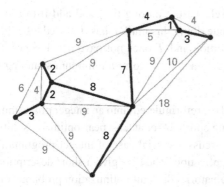

Figure 8.1 Example utility for minimum spanning tree.

how unfavorable it is, and use this to assign a weight to a spanning tree by computing the sum of the weights of the edges in that spanning tree. A minimum spanning tree or minimum weight spanning tree is then a spanning tree with weight less than or equal to the weight of every other spanning tree. More generally, any undirected graph has a minimum spanning forest.

One example is network design. If it is constrained to transmit along only certain paths, then there will be a graph representing which points are connected by those paths. Some of those paths might be more expensive, because they are longer; these paths will be represented by edges with larger weights. A spanning tree for that graph will be a subset of those paths that has no cycles but still connects to nodes. There might be several spanning trees possible. A minimum spanning tree will be one with the lowest total cost. One example is shown in Figure 8.1. There are other applications, such as visualizing multidimensional data and clustering by using a minimal spanning tree.

In case of a tie, there can be several minimum spanning trees; in particular, if all weights are the same, every spanning tree is minimum. However, one theorem states that, if each edge has a distinct weight, the minimum spanning tree is unique. This is true in many realistic situations, such as the preceding one, in which it is unlikely any two paths have exactly the same cost. This generalizes to spanning forests as well.

In general, there are two solutions of minimal spanning tree problems:

- Kruskal's algorithm:
  Let G be a network with n vertices. Sort the edges of G in increasing order by weight (ties can be in arbitrary order). Create a list T of edges by passing through the sorted list of edges and add to T any edge that does not create a circuit with the edges already in T. Stop when T contains $n - 1$ edges. If you run out of edges before reaching $n - 1$ in T, then the graph is disconnected and no spanning tree exists. If T contains $n - 1$ edges, then these edges form a minimum spanning tree of G.
- Prim's algorithm:
  Let G be a network with n vertices. Pick a vertex and place it in a list L. Among all the edges with one end point in the list L and the other not in L, choose one

with the smallest weight (arbitrarily select from among ties) and add it to a list T and put the second vertex in L. Continue doing this until T has $n - 1$ edges. If there are no more edges satisfying the condition and T does not have $n - 1$ edges, then G is disconnected. If T has $n - 1$ edges, then these form a minimum spanning tree of G.

Once the problem has been formulated (or reformulated) into an integer programming problem, solution approaches for obtaining optimal—or at least near-optimal—solutions must be found. In the next section, we discuss a special case of integer programming, knapsack problems, and its variety of application. Then we give a short description of solution approaches to these large, combinatorial difficult optimization problems in the following sections.

8.3   Knapsack Problem

In this section, we discuss one special case of integer programming and its applications. Then we list different formats of the problem formulations.

Suppose one wants to fill a knapsack that can hold a total weight of c with a certain combination of items from a list of n possible items each with weight w_i and value p_i so that the value of the items packed into the knapsack is maximized. This problem has a single linear constraint (that the weight of the items in the knapsack not exceed c), a linear objective function that sums the values of the items in the knapsack, and the added restriction that each item either be in the knapsack or not, i.e., a fractional amount of the item is not possible. Define a vector of binary variable x_j having the following meaning:

$$x_j = \begin{cases} 1, & \text{if item } j \text{ is selected} \\ 0, & \text{otherwise} \end{cases}. \tag{8.3}$$

The simplest knapsack problem is how to pack the packets with as much valuable items as possible, which has the following formulation:

$$\max_{x_j} \sum_{j=1}^{n} p_j x_j, \tag{8.4}$$

$$\text{s.t.} \sum_{j=1}^{n} w_j x_j \leq c.$$

In general the knapsack problems are NP-hard. For solution approaches specific to the knapsack problem, see [201].

Although this problem might seem almost too simple to have much applicability, the knapsack problem is important to cryptographers and to those interested in protecting computer files, electronic transfers of funds, and electronic mail. These applications use a "key" to allow entry into secure information. Often the keys are designed based on linear combinations of a certain collection of data items that must equal a certain value. This problem is also structurally important in that most integer programming problems are generalizations of this problem (i.e., there are many knapsack constraints that together comprise the problem). Approaches for the solution of multiple knapsack problems are often based on examining each constraint separately.

Next, we will list all types of knapsack problem formulations. For wireless networking and resource allocation, we can select the best fits for the specific problem.

- 0–1 knapsack problem
  The problem formulation is the same as (8.4). The problem itself attracts a lot of attention because of these facts: First, it can be viewed as the simplest integer optimization problem. Second, it appears as a subproblem of many complex problems. Third, it may represent many practical situations.
- Bounded knapsack problem
  The bounded knapsack problem is a generalized 0–1 knapsack problem because x_j can be a certain integer other than binary. The problem is formulated as

$$\max_{x_j} \sum_{j=1}^{n} p_j x_j, \tag{8.5}$$

$$\text{s.t.} \begin{cases} \sum_{j=1}^{n} w_j x_j \leq c \\ 0 \leq x_j \leq b_j \text{ and integer, } j \in N = \{1, \ldots, n\} \end{cases}.$$

- Subset–sum problem
  The subset–sum problem is also called the value-independent knapsack problem or stick-stacking problem. It is a particular case of the 0–1 knapsack problem with $p_j = w_j, \ \forall j$. The problem can be written as

$$\max_{x_j} \sum_{j=1}^{n} w_j x_j, \tag{8.6}$$

$$\text{s.t.} \begin{cases} \sum_{j=1}^{n} w_j x_j \leq c \\ x_j = 1, \text{ if item } j \text{ is selected; } x_j = 0, \text{ otherwise} \end{cases}.$$

- Change-making problem
  The problem is to make changes for a fixed amount of money and minimize the number of changes. The unit for each change is denoted by w_j and overall amount of money is c. The problem can be formulated as

$$\min_{x_j} \sum_{j=1}^{n} x_j, \tag{8.7}$$

$$\text{s.t.} \begin{cases} \sum_{j=1}^{n} w_j x_j = c \\ 0 \leq x_j \text{ and is an integer, } j \in N = \{1, \ldots, n\} \end{cases}.$$

- Multiple-knapsack problem

  If there is more than one knapsack, the problem is a multiple-knapsack problem. Suppose there are total of m knapsack and n items. The 0–1 multiple-knapsack problem is

  $$\max \sum_{i=1}^{m} \sum_{j=1}^{n} p_j x_{ij}, \tag{8.8}$$

  s.t. $\begin{cases} \sum_{j=1}^{n} w_j x_{ij} \leq c_i, \ \forall i \in M = \{1, \dots, m\} \\ \sum_{i=1}^{m} x_{ij} \leq 1, \ \forall j \in N = \{1, \dots, n\} \\ x_{ij} = 1, \ \text{if item } j \text{ is selected for knapsack } i; \ x_{ij} = 0, \ \text{otherwise} \end{cases}$.

- Generalized assignment problem

  In the generalized assignment problem, profit and weight are different for different knapsacks, i.e., we have p_{ij} and w_{ij}. The formulated problem is given by

  $$\max \sum_{i=1}^{m} \sum_{j=1}^{n} p_{ij} x_{ij}, \tag{8.9}$$

  s.t. $\begin{cases} \sum_{j=1}^{n} w_{ij} x_{ij} \leq c_i, \ \forall i \in M = \{1, \dots, m\} \\ \sum_{i=1}^{m} x_{ij} = 1, \ \forall j \in N = \{1, \dots, n\} \\ x_{ij} = 1, \ \text{if item } j \text{ is selected for knapsack } i; \ x_{ij} = 0, \ \text{otherwise} \end{cases}$.

- Bin-packing problem

  The bin-packing problem is to select the minimal number of knapsacks with capacity c to pack all the items. Suppose $y_i = 1$ if the ith knapsack is occupied; otherwise $y_i = 0$. The problem formulation is given by

  $$\min \sum_{i=1}^{n} y_i, \tag{8.10}$$

  s.t. $\begin{cases} \sum_{j=1}^{n} w_j x_{ij} \leq c y_i, \ \forall i \in N = \{1, \dots n\} \\ \sum_{i=1}^{m} x_{ij} = 1, \ \forall j \in N = \{1, \dots, n\} \\ x_{ij} = 1, \ \text{if item } j \text{ is selected for knapsack } i; \ x_{ij} = 0, \ \text{otherwise} \end{cases}$.

In the next three sections, we discuss three different approaches to solving the integer/combinatorial problem. Examples are given to clarify the approaches.

8.4   Relaxation and Decomposition

One approach to the solution to integer programming problems is to take a set of "complicating" constraints into the objective function in a Lagrangian fashion (with fixed multipliers that are changed iteratively). This approach is known as Lagrangian relaxation. By removing the complicating constraints from the constraint set, the resulting subproblem is frequently considerably easier to solve. The latter is a necessity for the approach to work because the subproblems must be solved repetitively until optimal values for the multipliers are found. The bound found by Lagrangian relaxation can be tighter than that found by LP, but only at the expense of solving subproblems in integers, i.e., only if the subproblems do not have the integrality property. (A problem has the integrality property if the solution to the Lagrangian problem is unchanged when the integrality restriction is removed.) Lagrangian relaxation requires that one understand the structure of the problem being solved in order to then relax the constraints that are "complicating" [74]. A related approach that attempts to strengthen the bounds of Lagrangian relaxation is called Lagrangian decomposition [99]. This approach consists of isolating sets of constraints so as to obtain separate, easy problems to solve over each of the subsets. The dimension of the problem is increased by creating linking variables that link the subsets. All Lagrangian approaches are problem dependent, and no underlying general theory has evolved.

Suppose an example of a constrained combinatorial optimization is

$$\max \sum_{j=1}^{n} p_j x_j, \tag{8.11}$$

$$\text{s.t.} \begin{cases} \sum_{j=1}^{n} w_j x_j = c \\ x_j = \{0, 1\}, \forall j \end{cases}.$$

The Lagrangian relaxation relaxes the complicated constraint to the objective function, which is given by

$$\max \sum_{j=1}^{n} p_j x_j + \lambda \left(c - \sum_{j=1}^{n} w_j x_j \right), \tag{8.12}$$

$$\text{s.t. } x_j = \{0, 1\}, \forall j,$$

where λ is the Lagrangian multiplier. The goal is that there is a certain simple solution to solve (8.12) with fixed λ. Consequently, the complexity can be reduced a lot. By adjusting λ, the feasibility and slackness of the complicated constraints are improved. This method is called the subgradient method and can be illustrated as follows.

1. Begin with each λ at 0. Let the step size be a certain (problem-dependent value) k.
2. Solve (8.12) to get current solution x.
3. For every constraint violated by x, increase the corresponding λ by k.
4. For every constraint with positive slack relative to x, decrease the corresponding λ by k.
5. If m iterations have passed since the best relaxation value has decreased, cut k in half.
6. Go to 2.

Next, we give an example for Lagrangian relaxation. The integer optimization problem is

$$\max_{x_i \in \{0,1\}} 4x_1 + 5x_2 + 6x_3 + 7x_4, \tag{8.13}$$

$$\text{s.t.} \begin{cases} 2x_1 + 2x_2 + 3x_3 + 4x_4 \leq 7 \\ x_1 - x_2 + x_3 - x_4 \leq 0 \end{cases}.$$

The Lagrangian relaxation is given by

$$\max_{x_i \in \{0,1\}} 4x_1 + 5x_2 + 6x_3 + 7x_4 + \lambda_1(7 - 2x_1 - 2x_2 - 3x_3 - 4x_4)$$
$$+ \lambda_2(-x_1 + x_2 - x_3 + x_4). \tag{8.14}$$

We select the initial value as $\lambda_1 = \lambda_2 = 0$ and step size 0.5. The solution is $x_j = 1, \forall j$. However, it violates the first constraint. We set $\lambda_1 = 0.5$ and $\lambda_2 = 0$; the solution and the constraint violation are the same. We set the new value until $\lambda_1 = 2$ and $\lambda_2 = 0$. There is slackness of the first constraint, and the second constraint is satisfied. Because we cannot go back to $\lambda_1 = 1.5$, which causes a constraint violation, we reduce the step size by setting $\lambda_1 = 1.75$ and $\lambda_2 = 0$. The process stops when $\lambda_1 = 1.83$ and $\lambda_2 = 0.33$ with optimal solution $x_1 = x_2 = x_3 = 1$ and $x_4 = 0$.

Most Lagrangian-based strategies provide approaches that deal with special row structures. Other problems may possess a special column structure, such that, when some subsets of the variables are assigned specific values, the problem reduces to one that is easy to solve. Benders' decomposition algorithm fixes the complicating variables and solves the resulting problem iteratively [24]. Based on the problem's associated duality, the algorithm must then find a cutting plane (i.e., a linear inequality) that "cuts off" the current solution point but no integer feasible points. This cut is added to the collection of inequalities and the problem is re-solved.

Because each of the decomposition approaches previously described provides a bound on the integer solution, they can be incorporated into a branch-and-bound algorithm, instead of the more commonly used LP relaxation. However, these algorithms are special-purpose algorithms in that they exploit the "constraint pattern" or special structure of the problem.

8.5   Enumerative Techniques

The simplest approach to solving a pure integer programming problem is to enumerate all finitely many possibilities. However, because of the "combinatorial explosion" resulting from the parameter "size," only the smallest instances can be solved by such an approach. Sometimes one can implicitly eliminate many possibilities by domination or feasibility arguments. Besides straightforward or implicit enumeration, the most commonly used enumerative approach is called branch-and-bound, in which the "branching" refers to the enumeration part of the solution technique and "bounding" refers to the fathoming of possible solutions by comparison with a known upper or lower bound on the solution value. Next, we will discuss the branch-and-bound approach in detail, and some examples are also given.

The general idea may be described in terms of finding the minimal value of a function $f(x)$ over a set of admissible values of the argument x called the feasible region. Both f and x may be of arbitrary nature. A branch-and-bound procedure requires two tools.

The first one is a smart way of covering the feasible region by several smaller feasible subregions (ideally, splitting into subregions). This is called branching, because the procedure is repeated recursively to each of the subregions and all produced subregions naturally form a tree structure, called a search tree, or branch-and-bound-tree, or something similar. Its nodes are the constructed subregions. Another tool is bounding, which is a fast way of finding upper and lower bounds for the optimal solution within a feasible subregion.

The core of the approach is a simple observation that (for a minimization task) if the lower bound for a subregion A from the search tree is greater than the upper bound for any other (previously examined) subregion B, then A may be safely discarded from the search. This step is called pruning. It is usually implemented by maintaining a global variable m that records the minimum upper bound seen among all subregions examined so far; any node whose lower bound is greater than m can be discarded.

It may happen that the upper bound for a node matches its lower bound; that value is then the minimum of the function within the corresponding subregion. Sometimes there is a direct way of finding such a minimum. In both these cases it is said that the node is solved. Note that this node may still be pruned as the algorithm progresses.

Ideally the procedure stops when all nodes of the search tree are either pruned or solved. At that point, all nonpruned subregions will have their upper and lower bounds equal to the global minimum of the function. In practice the procedure is often terminated after a given time; at that point, the optimal lower bound and the optimal upper bound, among all nonpruned sections, define a range of values that contains the global minimum.

The efficiency of the method depends critically on the effectiveness of the branching and bounding algorithms used; bad choices can lead to repeated branching, without any pruning, until the subregions become very small. In that case the method will be reduced to an exhaustive enumeration of the domain, which is often impractically large. There is no universal bounding algorithm that works for all problems, and there is little hope that one will ever be found; therefore the general paradigm needs to be implemented separately for each application, with branching and bounding algorithms that are specially designed for it.

In the following discussion, we give an example of a branch-and-bound algorithm. We maximize the following constrained integer optimization:

$$\max Z = 21x_1 + 11x_2, \tag{8.15}$$

$$\text{s.t.} \begin{cases} 7x_1 + 4x_2 \leq 13 \\ x_1 \geq 0, x_2 \geq 0 \\ x_1, x_2 \text{ are integers} \end{cases}.$$

1. Step 1: The first step relaxes (8.15) so that x_1 and x_2 are continuous variables. We have the solution $Z = 39$, $x_1 = 1.86$, and $x_2 = 0$. Then we try to branch on x_1 to step 2 and step 3, because it is not an integer value.

2. Step 2: $x_1 \geq 2$. In this case, there is no feasible solution.

3. Step 3: $0 \leq x_1 \leq 1$. The solution is $Z = 37.5$, $x_1 = 1$, and $x_2 = 1.5$. Then we try to branch on x_2 to step 4 and step 5, because it is not an integer value.

4. Step 4: $0 \leq x_1 \leq 1$ and $0 \leq x_2 \leq 1$. The solution is $Z = 32$, $x_1 = 1$, and $x_2 = 1$. Because all variables are integer, stop branching. Return one of the possible solutions.

5. Step 5: $0 \leq x_1 \leq 1$ and $x_2 \geq 2$. The solution is $Z = 37$, $x_1 = 0.71$, and $x_2 = 2$. Then we try to branch on x_1 to step 6 and step 7, because it is not an integer value.

6. Step 6: $x_1 = 1$ and $x_2 \geq 2$. No feasible solution.

7. Step 7: $x_1 = 0$ and $x_2 \geq 2$. The solution is $Z = 35.75$, $x_1 = 0$, and $x_2 = 3.25$. Then we try to branch on x_2 to step 8 and step 9, because it is not an integer value.

8. Step 8: $x_1 = 0$ and $2 \leq x_2 \leq 3$. The solution is $Z = 33$, $x_1 = 0$, and $x_2 = 3$. Because all variables are integer, stop branching. Return one of the possible solutions.

9. Step 9: $x_1 = 0$ and $x_2 \geq 4$. No feasible solution.

The steps are shown in Figure 8.2. Compared with the results in step 4 and step 8, the optimal solution is $Z = 32$, $x_1 = 1$, and $x_2 = 1$.

8.6   Cutting Planes

Cutting-plane algorithms are based on polyhedral combinatorics: Significant computational advances in exact optimization have taken place. Both the size and the complexity of the problems solved have been increased considerably when polyhedral theory, developed over the past 25 years, was employed in numerical problem solving. The underlying idea of polyhedral combinatorics is to replace the constraint set of an integer programming problem with an alternative convexification of the feasible points and extreme rays of the problem.

H. Weyl [331] established the fact that a convex polyhedron can alternatively be defined as the intersection of finitely many half-spaces or as the convex hull plus the conical hull of a certain finite number of vectors or points. If the data of the original problem formulation are rational numbers, then Weyl's theorem implies the existence of a finite system of linear inequalities whose solution set coincides with the convex hull of the mixed-integer points in S that we denote conv(S). Thus, if we can list the set of linear inequalities that completely define the convexification of S, then we can solve the integer programming problem by LP. Gomory [87] derived a "cutting-plane" algorithm for integer programming problems that can be viewed as a constructive proof of Weyl's theorem in this context.

Although Gomory's algorithm converges to an optimal solution in a finite number of steps, the convergence to an optimum is extraordinarily slow because of the fact that these algebraically derived cuts are "weak" in the sense that they frequently do not even define supporting hyperplanes to the convex hull of feasible points. Because one is

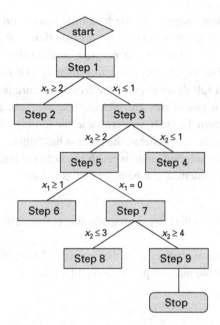

Figure 8.2 A branch-and-bound example.

interested in a linear constraint set for conv(S) that is as small as possible, one is led to the consider minimal systems of linear inequalities such that each inequality defines a facet of the polyhedron conv(S). When viewed as cutting planes for the original problem then the linear inequalities that define facets of the polyhedron conv(S) are the "best possible" cuts, i.e., they cannot be made "stronger" in any sense of the word without losing certain feasible integer or mixed-integer solutions to the problem. Considerable research activity has focused on identifying part (or all) of those linear inequalities for specific combinatorial optimization problems that are, however, derived from an underlying general theme that is due to Weyl's theorem that applies generally. Because for most interesting integer programming problems the minimal number of inequalities necessary to describe this polyhedron is exponential in the number of variables, one is led to wonder whether such an approach can ever be computationally practical. It is therefore all the more remarkable that the implementation of cutting-plane algorithms based on polyhedral theory has been successful in solving problems of sizes previously believed intractable. The numerical success of the approach can be explained, in part, by the fact that we are interested in proving optimality of a single extreme point of conv(S). We therefore do not require the complete description of S but rather only a partial description of S in the neighborhood of the optimal solution.

Thus a general cutting-plane approach relaxes in a first step the integrality restrictions on the variables and solves the resulting linear program over the set S. If the linear program is unbounded or infeasible, so is the integer program. If the solution to the linear program is an integer, then one has solved the integer program. If not, then one solves a facet-identification problem whose objective is to find a linear inequality that "cuts off" the fractional LP solution while ensuring that all feasible integer points satisfy the inequality, i.e., an inequality that "separates" the fractional point from the polyhedron conv(S). The algorithm continues until (1) an integer solution is found (we have successfully solved the problem); (2) the linear program is infeasible and therefore the integer problem is infeasible; or (3) no cut is identified by the facet-identification procedures either because a full description of the facial structure is not known or because the facet-identification procedures are inexact, i.e., one is unable to algorithmically generate cuts of a known form. If we terminate the cutting-plane procedure because of the third possibility, then, in general, the process has "tightened" the LP formulation so that the resulting LP solution value is much closer to the integer solution value. In general, the cutting-plane method can be described as follows:

1. Solve the LP relaxation.
2. If the solution to the relaxation is feasible in the integer programming problem, STOP with optimality.
3. Otherwise, find one or more cutting planes that separate the optimal solution to the relaxation from the convex hull of feasible integral points, and add a subset of these constraints to the relaxation.
4. Return to the first step.

Typically, the first relaxation is solved using the primal simplex algorithm. After the addition of cutting planes, the current primal iterate is no longer feasible. However, the dual problem is modified only by the addition of some variables. If these extra dual variables are given the value 0, the current dual solution is still dual feasible. Therefore subsequent relaxations are solved by the dual simplex method. Notice that the values of the relaxations provide lower bounds on the optimal value of the integer program. These lower bounds can be used to measure progress toward optimality and to give performance guarantees on integral solutions.

Consider an example of integer programming problem as

$$\min -2x_1 - x_2, \tag{8.16}$$

$$\text{s.t.} \begin{cases} x_1 + 2x_2 \leq 7 \\ 2x_1 - x_2 \leq 3 \\ x_1, x_2 \geq 0, \text{ integer} \end{cases} .$$

The feasible integer points are indicated in Figure 8.3. We obtain the LP relaxation by ignoring the integrality restrictions; this is given by the polyhedron contained in the solid lines. The boundary of the convex hull of the feasible integer points is indicated by dashed lines.

If a cutting-plane algorithm were used to solve this problem, the LP relaxation will first be solved, giving the point $x_1 = 2.6$, $x_2 = 2.2$, which has a value of -7.4. The inequalities $x_1 + x_2 \leq 4$ and $x_1 \leq 2$ are satisfied by all the feasible integer points but they are violated by the point $(2.6; 2.2)$. Thus these two inequalities are valid cutting planes. These two constraints can then be added to the relaxation, and when the relaxation is solved again, the point $x_1 = 2$, $x_2 = 2$ results, with a value of -6.

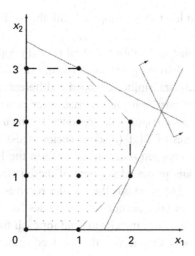

Figure 8.3 A cutting-plane example.

Notice that this point is feasible in the original integer program, so it must actually be optimal for that problem, because it is optimal for a relaxation of the integer program.

If, instead of adding both inequalities, just the inequality $x_1 \leq 2$ had been added, the optimal solution to the new relaxation will have been $x_1 = 2$, $x_2 = 2.5$, with a value of -6.5. The relaxation can then have been modified by adding a cutting plane that separates this point from the convex hull, for example $x_1 + x_2 \leq 4$. Solving this new relaxation will again result in the optimal solution to the integer program.

8.7 A Knapsack Example for Resource Allocation

Transmitting a real-time compressed video over CDMA networks is an emerging service. Compressed video exhibits a highly bursty rate because the various complexities of different video contents and intracoding/intercoding mode. Many recent research works have concentrated on different aspects of this issue [43, 44, 57, 126, 143, 326, 359, 365]. To provide all subscribers with satisfactory received qualities, we face a critical issue: The system's resources are limited. How to jointly perform the rate adaptation, code allocation, and power control to achieve required perceptual qualities of received video via distortion management becomes an important research problem.

In this example, we study the resource-allocation problem of transmitting real-time MPEG4 fine-granularity scalability (FGS) video sequences over downlink multicode CDMA systems. We first design a protocol for the video transmission system over wireless. Then the distortion management is formulated as a knapsack optimization problem to achieve minimal overall distortion received by all users subject to the available number of codes and maximal power for transmission. We develop a fast distortion-management algorithm to allocate resources to each user. Simulations show that the proposed algorithm reduces the distortion by at least 45%, compared with the modified greedy algorithm [111].

MPEG-4 FGS coding [243] and fine-granular scalability temporal (FGST) coding [271] are the new two-layer video techniques for delivering streaming video. FGS coding enables a video sequence to be encoded once and transmitted/decoded at different rates according to the available bandwidth. The encoder generates a base layer at a low bit rate by using a large quantization step and computes the residues between the original frame and the base layer. The bit planes of DCT (discrete cosine transform) transformed coefficients of these residues are encoded sequentially to form the FGS enhancement layer. The decoder can decode any truncated segment of the bit stream of the FGS layer corresponding to each frame. The more bits the decoder receives and decodes, the higher the video quality is. The encoder encodes all bit planes for each video frame and lets the video server determine how many bits to send for each frame according to the channel condition. The decoder can decode the received truncated bit stream.

Figure 8.4 shows a block diagram of our proposed distortion-management protocol to transmit FGS video over multicode CDMA. The protocol is implemented at the base station. The system resources, such as the number of codes and power, are managed to reduce the overall distortion. All users have their own FGS encoders to encode different real-time video programs. Those FGS encoders send the rate-distortion (R-D) information to the proposed protocol. The protocol assigns a variable number of codes to each user according to his/her resource needs and channel conditions. For example, an I frame requires more codes than a P frame. Also, according to the feedback of downlink channel estimations, the protocol assigns the channel-coding rates and power allocations to each code. In allocating resources, our goal is to maintain good video qualities, even when transmitting through a noisy channel with interference. Channel-induced errors affect qualities in an unpredictable way: For the same channel conditions, random errors may affect the received qualities in very different ways for different users. To avoid the uncertainty and maintain controllable video qualities, we use adaptive channel coding and power control to achieve a sufficiently small BER. Because the number of codes and the overall transmitted power are limited, the challenge for the proposed protocol is how to efficiently allocate these resources such that the overall system distortion can be minimized.

Consider a single-cell MC-CDMA system with N users and a total of C codes. We assume that the system is synchronous and that each user is assigned a set of unique pseudo-random codes. Because of the multipath effect [16], the orthogonality among codes may not be guaranteed. Consequently each mobile user is subject to the interference from other users in the cell. If the ith code is assigned to user j, the received SINR is

$$\Gamma_i = \frac{W}{R} \frac{P_i G_j}{G_j \displaystyle\sum_{k=1, k\neq i}^{C} \alpha_{ki} P_k + \sigma^2}, \tag{8.17}$$

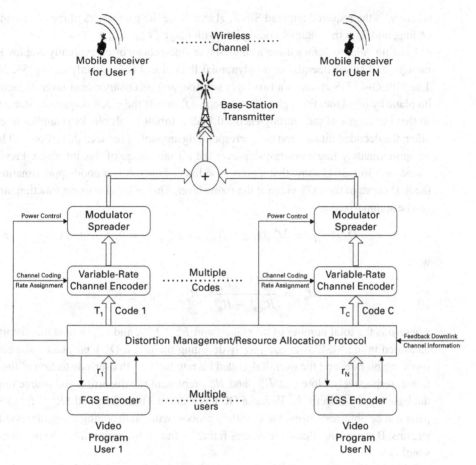

Figure 8.4 Block diagram for the proposed protocol.

where W is the total bandwidth and is fixed, R is the transmission rate, P_i is the transmitted power from the base station for code i, α_{ki} is the orthogonality factor between codes, G_j is the jth user's path loss, and σ^2 is the thermal-noise level that is assumed to be the same at all mobile receivers. The ratio W/R is the processing gain.

Our goal is to maintain the BER after channel decoding below a threshold such that the video quality is controllable by the MPEG-4 FGS video compression with error-resilient and concealment techniques. To reach this goal, the received SINR shall be no less than a targeted SINR, which is a function of channel-coding rate. Note that, for a fixed transmission rate, a reduction in source-encoding rate allows for inserting more channel protection. Through simulations using RCPC codes [101], we found that for our application the targeted SINR can be accurately approximated as a function of channel-coding rate by

$$\gamma_i = 2^{AT_i + B}, \qquad (8.18)$$

where γ_i is the required targeted SINR, A and B are the parameters of the error-control coding, and T_i is the channel-coding rate with range $[T_{\min}, T_{\max}]$.

Existing video schemes using a single-layer video codec often explicitly employ R-D models, and an exponential or a polynomial R-D model is frequently used [253, 360]. The MPEG-4 FGS codec is a two-layer scheme, and its enhancement layer is encoded bit plane by bit plane. For a given bit plane in a frame, if the video is spatially stationary so that the length of the entropy encoded FGS symbols in all blocks is similar to each other, the decoded bit rate and the corresponding amount of reduced distortion will have an approximately linear relationship over the bit rate range of this bit plane. Previous studies in [360, 361] show that a piecewise linear function is a good approximation to the R-D curve of the FGS video at the frame level. This piecewise linear function model can be summarized as

$$D_{n,j}(r_{n,j}) = M_{n,j}^k(r_{n,j} - R_{n,j}^k) + E_{n,j}^k, \ k = 0, \ldots, p-1, \tag{8.19}$$

with

$$M_{n,j}^k = \frac{E_{n,j}^{k+1} - E_{n,j}^k}{R_{n,j}^{k+1} - R_{n,j}^k}, \ R_{n,j}^k \le r_{n,j} \le R_{n,j}^{k+1},$$

where p is the total number of bit planes, and $E_{n,j}^k$, $R_{n,j}^k$, and $r_{n,j}$ denote the distortion measured in the MSE after completely decoding the first k DCT bit planes, the corresponding bit rate, and the overall decoded bit rate for the jth user's distortion of the nth frame, respectively. Note that $E_{n,j}^0$ and $R_{n,j}^0$ represent the distortion and source rate of the base layer, respectively. Because DCT is a unitary transform, all $(R_{n,j}^k, E_{n,j}^k)$ R-D pairs can be obtained during the encoding process without decoding the compressed bit streams. Because we allocate resources frame by frame, we omit n from the notation for simplicity.

In this multicode system, we denote a_{ij} as an indicator to specify whether the ith code is assigned to user j. The maximal overall power is $P_{\max}$, and each user's throughput, r_j, shall be larger than the base-layer rate R_j^0 to guarantee the baseline quality and smaller than the maximum source rate R_j^P. We formulate this optimization problem as

$$\min_{T_i, a_{ij}} \sum_{j=1}^N D_j, \tag{8.20}$$

$$\text{subject to} \begin{cases} \sum_{j=1}^N a_{ij} \le 1, a_{ij} \in \{0, 1\}, \forall i \\ P_{\text{sum}} = \sum_{i=1}^C P_i \le P_{\max} \\ R_j^0 \le r_j = R \sum_{i=1}^C a_{ij} T_i \le R_j^P, \forall j \end{cases}.$$

The difficulty in solving (8.20) lies in the power constraint and code constraint. Each user can reduce his/her distortion by increasing transmitted power. Because different users experience different channel conditions, users will need different increases of power to have the same distortion reduction. However, the overall power at the base station is limited. We need to allocate power to each user efficiently. Moreover, each user may transmit an I frame or a P frame and may encounter different complexities of

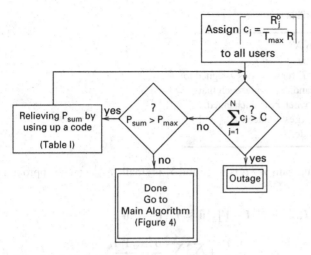

Figure 8.5 Initialization algorithm for the base layer.

the video contents at each time instance. Allocating a code to different users results in different reductions of the overall distortion. In the next section, we develop algorithms to efficiently allocate the limited power and codes to reduce the overall distortion.

There are two main stages in our proposed algorithm. At the first stage, we allocate the resources for delivering the base-layer data to provide the baseline video quality for each user. Some FGS layer data are then delivered to reduce the overall distortion by adjusting the system's resource allocation at the second stage. The goal for the adjustment is to fully utilize the code and power resources and to avoid exhausting only one resource first, while having the other resource left, which leads to local optima. The proposed algorithm can overcome the problem previously mentioned by keeping a balance between code and power allocation during the process of resource allocation.

Figure 8.5 shows the initialization procedure, in which codes, channel-coding rates, and power are assigned to each user so that all users can transmit the base-layer rate R_j^0 only, when the power constraint is satisfied. First, we use the maximal channel-coding rate $T_{\max}$ and assign the number of codes c_j for each user, such that all base layers can be transmitted. If there is no code left, an outage will be reported, indicating that there are too many users in the system and there are no resources even for accommodating the base layers only. If there are enough codes for the base layer, we judge whether the power constraint is violated. If not, the initialization is done and we go to the main algorithm for FGS layer. Otherwise, we relieve P_{sum} by assigning a code once a time to the user who can reduce the power most when the distortion is fixed, until the power constraint is satisfied.

We can approximate P_{sum} as follows. Depending on which user the code is assigned to, we define

$$Y_i = \begin{cases} 0, & \text{if code is not assigned} \\ \dfrac{2^{\frac{AT_i+B}{W}}R}{} = \dfrac{P_iG_j}{G_j\sum_{k\neq i}\alpha_{ki}P_k+\sigma^2}, & \text{for user } j \end{cases} \tag{8.21}$$

Table 8.1 P_{sum} relieve algorithm

1. For hypothesis $j = 1$ to N:
   - Assign a candidate code to user j.
   - Calculate the optimal T_i for all codes assigned to user j including the candidate code, such that P_{sum} is reduced most, when r_j is unchanged.
2. Pick the user with the largest reduced P_{sum} and assign him/her a real code.

Because the processing gain W/R is large and Y_i is small, P_{sum} can be approximated as

$$P_{sum} = \mathbf{1}^T[\mathbf{I} - \mathbf{F}]^{-1}\mathbf{u} \approx \mathbf{1}^T[\mathbf{I} + \mathbf{F}]\mathbf{u}$$

$$= \sum_{i=1}^{C} \frac{\sigma^2 Y_i}{G_i} + \sum_{i=1}^{C} \sum_{\substack{k=1 \\ k \neq i}}^{C} \frac{\sigma^2 \alpha_{ki} Y_i Y_k}{G_k}, \tag{8.22}$$

where $\mathbf{1} = [1 \ldots 1]^T$, $\mathbf{u} = [u_1, \ldots, u_C]^T$ with $u_i = \sigma^2 Y_i/G_i$, and $[\mathbf{F}]_{ij} = 0$ if $j = i$; $[\mathbf{F}]_{ij} = \alpha_{ji} Y_i$ if $j \neq i$.

The P_{sum} relieve algorithm is shown in Table 8.1. Before we assign a real code to a specific user, we make N hypotheses. For the jth hypothesis, we assign a candidate code to the jth user and the settings of the other users keep unchanged. The jth user will keep the amount of his/her source-coding rate, r_j, unchanged, but redistribute r_j to his/her existing assigned codes with the candidate code. Consequently, the channel-coding rates for those codes are reduced, so that the required SINR and the overall transmitted power are reduced. The problem can be formulated as an optimization problem to minimize P_{sum} subject to a fixed distortion by allocating the source-coding rates and channel-coding rates to the already assigned codes plus the candidate code. This problem can be solved using the water-filling method. First, we assign the candidate code with T_{max}, so the current throughput is larger than r_j. Because Y_i is a monotonic increasing function of T_i, we can reduce P_{sum} by searching the code with the largest $|g_i^T = \frac{\partial P_{sum}}{\partial T_i}|$ and reduce the channel-coding rate of this code. The algorithm repeats the searching procedure until the throughput is equal to r_j. From all hypotheses, the user reducing P_{sum} by the highest amount is selected and assigned a real code.

After initialization, we apply the distortion-management algorithm in Figure 8.6 to efficiently allocate resources for the FGS layer to reduce users' overall distortion. The algorithm first decides whether all codes are used up. If yes, we use the remaining power to reduce the distortion as described in Table 8.2. Otherwise, different subalgorithms are employed by assigning a new code to reduce the power or distortion depending on whether the system is power unbalanced. The whole distortion-management algorithm for the FGS layer is terminated under two conditions: First, all available codes are distributed and the power is within the range. This is the situation in which the system is heavy loaded. Second, all the users have the minimal distortion. This is the light-load situation.

Table 8.2 Distortion reduction by increasing power

---

1. For hypothesis $i = 1$ to C:
   - If T_i of code i is equal to T_{max}, do next hypothesis.
   - For code i, calculate the corresponding decrease in channel-coding rate of one discrete step, ΔT_i.
   - Given ΔT_i, calculate Δr_j, ΔD_j, and ΔP_{sum}. If $P_{sum} < P_{max}$, add hypothesis i to candidate list.
2. If no candidate left, exit. Otherwise, choose the code with the largest $|\Delta D_j / \Delta P_{sum}|$ and change the channel-coding rate to the chosen code.
3. Empty candidate list. Go to step 1.

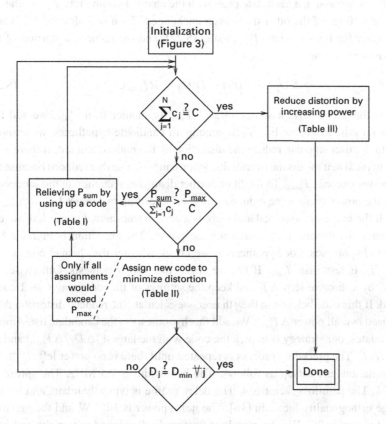

Figure 8.6 Distortion-management algorithm for FGS layer.

The criterion for judging whether the system is power unbalanced is the average consumed power per assigned code. If the current P_{sum} over the assigned code $\sum_{j=1}^{N} c_j$ is greater than P_{max} over the maximal number of codes C, the system is power unbalanced. Then the algorithm in Table 8.1 is employed to reduce the power. Otherwise, we apply the algorithm in Table 8.3 to reduce the distortion.

Table 8.3 Code assignment to reduce distortion

1. For hypothesis $j = 1$ to N:
   - Assign user j a candidate code, analyze $\Delta D_j, P_{sum}$.
   - If $P_{sum} < P_{max}$, add hypothesis j to candidate list.
2. If there is no candidate user, do not assign the code and go to the algorithm in Table 8.1.
3. Among the candidates, choose the one with the largest ΔD_j and assign a real code to user j.

In Table 8.3, we reduce the overall distortion by assigning one code to a user at a time. Before we assign a real code to a specific user, we make N hypotheses. For the jth hypothesis, we assign a candidate code with the channel-coding rate T_{max} to the jth user and the settings of the other users keep unchanged. Then we calculate the overall required power for transmission, P_{sum}, using (8.22), and the reduced distortion of the received video as follows:

$$\Delta D_j = D_j(r_j) - D_j(r_j + RT_{max}), \tag{8.23}$$

where r_j is the user's current source rate. If P_{sum} is smaller than P_{max}, we add this hypothesis into the candidate list. Then, among all candidate hypotheses, we assign a real code to the user who can reduce the distortion by the highest amount. If there is no candidate in the list, it means the overall distortion cannot be further reduced because the required power exceeds P_{max}. To facilitate further distortion reduction in the successive iterations, the power needs to be reduced using the algorithm in Table 8.1.

When all the codes are assigned and there is a certain transmitted power left, we can further reduce the distortion by increasing the power. The algorithm is in Table 8.2. We make C hypotheses. For hypothesis i, we check whether the channel-coding rate of code i, T_i, is less than T_{max}. If no, we check the next hypothesis. Otherwise, we increase T_i by a discrete step ΔT_i and keep the settings of the remaining $C - 1$ codes unchanged. If this code belongs to the jth user, we calculate the reduced distortion ΔD_j and increased overall power ΔP_{sum}. We add this hypothesis in the candidate list. Among these candidates, our strategy is to pick the code with the largest $|\Delta D_j / \Delta P_{sum}|$ and set $T_i = T_i + \Delta T_i$. The preceding process is repeated until there is no power left.

The simulations are set up as follows. The bandwidth is 7.5 MHz. The spreading factor is 64. The path-loss factor is 4. The delay profile is typically urban, and we use the average orthogonality factor in [16]. The noise power is 10^{-9} W, and the maximal transmitted power is 280 W. The mobile is uniformly distributed within the cell with radius from 20 to 700 m. We use RCPC codes with a memory 4, puncturing period 8, and mother code rate $1/4$ [101]. Our experimental results show that to achieve BER = 10^{-6} using the FGS codec (including base and FGS layer), parameters (A, B) in (8.18) are (4.4, -1.4). The video refresh rate is 15 frames per second. We concatenate 15 classic video sequences (*Akiyo, Carphone, Claire, Coastguard, Container, Foreman, Grandmother, Hall objects, Miss American, Mother and daughter, MPEG4 news, Salesman, Silent, Suzie,* and *Trevor*) with a temporal down sampling factor of 2 to form a basic testing

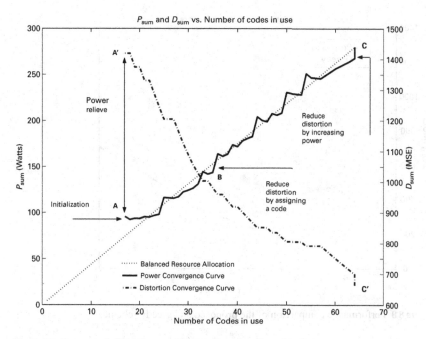

Figure 8.7 Power and distortion vs. the number of assigned codes.

video sequence source of 2775 frames. The base layer is generated by the MPEG-4 encoder with a fixed quantization step of 30 and the GOP (group of pictures) pattern is 14 P frames after one I frame. All frames of the FGS layer have up to six bit planes. The content program of the ith user is 100 frames long and starts from frame 173 × $(i - 1) + 1$ of the concatenated video source.

Figure 8.7 shows a simulation result for the convergence track of the overall power and distortion with the number of assigned codes by using the proposed algorithm. After initialization (shown at A), the base layer of each user is allocated and 17 codes are assigned. The overall distortion (shown at A′) is large because only the base layer is transmitted. When the system is power unbalanced, i.e., the operating point (such as position A) is above the balanced resource-allocation line, we apply the power relieve algorithm in Table 8.1 to reduce the power when keeping the distortion fixed. When the system is not power unbalanced (such as position B), we assign codes to reduce distortion (consequently, the required power is increased) using the algorithm in Table 8.3, until all the codes are used up. Finally, we use the algorithm in Table 8.2 to further reduce distortion by the remaining power quota (shown at C).

We compare the proposed algorithm with a modified greedy approach (see Figure 8.8) [111]. This modified approach is similar to our proposed framework, but uses a greedy approach for the code assignment in FGS layer. For each iteration, this greedy algorithm tries to assign a candidate code to every user, calculates $|\Delta D_j / \Delta P_{sum}|$, and assigns a new code to the user with the largest value. Figure 8.8 shows the number of users versus average of the total distortion D_{sum} ($D_{sum} = \sum_{j=1}^{N} D_j$) over 100 frames from 50

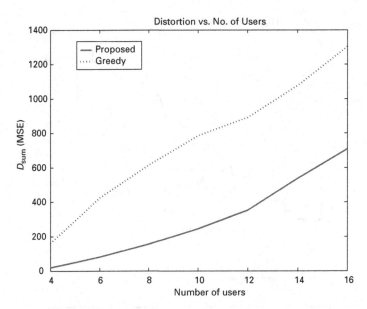

Figure 8.8 Performance comparison of the proposed and greedy scheme.

different mobiles' locations. The simulation results demonstrate that the average D_{sum} of the proposed algorithm outperforms that of the greedy algorithm by at least 45%.

8.8   Summary

We briefly note some topics related to formulating and solving the integer/combinatorial problems in wireless networking and resource allocation. In practice, there are many parameters that can have only integer or combinatorial value. The examples are modulation level, coding rate, route selection, and source-output rate. Moreover, even for some continuous parameters such as transmission power, the real implementation has finite granularity that leads to limited integer values as choices. We illustrate the basic problem formulation methods.

To solve the formulated problems, we mainly discuss three major methods: relaxation and decomposition, branch-and-bound, and cutting plane. Depending on different scenarios, the different algorithms can be employed to solve the problem efficiently and consider the available information for optimization. Beyond these methods, there are other methods such as dynamic programming, matroids, and greedy algorithm [178, 208, 382].

In most practical scenarios, integer programming and other types of continuous programming are usually combined together. The results are mixed programming. Some hybrid methods and heuristics can be utilized to solve the problems. In most cases, the solutions are problem oriented.

9   Game Theory

9.1   Introduction

Game theory is a branch of applied mathematics that uses models to study interactions with formalized incentive structures ("games"). It studies the mathematical models of conflict and cooperation among intelligent and rational decision makers. "Rational" means that each individual's decision-making behavior is consistent with the maximization of subjective expected utility. "Intelligent" means that each individual understands everything about the structure of the situation, including the fact that others are intelligent, rational, decision makers. It has applications in a variety of fields, including economics, international relations, evolutionary biology, political science, and military strategy. Game theorists study predicted and actual behavior, as well as optimal strategies, of individuals in games.

For the history of game theory, the basis of modern game theory can be considered an outgrowth of a few seminal works:

- Augustin Cournot, in his 1838 paper *Researches into the Mathematical Principles of the Theory of Wealth*, gives an intuitive explanation of what would eventually be formalized as the Nash equilibrium, as well as provides an evolutionary or dynamic notion of best responses to the actions of others.
- Francis Ysidro Edgeworth's *Mathematical Physics* (1881) demonstrated the notion of competitive equilibriums in a two-person (as well as two-type) economy.
- Emile Borel, in *Algebre et calcul des probabilites*, Comptes Rendus Academie des Sciences, Vol. 184, 1927, provided the first insight into mixed strategies that state that randomization may support a stable outcome.

Although many other contributors hold a place in the history of game theory, it is widely accepted that modern analysis began with John von Neumann's and Oskar Morgenstern's book, *Theory of Games and Economic Behavior*, which is the analytical base for the works [217] of John Nash.

Game theory has important applications in fields such as operations research, economics, collective action, political science, psychology, and biology. It has close links with economics in that it seeks to find rational strategies in situations in which the outcome depends not only on one's own strategy and "market conditions," but on the strategies chosen by other players with possibly different or overlapping goals. Applications in military strategy drove some of the early developments of game theory. Game theory has come to play an increasingly important role in logic and in computer science. Several logical theories have a basis in game semantics. And computer scientists have used games to model interactive computations. Computability logic attempts to develop a comprehensive formal theory (logic) of interactive computational tasks and resources, formalizing these entities as games between a computing agent and its environment.

Game-theoretic analysis can apply to simple games of entertainment or to more significant aspects of life and society. The Prisoner's Dilemma, as popularized by mathematician Albert W. Tucker, furnishes an example of the application of game theory to real life; it has many implications for the nature of human cooperation, and has even been used as the basis of a game show called *Friend or Foe?* Biologists have used game theory to understand and predict certain outcomes of evolution, such as the concept of evolutionarily stable strategy introduced by John Maynard Smith and George R. Price.[1]

In wireless networks, to obtain information such as channel conditions, signaling is performed so that resource allocation can be conducted in an optimal way. However, signaling has considerable overhead for communications. Most of the current wireless networks have more than 50% of overhead. Reducing the overhead can greatly increase the spectrum utilization, increase the number of users, and improve the network performances. One of the possible ways to reduce overhead is to do resource optimization by using only local information. This is very important, especially if the system topology is distributed.

In some wireless network scenarios, it is hard for an individual user to know the channel conditions of the other users. The users cannot cooperate with each other. They act selfishly to maximize their own performances in a distributive fashion. Such a fact motivates us to adopt game theory. Resource allocation can be modeled as a game that deals largely with how rational and intelligent individuals interact with each other in an effort to achieve their own goals. In this game, each mobile user is self-interested and trying to optimize his/her utility function, in which the utility function represents the user's performance and controls the outcomes of the game. There are many advantages of applying game theory to wireless networking and resource allocations:

1. Only local information and distributive implementation: The individual user observes the outcome of the game and adjusts only his/her resources in response to optimize his/her own benefit. As a result, there is no need for collecting all the information and conducting constrained optimization in a centralized way.
2. More robust outcome: For the optimization problems discussed in the other chapters in the second part of this book, if the information for optimization is not quite accurately obtained, the optimized results can be far from optimality. In contrast, local information is always accurate, so the outcome of distributed game approaches is always robust.
3. Combinatorial nature: For traditional optimization techniques, such as programming, it is hard to handle combinatorial problems. For game theory, it is natural to discuss the problem in a discrete form such as the strategic form. In problems such as discrete

---

[1] See also evolutionary game theory and behavioral ecology.

modulation levels and channel-coding rates, analyzing the combinatorial problems by game theory is considerably convenient.

4. Rich mathematics for optimization: There are many mathematics tools to analyze the outcome of the game. Specifically, if the game is played noncooperatively once, the static game can be studied. If the game is played multiple times, dynamic game theory is employed. If some contracts and mutual benefits can be obtained, cooperative game explains how to divide the profits. Auction theory studies the behaviors of both seller and bidder.

On the other hand, there are many challenges for using game theory:

1. Low-efficiency game outcome: The outcome of the game usually has lower efficiency than the centralized optimization. This lower efficiency exists because each individual greedily optimizes his/her own performance by consuming much more resources than necessary. The centralized results with perfect information can serve as a performance upper bound. Many techniques, such as pricing and repeated game, can improve the outcome of the games by giving incentives for distributed users to cooperate for resource usage.

2. Hard-to-model utility: One of the biggest challenges for applying game theory is how to formulate and obtain a reasonable and good utility function. First, the utility function must represent the performances of the users. The users may have multiple objectives, so that it is hard to describe in one function how to present the users' benefits. Second, the utility function must have some properties so that the game results can be nontrivial and make sense. Third, the utility function shall include the influences of the other users' strategies. Otherwise, there will be no conflict among users.

The goals of this chapter are as follows: First, we explain the basic ideas for game theory and give the readers some insight on how to model resource-allocation problems by means of game theory. Then we discuss four different existing game-theory approaches. The basic concepts and properties are illustrated. Finally, we connect these approaches to the real problems in wireless networking.

This chapter is organized as follows: In Section 9.2, we explain the basic components of a game. In Section 9.3, the noncooperative static game is studied. In Section 9.4, the game is played multiple times and the behaviors are analyzed. In Section 9.5, another type of game, the cooperative game, is examined and a resource-allocation framework is proposed. In Section 9.6, auction theory is studied and some types of auctions are illustrated. Then mechanism design concepts are illustrated. A summary is given in 9.7. Some wireless networking and resource-allocation examples are also given in the different sections.

9.2   Game-Theory Basics

In this section, we discuss the basic concepts and elements of game theory. Then we give an example of the Prisoner's Dilemma to explain the players' behaviors. Finally, we study how to write utility functions in wireless networking and resource-allocation problems to represent distributed users' own interests. We also explain one example [268] that is well known in wireless resource allocation.

A game can be roughly defined as each user adjusts his/her strategy to optimize his/her own utility to compete with others. Strategy and utility can be defined as follows:

DEFINITION 12 *A **strategy** r is a complete contingent plan, or a decision rule, that defines the action an agent will select in every distinguishable state Ω of the world.*

DEFINITION 13 *In any game, **utility** (payoff) u represents the motivations of players. A utility function for a given player assigns a number for every possible outcome of the game with the property that a higher (or lower) number implies that the outcome is more preferred.*

One of the most common assumptions made in game theory is rationality. In its mildest form, rationality implies that all players are motivated by maximizing their own utilities. In a stricter sense, it implies that every player always maximizes his or her utility, thus being able to perfectly calculate the probabilistic result of every action. A game can be defined as follows.

DEFINITION 14 *A **game** G in the strategic form has three elements: the set of players $i \in \mathcal{I}$, which is a finite set $\{1, 2, \ldots, N\}$; the strategy space Ω_i for each player i; and utility function u_i, which measures the outcome of the ith user for each strategy profile $\mathbf{r} = (r_1, r_2, \ldots, r_N)$. We define $\mathbf{r}_{-i}$ as the strategies of player i's opponents, i.e., $r_{-i} = (r_1, \ldots, r_{i-1}, r_{i+1}, \ldots, r_N)$. In static games, the interaction between users occurs only once, where as in dynamic games the interaction occurs several times.*

The basics of game theory can be demonstrated by a game called the "Prisoner's Dilemma." The name comes from a hypothetical situation: Two criminals are arrested for committing a crime in unison, but the police do not have enough proof to convict either. Thus, the police separate the two and offer a deal: If one testifies to convict the other, he will get a reduced sentence or go free. Here the prisoners do not have information about the other's "move" as they would in chess. The payoff if they both say nothing (and thus cooperate with each other) is good, as neither can be convicted without further proof. If one of them betrays but the other remains silent, then he benefits because he goes free while the other is imprisoned because there is sufficient proof to convict the silent one. If they both betray the each other, they both get reduced sentences, which can be described as a null result. The obvious dilemma is the choice between two options, in which a good decision cannot be made without information.

This decision in terms of the outcomes of the decisions of the prisoner may be assigned arbitrary point values and is described in Table 9.1. The Prisoner's Dilemma is a two-player game. One player plays as the row player and one plays as the column player. Both have the strategy option of cooperating (C) or defecting (D). Thus there are four possible outcomes to the game: $\{(C, C), (D, D), (C, D), (D, C)\}$. Under mutual cooperation, $\{(C, C)\}$, both players will receive the reward payoff, 3. Under mutual defection, $\{(D, D)\}$, both players receive the punishment of defection, 1. When one player

Table 9.1 Prisoner's Dilemma

	Cooperate	Defect
Cooperate	(3,3)	(0,5)
Defect	(5,0)	(1,1)

cooperates while the other defects {(C, D), (D, C)}, the cooperating player receives the payoff, 0, and the defecting player receives the temptation to defect, 5.

In the Prisoner's Dilemma example, if one player cooperates, the other player will have a better payoff (5 instead of 3) if he defects; if one player defects, the other player will still have a better payoff (1 instead of 0) if he defects. Regardless of the other player's strategies, the player always selects defect and {(D, D)} is an equilibrium. Although cooperation will give each player a better payoff of 3, greediness leads to an inefficient outcome. As a result, the outcome of the noncooperative game can be less efficient.

In the rest of this section, we study how to formulate the game in wireless networking and resource-allocation problems. The utility function represents the real interests of the mobile users. For example, the utility can be the communication rate, power, energy, or a combination of these. The difficulty is how to write the utility function that has a physical meaning and the game outcome is not trivial. Typically, if the utility function is monotony, the game outcome occurs in the boundary situation, which is trivial. Therefore the utility function shall be designed so that the utility is either quasi-concave or quasi-convex. In addition, the optimal point is selected somewhere within the practical parameter range. Moreover, the location of this optimal point also depends on the other users' behaviors. If the utility function has the preceding properties, the analysis and improvement of the game outcomes can be further studied. In the next section we study a commonly used utility function in wireless networking and resource-allocation problems proposed in [268].

Users access a wireless system through the air interface that is a common resource, and they transmit information expending battery energy. Because the air interface is a shared medium, each user's transmission is a source of interference for others. The signal-to-interference ratio (SIR) is a measure of the quality of signal reception for the wireless user. Typically, a user intends to achieve a high quality of reception (high SIR) while at the same time expending a small amount of energy. Thus it is possible to view both SIR and battery energy (or, equivalently, transmitted power) as commodities that a wireless user desires. There exists a trade-off relationship between obtaining high SIR and low energy consumption. Finding a good balance between the two conflicting objectives is the primary focus of the power-control component of radio resource management.

An optimum power-control algorithm for wireless voice systems maximizes the number of conversations that can simultaneously achieve a certain quality-of-service (QoS) objective. Typically, the QoS objective for a voice terminal is to achieve a minimum acceptable SIR. However, this approach is not appropriate for the efficient operation of a wireless data system. This is because the QoS objective for data signals differs from

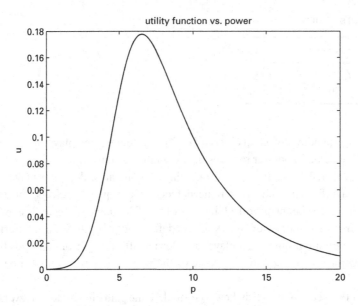

Figure 9.1 Example utility for game theory over wireless networks.

the QoS objective for telephones. In a data system, error-free communication has high priority. The SIR is an important quantity because there is a direct relationship between the SIR and the probability of transmission errors.

Suppose the frame length is M bits with L information bits and $M - L$ overhead or channel protections. The rate is R bit/s and power per bit is p. Let P_c denote the probability of correct reception of a frame at the receiver. P_c is a function of the SINR. The utility function proposed in [268] is given by

$$u = \frac{LRP_c}{Mp} \text{ bits/J.} \tag{9.1}$$

The utility has the physical meaning of the number of successfully received information bits per joule of energy cost. To avoid the problem of infinite utility with $p = 0$, an efficiency function is proposed to approximate P_c as

$$f(\text{SNR}) = [1 - 2P_e(\text{SNR})]^M, \tag{9.2}$$

where P_e depends on the modulation and coding scheme of the system.

Suppose $M = L = 10$, $R = 1$, and $p = \text{SNR}$. In Figure 9.1, we show the utility function with differential PSK (DPSK) schemes where $P_e = \frac{1}{2}e^{-\text{SNR}}$. We can see that the curve is a quasi-concave function of the SNR. When the SNR is small, the frame error rate in (9.2) is large so that the utility in (9.1) is small. On the other hand, when the SNR is high, the frame error rate is always approaching 1. Therefore, increasing power cannot improve the frame error rate, but only waste energy. Therefore the utility decreases. We will use this example in some of the sections in the rest of this chapter to illustrate the game-theory ideas.

9.3 Noncooperative Static Game

In this section, we first define the noncooperative static game. Then we study two ways to present a game. Then we give some properties of the game, such as dominance, Nash equilibrium, Pareto optimality, and mixed strategies. Some basic examples are employed for illustration. Finally, we discuss the low efficiency of the outcome for noncooperative static games employed to wireless networking and resource-allocation problems. The ideas of two methods, pricing and referee, are briefly discussed to improve the game outcomes.

DEFINITION 15 *A **noncooperative game** is one in which players are unable to make enforceable contracts outside of those specifically modeled in the game. Hence it is not defined as games in which players do not cooperate, but as games in which any cooperation must be self-enforcing.*

DEFINITION 16 *A **static game** is one in which all players make decisions (or select a strategy) simultaneously, without knowledge of the strategies that are being chosen by other players. Even though the decisions may be made at different points in time, the game is simultaneous because each player has no information about the decisions of others; thus it is as if the decisions are made simultaneously.*

Static games can be represented by the normal form defined as follows:

DEFINITION 17 *The **strategic (or normal) form** is a matrix representation of a simultaneous game. For two players, one is the "row" player and the other is the "column" player. Each row or column represents a strategy, and each box represents the payoffs to each player for every combination of strategies.*

One example is the "battle of the sexes," shown in Table 9.2 with the following scenario: A husband and wife have agreed to attend a rare entertainment event in the evening. Unfortunately, neither remembers which of the two special events in town they had agreed on: the boxing match or the opera. The husband prefers the boxing match whereas the wife prefers the opera; yet both prefer being together to being apart. They must decide simultaneously and without communication which event to attend. There are two pure strategy equilibria. Different pure strategy equilibria are preferred by each player. However, either equilibrium is preferred by both players to any of the nonequilibrium outcomes.

For each user, there are some strategy spaces. Some of the strategies are superior to the others for the user's interests. To define such superiority, we have the following two definitions:

Table 9.2 Battle of the sexes (wife, husband)

	Boxing	Opera
Boxing	(1,2)	(0,0)
Opera	(0,0)	(2,1)

DEFINITION 18 *Dominant Strategies: A strategy is dominant if, regardless of what any other players do, the strategy earns a player a larger payoff than any other. Hence a strategy is dominant if it is always better than any other strategy, regardless of what opponents may do. If a player has a dominant strategy, then he or she will always play it in equilibrium. Also, if one strategy is dominant, then all others are dominated.*

DEFINITION 19 *Dominated Strategies: A strategy is dominated if, regardless of what any other players do, the strategy earns a player a smaller payoff than a certain other strategy. Hence a strategy is dominated if it is always better to play some other strategy, regardless of what opponents may do. A dominated strategy is never played in equilibrium.*

For example, in the Prisoner's Dilemma in Table 9.1, each player has a dominated strategy of cooperation and a dominant strategy of defection. This is because, no matter what other players' strategies, the defection for each user always yields higher utility. Notice that here domination does not imply higher payoff. In this case, the dominated strategy for both players has a better payoff than that of the dominating strategy.

To analyze the outcome of the game, the Nash equilibrium is a well-known concept that states that in the equilibrium every agent will select a utility-maximizing strategy given the strategies of every other agent.

DEFINITION 20 *Define a strategy vector* $\mathbf{r} = [r_1 \ldots r_K]$ *and define the strategy vector of the ith player's opponents as* $\mathbf{r}_i^{-1} = [r_1 \ldots r_{i-1}\ r_{i+1} \ldots r_K]$, *where K is number of users and r_i is the ith user's strategy. u_i is the ith user's utility. Nash equilibrium point (NEP) $\mathbf{r}$ is defined as*

$$u_i(r_i, \mathbf{r}_i^{-1}) \ge u_i(\tilde{r}_i, \mathbf{r}_i^{-1}), \ \forall i, \ \forall \tilde{r}_i \in \mathbf{\Omega}, \ \mathbf{r}_i^{-1} \in \mathbf{\Omega}^{K-1}, \tag{9.3}$$

i.e., given the other users' resource allocations, no user can increase his/her utility alone by changing his/her own resource allocation.

In other word, a Nash equilibrium, named after John Nash, is a set of strategies, one for each player, such that no player has the incentive to unilaterally change his/her action. Players are in an equilibrium if a change in strategies by any one of them will lead that player to earn less than if he/she remained with his/her current strategy.

The existence of the Nash equilibrium is a difficult problem to prove. There are some existing theorems to show the existence. We need to prove only that the proposed game satisfies the requirements of the theorems. For example, in [80], it has been shown a NEP exists if, $\forall i$,

1. Ω, the support domain of $u_i(\mathbf{r}_i)$, is a nonempty, convex, and compact subset of a certain Euclidean space $\Re^L$;
2. $u_i(\mathbf{r}_i)$ is continuous and quasi-convex in $\mathbf{r}_i$.

There might be an infinite number of Nash equilibria. Among all these equilibria, we need to select the optimal one. There are many criteria by which to judge whether the equilibrium is optimal. Among these criteria, Pareto optimal is one of the most important definitions.

Table 9.3 Chicken (driver 1, driver 2)

	Stay	Swerve
Stay	(−100,−100)	(1,−1)
Swerve	(−1,1)	(0,0)

DEFINITION 21 *Pareto optimal: Named after Vilfredo Pareto, Pareto optimality is a measure of efficiency. An outcome of a game is Pareto optimal if there is no other outcome that makes every player at least as well off and at least one player strictly better off. That is, a Pareto optimal outcome cannot be improved on without hurting at least one player. Often, a Nash equilibrium is not Pareto optimal, implying that the players' payoffs can all be increased.*

Until now, we have discussed only the strategy that is deterministic, or pure strategy. A pure strategy defines a specific move or action that a player will follow in every possible attainable situation in a game. Such moves may not be random or drawn from a distribution, as in the case of mixed strategies.

DEFINITION 22 *Mixed Strategy: A strategy consists of possible moves and a probability distribution (collection of weights) that corresponds to how frequently each move is about to play. A player will use a mixed strategy only when she is indifferent about several pure strategies. Moreover, if the opponent can benefit from knowing the next move, the mixed strategy is preferred because keeping the opponent guessing is desirable.*

To illustrate the idea of Pareto optimality and mixed strategy, we give an example of the "game of Chicken" shown in Table 9.3 (or it can be called the hawk–dove game.). The scenario is as follows: Two hooligans with something to prove drive at each other on a narrow road. The first to swerve loses face among his peers. If neither swerves, however, a terminal fate plagues both. There are two pure strategy equilibria. A different pure strategy equilibrium is preferred by each player. Both equilibria (1,−1) and (−1,1) are Pareto optimal. A mixed strategy equilibrium also exists, in which each user plays swerve with probability 0.99 and plays stay with probability 0.01.

A zero-sum game is a special case of a constant-sum game in which all outcomes involve a sum of all players' payoffs of 0. Hence a gain for one participant is always at the expense of another, such as in most sporting events. Therefore one of the strategies to gain one's own benefit is to suppress the other's play. Given the conflicting interests, the equilibrium of such games is often in mixed strategies.

One example of a zero-sum game is the game of "matching pennies" shown in Table 9.4. The scenario determines who is required to do the nightly chores; two children first select who will be represented by "same" and who will be represented by "different." Then, each child conceals in the palm a penny either with its face up or face down. Both coins are revealed simultaneously. If they match (both are heads or both are tails), the child who claims "same" wins. If they are different (one heads and one tails), the child who claims "different" wins. The game is equivalent to "odds or evens" and quite

Table 9.4 Matching pennies (different, same)

	Heads	Tails
Heads	$(-1,1)$	$(1,-1)$
Tails	$(1,-1)$	$(-1,1)$

similar to a three-strategy version—rock, paper, scissors. The game is zero sum. The only equilibrium is in mixed strategies. Each plays each strategy with equal probability, resulting in an expected payoff of zero for each player.

Unfortunately, the Nash equilibria of noncooperative static games often have low efficiency. For example, in the Prisoner's Dilemma example in Table 9.1, if one player cooperates, the other player will have a better payoff, 5 rather than 3, if he/she defects; if one player defects, the other player will still have a better payoff, 1 rather than 0, if he/she defects. Therefore, regardless of the other players' strategies, the player always selects defect and $\{(D, D)\}$ is a Nash equilibrium. Although cooperation will give each player a better payoff of 3, greediness leads to an inefficient outcome, so the outcome of the noncooperative static game can be less efficient. To overcome this low efficiency, there are many schemes in the literature. In the following discusion, we give two examples, namely, pricing- and referee-based approaches.

Because the individual user has no incentive to cooperate with the other users in the system and imposes harm to the other users' resources, the outcome of the game might not be optimal from the system point of view. To overcome this problem, pricing (or taxation) has been used as an effective tool both by economists and researchers in computer networks. The pricing technique is motivated by the following two objectives:

1. The revenue for the system is optimized.
2. The cooperation for resource usage is encouraged.

An efficient pricing mechanism can make distributed decisions compatible with the system efficiency obtained by centralized control. A pricing policy is called *incentive compatible* if pricing enforces a Nash equilibrium that achieves the system optimum. In [268], a policy is proposed by usage-based pricing, in which the price a user pays for using the resources is proportional to the amount of resources consumed by the user. Specifically, for the utility function in (9.1), the new utility with pricing is

$$u' = \frac{LRP_c}{Mp} - \alpha p, \tag{9.4}$$

where α is the price for transmitted power p and the price can be different for different users. It has been shown from [268] that the preceding utility function can achieve Pareto optimality.

The basic idea for the referee-based scheme is to introduce a referee for the non-cooperative game. The game may have multiple Nash equilibria. A referee is in charge of detecting these less efficient Nash equilibria and then changes the game rule to prevent the players from falling into undesirable game outcomes. It is worth mentioning that the noncooperative game is still played in a distributive way. The referee intervenes only when it is necessary. In [105, 106], the preceding idea is employed in a multicell OFDMA network to have an efficiently distributed resource allocation.

9.4   Dynamic/Repeated Game

When players interact by playing a similar stage game numerous times, the game is called a dynamic, or repeated, game. Unlike simultaneous games, players have at least some information about the strategies chosen by others and thus their play may be contingent on past moves. The cooperation among autonomous users can be encouraged by considering the long-term benefit and threat from others. In this section, we first describe how to represent a dynamic game. Then we discuss the information available to the players. We further our study for the properties of the dynamic games. Finally, two practical strategies, Tit-for-Tat and Cartel Maintenance, are proposed using repeated games.

First we define the concept of a sequential game.

DEFINITION 23 *A **sequential game** is one in which players make decisions (or select a strategy) following a certain predefined order and in which at least some players can observe the moves of players who preceded them. If no players observe the moves of previous players, then the game is simultaneous. If every player observes the moves of every other player who has gone before her, the game is one of perfect information. If some (but not all) players observe prior moves, when others move simultaneously, the game is one of imperfect information.*

For the information to a game, there are three concepts:

DEFINITION 24 *A **game** is one of **complete information** if all factors of the game are common knowledge. Specifically, each player is aware of all other players, the timing of the game, and the set of strategies and payoffs for each player.*

DEFINITION 25 *A sequential game is one of **perfect information** if only one player moves at a time and if each player knows every action of the players that moved before him at every point. Technically, every information set contains exactly one node. Intuitively, if it is my turn to move, I always know what every other player has done up to that point. All other games are games of imperfect information.*

DEFINITION 26 *A sequential game is one of **imperfect information** if a player does not know exactly what actions other players took up to that point. Technically, there exists at least one information set with more than one node. If every information set contains exactly one node, the game is one of perfect information. Intuitively, if it is my turn to move, I may not know what every other player has done up to now. Therefore I have to infer from their likely actions and from Bayes' rule which actions likely led to my current decision.*

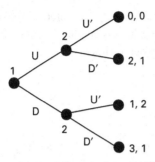

Figure 9.2 Extensive form (game tree) example.

Sequential games are represented by game trees (the extensive form).

DEFINITION 27 *A **game tree** (also called the **extensive form**) is a graphical representation of a sequential game. It provides information about the players, payoffs, strategies, and the order of moves. The game tree consists of nodes, which are points at which players can take actions, connected by vertices, that represent the actions that may be taken at that node. An initial (or root) node represents the first decision to make. Every set of vertices from the first node through the tree eventually arrives at a terminal node, representing an end to the game. Each terminal node is labeled with the payoffs earned by each player if the game ends at that node.*

To illustrate the idea, one example is shown in Figure 9.2. The game has two players: 1 and 2. The numbers by every nonterminal node indicate to which player that decision node belongs. The numbers by every terminal node represent the payoffs to the players (e.g., 2,1 represents a payoff of 2 to player 1 and a payoff of 1 to player 2). The labels by every edge of the graph are the name of the action that that edge represents.

The initial node belongs to player 1, indicating that that player moves first. Play according to the tree is as follows: Player 1 chooses between U and D; player 2 observes player 1's choice and then chooses between U′ and D′. The payoffs are as specified in the tree. There are four outcomes represented by the four terminal nodes of the tree: (U,U′), (U,D′), (D,U′), and (U,U′). The payoffs associated with each outcome respectively are as follows: (0,0), (2,1), (1,2), and (3,1).

One can solve a sequential game by using the concept of subgame perfect equilibrium.

DEFINITION 28 *A **subgame perfect** Nash equilibrium is an equilibrium such that players' strategies constitute a Nash equilibrium in every subgame of the original game. It may be found by backward induction, an iterative process for solving finite extensive form or sequential games. First, one determines the optimal strategy of the player who makes the last move of the game. Then the optimal action of the next-to-last moving player is determined taking the last player's action as given. The process continues in this way backwards in time until all players' actions have been determined.*

In the example shown in Figure 9.2, if player 1 plays D, player 2 will play U′ to maximize his payoff and so player 1 will receive only 1. However, if player 1 plays U, player 2 maximizes his payoff by playing D′ and player 1 receives 2. Player 1 prefers 2 to 1 and so will play U and player 2 will play D′ . This is the subgame perfect equilibrium.

Subgame perfect equilibria eliminate noncredible threats. A noncredible threat is a threat made by a player in a sequential game that will not be in the best interest for the player to carry out. The hope is that the threat is believed in which case there is no need to carry it out. Even though Nash equilibria may depend on noncredible threats, backward induction eliminates them.

For repeated games, the technical definition can be written as follows:

DEFINITION 29 *Let G be a static game and β be a discount factor. The T-period* **repeated game**, *denoted as G(T, β), consists of game G repeated T times. The payoff for such a game is given by*

$$V_i = \sum_{t=1}^{T} \beta^{t-1} \pi_i^t, \tag{9.5}$$

where π_i^t denotes the payoff to player i in period t. If T goes to infinity, then G(∞, β) is referred to as the infinitely repeated game. In this section, we employ the infinitely repeated game.

From Folk Theorem, we know that, in an infinitely repeated game, any feasible outcome that gives each player a better payoff than the Nash equilibrium can be obtained.

THEOREM 18 *Folk Theorem: Let $(e_1, \ldots, e_n)$ be the payoffs from a Nash equilibrium of G and let $(x_1, \ldots, x_n)$ be any feasible payoffs from G. If $x_i > e_i$ for every player i, then there exists an equilibrium (it is also subgame perfect) of G(∞, β) that attains $(x_1, \ldots, x_n)$ as the average payoff, provided that β is sufficiently close to 1.*

Now we know that, by using the repeated game, any feasible payoffs better than the Nash equilibrium can be obtained. The reason is that, with enough patience, a player's noncooperative behavior will be punished by the future revenge from other cooperative users, or, on the other hand, a player's cooperation can be rewarded in the future by others' cooperation. The remaining problem is how to define a good rule to achieve these better payoffs by enforcing cooperation among users. In what follows, we propose two approaches, namely Tit-for-Tat and Cartel Maintenance.

Tit-for-Tat is a type of trigger strategy in which a player responds in one period with the same action her opponent used in the last period. Many research works have been proposed by this method [10, 287]. The advantage of Tit-for-Tat is that it is easy to implement. However, there are some potential problems. The best response of a user is not the same action of the other opponent. The information of others' actions is hard to obtain. All these limit the possible applications for Tit-for-Tat. There are many less harsh and more optimal variation trigger strategies than the Tit-for-Tat. One of the optimal design criteria is Cartel Maintenance [238].

The basic idea for the Cartel Maintenance repeated game framework is to provide enough threat to greedy users so as to prevent them from deviating from cooperation. First, the cooperative point is obtained so that all users have better performances than those of noncooperative NEPs. However, if any user deviates from cooperation while others still play cooperatively, this deviating user has a better utility, whereas others have relatively worse utilities. If no rule is employed, the cooperative users will also have incentives to deviate. Consequently, the network deteriorates to the noncooperation results with inefficient performances. The proposed framework provides a mechanism so that the current defecting gains of the selfish user will be outweighed by future punishment strategies from other users. For any rational user, this threat of punishment prevents them from deviation, so the cooperation is enforced.

The proposed trigger strategy is a strategy to introduce punishment on the defecting users. In the trigger strategy, the players start with cooperation. Assume each user can observe the public information (e.g., the outcome of the game) P_t at time t. Examples of this public information can be the successful transmission rate and network throughput. Notice that such public information is mostly imperfect or simply partial information about the users' strategies. Here we assume a larger P_t stands for a higher cooperation level, resulting in higher performances for all users. Let the cooperation strategies be $\bar{\lambda} = [\lambda_1, \lambda_2, \ldots, \lambda_N]^\mathsf{T}$ and the noncooperative strategies be $\bar{s} = [s_1, s_2, \ldots, s_N]^\mathsf{T}$, respectively. The trigger–punishment game rule is characterized by three parameters: the optimal punishment time T, trigger threshold P^*, and the cooperation strategy $\bar{\lambda}$. The trigger–punishment strategy $(\bar{\lambda}, P^*, T)$ for distributed user i is given as follows:

1. User i plays the strategy of the cooperation phase, $\bar{\lambda}$, in period 0;
2. if the cooperation phase is played in period t and $P_t > P^*$, user i plays the cooperation phase in period $t + 1$;
3. if the cooperation phase is played in period t and $P_t < P^*$, user i switches to a punishment phase for $T - 1$ periods, in which the players play a static Nash equilibrium $\bar{s}$ regardless of the realized outcomes. At the Tth period, play returns to the cooperative phase.

Note that $\bar{s}$ generates the noncooperation outcome, which is much worse than that generated by the cooperation strategy $\bar{\lambda}$. Therefore the selfish users who deviate will have much lower utilities in the punishment phase. Moreover, the punishment time T is designed to be long enough to let all cheating gains of the selfish users be outweighed by the punishment. Therefore the users have no incentive to deviate from cooperation, because the users aim to maximize the long-run payoffs over time.

The rest of the problem is how to calculate the optimal parameters of $(\bar{\lambda}, P^*, T)$, i.e., how to construct a Cartel so that the benefit is optimized and incentive for deviation is eliminated. We define $\mathcal{P}_{\text{trig}} = \Pr(P_t < P^*)$, which is the trigger probability that the realization of public information is less than the trigger threshold. If we discuss the different future situations in (9.5), the expected payoff V_i is given as follows:

$$V_i(\bar{\lambda}, P^*, T) = \pi_i(\bar{\lambda}) + (1 - \mathcal{P}_{\text{trig}})\beta V_i(\bar{\lambda}, P^*, T) \qquad (9.6)$$

$$+ \mathcal{P}_{\text{trig}}[\sum_{t=1}^{T-1} \beta^t \pi_i(\bar{s}) + \beta^T V_i(\bar{\lambda}, P^*, T)], \forall i,$$

where $\pi_i(\bar{\lambda})$ and $\pi_i(\bar{s})$ are the cooperation and noncooperation payoffs, respectively. The first term on the right-hand side of (9.7) is the current expected payoff if cooperation is played, and the second term and third term are the payoffs for two different results depending on whether or not the punishment is triggered, respectively. Notice that V_i is a function of not only users' strategies $\bar{\lambda}$ but also the game parameters P^* and T. Our objective is to maximize the expected payoff V_i for each user while the optimal strategy yields a NEP for the proposed algorithm. To achieve the NEP given P^* and T, the optimal strategies of the repeated games can also be characterized by the first-order necessary conditions,

$$\frac{\partial V_i(\lambda_i, \lambda_{-i})}{\partial \lambda_i} = 0, \forall i. \tag{9.7}$$

If all users have the same utility and the game outcome is symmetric for all users, the solution λ^* of the first-order conditions is the same for all users. This solution is also a function of parameter P^* and T. To obtain the optimal P^* and T for maximizing the expected payoff V_i, we have the following differential equations:

$$\frac{\partial V_i(P^*, T)}{\partial P^*} = 0, \quad \frac{\partial V_i(P^*, T)}{\partial T} = 0, \forall i. \tag{9.8}$$

In general, (9.7) and (9.8) need to be solved by numerical methods. For a certain structure of the payoff function, we are able to derive the closed-form optimal configuration $\{\bar{\lambda}, P^*, T\}$.

9.5   Cooperative Game

A cooperative game is one in which players are able to make enforceable contracts. Hence it is not defined as the game in which players actually do cooperate, but as games in which any cooperation is enforceable by an outside party (e.g., a judge and police). There are two major components of cooperative game theory: the bargaining solution and coalition concepts.

The bargaining problem of cooperative game theory can be described as follows [95, 227, 343]: Let $\mathbf{K} = \{1, 2, \ldots, K\}$ be the set of players. Let $\mathbf{S}$ be a closed and convex subset of $\Re^K$ to represent the set of feasible payoff allocations that the players can get if they all work together. Let u_{min}^i be the minimal payoff that the ith player will expect; otherwise, he/she will not cooperate. Suppose $\{u_i \in \mathbf{S} | u_i \geq u_{min}^i, \forall i \in \mathbf{K}\}$ is a nonempty bounded set. Define $\mathbf{u}_{min} = (u_{min}^1, \ldots, u_{min}^K)$; then the pair $(\mathbf{S}, \mathbf{u}_{min})$ is called a K-person bargaining problem.

Within the feasible set $\mathbf{S}$, we define the notion of Pareto optimal as a selection criterion for the bargaining solutions.

DEFINITION 30 *The point $(u_1, \ldots u_K)$ is said to be **Pareto optimal** if and only if there is no other allocation u_i' such that $u_i' \geq u_i, \forall i$, and $u_i' > u_i, \exists i$, i.e., there exists no other allocation that leads to superior performance for some users without inferior performance for some other users.*

There might be an infinite number of Pareto optimal points. We need further criteria to select a bargaining result. A possible criterion is the fairness. One commonly used fairness criterion for wireless resource allocation is max–min [252], in which the performance of the user with the worst channel conditions is maximized. This criterion penalizes the users with good channels and as a result generates inferior overall system performance. Here, we use the criterion of the fairness Nash bargaining solution (NBS) [227]. The intuitive idea is that, after the minimal requirements are satisfied for all users, the rest of the resources are allocated proportionally to users according to their conditions. There exist many kinds of bargaining solutions [227]. Among them, the NBS provides a unique and fair Pareto optimal operation point under the following conditions.

DEFINITION 31 $\bar{\mathbf{u}}$ *is said to be a **Nash Bargaining Solution** in* $\mathbf{S}'$ *for* $\mathbf{u}_{\min}$, *i.e.,* $\bar{u} = \phi(\mathbf{S}, \mathbf{u}_{\min})$, *if the following axioms are satisfied:*

1. *Individual rationality:* $\bar{u}_i \geq \mathbf{u}_{\min}^i, \forall i$.
2. *Feasibility:* $\bar{\mathbf{u}} \in \mathbf{S}$.
3. *Pareto optimality: For every* $\hat{\mathbf{u}} \in \mathbf{S}$, *if* $u_i \geq \bar{u}_i, \forall i$, *then* $\hat{u}_i = \bar{u}_i, \forall i$.
4. *Independence of irrelevant alternatives: If* $\bar{\mathbf{u}} \in \mathbf{S}' \subset \mathbf{S}$, $\bar{\mathbf{u}} = \phi(\mathbf{S}, \mathbf{u}_{\min})$, *then* $\bar{\mathbf{u}} = \phi(\mathbf{S}', \mathbf{u}_{\min})$.
5. *Independence of linear transformations: For any linear scale transformation* ψ, $\psi[\phi(\mathbf{S}, \mathbf{u}_{\min})] = \phi[\psi(\mathbf{S}), \psi(\mathbf{u}_{\min})]$.
6. *Symmetry: If* $\mathbf{S}$ *is invariant under all exchanges of agents,* $\phi_j(\mathbf{S}, \mathbf{u}_{\min}) = \phi_{j'}(\mathbf{S}, \mathbf{u}_{\min}), \forall j, j'$.

Axioms 4–6 are called axioms of fairness. The irrelevant alternative axiom asserts that eliminating the feasible solutions that would not have been chosen shall not affect the NBS solution. Axiom 5 asserts that the bargaining solution is scale invariant. The symmetry axiom asserts that, if the feasible ranges for all users are completely symmetric, then all users have the same solution.

The following optimization has been shown to have the NBS that satisfies the preceding axioms [227].

THEOREM 19 ***Existence of NBS:*** *There is a solution function* $\phi(\mathbf{S}, \mathbf{u}_{\min})$ *that satisfies all six axioms in Definition 31. And this solution satisfies*

$$\phi(\mathbf{S}, \mathbf{u}_{\min}) \in \arg \max_{\bar{u} \in S, \bar{u}_i \geq u_{\min}^i, \forall i} \prod_{i=1}^{K} \left(\bar{u}_i - u_{\min}^i \right). \tag{9.9}$$

Until now, we have discussed how to play the cooperative game by bargaining. Other analysis tools for the game are coalition, core, Shapley function, and nucleolus. In the following discussion, we explain these concepts and give examples.

DEFINITION 32 *A **coalition** S is defined to be a subset of the total set of player N,* $S \in N$. *The users in a coalition try to cooperate with each other. The **coalition form*** *of a game is given by the pair (N, v), where v is a real value function, called the* **characteristic function**. *$v(S)$ is the value of the cooperation for coalition S with the following properties:*

1. $v(\emptyset) = 0$,
2. (superadditivity) if S and T are disjoint coalitions $(S \cap T = \emptyset)$, then $v(S) + v(T) \leq v(S \cup T)$.

The coalition states the benefit obtained via a cooperation agreement. But we still need to study how to divide the benefit to the cooperative users. This division should also be stable because no coalition shall have the incentive and power to upset the cooperative agreement. The set of such division of v is called the core, defined as follows:

DEFINITION 33 *A payoff vector $\mathbf{x} = (x_1, \ldots, x_N)$ is said to be **group rational** or **efficient** if $\sum_{i=1}^{N} x_i = v(N)$. A payoff vector $\mathbf{x}$ is said to be **individually rational** if the user can obtain the benefit no less than acting alone, i.e., $x_i \geq v(\{i\})$, $\forall i$. An **imputation** is a payoff vector satisfying the preceding two conditions.*

DEFINITION 34 *An imputation $\mathbf{x}$ is said to be unstable through a coalition S if $v(S) > \sum_{i \in S} x_i$, i.e., the users have incentive for coalition S and upset the proposed $\mathbf{x}$. The set C of a stable imputation is called the core, i.e.,*

$$C = \{\mathbf{x} : \sum_{i \in N} x_i = v(N) \text{ and } \sum_{i \in S} x_i \geq v(S), \forall S \subset N\}. \tag{9.10}$$

Core gives a reasonable set of possible shares. A combination of shares is in a core if there exists no subcoalition in which its members may gain a higher total outcome than the share of concern. If the share is not in a core, some members may be frustrated and may think of leaving the whole group with some other members and forming a smaller group.

To illustrate the idea of core, we give the following example. Suppose the game has the following characteristic functions:

$$v(\emptyset) = 0, v(\{1\}) = 1, v(\{2\}) = 0, v(\{3\}) = 1, \tag{9.11}$$

$$v(\{1, 2\}) = 4, v(\{1, 3\}) = 3, v(\{2, 3\}) = 5, v(\{1, 2, 3\}) = 8.$$

By using $v(\{2, 3\}) = 5$, we can eliminate the payoff vector [such as $(4, 3, 1)$], because user 2 and user 3 can achieve better payoff by forming a coalition themselves. Using the same analysis, the final core of the game is $(3,4,1)$, $(3,3,2)$, $(3,2,3)$, $(3,1,4)$, $(2,5,1)$, $(2,4,2)$, $(2,3,3)$, $(2,2,4)$, $(1,5,2)$, $(1,4,3)$, and $(1,3,4)$.

Core concept defines the stability of an allocation of payoff. Next, we study each individual player's power in the coalition by defining a value called the Shapley function.

DEFINITION 35 *A **Shapley function** ϕ is a function that assigns to each possible characteristic function v a real number, i.e.,*

$$\phi(v) = [\phi_1(v), \phi_2(v), \ldots, \phi_N(v)], \tag{9.12}$$

where $\phi_i(v)$ represents the worth or value of player i in the game. The Shapley axioms for $\phi(v)$ are

1. **Efficiency:** $\sum_{i \in N} \Phi_i(v) = v(N)$.
2. **Symmetry:** If i and j are such that $v(S \bigcup \{i\}) = v(S \bigcup \{j\})$ for every coalition S not containing i and j, then $\phi_i(v) = \phi_j(v)$.
3. **Dummy axiom:** If i is such that $v(S) = v(S \bigcup \{i\})$ for every coalition S not containing i, then $\phi_i(v) = 0$.
4. **Additivity:** If u and v are characteristic functions, then $\phi(u + v) = \phi(v + u) = \phi(u) + \phi(v)$.

It can be proved that there exists a unique function ϕ satisfying the Shapley axioms. To calculate the Shapley function, suppose we form the grand coalition by entering the players into this coalition one at a time. As each player enters the coalition, he receives the amount by which his entry increases the value of the coalition he enters. The amount a player receives by this scheme depends on the order in which the players are entered. The Shapley value is just the average payoff to the players if the players are entered in completely random order, i.e.,

$$\phi_i(v) = \sum_{S \subset N, i \in S} \frac{(|S| - 1)!(N - |S|)!}{N!}[v(S) - v(S - \{i\})]. \qquad (9.13)$$

For the example in (9.11), it can be shown that the Shapley value is $\phi = (14/6, 17/6, 17/6)$.

Another concept for multiple cooperative games is nucleolus. For a fixed characteristic function, an imputation **x** is found such that the worst inequity is minimized, i.e., for each coalition S and its associated dissatisfaction, an optimal imputation is calculated to minimize the maximum dissatisfaction. First we define the concept of excess that measures the dissatisfactions.

DEFINITION 36 *The measure of the inequity of an imputation* **x** *for a coalition S is defined as the excess:*

$$e(\mathbf{x}, S) = v(S) - \sum_{j \in S} x_j. \qquad (9.14)$$

Obviously, any imputation **x** is in the core if and only if all its excesses are negative or zero.

Among all allocations, a kernel is a fair allocation, defined as follows:

DEFINITION 37 *A **kernel** of v is the set of all allocations* **x** *such that*

$$\max_{S \subseteq N - j, i \in S} e(\mathbf{x}, S) = \max_{T \subseteq N - i, j \in T} e(\mathbf{x}, T). \qquad (9.15)$$

If players i and j are in the same coalition, then the highest excess that i can make in a coalition without j is equal to the highest excess that j can make in a coalition without i.

Finally, we define a nucleolus as follows:

DEFINITION 38 *Nucleolus is the allocation **x** that minimizes the maximum excess:*

$$\mathbf{x} = \arg\min_{\mathbf{x}}[\max e(\mathbf{x}, S), \ \forall S]. \tag{9.16}$$

The nucleolus has the following property: The nucleolus of a game in coalitional form exists and is unique. The nucleolus is group rational, individually rational, and satisfies the symmetry axiom and the dummy axiom. If the core is not empty, the nucleolus is in the core and kernel. In other words, the nucleolus is the best allocation with the min–max criteria.

9.6 Auction Theory and Mechanism Design

In this section, we first discuss the basics of auction theory. Then mechanism design is illustrated to show how to control the game outcome by cleverly designing the game rule. Finally, some applications of auction theory and mechanism design are explained.

Auction theory is important for practical, empirical, and theoretical reasons. First, a large amount of wireless networking and resource-allocation problems can be formulated as auction theory, for example, the routing problem for self-interested users [11]. Second, the auction theory has a simple game setup and many theoretical results are available for analysis. The following are the definitions of auction.

DEFINITION 39 *A market mechanism in which an object, service, or set of objects is exchanged on the basis of bids submitted by participants.* **Auction** *provides a specific set of rules that will govern the sale or purchase (procurement auction) of an object to the submitter of the most favorable bid. The specific mechanisms of the auction include first- and second-price auctions and English and Dutch auctions.*

Some basic types of auctions are listed as follows.

DEFINITION 40 **First-Price Auction:** *An auction in which the bidder who submitted the highest bid is awarded the object being sold and pays a price equal to the amount bid. Alternatively, in a procurement auction, the winner is the bidder who submits the lowest bid and is paid an amount equal to his or her bid. In practice, first-price auctions are either sealed bid, in which bidders submit bids simultaneously, or Dutch. Alternatively, second-price auctions also award the object to the highest bidder, but the payment is equal to the second-highest bid. Unlike second-price auctions, in which bidding one's true value is a dominant strategy, in first-price auctions, bidders shade their bids below their true value.*

DEFINITION 41 **Second-Price Auction:** *An auction in which the bidder who submitted the highest bid is awarded the object being sold and pays a price equal to the second highest amount bid. Alternatively, in a procurement auction, the winner is the bidder who submits the lowest bid, and is paid an amount equal to the next lowest submitted bid. In practice, second-price auctions are either sealed bid, in which bidders submit* *bids simultaneously, or English auctions, in which bidders continue to raise each other's bids until only one bidder remains. The theoretical nicety of second-price auctions, first pointed out by William Vickrey, is that bidding one's true value is a dominant strategy. Alternatively, first-price auctions also award the object to the highest bidder, but the payment is equal to the amount bid.*

DEFINITION 42 **English Auction:** *A type of sequential second-price auction in which an auctioneer directs participants to beat the current, standing bid. New bids must increase the current bid by a predefined increment. The auction ends when no participant is willing to outbid the current standing bid. Then the participant who placed the current bid is the winner and pays the amount bid. A second-price auction is also known as a Vickrey auction after William Vickrey, who first described it and pointed out that bidders have a dominant strategy to bid their true values. Although the highest bidder pays the amount bid, an English auction is termed second price because the winning bidder need outbid only the next highest bidder by the minimum increment. Thus the winner, effectively, pays an amount equal to (slightly higher than) the second highest bid.*

DEFINITION 43 **Dutch Auction:** *A type of first-price auction in which a "clock" initially indicates a price for the object for sale substantialy higher than any bidder is likely to pay. Then the clock gradually decreases the price until a bidder "buzzes in" or indicates his or her willingness to pay. The auction is then concluded, and the winning bidder pays the amount reflected on the clock at the time he or she stopped the process by buzzing in. These auctions are named after a common market mechanism for selling flowers in Holland, but also reflect stores successively reducing prices on sale items.*

DEFINITION 44 **Japanese Auction:** *A type of sequential second-price auction, similar to an English auction in which an auctioneer regularly raises the current price. Participants must signal at every price level their willingness to stay in the auction and pay the current price. Thus, unlike an English auction, each participant must bid at each level to stay in the auction. The auction concludes when only one bidder indicates a willingness to stay in. This auction format is also known as the button auction.*

To implement auction theory in wireless networking and resource allocation, the credit-based system is usually proposed. The individual user can select to pay for some kind of services such as a route. The payment can be implemented via a certain central "bank" system. However, this requires more control than the other game-theory approaches, such as noncooperative games. Moreover, to achieve different design goals such as the network total benefit, the auction method shall be designed according to different available information. Mechanism design is the tool for game and auction design.

Mechanism design is the subfield of microeconomics and game theory that considers how to implement good system-wide solutions to problems that involve multiple self-interested agents, each with private information about their preferences.

The goal is to achieve a social-choice function implemented in distributed systems with private information and rational agents. The design criteria can be different as follows:

1. Efficiency: Select the outcome that maximizes total utility.
2. Fairness: Select the outcome that minimizes the variance in utility.
3. Revenue maximization: Select the outcome that maximizes revenue to a seller (or more generally, utility to one of the agents).
4. Budget balance: Implement outcomes that have balanced transfers across agents.
5. Pareto optimality.

There are some basic concepts and theorems for mechanism design. The Revelation Principle states that under quite weak conditions any mechanism can be transformed into an equivalent incentive-compatible direct-revelation mechanism [167] that implements the same social-choice function. One of the most important families of mechanisms is Vickrey–Clarke–Groves mechanisms, which are the only mechanisms that are allocatively efficient and strategy proof [167]. A detailed discussion of mechanism design can be found at [231].

9.7 Summary

In this chapter, we discussed four different types of games, namely, the noncooperative game, repeated game, cooperative game, and auction theory. The basic concepts are listed, and simple examples are illustrated. The goal is to let the readers understand the basic problems and basic approaches. As a result, the authors hope the readers can formulate the problem and find solutions in their research areas.

Depending on the nature of the different approaches, there are different possible applications. If the information is strictly limited to local information, the noncooperative game might be the only choice for each individual to play. However, such a game might have a very low-efficiency outcome. To overcome the problem, pricing or referee approaches are proposed. If the users care about long-term benefit, the repeated game can be employed to enforce cooperation by the threat of the future punishment from others. Tit-for-Tat and Cartel Maintenance approaches are proposed. If the signaling is allowed and the problem is about the contract on efficient usage of limited resources, cooperative game theory can be employed to have mutual benefits. Finally, if the outcome is binary and signaling is allowed, auction theory and mechanism design can be implemented for resource allocation.

There are other game-theory approaches. For example, if the information is not deterministic but statistical, statistic games [204, 276] can be employed to solve the problem and analyze the stability. Even if the distributed individual users have the incentive to cooperate, they might not know what the best point for cooperation is. Under this condition, reinforcement learning or artificial intelligence approaches [302] can also be studied to find the better cooperation point. Overall, depending on different applications, various game-theory approaches can be employed to optimize the performances. The game-theory approaches bring a new perspective to the traditional wireless resource allocation and networking optimization.

We hope readers know how to formulate the game-theory problems in their research areas after reading this chapter. Our purpose is not to teach game theory, so the mathematics is limited to necessity. For details on game theory, readers can refer to [80, 167, 227, 388].

Part III

Advanced Topics

10 Resource Allocation with Antenna-Array Processing

10.1 Introduction

Because different users have different channels and locations at different times, resource allocation can take advantage of the time, frequency, multiuser, and spatial diversity. Specifically for spatial diversity, transceivers employ antenna arrays and adjust their beam patterns such that they have good channel gain toward the desired directions, whereas the aggregate interference power is minimized at their output. Antenna-array processing techniques such as beam forming can be applied to receive and transmit multiple signals that are separated in space. Hence multiple cochannel users can be supported in each cell to increase the capacity by exploring the spatial diversity. Many works have been reported in the literature. Traditional beam formers such as minimum-mean-square-error (MMSE) and minimum-variance-distortion-response (MVDR) methods have been commonly employed [130]. Many joint power-control and beam forming algorithms are proposed in [183, 203, 248, 250]. The application of antenna arrays has been proposed in [214] to increase the network capacity in CDMA systems.

In this chapter, we first consider the resource-allocation example that jointly considers the antenna-array processing. We consider a system with beam forming capabilities in the receiver. An iterative algorithm is proposed to jointly update the transmission powers and the beam-former weights so that it converges to the joint optimal beam-forming and transmission power vector. The algorithm is distributed and uses only local interference measurements. In an uplink transmission scenario it is shown how base assignment can be incorporated in addition to beam forming and power control such that a globally optimum solution is obtained. The network capacity and the saving in mobile power are then evaluated through numerical study.

The second example in this chapter focuses on blind beam forming over multiple users in the wireless networks. As a majority of communication systems often struggle with a limited bandwidth constraint, it is desirable for the receiver with multiple antennas to steer to the desired direction and to estimate the transmit signals without consuming much channel bandwidth. By eliminating the training sequence overhead used for estimation and maximizing the channel capacity for information transmission, blind estimation and beam forming [58, 182, 184, 222, 266, 303, 304, 310] offer a bandwidth-efficient solution to signal separation and estimation. The importance of blind techniques also lies in the practical need for some communication receivers to equalize unknown channels without the assistance and the expense of training sequences.

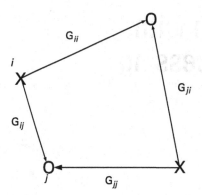

Figure 10.1 A pair of cochannel links i and j is depicted.

10.2   Joint Optimization with Antenna-Array Processing

The interference reduction capability of antenna arrays and the power-control algorithms have been considered separately as means to increase the capacity in wireless communication networks. In this section, we consider the joint problem of power control and beam forming. We consider a set of cochannel links, such as a set of cochannel uplinks in a cellular network, in which only receivers employ antenna arrays. An algorithm is provided for computing the transmission powers and the beam-forming weight vectors, such that a target SINR is achieved for each link (if it is achievable) with minimal transmission power. The algorithm is decentralized and amenable to a distributed implementation. It operates as follows: For a fixed power allocation, each base station (BS) maximizes the SINR using the MVDR beam former. Next, the mobile powers are updated to reduce the cochannel interference (CCI). This operation is done iteratively until the vector of transmitter powers and the weight coefficients of the beam formers converge to the jointly optimal value. For the case that each transmitter can select its BS among a set of possible options, the algorithm easily extends to find the joint optimum power, BS, and beam forming.

For the system model, a set of M transmitter–receiver pairs that share the same channel is considered. The shared channel could be a frequency band in FDMA, a time slot in TDMA, or even CDMA spreading codes. The link gain between transmitter i and receiver j is denoted by G_{ij}, and the ith transmitter power by P_i. For an isotropic antenna with unity gain in all directions, the signal power received at receiver i from transmitter j is $G_{ji}P_j$, as illustrated in Figure 10.1. It is assumed that transmitter i communicates with receiver i. Hence the desired signal at receiver i is equal to $G_{ii}P_i$, and the interfering signal power from other transmitters to receiver i is $I_i = \sum_{j \neq i} G_{ji}P_j$. If we neglect thermal noise, the carrier-to-interference ratio (CIR) at the ith receiver is given by

$$\Gamma_i = \frac{G_{ii}P_i}{\sum_{j \neq i} G_{ji}P_j}. \tag{10.1}$$

The quality of the link from transmitter i to receiver j depends solely on Γ_i. The quality is acceptable if Γ_i is above a certain threshold γ_0, the *minimum protection ratio*. The minimum protection ratio is determined based on the signaling scheme and the link-quality requirements (target BER). Hence, for acceptable link quality,

$$\frac{G_{ii} P_i}{\sum_{j \neq i} G_{ji} P_j} \geq \gamma_0. \tag{10.2}$$

In matrix form, (10.2) can be written as

$$\mathbf{P} \geq \gamma_0 \mathbf{F} \mathbf{P}, \tag{10.3}$$

where $\mathbf{P} = [P_1, P_2, \ldots, P_M]^T$ is the power vector and $\mathbf{F}$ is a nonnegative matrix defined as

$$[\mathbf{F}]_{ij} = \begin{cases} 0 & \text{if } j = i \\ \frac{G_{ji}}{G_{ii}} > 0 & \text{if } j \neq i \end{cases}. \tag{10.4}$$

The objective of a power-control scheme is to maintain the link quality by keeping the CIR above the threshold γ_0, that is, to adjust the power vector $\mathbf{P}$ such that (10.3) is satisfied. This problem has been studied recently in [2, 125]. Given that $\mathbf{F}$ is irreducible, it is known by Perron–Frobenius theorem, in which the maximum value of γ_0 for which there exists a positive $\mathbf{P}$ such that (10.3) is satisfied is $\frac{1}{\rho(\mathbf{F})}$, where $\rho(\mathbf{F})$ is the spectral radius of $\mathbf{F}$ [82]. According to this theorem, the power vector that satisfies (10.3) is the eigenvector corresponding to $\rho(\mathbf{F})$ and is positive. Now we consider thermal noise at the receivers. The SINR at the ith receiver is then expressed as

$$\Gamma_i = \frac{G_{ii} P_i}{\sum_{j \neq i} G_{ji} P_j + N_i}, \tag{10.5}$$

where N_i is the noise power at the ith receiver. The requirement for acceptable link quality is again

$$\Gamma_i \geq \gamma_0, \quad 1 \leq i \leq M, \tag{10.6}$$

or in matrix form,

$$[\mathbf{I} - \gamma_0 \mathbf{F}]\mathbf{P} \geq \mathbf{u}, \tag{10.7}$$

where $\mathbf{I}$ is an $M \times M$ identity matrix, and $\mathbf{u}$ is an element-wise positive vector with elements u_i defined as

$$u_i = \frac{\gamma_0 N_i}{G_{ii}}, \quad 1 \leq i \leq M. \tag{10.8}$$

The SINR threshold γ_0 is achievable if there exists at least one solution vector $\mathbf{P}$ that satisfies (10.7). The power-control problem is defined as follows:

$$\text{minimize} \quad \sum_i P_i, \tag{10.9}$$

$$\text{subject to} \quad [\mathbf{I} - \gamma_0 \mathbf{F}]\mathbf{P} \geq \mathbf{u}.$$

It can be shown that if the spectral radius of $\mathbf{F}$ is less than $\frac{1}{\gamma_0}$, the matrix $\mathbf{I} - \gamma_0 \mathbf{F}$ is invertible and positive [82]. In this case the power vector

$$\hat{\mathbf{P}} = [\mathbf{I} - \gamma_0 \mathbf{F}]^{-1} \mathbf{u} \tag{10.10}$$

solves the optimization problem.

A centralized power-control algorithm [91, 351] solves (10.10) by requiring all link gains in the network and noise levels at receivers. In [79, 352], a decentralized solution to the power-control problem is proposed that solves (10.10) by performing the following iterations:

$$P_i^{n+1} = \frac{\gamma_0}{G_{ii}} \left(\sum_{j \neq i} G_{ji} P_j^n + N_i \right) = \frac{\gamma_0}{G_{ii}} I_i, \tag{10.11}$$

where P_i^n is the ith mobile power at the nth iteration step. The right-hand side of (10.11) is a function of the interference at the ith receiver, denoted by I_i, as well as the link gain between each receiver and its transmitter (G_{ii}). That is, there is no need to know all the existing path gains and transmitter power in order to update the power. At each iteration, transmitters update their power based on the interference measured at the receivers and the link gain between each transmitter and its own receiver. The link gain can be measured from the information sent in the control channel. It has been shown in [79, 352] that, starting from any arbitrary power vector, this solution converges to the optimal solution $\hat{\mathbf{P}}$.

An adaptive antenna array consists of a set of antennas, designed to receive signals radiating from some specific directions and attenuate signals radiating from other directions of no interest. The outputs of array elements are weighted and added by a beam former as shown in Figure 10.2 to produce a directed main beam and adjustable nulls. To reject the interference, the beam former has to place its nulls in the directions of sources of interference, and steer to the direction of the target signal by maintaining constant gain at this direction. A sample antenna-array pattern, which is depicted in Figure 10.3, shows this effect.

Now consider a cochannel set consisting of M transmitter and receiver pairs, and assume that antenna arrays with K elements are used at the receivers. Denote the array response to the direction of arrival θ by $\mathbf{v}(\theta)$, defined as $\mathbf{v}(\theta) = [v^1(\theta), v^2(\theta), \ldots, v^K(\theta)]$, where $v^k(\theta)$ is the response of the kth antenna element at the direction θ. We consider multipath channels with negligible delay spreads. That is, the propagation delays in different paths are much smaller than a fraction of a symbol. Also we assume slow-fading channels in which the channel response can be assumed constant over several symbol intervals. Under the preceding assumptions the received vector at the ith array can be written as

$$\mathbf{x}_i(t) = \sum_{j=1}^{M} \sqrt{P_j G_{ji}} \sum_{l=1}^{L} \alpha_{ji}^l \mathbf{v}_j(\theta_l) s_j(t - \tau_j) + \mathbf{n}_i(t), \tag{10.12}$$

where $s_j(t)$ is the message signal transmitted from the jth user, τ_j is the corresponding time delay, $\mathbf{n}_i(t)$ is the thermal-noise vector at the input of antenna array at the ith

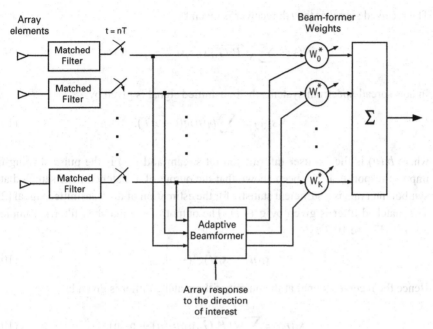

Figure 10.2 Antenna array and beam former.

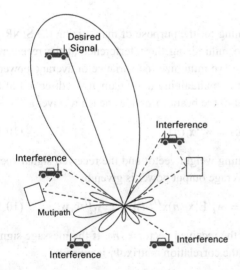

Figure 10.3 Sample antenna-array pattern.

receiver, and P_j is the power of the jth transmitter. $\mathbf{v}_j(\theta_l)$ is the response of the jth receiver array to the direction θ_l. The attenuation that is due to shadowing in the lth path is denoted by α_{ji}^l. Define the $K \times 1$ vector $\mathbf{a}_{ji}$, called the *spatial signature* or *array response* of the ith antenna array to the jth source, as

$$\mathbf{a}_{ji} = \sum_{l=1}^{L} \alpha_{ji}^l \mathbf{v}_j(\theta_l). \tag{10.13}$$

The received signal at the ith receiver is given by

$$\mathbf{x}_i(t) = \sum_{j=1}^{M} \sqrt{P_j G_{ji}} \mathbf{a}_{ji} s_j(t - \tau_j) + \mathbf{n}_i(t). \tag{10.14}$$

In non-spread-spectrum systems the transmitted signal is given by

$$s_i(t) = \sum_n b_i(n) g(t - nT), \tag{10.15}$$

where $b_i(n)$ is the ith user information bit stream and $g(t)$ is the pulse-shaping filter impulse response. It has been shown that the output of a matched filter sampled at the symbol intervals is a sufficient statistic for the estimation of the transmitted signal [239]. The matched filter is given by $g^*(-t)$. The output of the matched filter is sampled at $t = nT$ (Figure 10.2):

$$\mathbf{x}_i(n) = \mathbf{x}_i(t) g^*(-t)|_{t=nT}. \tag{10.16}$$

Hence the received signal at the output of the matched filter is given by

$$\mathbf{x}_i(n) = \sum_{j=1}^{M} \sqrt{P_j G_{ji}} \mathbf{a}_{ji} b_j(n) + \mathbf{n}_i(n), \tag{10.17}$$

where $\mathbf{n}_i(n) = \mathbf{n}_i(t) g^*(-t)|_{t=nT}$.

Consider the problem of beam forming for the purpose of maximizing the SINR for a specific link, which is equivalent to minimizing the interference at the receiver of that link. To minimize the interference, we minimize the variance or average power at the output of the beam former subject to maintaining unity gain at the direction of the desired signal. We can write the output of the beam former at the ith receiver as

$$\mathbf{e}_i(n) = \mathbf{w}_i^H \mathbf{x}_i(n), \tag{10.18}$$

where $\mathbf{w}_i$ and $\mathbf{x}_i(n)$ are the beam-forming weight vector and the received signal vector at the ith receiver, respectively. The average output power is given by

$$\mathcal{E}_i = \mathrm{E}\{\mathbf{w}_i^H \mathbf{x}_i(n) \mathbf{x}_i^H(n) \mathbf{w}_i\} = \mathbf{w}_i^H \mathrm{E}\{\mathbf{x}_i(n) \mathbf{x}_i^H(n)\} \mathbf{w}_i = \mathbf{w}_i^H \Phi_i \mathbf{w}_i, \tag{10.19}$$

where Φ_i is the correlation matrix of the received vector $\mathbf{x}_i(n)$. If the message signals $s_j(t)$ are uncorrelated and zero mean, the correlation matrix Φ_i is given by

$$\Phi_i = \sum_{j \neq i} P_j G_{ji} \mathbf{a}_{ji} \mathbf{a}_{ji}^H + N_i \mathbf{I} + P_i G_{ii} \mathbf{a}_{ii} \mathbf{a}_{ii}^H$$
$$= \Phi_{in} + P_i G_{ii} \mathbf{a}_{ii} \mathbf{a}_{ii}^H, \tag{10.20}$$

where

$$\Phi_{in} = \sum_{j \neq i} P_j G_{ji} \mathbf{a}_{ji} \mathbf{a}_{ji}^H + N_i \mathbf{I} \tag{10.21}$$

is the correlation matrix of unwanted signals and N_i is the noise power at the input of each array element. Combining (10.19) and (10.20), we obtain the received signal plus interference power as a function of weight vector $\mathbf{w}_i$:

$$\mathcal{E}_i = P_i G_{ii} + \sum_{j \neq i} P_j G_{ji} \mathbf{w}_i^H \mathbf{a}_{ji} \mathbf{a}_{ji}^H \mathbf{w}_i + N_i \mathbf{w}_i^H \mathbf{w}_i. \tag{10.22}$$

Here we use the fact that the gain at the direction of interest is unity, i.e, $\mathbf{w}_i^H \mathbf{a}_{ii} = 1$. The first term in (10.22) is the received power from the signal of interest, whereas the other terms are related to the interference and noise. That is, the total interference is written as

$$I_i = \sum_{j \neq i} G_{ji} \mathbf{w}_i^H \mathbf{a}_{ji} \mathbf{a}_{ji}^H \mathbf{w}_i P_j + N_i \, \mathbf{w}_i^H \mathbf{w}_i. \tag{10.23}$$

The goal of beam forming is to find a weight vector $\mathbf{w}_i$ that minimizes the interference I_i subject to $\mathbf{w}_i^H \mathbf{a}_{ii} = 1$. It can be shown that the unique solution to this problem is given by [211]

$$\hat{\mathbf{w}}_i = \frac{\Phi_{in}^{-1} \mathbf{a}_{ii}}{\mathbf{a}_{ii}^H \Phi_{in}^{-1} \mathbf{a}_{ii}}. \tag{10.24}$$

The antenna gain for the signal of interest is unity. As a result, the desired signal is unaffected by beam forming, and the SINR at the ith receiver is then given by

$$\Gamma_i = P_i G_{ii} \mathbf{a}_{ii}^H \Phi_{in}^{-1} \mathbf{a}_{ii}. \tag{10.25}$$

In a spread-spectrum system, the message signal is given by

$$s_i(t) = \sum_n b_i(n) c_i(t - nT), \tag{10.26}$$

where $c(t)$ is the spreading sequence. The matched filter in a spread-spectrum receiver is given by $c^*(-t)$. The received signal, sampled at the output of matched filter, is expressed as

$$y_i(n) = \int_{(n-1)T}^{nT} c_i(t - nT - \tau_i) \left[\sum_j \sqrt{P_j G_{ji}} \sum_m b_j(m) c_j(t - mT - \tau_j) \mathbf{a}_{ji} + \mathbf{n}_i(t) \right] dt. \tag{10.27}$$

We assume the signature sequences of the interfering users appear as mutually uncorrelated noise. The correlation matrix of the signal at the output of correlator is then given by [299]

$$\Phi_i = E\{\mathbf{y}_i(n) \mathbf{y}_i^H(n)\} = \sum_{j \neq i} P_j G_{ji} \mathbf{a}_{ji} \mathbf{a}_{ji}^H + N_i \mathbf{I} + L P_i G_{ii} \mathbf{a}_{ii} \mathbf{a}_{ii}^H$$

$$= \Phi_{in} + L P_i G_{ii} \mathbf{a}_{ii} \mathbf{a}_{ii}^H, \tag{10.28}$$

where L is the processing gain and Φ_{in} is defined as in (10.21). The optimum beamforming weight vector is similarly given by (10.24), and the maximum SNR can be written as

$$\Gamma_i = LP_i G_{ii} \mathbf{a}_{ii}^H \Phi_{in}^{-1} \mathbf{a}_{ii}. \tag{10.29}$$

Equations (10.25) and (10.29) are similar, but the latter includes the processing gain. For simplicity of notation henceforth we assume the processing gain is absorbed in Γ_i. Therefore (10.25) can be used to express the SINR in both cases.

To calculate the received power for transmitter j, we have to multiply the transmitter power by the antenna power gain in addition to the propagation path gain, i.e.,

$$G_{ji} G_{a_i}(\mathbf{w}_i, \mathbf{a}_{ji}),$$

where $G_{a_i}(\mathbf{w}_i, \mathbf{a}_{ji}) = |\mathbf{w}_i^H \mathbf{a}_{ji}|^2$. Then the maximum SINR at the ith receiver can be written as

$$\Gamma_i = \frac{P_i G_{ii}}{\sum_{j \neq i} G_{ji} G_{a_i}(\hat{\mathbf{w}}_i, \mathbf{a}_{ji}) P_j + N_i \, \hat{\mathbf{w}}_i^H \hat{\mathbf{w}}_i}, \tag{10.30}$$

where it is assumed that the array response to the source of interest, given by (10.13), is known. Knowing the response vector $\mathbf{v}_j(\theta)$ and the *direction of arrival* (DOA) for the signal of interest and its multipaths, we can calculate the array response from (10.13). In wireless networks usually the number of cochannels and multipath signals are much larger than the number of array elements. As a result, conventional DOA estimation methods like ESPRIT and MUSIC are not applicable. However, there exist some schemes that can be used to estimate the array response in non-spread-spectrum [226, 341] and spread-spectrum systems [299], without the need to estimate the DOA. Further, as we will see later when we use a training sequence, there is no need to estimate the array response.

The level of CCI at each receiver depends both on the gain between interfering transmitters and receivers and also on the level of transmitter power, i.e., the optimal beam-forming vector may vary for different power. Hence beam forming and power control should be considered jointly.

In the joint power-control and beam-forming problem the objective is to find the optimal weight vector and power allocations such that the SINR threshold is achieved by all links, while each transmitter keeps the transmission power at the minimum required level to reduce the interference to other users. The SINR at the ith receiver is given by

$$\Gamma_i = \frac{P_i G_{ii}}{\sum_{j \neq i} G_{ji} G_{a_i}(\mathbf{w}_i, \mathbf{a}_{ji}) P_j + N_i \, \mathbf{w}_i^H \mathbf{w}_i}. \tag{10.31}$$

The optimization problem now is defined as

$$\min_{\mathbf{W}, \mathbf{P}} \sum_{i=1}^{M} P_i,$$

$$\text{subject to} \quad \Gamma_i \geq \gamma_i, \quad (i = 1, 2, \ldots, M), \tag{10.32}$$

where $\mathbf{W} = \{\mathbf{w}_1, \mathbf{w}_2, \ldots, \mathbf{w}_M\}$ is a set of beam-forming vectors and γ_i is the minimum protection ratio for the ith link. This constraint can be presented in matrix form as

$$[\mathbf{I} - \mathbf{F}^w]\mathbf{P} \geq \mathbf{u}^w, \tag{10.33}$$

where

$$[\mathbf{F}^w]_{ij} = \begin{cases} 0 & \text{if } j = i \\ \frac{\gamma_i G_{ji} G_{a_i}(\mathbf{w}_i, \mathbf{a}_{ji})}{G_{ii}} > 0 & \text{otherwise} \end{cases}, \tag{10.34}$$

and $\mathbf{u}^w$ is an element-wise positive vector with elements u_i defined as

$$u_i^w = \frac{\gamma_i N_i \mathbf{w}_i^H \mathbf{w}_i}{G_{ii}}, \quad (i = 1, 2, \ldots, M). \tag{10.35}$$

Assume that there is a set of weight vectors $\mathbf{W}$ for which $\rho(\mathbf{F}^w) < 1$. The matrix $\mathbf{I} - \mathbf{F}^w$ is then invertible and $\mathbf{P}_w = [\mathbf{I} - \mathbf{F}^w]^{-1}\mathbf{u}^w$ minimizes the objective function in the optimization problem for the fixed-weight-vector set $\mathbf{W}$. For any feasible $\mathbf{W}$, the vector $\mathbf{P}_w$ can be computed as the limit of the following iteration:

$$P_i^{n+1} = \sum_{j \neq i} \frac{\gamma_i G_{ji} G_{a_i}(\mathbf{w}_i, \mathbf{a}_{ji})}{G_{ii}} P_j^n + \frac{\gamma_i N_i \mathbf{w}_i^H \mathbf{w}_i}{G_{ii}} \quad (i = 1, 2, \ldots, M). \tag{10.36}$$

The preceding iteration is similar to the distributed power-control algorithm (see [209, 352]), in which the link gain G_{ji} is replaced with the multiplication of the path loss and antenna gain, and the noise power is replaced with the weighted sum of the noise power at the inputs of array elements. Denote the iteration in (10.36) as

$$\mathbf{P}^n = m^w(\mathbf{P}^{n-1}). \tag{10.37}$$

Starting from any initial power vector $\mathbf{P}^0$, the mapping m^w will converge to the optimal power vector $\mathbf{P}_w$ that is the fixed point of the mapping, i.e., $\lim_{n \to \infty} \mathbf{P}^n = \mathbf{P}_w$, $\mathbf{P}_w = m^w(\mathbf{P}_w)$. The objective in the joint beam-forming and power-control problem is to find the beam-forming set $\mathbf{W}$ among all feasible beam-forming sets, in such a way that $\mathbf{P}_w$ is minimal.

To find the optimal solution for the minimization problem $\hat{\mathbf{P}}$, we define the ith element of the mapping m as

$$m_i(\hat{\mathbf{P}}) = \min_{\mathbf{w}_i} \{ \sum_{j \neq i} \frac{\gamma_i G_{ji} G_{a_i}(\mathbf{w}_i, \mathbf{a}_{ji})}{G_{ii}} \hat{P}_j + \frac{\gamma_i N_i \mathbf{w}_i^H \mathbf{w}_i}{G_{ii}} \},$$

$$\text{subject to } \mathbf{w}_i^H \mathbf{a}_{ii} = 1 \ (i = 1, 2, \ldots, M). \tag{10.38}$$

In the following we show that the optimum power allocation is the fixed point of the mapping m, i.e.,

$$\hat{\mathbf{P}} = m(\hat{\mathbf{P}}). \tag{10.39}$$

The following lemma holds for the mapping m:

LEMMA 1 *The fixed point of mapping m and the optimal beam-forming weight vectors are unique.*

Proof 1 The uniqueness can be shown by a similar approach to that in [348]. Assume positive power vectors $\hat{\mathbf{P}}$ and $\mathbf{P}^*$ are the fixed points of the mappings. Without loss of generality, we assume for the kth element of these two vectors that the following relationship holds: $\hat{P}_k > P_k^*$. Let $\alpha = \max_l(\hat{P}_l/P_l^*) > 1$, such that $\alpha \mathbf{P}^* \geq \hat{\mathbf{P}}$. We can find an index i such that $\alpha P_i^* = \hat{P}_i$. Because both $\hat{\mathbf{P}}$ and $\mathbf{P}^*$ are the fixed points of mapping m,

$$\hat{P}_i = \min_{\mathbf{w}_i} \left\{ \sum_{j \neq i} \frac{\gamma_i G_{ji} G_{a_i}(\mathbf{w}_i, \mathbf{a}_{ji})}{G_{ii}} \hat{P}_j + \frac{\gamma_i N_i \mathbf{w}_i^H \mathbf{w}_i}{G_{ii}} \right\},$$

$$\text{subject to} \qquad \mathbf{w}_i^H \mathbf{a}_{ii} = 1$$

$$\leq \min_{\mathbf{w}_i} \left\{ \sum_{j \neq i} \frac{\gamma_i G_{ji} G_{a_i}(\mathbf{w}_i, \mathbf{a}_{ji})}{G_{ii}} \alpha P_j^* + \frac{\gamma_i N_i \mathbf{w}_i^H \mathbf{w}_i}{G_{ii}} \right\},$$

$$\text{subject to} \qquad \mathbf{w}_i^H \mathbf{a}_{ii} = 1$$

$$< \alpha \left(\min_{\mathbf{w}_i} \left\{ \sum_{j \neq i} \frac{\gamma_i G_{ji} G_{a_i}(\mathbf{w}_i, \mathbf{a}_{ji})}{G_{ii}} P_j^* + \frac{\gamma_i N_i \mathbf{w}_i^H \mathbf{w}_i}{G_{ii}} \right\} \right),$$

$$\text{subject to} \qquad \mathbf{w}_i^H \mathbf{a}_{ii} = 1$$

$$= \alpha P_i^*. \tag{10.40}$$

The preceding contradiction implies that the fixed point of mapping m is unique. The optimal weight vectors are given by

$$\hat{\mathbf{w}}_i = \arg\min_{\mathbf{w}_i} \left\{ \sum_{j \neq i} \frac{\gamma_i G_{ji} G_{a_i}(\mathbf{w}_i, \mathbf{a}_{ji})}{G_{ii}} \hat{P}_j + \frac{\gamma_i N_i \mathbf{w}_i^H \mathbf{w}_i}{G_{ii}} \right\},$$

$$\text{subject to} \qquad \mathbf{w}_i^H \mathbf{a}_{ii} = 1 \qquad (i = 1, 2, \ldots, M). \tag{10.41}$$

Because the solution to the optimal beam-forming problem, given by (10.24), is unique [211], the optimal weight vectors, which are denoted by a set $\hat{\mathbf{W}} = \{\hat{\mathbf{w}}_1, \ldots, \hat{\mathbf{w}}_M\}$, are also unique.   **QED**

Let $(\hat{\mathbf{P}}, \hat{\mathbf{W}})$ be the power vector and the weight vector sets that achieve the minimum in (10.32). In the following, we present an iterative algorithm for adjusting $\mathbf{P}$ and $\mathbf{W}$ simultaneously, and we show that, starting from any arbitrary power vector, it converges to the optimal solution $(\hat{\mathbf{P}}, \hat{\mathbf{W}})$. The iteration step for obtaining $(\mathbf{P}^{n+1}, \mathbf{W}^{n+1})$ given $\mathbf{P}^n$ is as follows:

Algorithm A:

1. $\mathbf{w}_i^{n+1}$ is computed at each receiver i such that the CCI is minimized under the constraint of maintaining constant gain for the direction of interest, i.e.,

$$\mathbf{w}_i^{n+1} = \arg\min_{\mathbf{w}_i} \left\{ \sum_{j \neq i} G_{ji} G_{a_i}(\mathbf{w}_i, \mathbf{a}_{ji}) P_j^n + N_i \mathbf{w}_i^H \mathbf{w}_i \right\} \quad (i = 1, 2, \ldots, M), \tag{10.42}$$

$$\text{subject to} \qquad \mathbf{w}_i^H \mathbf{a}_{ii} = 1,$$

where $\mathbf{P}^n$ is the power vector updated at the $(n-1)$th step.

2. The updated power vector, $\mathbf{P}^{n+1}$, is then obtained by

$$P_i^{n+1} = \sum_{j \neq i} \frac{\gamma_i G_{ji} G_{a_i}(\mathbf{w}_i^{n+1}, \mathbf{a}_{ji})}{G_{ii}} P_j^n + \frac{\gamma_i N_i (\mathbf{w}_i^{n+1})^H \mathbf{w}_i^{n+1}}{G_{ii}}, \quad (10.43)$$

by performing one iteration with the mapping $m^{\mathbf{w}_i^{n+1}}$ on the power vector $\mathbf{P}^n$.

Combining two iteration steps in the algorithm, we obtain the power vector update in a single step:

$$P_i^{n+1} = \min_{\mathbf{w}_i} \left\{ \sum_{j \neq i} \frac{\gamma_i G_{ji} G_{a_i}(\mathbf{w}_i, \mathbf{a}_{ji})}{G_{ii}} P_j^n + \frac{\gamma_i N_i \mathbf{w}_i^H \mathbf{w}_i}{G_{ii}} \right\},$$

$$\text{subject to} \quad \mathbf{w}_i^H \mathbf{a}_{ii} = 1, \quad (10.44)$$

which is expressed as

$$\mathbf{P}^{n+1} = m(\mathbf{P}^n). \quad (10.45)$$

THEOREM 20 *The sequence* $(\mathbf{P}^n, \mathbf{W}^n)$, $(n = 1, 2 \ldots)$ *produced by iteration (10.44), starting from an arbitrary power* $\mathbf{P}^0$, *converges to the optimal pair* $(\hat{\mathbf{P}}, \hat{\mathbf{W}})$.

To prove Theorem 20, first we present a lemma and then we show that the theorem holds when the iteration starts from the power vector $\mathbf{P}^0 = 0$.

LEMMA 2 *For any two power vectors* $\mathbf{P}_1$ *and* $\mathbf{P}_2$ *such that* $\mathbf{P}_1 \leq \mathbf{P}_2$ *the following holds:*
    (a) $m(\mathbf{P}_1) \leq m^w(\mathbf{P}_1)$, $\forall \mathbf{W}$,
    (b) $m^w(\mathbf{P}_1) \leq m^w(\mathbf{P}_2)$, $\forall \mathbf{W}$,
    (c) $m(\mathbf{P}_1) \leq m(\mathbf{P}_2)$.

Proof 2 (a) holds because in the mapping m, we are minimizing the power vector $\mathbf{P}$ over all possible weight vectors $\mathbf{W}$. (b) can be concluded immediately from the fact that the coefficients in the mapping m^w are positive. (c) can be shown as follows:

$$m(\mathbf{P}_2) = m^{\hat{w}}(\mathbf{P}_2). \quad (10.46)$$

Because $\mathbf{P}_1 \leq \mathbf{P}_2$, from (b) we conclude

$$m(\mathbf{P}_2) \geq m^{\hat{w}}(\mathbf{P}_1), \quad (10.47)$$

and from (a)

$$m(\mathbf{P}_2) \geq m(\mathbf{P}_1). \quad (10.48)$$

$$\text{QED}$$

THEOREM 21 *The sequence* $\mathbf{P}^n$, *generated by iteration (10.44) and initial condition* $\mathbf{P}^0 = 0$, *converges to the fixed point of the mapping* m, $\hat{\mathbf{P}}$.

Proof 3 We define two power vector sequences $\mathbf{P}^n$ and $\mathbf{P}_{\hat{w}}^n$ produced by the mappings m and $m^{\hat{w}}$, respectively, with zero initial condition. That is,

$$\mathbf{P}^{n+1} = m(\mathbf{P}^n), \quad \mathbf{P}^0 = 0, \quad (10.49)$$

and

$$P_{\hat{w}}^{n+1} = m^{\hat{w}}(P_{\hat{w}}^n), \qquad P_{\hat{w}}^0 = 0. \tag{10.50}$$

The power vector sequence $\mathbf{P}^n$ is nondecreasing. To show this we observe that $\mathbf{P}^1 = m(\mathbf{P}^0) = \mathbf{u}^w \geq 0$, i.e., $\mathbf{P}^0 \leq \mathbf{P}^1$, and if $\mathbf{P}^{n-1} \leq \mathbf{P}^n$ Lemma 2(c) implies $m(\mathbf{P}^{n-1}) \leq m(\mathbf{P}^n)$ or $\mathbf{P}^n \leq \mathbf{P}^{n+1}$. By induction we conclude that $\mathbf{P}^n$ is a nondecreasing sequence.

We start the mappings m and $m^{\hat{w}}$ from the same starting vector $\mathbf{P}^0 = \mathbf{P}_{\hat{w}}^0 = 0$. We can follow the same steps to prove that the sequence $\mathbf{P}_{\hat{w}}^n$ is also nondecreasing. Because $\hat{\mathbf{W}}$ is the optimal beam-forming set, the sequence $\mathbf{P}_{\hat{w}}^n$ will converge to the optimal power vector $\hat{\mathbf{P}}$, i.e.,

$$\lim_{n \to \infty} \mathbf{P}_{\hat{w}}^n = \hat{\mathbf{P}}. \tag{10.51}$$

By Lemma 2(a) $m(\mathbf{P}^0) \leq m^{\hat{w}}(\mathbf{P}_{\hat{w}}^0)$ or $\mathbf{P}^1 \leq \mathbf{P}_{\hat{w}}^1$, and if $\mathbf{P}^n \leq \mathbf{P}_{\hat{w}}^n$, by Lemma 2(a) and 2(b), $m(\mathbf{P}^n) \leq m^{\hat{w}}(\mathbf{P}_{\hat{w}}^n)$ or $\mathbf{P}^{n+1} \leq \mathbf{P}_{\hat{w}}^n$ for all n. That is, by induction we may write $\mathbf{P}^n \leq \mathbf{P}_{\hat{w}}^n$ (for $n = 1, 2, \ldots$). Hence, $\mathbf{P}^n$ is a nondecreasing sequence and bounded from above by $\hat{\mathbf{P}}$, so it has a limit denoted by $\mathbf{P}^*$. Because the mapping m is continuous, $\mathbf{P}^* = \lim_{n \to \infty} \mathbf{P}^n = m(\lim_{n \to \infty} \mathbf{P}^n) = m(\mathbf{P}^*)$. That is, the power vector $\mathbf{P}^*$ is the fixed point of the mapping m. It is shown in the following that $\mathbf{P}^* = \hat{\mathbf{P}}$. Let

$$P_i^* = \min_{\mathbf{w}_i} \left\{ \sum_{j \neq i} G_{ji} G_{a_i}(\mathbf{w}_i, \mathbf{a}_{ji}) P_j^* + N_i \, \mathbf{w}_i^H \mathbf{w}_i \right\} \quad (i = 1, 2, \ldots, M), \tag{10.52}$$

$$\text{subject to} \qquad \mathbf{w}_i{}^H \mathbf{a}_{ii} = 1.$$

By definition $\mathbf{P}^* = \hat{\mathbf{P}}$. That is, the sequence $\mathbf{P}^n$ converges to the optimal power vector $\hat{\mathbf{P}}$. Because the power vector is converging to $\hat{\mathbf{P}}$, beam-forming vectors are also converging to $(\check{\mathbf{w}}_i, i = 1, \ldots, M)$, given by

$$\check{\mathbf{w}}_i = \min_{\mathbf{w}_i} \left\{ \sum_{j \neq i} G_{ji} G_{a_i}(\mathbf{w}_i, \mathbf{a}_{ji}) \hat{P}_j + N_i \, \mathbf{w}_i^H \mathbf{w}_i \right\} \quad (i = 1, 2, \ldots, M). \tag{10.53}$$

The uniqueness of the optimal beam-forming weight vectors implies $\check{\mathbf{w}}_i = \hat{\mathbf{w}}_i, (i = 1, 2, \ldots, M)$.   **QED**

Now we give the proof of Theorem 20:

Proof 4 Now we show that a power vector sequence starting from any initial power vector converges to the optimal power vector $\hat{\mathbf{P}}$. We consider the sequence $\tilde{\mathbf{P}}^{n+1} = m(\tilde{\mathbf{P}}^n)$ with the arbitrary initial power vector $\tilde{\mathbf{P}}^0$.

Assume there exists a feasible pair $(\hat{\mathbf{P}}, \hat{\mathbf{W}})$. The power vector iteration for this pair is given by

$$\mathbf{P}_{\hat{w}}^{n+1} = m^{\hat{w}}(\mathbf{P}_{\hat{w}}^n) \qquad (n = 0, 1, \ldots). \tag{10.54}$$

The optimality of $\hat{\mathbf{W}}$ implies that $\lim_{n \to \infty} \mathbf{P}_{\hat{w}}^n = \hat{\mathbf{P}}$, where $\hat{\mathbf{P}}$ is the fixed point of the mapping defined in (10.54). Assume that both sequences start from the same point, i.e., $\tilde{\mathbf{P}}^0 = \mathbf{P}_{\hat{w}}^0$. Lemma 2(a) implies $m(\tilde{\mathbf{P}}^0) \leq m^{\hat{w}}(\mathbf{P}_{\hat{w}}^0)$ or $\tilde{\mathbf{P}}^1 \leq \mathbf{P}_{\hat{w}}^1$. If $\tilde{\mathbf{P}}^n \leq \mathbf{P}_{\hat{w}}^n$, then $m(\tilde{\mathbf{P}}^n) \leq m^{\hat{w}}(\mathbf{P}_{\hat{w}}^n)$ or $\tilde{\mathbf{P}}^{n+1} \leq \mathbf{P}_{\hat{w}}^{n+1}$. Hence, by induction we have

$$\tilde{\mathbf{P}}^n \leq \mathbf{P}_{\hat{w}}^n \qquad (n = 0, 1, \ldots), \tag{10.55}$$

and because $\lim_{n \to \infty} \mathbf{P}_{\check{w}}^n = \hat{\mathbf{P}}$, we have $\tilde{\mathbf{P}}^n \leq \hat{\mathbf{P}}$ $(n = 0, 1, \ldots)$. That is, the sequence $\tilde{\mathbf{P}}^n$ is bounded, therefore it has accumulation points. For any accumulation point $\tilde{\mathbf{P}}^*$ the following inequality holds:

$$\tilde{\mathbf{P}}^* \leq \hat{\mathbf{P}}. \tag{10.56}$$

Let the sequence $\check{\mathbf{P}}^n$ defined by the iteration $\check{\mathbf{P}}^n = m(\check{\mathbf{P}}^{n-1})$ start from $\check{\mathbf{P}}^0 = 0$. Lemma 2(c) implies $m(\check{\mathbf{P}}^0) \leq m(\tilde{\mathbf{P}}^0)$, that is, $\check{\mathbf{P}}^1 \leq \tilde{\mathbf{P}}^1$. If $\check{\mathbf{P}}^n \leq \tilde{\mathbf{P}}^n$, then $m(\check{\mathbf{P}}^n) \leq m(\tilde{\mathbf{P}}^n)$ or $\check{\mathbf{P}}^{n+1} \leq \tilde{\mathbf{P}}^{n+1}$. By induction we may write

$$\check{\mathbf{P}}^n \leq \tilde{\mathbf{P}}^n, \qquad (n = 0, 1 \ldots). \tag{10.57}$$

From Theorem 21, it follows that the sequence $\check{\mathbf{P}}$ converges to $\hat{\mathbf{P}}$; therefore for the accumulation points we have

$$\check{\mathbf{P}}^* = \hat{\mathbf{P}} \leq \tilde{\mathbf{P}}^*. \tag{10.58}$$

Inequalities (10.56) and (10.58) imply that $\tilde{\mathbf{P}}^* = \hat{\mathbf{P}}$.   **QED**

The proofs of Theorems 20 and 21 can be done by the standard function approach [347]. In practice Algorithm A is implemented as follows:

1. The received signal correlation matrix is calculated at the BS: $\Phi_i = E\{\mathbf{x}_i \mathbf{x}_i^H\}$.
2. The optimal weight vectors $\mathbf{w}_i$, $i = 1, \ldots, N$, are calculated and the total interference is sent to the mobile.
3. Mobiles update their power based on the total interference and link gain:

$$P_i^{n+1} = \frac{\gamma_i}{G_{ii}} \{\mathbf{w}_i^H(\Phi_i)\mathbf{w}_i - x P_i G_{ii}\}, \tag{10.59}$$

where $x = 1$ for non-spread-spectrum systems, and $x = L$ (the processing gain) in spread-spectrum systems.

To calculate the optimal weight vector we need to estimate the array response from each mobile to its BS. Assume that, because of estimation errors, the array response from the ith mobile to the ith BS is estimated as $\tilde{\mathbf{a}}_{ii}$. Note that there is no need to estimate the array response $\mathbf{a}_{ij}$, $i \neq j$, for those terms appear in only the interference measured at each BS. The optimal weight vector is given by

$$\tilde{\mathbf{w}}_i = \frac{\Phi_{in}^{-1} \tilde{\mathbf{a}}_{ii}}{\tilde{\mathbf{a}}_{ii}^H \Phi_{in}^{-1} \tilde{\mathbf{a}}_{ii}}. \tag{10.60}$$

Replacing Φ_i from (10.20) or (10.28), we express the algorithm as

$$P_i^{n+1} = \min_{\mathbf{w}_i}\left\{\gamma_i \left[\sum_{j \neq i} \frac{G_{ji} G_{a_i}(\mathbf{w}_i, \mathbf{a}_{ji})}{G_{ii}} P_j^n + \frac{N_i \mathbf{w}_i^H \mathbf{w}_i}{G_{ii}} + x P_i(|\mathbf{w}_i^H \mathbf{a}_{ii}|^2 - 1)\right]\right\},$$

$$\text{subject to} \qquad \mathbf{w}_i^H \tilde{\mathbf{a}}_{ii} = 1, \tag{10.61}$$

and the SNR at each link would be given by

$$\Gamma_i = \frac{|\tilde{\mathbf{w}}_i^H \tilde{\mathbf{a}}_{ii}|^2 P_i G_{ii}}{\sum_{j \neq i} G_{ji} P_j |\tilde{\mathbf{w}}_i^H \tilde{\mathbf{a}}_{ji}|^2 + N_i |\tilde{\mathbf{w}}_i|^2}. \tag{10.62}$$

Note that, in spread-spectrum systems, the processing gain is also included in Γ_i and γ_i. The array response estimation error will change the gain matrix, and it may affect the feasibility of the network if the number of users is close to the maximum capacity of the network. It will also degrade the SNR at each link.

If the array response is not available, or the estimation error is large, we use a training sequence that is correlated with the desired signal. The weight vector is obtained by minimizing the difference of the estimated signal and the training sequence [211]. The minimization problem is defined as

$$\hat{\mathbf{w}}_i = \arg \min_{\mathbf{w}_i} E\{|d_i - \mathbf{w}_i^H \mathbf{x}_i|^2\}, \tag{10.63}$$

$$E_{i,\min} = \min_{\mathbf{w}_i} E\{|d_i - \mathbf{w}_i^H \mathbf{x}_i|^2\}. \tag{10.64}$$

The solution to the preceding minimization problem is given by [211]

$$\hat{\mathbf{w}}_i = \Phi_i^{-1} \mathbf{p}_i, \tag{10.65}$$

where the cross correlation $\mathbf{p}_i$ is given by

$$\mathbf{p}_i = E\{\mathbf{x}_i d_i^*\}. \tag{10.66}$$

If we assume that the training sequence is simply chosen as a copy of the message signal, the cross-correlation vector $\mathbf{p}_i$ is expressed as

$$\mathbf{p}_i = \sqrt{P_i G_{ii}} \mathbf{a}_{ii}. \tag{10.67}$$

The optimal weight vector is then given by

$$\hat{\mathbf{w}}_i = \sqrt{P_i G_{ii}} \Phi_i^{-1} \mathbf{a}_{ii}. \tag{10.68}$$

The preceding method, known as optimum combining, will result in a similar solution as MVDR. It can be shown [211] that the preceding method also maximizes the SINR. As a result, using the same approach, we can prove the convergence of the joint power control and optimum combining [249]. However, in this method there is no need to estimate the array response. The power-control update is given by [249]

$$P_i^{n+1} = P_i^n \frac{\gamma_i}{\Gamma_i} = \gamma_i P_i^n \frac{E_{i,\min}}{1 - E_{i,\min}}. \tag{10.69}$$

Therefore, to update the transmitted power, $E_{i,\min}$ is evaluated at BS (measured locally) and sent to the assigned mobile. Knowing its previous transmitted power and the target SINR, the mobile will update its power according to (10.69).

So far we have considered the power-control problem for a number of transmitter–receiver pairs with fixed assignments, which can be used in uplink or downlink in mobile communication systems. In the uplink power-control problem without beam forming, the power-allocation and BS assignment can be integrated to attain a higher capacity while achieving smaller power allocated to each mobile, as has been demonstrated in previous studies [125, 348].

In the joint power-control and BS assignment, a number of BSs are potential receivers of a mobile transmitter. Here the objective is determining the assignment of users to BSs that minimizes the allocated mobile power. Iterative algorithms that compute the joint optimal BS and power assignment were proposed in [125, 348].

In an uplink scenario in which BSs are equipped with antenna arrays, the problem of joint power control and beam forming as well as BS assignment naturally arises. We modify Algorithm A to support BS assignment as well. The modified algorithm can be summarized as follows:

Algorithm B

1. Each BS in the allowable set of a mobile i minimizes the total interference subject to maintaining unity gain toward the direction of the ith mobile:

$$\mathbf{w}_{im}^{n+1} = \arg\min_{\mathbf{w}_i} \left\{ \sum_{j \neq m} G_{jm} G_{a_i}(\mathbf{w}_i, \mathbf{a}_{jm}) P_j^n + N_m \mathbf{w}_i^H \mathbf{w}_i \right\}, \, m \in B_i, \qquad (10.70)$$

$$\text{subject to} \qquad \mathbf{w}_i^H \mathbf{a}_{im} = 1,$$

where $\mathbf{w}_{im}^{n+1}$ is the optimal beam-forming weight vector at the mth BS for the ith mobile, and B_i is the set of allowable BSs for the ith mobile.

2. Each mobile finds the optimal BS such that the allocated power for the next iteration is minimized:

$$b_i = \arg\min_{j \in B_i} \left\{ \gamma_i \sum_{k \neq i} \frac{G_{kj} G_{a_i}(\mathbf{w}_{ij}^{n+1}, \mathbf{a}_{kj})}{G_{ij}} P_k^n + \frac{\gamma_i N_j \mathbf{w}_{ij}^{n+1}{}^H \mathbf{w}_{ij}^{n+1}}{G_{ij}} \right\}, \qquad (10.71)$$

where b_i is the optimal assignment for mobile i.

3. Each mobile updates its transmitted power based on the optimum beam-forming and BS assignment:

$$P_i^{n+1} = \gamma_i \sum_{k \neq i} \frac{G_{kb_i} G_{a_i}(\mathbf{w}_{ib_i}^{n+1}, \mathbf{a}_{kb_i})}{G_{ib_i}} P_k^n + \frac{\gamma_i N_{b_i} \mathbf{w}_{ib_i}^{n+1}{}^H \mathbf{w}_{ib_i}^{n+1}}{G_{ib_i}}. \qquad (10.72)$$

The preceding steps are combined in one iteration, denoted by $\tilde{m}_i$:

$$P_i^{n+1} = \tilde{m}_i(\mathbf{P}_i^n) = \min_{\mathbf{w}_{ij}, j \in B_i} \left\{ \gamma_i \sum_{k \neq i} \frac{G_{kj} G_{a_i}(\mathbf{w}_{ij}, \mathbf{a}_{kj})}{G_{ij}} P_k^n + \frac{\gamma_i N_j \mathbf{w}_{ij}^H \mathbf{w}_{ij}}{G_{ij}} \right\},$$

$$\text{subject to} \qquad \mathbf{w}_i^H \mathbf{a}_{ij} = 1. \qquad (10.73)$$

Consider a set of BS assignments by $\mathbf{B} = \{b_1, \ldots, b_M\}$. Define the ith element of the mapping $\tilde{m}^{w,b}$ as

$$P_i^{n+1} = \tilde{m}_i^{w,b}(\mathbf{P}_i^n) = \left\{ \gamma_i \sum_{k \neq i} \frac{G_{kb_i} G_{a_i}(\mathbf{w}_{ib_i}, \mathbf{a}_{kb_i})}{G_{ib_i}} P_k^n + \frac{\gamma_i N_{b_i} \mathbf{w}_{ib_i}^H \mathbf{w}_{ib_i}}{G_{ib_i}} \right\},$$

$$\text{subject to} \qquad \mathbf{w}_i^H \mathbf{a}_{ib_i} = 1. \qquad (10.74)$$

The following lemma holds for $\tilde{m}$ and $\tilde{m}^{w,b}$:

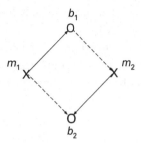

Figure 10.4 A simple degenerate network.

LEMMA 3   *For any two power vectors* $\mathbf{P}_1$ *and* $\mathbf{P}_2$ *such that* $\mathbf{P}_1 \leq \mathbf{P}_2$ *the following holds:*

(a) $\tilde{m}(\mathbf{P}_1) \leq \tilde{m}^{w,B}(\mathbf{P}_1),$ $\forall \mathbf{W}, \mathbf{B};$

(b) $\tilde{m}^{w,b}(\mathbf{P}_1) \leq \tilde{m}^{w,B}(\mathbf{P}_2),$ $\forall \mathbf{w}, \mathbf{B};$

(c) $\tilde{m}(\mathbf{P}_1) \leq \tilde{m}(\mathbf{P}_2).$

Similar to the joint beam-forming and power-control case, we can show that Theorems 20 and 21 hold for mapping $\tilde{m}$ and Algorithm B converges to the optimal power allocation starting from any initial power vector.

In practice, each mobile can be assigned to a set of BSs, denoted by B_i, for the ith mobile. At each iteration all of the BSs in the set will perform beam forming, and the mobile transmitted power for the next iteration is calculated. The BS assignment, or in other words the "handoff," is performed by comparing the power requirements for different BS assignments. The BS with the least required power will be chosen for the mobile. It is worthwhile to note that the beam formings at the BSs are done independently, without the knowledge of other channel responses.

We have shown that the solution to the joint power control and beam forming is unique. In the joint problem with BS assignment, using the same approach as in Lemma 1, we can show that the optimal power allocation is also unique. However, the optimal base station and beam-forming vectors may not be unique. In practice the probability of nonuniqueness is almost zero, and if it happens, it will be lost by a slight variation in parameters. As a simple example, consider Figure 10.4. Assume that mobile m_1 and m_2 are assigned to b_1 and b_2, respectively. In this case the optimal power allocation is given by $P_1 = P_2 = P$. Because of symmetry of the network, the same power vector can achieve the required SNR at each link when m_1 is assigned to b_2 and m_2 is assigned to b_1. In the latter case the beam-forming vectors are different, although the same optimal power vector can be achieved.

We evaluate the performance of our algorithm by simulating the same system as in [125]. The quality constraint is considered to be 0.0304, which is equivalent to a SINR of -14 dB. This threshold results in an acceptable BER only in CDMA systems where there is a processing gain of the order of 128 or more. However, the same methodology can be applied to any wireless network such as TDMA and FDMA. In the latter cases the interference rejection capability of antenna arrays can be utilized to decrease the reuse distance or support more than one user with the same time slot or frequency in each cell. Both of these effects will increase the capacity significantly.

Figure 10.5 shows a network with 36 BSs with 400 users randomly distributed in the area $[0.5, 6.5] \times [0.5, 6.5]$ with uniform distribution. The link gain is modeled as $G_{ij} = 1/d_{ij}^4$, where d_{ij} is the distance between base i and mobile j. Throughout the simulations, we consider two system setups: in System Setup I we use omnidirectional antennas; in System Setup II we use an antenna array with four elements.

Figure 10.5(a) illustrates the use of System Setup I. Traditionally, the mobiles are assigned to the BSs with the largest path gains, and the mobile power is obtained by iterative fixed-assignment power-control algorithm as given by (10.11). In Figure 10.6(a), the dash–dot curve shows the total mobile power at each iteration. This algorithm converges in about 16 iterations. In Figure 10.5(b), using the same system setup, the BS assignment is done by the jointly optimal BS assignment and power-control algorithm, and mobiles have the option to select among the four closest BSs [125]. The total mobile power is depicted in Figure 10.6(a). The dashed curves show that the total power is slightly less than that of the first algorithm considered in Figure 10.5(a). This algorithm converges in about 15 iterations. In Figure 10.5(c) we use System Setup II, i.e., the BSs are equipped with four-element antenna arrays. We apply our joint power control, BS assignment, and beam-forming algorithm to the same configuration of users as in Figures 10.5(a) and 10.5(b). The solid curve in Figure 10.6(a) shows that the total mobile power is an order of magnitude smaller than that of the previous algorithms. Furthermore, the convergence of this algorithm is much faster; it converges in about five iterations in our simulation study.

The capacity of the system is defined as the maximum number of users for which there exists a feasible power vector. As the number of users grow, the maximum eigenvalue of the gain matrix $\rho(\mathbf{F}^w)$ approaches unity, and the total sum of mobile power is increased. At the same time the number of iterations needed to achieve convergence is also increased. In our simulations, we set a maximum value for the number of iterations required for convergence. That is, if the power vector does not converge in 100 iterations, we consider the network an infeasible system.

Using an antenna array with four elements and our algorithm, we can increase the capacity of the network significantly. In Figure 10.6(b) the total mobile power versus the number of users is depicted. Using omnidirectional antennas and the power-control algorithm with a fixed base assignment, we can tolerate at most 660 users. In the same configuration, using the joint BS assignment and power-control algorithm proposed in [125], we can increase the capacity to 800 users. If we use an antenna array with four elements, using our algorithm, the network can tolerate 2800 users. Figure 10.7 illustrates the BS assignments for the preceding three cases. Figure 10.6(b) shows that, for a fixed number of users in our system, the total mobile power is an order of magnitude less than that of a power-controlled network with omnidirectional antennas.

Table 10.1 shows the maximum number of users for different system settings. In the first row of the table the maximum capacity of the network for a fixed power allocation and the same target SINR is shown. The capacity of the same network with fixed power allocations and where each BS uses four-element antennas is three times better than that of the fixed power network with omnidirectional antennas. However, it is significantly less than the capacity of a power-controlled network.

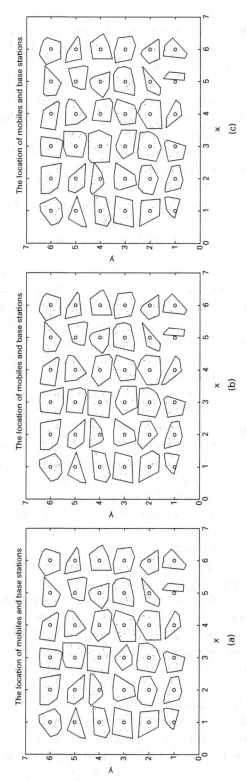

Figure 10.5 Mobile and BSs locations for 400 users: (a) traditional assignment; (b) optimal BS and power control; (c) optimal BS, beam forming, and power control.

Table 10.1 Maximum number of users

System setup	Maximum number of users
Fixed power	30
Fixed power and beam forming	90
Power control	660
Power control and base assignment	800
Power control and beam forming	2800

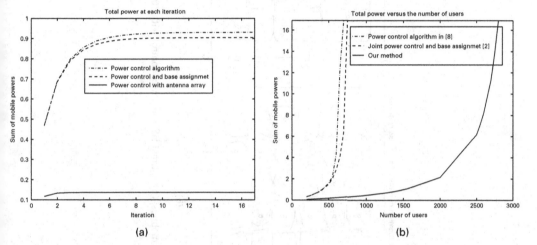

Figure 10.6 (a) Total mobile power versus the iteration number; (b) total mobile power versus the number of users.

It has been observed in [125] that the integration of BS assignment and power control significantly increases the local capacity, i.e., handling more users when we have a hot spot in a network. To demonstrate the effectiveness of our proposed approach, in Figure 10.8, 400 users are dispersed randomly around the network. We then added users randomly in the local area of $[3.5, 4.5] \times [3.5, 4.5]$. When we add 22 users to System Setup I, the traditional fixed BS assignment reaches its limit. Using the power allocation and BS assignment [125] and the same system setup, when we add 57 users we get overload. Using System Setup II and our method, we can add 150 users prior to overload.

In summary, when we have the same configuration of users, the use of adaptive antenna arrays in the BSs and our algorithm significantly reduces the mobile power by almost an order of magnitude, which is very critical in terms of battery life in mobile sets. Second, it provides faster convergence compared with the existing power-control algorithms. And third, it can increase the capacity of systems significantly.

We have introduced the consideration of joint optimal beam forming and power control. We provided an iterative algorithm amenable to distributed implementation that converges to the optimal beam forming and BS assignment if there exists at least one

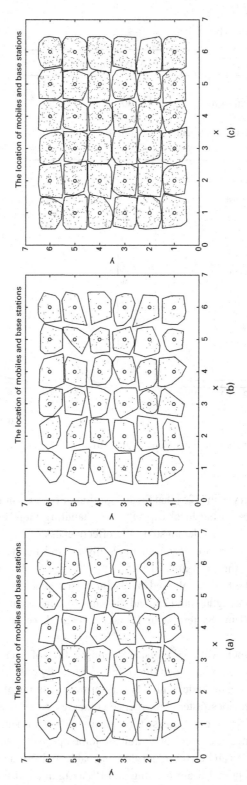

Figure 10.7 Mobile and BS locations: (a) traditional assignment with 660 mobiles; (b) optimal BS and power control with 800 mobiles; (c) optimal BS, beam forming, and power control with 2800 mobiles.

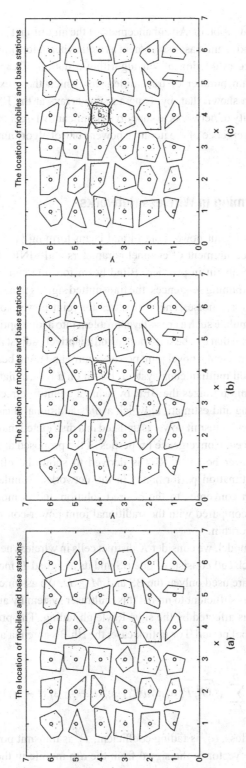

Figure 10.8 Mobile and BS locations with local congested area: (a) traditional assignment with 22 additional users; (b) optimal BS and power control with 57 additional users; (c) optimal BS, beam forming, and power control with 150 additional users.

solution to the joint problem. An enhancement of the algorithm that makes it appropriate for joint power control and base assignment as well as beam forming was also considered.

For performance evaluation of our algorithm a notion of capacity was considered to be the maximum number of transmitters for which there exists a feasible power vector. It has been shown that, by using antenna arrays at the BSs, the algorithm will improve the capacity of networks to support a significantly larger number of users. It also speeds up the convergence of the iterative power-control algorithm and saves the mobile power.

10.3   Blind Beam Forming in Wireless Networks

Current methods of joint power control and beam forming [183, 203, 248, 250, 314] assume perfect measurement of channel parameters and SINR at the receivers, which is very difficult to obtain in practice. Blind beam forming can estimate and separate, without the use of training sequences, the transmitted signals that suffer from the channel distortion and additive noise. The difficulties for joint power control and blind beam forming are to formulate such a cross-layer problem into a joint optimization problem and develop an algorithm that can be self-trained and adaptively adjust the system parameters. In this section, we present a novel joint power-control and blind beam forming algorithm [102] for a multicell multiantenna system. Based on a reformulated joint problem, our proposed algorithm optimizes the BER by using a quantity directly available from the blind beam forming and estimation, which avoids additional measurements previously mentioned. Mobiles' transmit power is updated in a distributed manner such that the CCI is effectively reduced. Convergence properties of the proposed algorithm are discussed. A Cramer–Rao lower bound (CRB) is derived to show the effect of power control on the symbol estimation performance in the networks. Simulation results illustrate that our algorithm converges to the desired solution and is more robust to channel-estimation errors compared with the traditional joint power-control and training-based beam-forming algorithm.

For the system model, we consider K distinct cells in wireless networks where cochannel links exist. Each cell consists of one BS and its assigned D mobiles. Antenna arrays with M elements are used only at the BS and $M \geq D$. We assume coherent detection is possible so that it is sufficient to model this multiuser system by an equivalent baseband model. Each link is affected by the slow Rayleigh fading. The propagation delay is far less than one symbol period. For uplink case, the ith BS antenna array's output vector is given by

$$\mathbf{x}_i(t) = \sum_{k=1}^{K} \sum_{d=1}^{D} \sqrt{G_{ki}^d P_k^d} \alpha_{ki}^d \mathbf{a}_{ki}^d(\theta_{ki}^d) \cdot g_k^d(t - \tau_{ki}) s_k^d(t - \tau_{ki}) + \mathbf{n}_i(t), \qquad (10.75)$$

where G_{ki}^d is path loss, α_{ki}^d is fading coefficient, P_k^d is transmit power, $\mathbf{a}_{ki}^d(\theta_{ki}^d)$ is the ith BS array response vector to the signal from the dth mobile in the kth cell at direction θ_{ki}^d, $g_k^d(t)$ is the shaping function, $s_k^d(t)$ is the message symbol, τ_{ki} is the delay, and $\mathbf{n}_i(t)$

is the thermal-noise vector. We assume the synchronous transmission for all the users within the same cell, i.e., $\tau_{ii} = 0$, $\forall i$. The synchronous assumption is reasonable because the symbol timing can be effectively controlled within each cell. We assume the CCI from other cells is asynchronous for the desired signals within the cell and τ_{ki}, $k \neq i$, is uniformly distributed within the symbol duration. We assume the channels are flat fading and stable within a frame of hundreds of symbols. Define the impulse response from the dth mobile in the kth cell to the pth element of the ith BS as $h_{ki}^{dp} = \alpha_{ki}^d a_{ki}^{dp}(\theta_{ki}^d) r_{ki}^{dp}$, where r_{ki}^{dp} includes the effect of the transmitter, receiver filter, and shaping function $g_k^d(t - \tau_{ki})$. In vector form, it is given by $\mathbf{h}_{ki}^d = [h_{ki}^{1d}, \ldots, h_{ki}^{Md}]^T$. The sampled received vector for these DK users and MK antenna output multicell system at time n is given by

$$\mathbf{X}(n) = \mathbf{A}\mathbf{S}(n) + \mathbf{n}(n), \tag{10.76}$$

where $\mathbf{X}(n) = [\mathbf{x}_1^T(n), \mathbf{x}_2^T(n) \ldots \mathbf{x}_K^T(n)]^T$, $\mathbf{S}(n) = [\mathbf{S}_1^T(n), \mathbf{S}_2^T(n) \ldots \mathbf{S}_K^T(n)]^T$, $\mathbf{S}_i(n) = [s_i^1(n), \ldots s_i^D(n)]^T$, $\mathbf{n}(n)$ is the sampled thermal-noise vector, and

$$\mathbf{A} = \begin{bmatrix} \mathbf{A}_{11} & \mathbf{A}_{21} & \cdots & \mathbf{A}_{K1} \\ \mathbf{A}_{12} & \mathbf{A}_{22} & \cdots & \mathbf{A}_{K2} \\ \vdots & \vdots & \vdots & \vdots \\ \mathbf{A}_{1K} & \mathbf{A}_{2K} & \cdots & \mathbf{A}_{KK} \end{bmatrix}_{MK \times DK}, \tag{10.77}$$

where $\mathbf{A}_{ij} = \left[\sqrt{P_i^1 G_{ij}^1} \mathbf{h}_{ij}^1 \cdots \sqrt{P_i^D G_{ij}^D} \mathbf{h}_{ij}^D \right]$.

Let $\mathbf{w}_i^d$ be the beam-forming weight vector for the dth mobile in the ith cell. Without loss of generality, we normalize the beam-former weight vector $\|(\mathbf{w}_i^d)^H \mathbf{h}_{ii}^d\|^2 = 1$, which will not change the receivers' SINRs. We assume the transmitted signals from different sources are uncorrelated and zero mean, and the additive noise is spatially and temporally white with variance $\mathbf{N}_i = \sigma^2 \mathbf{I}_{M \times M}$, where σ^2 is the thermal-noise variance. The dth user's SINR at its associated ith base station's beam-former output is

$$\Gamma_i^d = \frac{P_i^d G_{ii}^d}{\sum \sum_{(k,j) \neq (i,d)} P_k^j G_{ki}^j \|(\mathbf{w}_i^d)^H \mathbf{h}_{ki}^j\|^2 + (\mathbf{w}_i^d)^H \mathbf{N}_i \mathbf{w}_i^d}. \tag{10.78}$$

The issue in question here is how to find the users' beam-forming vectors and transmit power such that each user has the desired link quality and does not introduce unnecessary CCI to other users. Next, we briefly illustrate traditional joint power control and beam forming.

An adaptive antenna array is designed to receive the signals from the desired directions and attenuate signals' radiations from other directions of no interest. The outputs of the array elements are weighted by a beam former. To suppress the interferences, the beam former places its nulls in the directions of interference sources and steers in the direction of the target signal. Some of the most popular beam formers are MMSE and MVDR beam formers [130].

If the channel responses $\mathbf{h}_{ii}^d$ can be estimated, the beam-forming vector can be calculated by the MVDR method, which minimizes the total interferences at the output of a beam former, while the gain for the desired dth user in the ith cell is kept as a constant. The MVDR problem can be defined as

$$\min_{\mathbf{w}_i^d} \|(\mathbf{w}_i^d)^H \mathbf{x}_i\|^2 , \qquad (10.79)$$

$$\text{subject to } \|(\mathbf{w}_i^d)^H \mathbf{h}_{ii}^d\|^2 = 1, \; i = 1, ..., M.$$

Define correlation matrix as $\Phi_i = E[\mathbf{x}_i \mathbf{x}_i^H]$. The optimal weight vector is given by

$$\hat{\mathbf{w}}_i^d = \frac{\Phi_i^{-1} \mathbf{h}_{ii}^d}{(\mathbf{h}_{ii}^d)^H \Phi_i^{-1} \mathbf{h}_{ii}^d}. \qquad (10.80)$$

In traditional power-control schemes, the overall transmit power of all links is minimized, while each link's transmit power is selected so that its SINR is equal to or larger than a fixed and predefined targeted SINR threshold γ_i^d required to maintain the link quality. The power-control problem can be defined as

$$\min_{P_i^d} \sum_{i=1}^{K} \sum_{d=1}^{D} P_i^d , \qquad (10.81)$$

$$\text{subject to } (\mathbf{I} - \mathbf{BF})\mathbf{P} \geq \mathbf{u},$$

where $\mathbf{u} = [u_1^1, \dots, u_1^D, \dots, u_K^1, \dots, u_K^D]^T$, $\mathbf{P} = [P_1^1, \dots P_1^D, \dots P_K^1, \dots, P_K^D]^T$, $\mathbf{I}$ is the identical matrix, $\mathbf{B} = \text{diag}\{\gamma_1^1, \dots \gamma_1^D, \dots \gamma_K^D, \dots, \gamma_K^D\}$, and

$$[\mathbf{F}]_{kj} = \begin{cases} 0 & \text{if } j = k \\ \frac{G_{i'i}^{d'} \|(\mathbf{w}_i^d)^H \mathbf{h}_{i'i}^{d'}\|^2}{G_{ii}^d} & \text{if } j \neq k \end{cases}, \qquad (10.82)$$

where $i = \lfloor k/D \rfloor$, $d = \text{mod}(k, D)$, $i' = \lfloor j/D \rfloor$, $d' = \text{mod}(j, D)$, and $k, j = 1 \dots KD$.

If the spectral radius $\rho(\mathbf{BF})$ [85], i.e., the maximum eigenvalue of BF, is inside the unit circle, the system has feasible solutions, and there exists a positive power-allocation vector to achieve the desired targeted SINRs. By the Perron–Frobenius theorem [85], the optimum power vector for this problem is $\hat{\mathbf{P}} = (\mathbf{I} - \mathbf{BF})^{-1}\mathbf{u}$. Many adaptive algorithms [79, 250, 347] have been developed to reduce the system complexity by the following distributed iteration:

$$P_i^d(n + 1) = \frac{\gamma_i}{G_{ii}^d} I_i^d, \qquad (10.83)$$

where $I_i^d = (\mathbf{w}_i^d)^H \mathbf{N}_i \mathbf{w}_i^d + \sum_{(k,j) \neq (i,d)}^{K,D} \|(\mathbf{w}_i^d)^H \mathbf{h}_{ki}^j\|^2 P_k^j G_{ki}^j$ and I_i^d can be easily estimated at the receivers. The power allocation is balanced at the equilibrium when the power update in (10.83) has converged.

The level of CCI depends on both channel gain and transmit power. The optimal beam-forming vector may vary for different powers. Hence the beam forming and power control should be considered jointly. In [250], a joint power-control and beam-forming scheme has been proposed. An iterative algorithm is developed to jointly update the transmit power and beam-former weight vectors. The algorithm converges to the joint optimal transmit power and beam-forming solution. The joint iterative algorithm can be summarized by the following two steps:

1. Beam forming in physical layer: MVDR algorithm,
2. Power update in MAC layer: $\mathbf{P}^{n+1} = \mathbf{BFP}^n + \mathbf{u}$,

where the power-update step can be implemented by using only a local interference measurement. But the algorithm assumes the knowledge of SINR, directions of the desired signals, and the perfect measurements of channel responses, which are very difficult to get in practice.

In what follows, we first consider how to choose a blind beam-forming algorithm that can be used for joint optimization with power control. Then we reformulate the joint power control and blind beam-forming problem as a cross-layer approach. Finally, an adaptive iterative algorithm is developed.

The traditional beam forming needs the measurement of spatial responses of the array. A common practice is the use of training sequences [130]. However, it costs bandwidth, which is very precious and limited in wireless networks. Moreover, the measurement errors can greatly reduce the performance of beam forming. This gives us the motivation to use a blind beamforming method to separate and estimate the multiple signals arriving at the antenna array. Because beam forming and power control are two different layer techniques, we need to find the blind beam-forming algorithms that allow us to have joint optimization across the layers. In [303, 304], a maximum-likelihood approach called the iterative least-squares projection (ILSP) algorithm is proposed. The algorithm explores the finite alphabet property of digital signals. Channel estimation and symbol detection can be implemented at the same time. In addition, a quantity is available for BER performance and can be used for power-control optimization. Next, we briefly review the ILSP algorithm.

Consider the same channel module in (10.76). The dth mobile inside the ith cell generates binary data $s_i^d(n)$ with power P_i^d transmitted over a low-delay-spread Rayleigh fading channel. The channel and antenna-array response is $\mathbf{h}_{ii}^d$. The sampled antenna output at the ith BS is given by

$$\mathbf{x}_i(n) = \sum_{d=1}^{D} \mathbf{h}_{ii}^d \sqrt{P_i^d G_{ii}^d} s_i^d(n) + \mathbf{v}_i(n), \tag{10.84}$$

where $\mathbf{v}_i(n)$ includes the ith BS antenna thermal noise and all the CCI from the other cells, i.e.,

$$\mathbf{v}_i(n) = \mathbf{n}_i(n) + \sum_{k=1,k\neq i}^{K} \sum_{d=1}^{D} \mathbf{h}_{ki}^d \sqrt{P_k^d G_{ki}^d} s_k^d(n), \tag{10.85}$$

where $\mathbf{n}_i(n)$ is the $M \times 1$ sampled thermal-noise vector.

The ILSP algorithm works with a shifting window on data blocks of size N. Assume that the channel is constant over the N symbol periods. In the ith cell, we obtain the following formulation of the lth data block:

$$\mathbf{X}_i(l) = \mathbf{A}_i \mathbf{S}_i(l) + \mathbf{V}_i(l), \tag{10.86}$$

Table 10.2 ILSP Algorithm

1. Initial $\hat{\mathbf{A}}_{i,0}$, Step $m = 0$;
2. $m = m + 1$
   a. $\bar{\mathbf{S}}_{i,m} = \mathbf{A}_{i,m-1}^{+}\mathbf{X}_i$,
      where $\mathbf{A}_{i,m-1}^{+} = (\hat{\mathbf{A}}_{i,m-1}^{H}\hat{\mathbf{A}}_{i,m-1})^{-1}\hat{\mathbf{A}}_{i,m-1}^{H}$
   b. projection onto finite alphabet
      $\hat{\mathbf{S}}_{i,m} = \text{proj}[\bar{\mathbf{S}}_{i,m}]$
   c. $\hat{\mathbf{A}}_{i,m} = \mathbf{X}_i\hat{\mathbf{S}}_{i,m}^{+}$,
      where $\hat{\mathbf{S}}_{i,m}^{+} = \hat{\mathbf{S}}_{i,m}^{H}(\hat{\mathbf{S}}_{i,m}\hat{\mathbf{S}}_{i,m}^{H})^{-1}$
3. Repeat until $(\hat{\mathbf{A}}_{i,m}, \hat{\mathbf{S}}_{i,m}) \approx (\hat{\mathbf{A}}_{i,m-1}, \hat{\mathbf{S}}_{i,m-1})$.

where l is the block number, $\mathbf{X}_i(l) = \{\mathbf{x}_i(lN+1)\ \mathbf{x}_i(lN+2)\ldots\mathbf{x}_i[(l+1)N]\}$, $\mathbf{V}_i(l) = \{\mathbf{v}_i(lN+1)\ \mathbf{v}_i(lN+2)\ldots\mathbf{v}_i[(l+1)N]\}$, $\mathbf{S}_i(l) = \{\mathbf{s}_i(lN+1)\ \mathbf{s}_i(lN+2)\ldots$ $\mathbf{s}_i[(l+1)N]\}$, $\mathbf{s}_i(n) = [s_i^1(n)\ldots s_i^D(n)]^T$, and $\mathbf{A}_i = [\sqrt{P_i^1 G_{ii}^1}\mathbf{h}_{ii}^1 \ldots \sqrt{P_i^D G_{ii}^D}\mathbf{h}_{ii}^D]$. We assume that the number of users is known or has been estimated.

The ILSP algorithm uses the finite alphabet property of the input to implement a least-squares algorithm that has good convergence properties for the channel with low delay spread. The algorithm is carried out in two steps to alternatively estimate $\mathbf{A}_i$ and $\mathbf{S}_i$ as

$$\min_{\mathbf{A}_i, \mathbf{S}_i} f(\mathbf{A}_i, \mathbf{S}_i; \mathbf{X}_i) = \|\mathbf{X}_i(l) - \mathbf{A}_i\mathbf{S}_i(l)\|^2. \tag{10.87}$$

The first step is a least-squares minimization problem, where $\mathbf{S}_i$ is unstructured and its amplitude is continuous without considering the discrete nature of modulations, and $\mathbf{A}_i$ is fixed and equal to estimated $\hat{\mathbf{A}}_i$. In the second step, each element of the solution $\mathbf{S}_i$ is projected to its closest discrete values $\hat{\mathbf{S}}_i$. Then we obtain a better estimate of $\hat{\mathbf{A}}_i$ by minimizing $f(\mathbf{A}_i, \hat{\mathbf{S}}_i; \mathbf{X}_i)$ with respect to $\mathbf{A}_i$, keeping $\hat{\mathbf{S}}_i$ fixed. We continue this process until estimates of $\hat{\mathbf{A}}_i$ and $\hat{\mathbf{S}}_i$ converge. The ILSP algorithm is given in Table 10.2.

In traditional joint power control and beam forming, the user's received SINR is larger than or equal to a targeted value to maintain the link quality such as the desired BER. We proposed another quantity available from the ILSP algorithm to directly ensure each user's BER. For simplicity, we use BPSK modulation for the analysis and simulation. The other PAM or MQAM modulation methods can be easily extended in a similar way. It has been shown in [303] that the error probability of the ILSP algorithm is approximated by

$$P_r(s_i^d) = Q\left(\sqrt{\frac{2}{\text{Var}[\hat{s}_i^d(n)]}}\right), \tag{10.88}$$

where each estimated signal $\hat{s}_i^d(n)$ has $E[\hat{s}_i^d(n)] = s_i^d(n)$, i.e, ILSP is an unbiased estimator with variance

$$\text{Var}[\hat{s}_i^d(n)] = 2\sigma_i^2(\mathbf{A}_i^H\mathbf{A}_i)_{dd}^{-1}, \tag{10.89}$$

where, in our case, $\sigma_i^2 = E[\mathbf{v}_i(n)^H\mathbf{v}_i(n)]$ and can be estimated by

$$\sigma_i^2 \approx \frac{1}{N}\|\mathbf{X}_i - \hat{\mathbf{A}}_i\hat{\mathbf{S}}_i\|^2 = \frac{1}{N}\|\mathbf{V}_i\|^2. \tag{10.90}$$

In [303], (10.89) is developed for a single-cell environment with AWGNs. In our case, we need to perform optimization in a multicell scenario with CCI. Because there are a large number of CCI sources with similar received power, by the central-limit theorem, we can assume $v_i(n)$ approaches a zero-mean Gaussian vector. So (10.89) still holds in our case. From the simulation results later, we can show that this assumption is valid.

In our proposed joint power-control and blind beam-forming scheme, the key issue is the quantity $\text{Var}[\hat{s}_i^d(n)]$ that is directly related to error performance. $\text{Var}[\hat{s}_i^d(n)]$ is a function of σ_i^2 and $\mathbf{A}_i$, so it is also a function of all $P_i^d, \forall i, d$. We want the maximum variance for each user's $\text{Var}[\hat{s}_i^d(n)]$ to be less than or equal to a predefined value var_0, so that each user's BER is less than the desired value. However, if var_0 is too small, each user's transmit power will be too large and cause too much CCI. Under this condition, the system may not be feasible, i.e., no matter how large the transmit power is, the receivers cannot achieve the desired BER. So we need a feasibility constraint for var_0. The reformulated joint power-control and blind beam-forming problem is given by

$$\min_{P_i^d} \sum_{i=1}^{K} \sum_{d=1}^{D} P_i^d, \tag{10.91}$$

$$\text{subject to} \begin{cases} \text{Var}[\hat{s}_i^d(n)] \leq \text{var}_0, \ \forall i, d \\ \text{var}_0 \text{ is feasible} \end{cases}.$$

To solve this problem, we need to develop a distributed algorithm such that each user can adapt its transmit power by using only local information. We need to evaluate the feasible range of var_0 such that the system is feasible, i.e., there exists a possible power-allocation vector. The convergence and optimality of the adaptive algorithm will be considered later.

To solve the formulated problem, we propose adaptive iterative algorithms next. We assume that var_0 is feasible for the system. We will discuss the feasibility issue later. In the ILSP algorithm, the iteration stops when the estimated channel-response matrix and symbol matrix have converged. In the algorithm, we use the final channel-response matrix $\hat{\mathbf{A}}_i$ to substitute $\mathbf{A}_i$ in (10.89). Then the estimation of $\text{Var}[\hat{s}_i^d(n)]$ is calculated by

$$\text{var}_i^d = 2\sigma_i^2 (\hat{\mathbf{A}}_i^H \hat{\mathbf{A}}_i)_{dd}^{-1}. \tag{10.92}$$

In the uplink, the value of var_i^d is obtained in the BS and compared with the desired var_0. If var_i^d is too large, it means that the BER for the dth user is too large and consequently the dth user's power needs to be increased. If var_i^d is too small, it is unnecessary to have such a high power for the dth user. Consequently, the power needs to be reduced. The power update stops when transmit power has converged in the consecutive iterations, i.e., $\text{var}_i^d \approx \text{var}_0$. Each user's power is updated by the simple feedback of $\lambda = \text{var}_i^d / \text{var}_0$ from the BS. The power-update scheme can be easily implemented in a distributed manner. In each iteration, the power is updated by

$$P_i^d(m+1) = \lambda P_i^d(m), \tag{10.93}$$

where m is the iteration number.

Table 10.3 Joint power-control and blind beam-forming
algorithm

---

1. Given $\mathbf{P}(0)$, var_0, $m = 0$ and $\hat{\mathbf{A}}_i = \hat{\mathbf{A}}_{i,0}$.
2. Received data block at base station i,
     i. ILSP blind estimation to get $\hat{\mathbf{A}}_i$
     ii. For each mobile d inside ith cell,
         $\text{var}_i^d = 2\hat{\sigma}_i^2 (\hat{\mathbf{A}}_i^H \hat{\mathbf{A}}_i)_{dd}^{-1}$
         $\lambda = \frac{\text{var}_i^d}{\text{var}_0}$
         $P_i^d(m+1) = \lambda P_i^d(m)$
     iii. $\hat{\mathbf{A}}_{i,0} = \hat{\mathbf{A}}_i$
3. $m = m + 1$. Go to step 2;
     Repeat until $\mathbf{P}_i(m) \approx \mathbf{P}_i(m-1)$, $\forall i$.

With the preceding power-update equation, we develop the following joint adaptive power-control and blind beam-forming algorithm. The algorithm is initialized by some feasible power-allocation vector $\mathbf{P}(0)$ and approximate channel estimation $\hat{\mathbf{A}}_{i,0}$ [304]. The user's BER may be larger than the desired value during the initialization. In each iteration, first, the ILSP blind estimate algorithm is applied to estimate the antenna-array responses and the transmitted signals. Then var_i^d is calculated. The new transmit power is updated by (10.93). The iteration is stopped by comparing the power vector of the two consecutive iterations. When the algorithm stops, each user's desired BER will be satisfied. The adaptive algorithm is summarized in Table 10.3.

With the adaptive algorithm, we can construct a joint power-control and blind beam-forming system, as shown in Figure 10.9. The variance calculator module calculates the estimation var_i^d from the ILSP module. The updating information of transmit power is computed by the power-update module. Then the simple power-update information is sent back to mobiles via the feedback channels. When the algorithm converges, the output data from the ILSP module will have the desired BER.

Next, we analyze the condition for our proposed algorithm to converge, i.e., we find the feasible range for var_0. Then we prove that the power update converges to a unique solution when the system is feasible, whereas the blind beam forming may not converge to a unique solution. So our proposed joint power-control and blind beam-forming algorithm may have local minima because of the inherited characteristics of the blind estimation. We will propose a method to avoid the local minima. From the simulation results later, we can show that, even with the possible local minima, the proposed algorithm performs comparably well with the traditional joint power-control and beam-forming algorithm.

Consider the transmission from the dth mobile to its associated ith BS with $\mathbf{h}_{ii}^d$ and G_{ii}^d being the channel response and link gain, respectively, and $\mathbf{A}_i$ being the channel-response matrix. We want to find the expression $\text{Var}[\hat{s}_i^d(n)]$ in (10.89). Then we analyze the conditions for the convergence of our algorithm. We have

$$[\mathbf{A}_i^H \mathbf{A}_i]_{jk} = \sqrt{P_i^j P_i^k G_{ii}^j G_{ii}^k} (\mathbf{h}_{ii}^j)^H \mathbf{h}_{ii}^k. \tag{10.94}$$

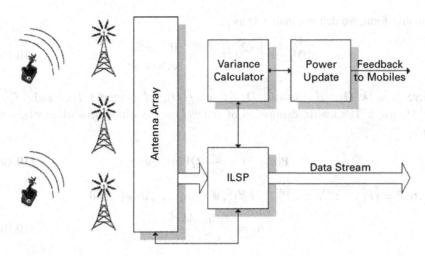

Figure 10.9 Joint power-control and blind beam-forming system.

The $\det(\mathbf{A}_i^H \mathbf{A}_i)$ can be expanded by the following alternative sum form:

$$\det(\mathbf{A}_i^H \mathbf{A}_i) = P_i^1 G_{ii}^1 \dots P_i^D G_{ii}^D f_1(\mathbf{h}_{ii}), \qquad (10.95)$$

where $\mathbf{h}_{ii} = [\mathbf{h}_{ii}^1, \dots, \mathbf{h}_{ii}^D]$, and $f_1(\mathbf{h}_{ii})$ is a real function of channel responses $\mathbf{h}_{ii}^d$, $\forall d$. Then it follows from the cofactor method of matrix inverse [85] that

$$(\mathbf{A}_i^H \mathbf{A}_i)_{dd}^{-1} = \frac{f_2^d(\mathbf{h}_{ii}) \prod_{j=1, j\neq d}^{j=D} P_i^j G_{ii}^j}{f_1(\mathbf{h}_{ii}) \prod_{j=1}^{j=D} P_i^j G_{ii}^j} = \frac{f_3(\mathbf{h}_{ii})}{P_i^d G_{ii}^d}, \qquad (10.96)$$

where $f_2^d(\mathbf{h}_{ii})$ is a real function of channel responses $\mathbf{h}_{ii}^j$, $j \neq d$, and $f_3(\mathbf{h}_{ii}) = f_2^d(\mathbf{h}_{ii})/f_1(\mathbf{h}_{ii})$.

Because the channels are not reused in the adjacent cells in most of the communication system, we assume the CCI plus thermal noise in (10.85) is Gaussian noise with the variance

$$\sigma_i^2 = \sum_{j\neq i}^{K} \sum_{d=1}^{D} \|\mathbf{h}_{ji}^d\|^2 G_{ji}^d P_j^d + M\sigma^2. \qquad (10.97)$$

Now we can calculate $\mathrm{Var}[\hat{s}_i^d(n)]$ as

$$\mathrm{Var}[\hat{s}_i^d(n)] = \frac{2\sigma_i^2}{(\mathbf{A}_i^H \mathbf{A}_i)_{dd}} = \frac{2\sigma_i^2}{P_i^d G_{ii}^d} f_3(\mathbf{h}_{ii}). \qquad (10.98)$$

An interesting result is that $\mathrm{Var}[\hat{s}_k^d(n)]$ is independent of the transmit power of the other mobiles in the same cell. So the main concern for power control is intercell CCI. Substituting into (10.93), we can express the power-update equation as

$$P_i^d(n+1) = \frac{\sum_{j\neq i}^{K} \sum_{d=1}^{D} \|\mathbf{h}_{ji}^d\|^2 G_{ji}^d P_j^d + M\sigma^2}{G_{ii}^d \mathrm{var}_0} f_3(\mathbf{h}_{ii}). \qquad (10.99)$$

In matrix form, we define a matrix $\mathbf{Q}$ as

$$[\mathbf{Q}]_{kj} = \begin{cases} G_{i'i}^{d'} f_4^{kj} / G_{ii}^d & \text{if } i' \neq i \\ 0 & \text{otherwise} \end{cases}, \tag{10.100}$$

where $i = \lfloor k/D \rfloor$, $d = \text{mod}(k, D)$, $i' = \lfloor j/D \rfloor$, $d' = \text{mod}(j, D)$, and $f_4^{kj} = \|\mathbf{h}_{ji}^d\|^2 f_3(\mathbf{h}_{ii})$. The matrix expression of (10.99) for the whole network can be written as

$$\mathbf{P}(n+1) = \frac{1}{\text{var}_0} \mathbf{Q}\mathbf{P}(n) + \mathbf{u}, \tag{10.101}$$

where $\mathbf{P} = [P_1^1 \dots P_1^D, \dots, P_K^1 \dots P_K^D]^T$, $\mathbf{u} = [u_1, \dots, u_{DK}]^T$, and

$$u_j = \frac{f_3(\mathbf{h}_{ii}) M \sigma^2}{G_{ii}^d \text{var}_0}. \tag{10.102}$$

By the Perron–Frobenius theorem [85], the power update in (10.101) has the equilibrium

$$\mathbf{P} = \left(\mathbf{I} - \frac{1}{\text{var}_0}\mathbf{Q}\right)^{-1}\mathbf{u}. \tag{10.103}$$

If $(\mathbf{I} - \frac{1}{\text{var}_0}\mathbf{Q})$ is positive definite, i.e., the spectrum radius $|\rho(\mathbf{Q})| < \text{var}_0$, the positive power vector exists and the power update converges. Under this condition, the system converges when $\text{Var}[\hat{s}_i^d(n)] = \text{var}_0$. From the simulation results later, we will see that our algorithm converges rapidly to the desired var_0 if $|\rho(\mathbf{Q})| < \text{var}_0$.

When var_0 is too small and less than $\rho(\mathbf{Q})$, the system is not feasible and the adaptive algorithm diverges. To prevent the algorithm from diverging, the system will detect the severity of CCI. If the system detects that $\rho(\mathbf{Q})$ approaches var_0 or the transmit power increases very fast, var_0 will be increased so that users will reduce their transmit power and CCI will be alleviated.

Following the same proof in [79], we can prove that the power update in (10.99) converges to a unique solution. Suppose $\hat{\mathbf{P}}$ and $\mathbf{P}^*$ are two different converge power-allocation vectors. Without loss of generality, we assume that $\beta = \max_l(\hat{P}_l^d / P_l^{d*}) > 1$, such that $\beta \mathbf{P}^* \geq \hat{\mathbf{P}}$. We can find an index i such that $\beta P_i^{d*} = \hat{P}_i^d$. We have

$$\begin{aligned}
\hat{P}_i^d &= \frac{\sum_{j \neq i}^K \sum_{d=1}^D \|\mathbf{h}_{ji}^d\|^2 G_{ji}^d \hat{P}_j^d + M\sigma^2}{G_{ii}^d \text{var}_0} f_3(\mathbf{h}_{ii}) \\
&\leq \frac{\sum_{j \neq i}^K \sum_{d=1}^D \|\mathbf{h}_{ji}^d\|^2 G_{ji}^d \beta P_j^{d*} + M\sigma^2}{G_{ii}^d \text{var}_0} f_3(\mathbf{h}_{ii}) \\
&< \beta \frac{\sum_{j \neq i}^K \sum_{d=1}^D \|\mathbf{h}_{ji}^d\|^2 G_{ji}^d P_j^{d*} + M\sigma^2}{G_{ii}^d \text{var}_0} f_3(\mathbf{h}_{ii}) \\
&= \beta P_i^{d*}.
\end{aligned} \tag{10.104}$$

The preceding contradiction implies that power-update equation (10.93) will converge to a unique solution. However, because the solution of blind beam forming may not be unique [303], our proposed joint scheme may fall into local minima. To prevent such local minima, we propose the following scheme to avoid the local minima.

When the two users are not well separated in the angle, i.e., the array response $\mathbf{A}_i$ is ill-conditioned, the ILSP algorithm can converge to some fixed points that are not the global minima. In this case, instead of projecting unstructured continuous estimated symbols to the closest discrete values in the ILSP algorithm, we enumerate over all Ω^D possible vectors $\mathbf{S}_i^j \in \Omega^D$ and choose the one that minimizes

$$\hat{\mathbf{S}}_i(n) = \arg \min_{\mathbf{S}_i^j \in \Omega^D} \|\mathbf{X}_i(n) - \mathbf{A}_i \mathbf{S}_i^j\|^2, \forall j, \tag{10.105}$$

where Ω is the modulation constellation alphabet. This enumerating method has a better performance but a higher complexity. If the global minimum is still not achieved, it has been shown in [304] that usually one or two random reinitializations are sufficient to yield the global minimum. So we can have two or three parallel structures with different initial values to calculate the ILSP algorithm. Then we select the minimal one. The probability of staying in a local minimum will be greatly reduced.

In our proposed joint power-control and blind beam-forming system, the performance of each user's BER is determined by the noise variance, channel conditions, and power allocation. When the additive noise is a zero-mean Gaussian random process, the estimation performance of the unbiased estimator is bounded by the CRB. Next, we derive the covariance matrix for the parameters of the thermal-noise variance, the input symbols, and the power-allocation vector for the CRB. The results will help us analyze the effects of power control on the users' symbol estimation performances in this multicell system.

For simplicity, we assume the data are modulated as BPSK, i.e., $\mathbf{S}(n) \in \Omega^{KD}$, where $\Omega = \{\pm 1\}$. Similar to the performance analysis of the ILSP in [303], we assume the channel responses are known (the algorithm itself does not need such information). The parameter for the Fisher information matrix is $\vartheta = [\sigma^2, \mathbf{S}(1), \ldots, \mathbf{S}(N), \mathbf{P}]$. The likelihood function L of the received data $\mathbf{X}(n)$ is given by

$$L[\mathbf{X}(1)\ldots\mathbf{X}(N)] = \frac{1}{(\pi\sigma^2)^{MKN}} \exp\{-\frac{1}{\sigma^2} \sum_{n=1}^{N} [\mathbf{X}(n) - \mathbf{A}\mathbf{S}(n)]^H [\mathbf{X}(n) - \mathbf{A}\mathbf{S}(n)]\}. \tag{10.106}$$

The Fisher information matrix is calculated by

$$\mathbf{I}(\theta)_{ij} = -E\left[\frac{\partial^2 ln(L)}{\partial \theta_i \partial \theta_j}\right]$$

$$= \begin{bmatrix} \frac{MKN}{\sigma^4} & 0 & \cdots & 0 & 0 \\ 0 & \mathbf{Q} & \cdots & 0 & \mathbf{R}(1) \\ \vdots & \vdots & \ddots & \vdots & \vdots \\ 0 & 0 & \cdots & \mathbf{Q} & \mathbf{R}(N) \\ 0 & \mathbf{R}(1) & \cdots & \mathbf{R}(N) & \mathbf{R}_P \end{bmatrix}, \tag{10.107}$$

where $\mathbf{Q}$, $\mathbf{R}(n)$, and $\mathbf{R}_P$ can be derived straightforwardly by use of the definition of the Fisher information matrix.

To see the effect of the proposed power control on the symbol estimation errors, we define the average mean square error (AMSE) as a performance measure of the symbol estimation:

$$\text{AMSE} = \frac{1}{N} \sum_{n=1}^{N} \frac{\|\hat{\mathbf{S}}(n) - \mathbf{S}(n)\|^2}{\|\mathbf{S}(n)\|^2}. \tag{10.108}$$

Because we use BPSK modulation, $\|\mathbf{S}(n)\|^2 = DK \ \forall \ n$, and the AMSE is the variance bounded by CRB. The CRB for the symbol estimation can be obtained directly from the inverse of the Fisher information matrix, i.e.,

$$\text{AMSE} \geq \frac{1}{NDK} \sum_{n=1}^{N} \sum_{j=1}^{DK} [\mathbf{I}^{-1}(\theta)]_{\mathbf{S}^j(n)\mathbf{S}^j(n)}, \tag{10.109}$$

where $\mathbf{S}^j(n)$ is the jth element of $\mathbf{S}(n)$. How close AMSE is to the CRB will show the relative efficiency of our proposed algorithm.

A network with 50 cells is simulated. Each hexagonal cell's radius is 1000 m. Two adjacent cells do not share the same channel. In each cell, one BS is placed at the center. Two mobiles are placed randomly with uniform distribution. Each mobile transmits BPSK data over Rayleigh fading channels. Each BS employs a four-element antenna array. The noise level is $\sigma = 1$. The transmit frame has $N = 1000$ data symbols. Our shaping function is a raised cosine function.

Path loss is due to the decay of the intensity of a propagating radio wave. In our simulations, we use the two-slope path-loss model [128] to obtain the average received power as a function of distance. According to this model, the average path loss is given by

$$G = \frac{C}{r^a[1 + r\lambda_c/(4h_b h_m)]^b}, \tag{10.110}$$

where C is a constant, r is the distance between the mobile and the BS, a is the basic path-loss exponent (approximately two), b is the additional path-loss component (ranging from two to six), h_b is the BS antenna height, h_m is the mobile antenna height, and λ_c is the wavelength of the carrier frequency. We assume the mobile antenna height is 2 m and the BS antenna height is 50 m. The carrier frequency is 900 MHz.

In Figure 10.10, we show the analytical and numerical performance of the ILSP compared with the MVDR with perfect channel estimation. The numerical results with CCI match the analytical results well, especially at a high SINR range, which proves our assumption that $\mathbf{V}_i(n)$ can be treated as Gaussian noise when the number of CCI is large. Our proposed joint power control and blind beam forming has only about a 1–2-dB performance loss over traditional power control and MVDR beam forming with perfect channel estimation. However, MVDR beam forming needs an additional training sequence to estimate the channel and SINR with prior information that may not be available in practice.

In reality, perfect channel estimation is hard to obtain. In Figure 10.11, we show the effects of DOA estimation errors on the traditional joint power-control algorithm, the MVDR beam-forming algorithm, and our algorithm. In Figure 10.11(a), we compare

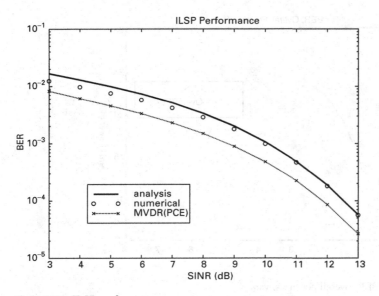

Figure 10.10 ILSP performance.

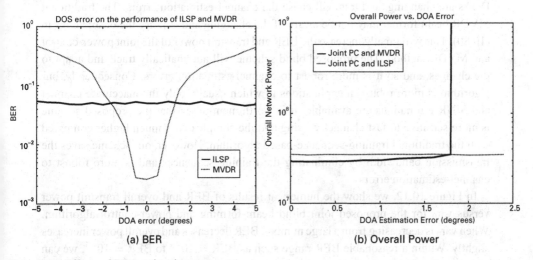

(a) BER   (b) Overall Power

Figure 10.11 Effects of DOA estimation error.

the BER performance; the transmit power allocation is the same for both algorithms. We can see from the curves that, when the channel estimation error for DOA is greater than about 2 deg, the blind beam-forming algorithm outperforms the traditional MVDR. In Figure 10.11(b), we compare the overall transmit power; BER performance is the same for both algorithms. We can see that the blind beam-forming algorithm needs a little bit more transmit power when the DOA estimation error is small. However, the traditional power control with the MVDR method will diverge when the DOA estimation error is about 2 deg. Our proposed joint power-control and beam-forming algorithm will always converge regardless of the DOA variations. When the mobiles are moving,

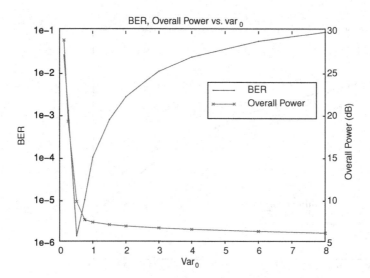

Figure 10.12 BER, overall power vs. var_0.

DOAs are changing, and this will cause the channel-estimation errors. The traditional MVDR beam former may not be aware of the changing and still use the obsolete $\mathbf{h}_{ii}^d$ in (10.80). This will greatly increase the BER and transmit power of the joint power-control and MVDR method. The proposed blind scheme will automatically track and adapt to the changes, and so it is more robust to channel-estimation errors. Consequently, our algorithm is more robust in applications in which usually only the inaccurate channel and SINR estimations are available. It is worth mentioning that the proposed scheme is more sensitive to fast channel varying and the complexity is much higher compared with the traditional training-sequence-based algorithm. However, our scheme saves the transmission bandwidth by eliminating the training sequences and is more robust to channel-estimation errors.

In Figure 10.12, we show the numerical results of BER and overall transmit power versus var_0 for the proposed joint blind beam-forming and power-control algorithm. When var_0 is decreasing from a large number, BER decreases and overall power increases slightly. Within a reasonable BER range such as BER = 10^{-3} to BER = 10^{-5}, we can calculate the threshold of var_0 for the desired BER. After var_0 decreases to a specific value, overall transmit power increases and the BER decreases quickly. This is because the CCI is too large and $var_0 \rightarrow \rho(\mathbf{Q})$. After var_0 is smaller than some value, the algorithm diverges. Consequently, there is no feasible power-control solution, i.e., no matter how large the transmit power is, the receivers cannot ensure the desired BER. This proves that our algorithm behaves exactly the same as the traditional power-control algorithm, except that our algorithm directly ensures the BER instead of each user's SINR. There is a trade-off between the overall transmit power and the BER, and var_0 is the bridge between the two quantities.

In Figure 10.13, we show the distribution of the number of iterations required for the convergence of our proposed algorithm with different values of var_0. The convergence criterion is that the maximum difference of users' transmit power between two

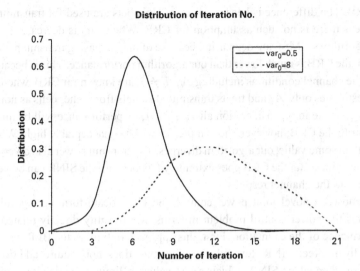

Figure 10.13 Convergence of the algorithm.

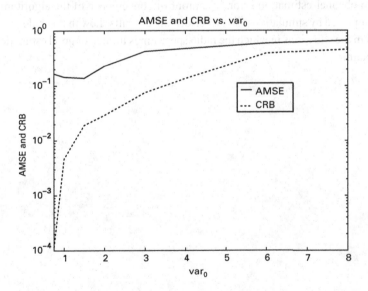

Figure 10.14 AMSE and CRB vs. var_0.

consecutive iterations is less than 3%. When var_0 is within the range such that the system is feasible, we can see that our algorithm converges within a small number of iterations, which demonstrates that our algorithm is robust in the wireless communication systems if the channel gains and topologies have been changed. When var_0 is large (i.e., the desired BER is large), the algorithm converges slower. This is because the transmit power is small when var_0 is large. Consequently, the var_i^d estimation is poor, and more iterations are needed for the convergence.

In Figure 10.14, we compare the AMSE and CRB with var_0. When var_0 is large and the transmit power of users is small, the CCI is small. The performance of the ILSP is close to the CRB. The difference is because discrete alphabets are used for transmitted symbols, whereas there is no such assumption for CRB. When var_0 is decreasing, the CCI and our algorithm's AMSE are decreasing because of the increasing transmit power. In this situation, the CRB is much lower than our algorithm performance. This is because we assume all the channel conditions including A_{ij}, $i \neq j$, are known for CRB, whereas our algorithm estimates only A_{ii} and treats transmitted signals from other cells as noise. If an algorithm can take into consideration all A_{ij}, $\forall i, j$, its performance will be much better and closer to the CRB; however, the complexity will be unacceptably high. When var_0 is smaller than some value, our algorithm diverges. The transmit power also diverges to arbitrary large values. But the CRB goes extremely low because the SINR can be very high, if we know all the channel responses.

We have proposed a novel joint power-control and blind beam-forming algorithm that reformulates the power-control problem in terms of a quantity directly related to the error performance of the estimation. First, this approach optimizes the BER instead of a theoretically indirect SINR. Second, the algorithm does not require additional measurements of channel or SINR, which saves valuable limited bandwidth. Third, our scheme can be easily implemented in a distributed manner. Fourth, our scheme is more robust to channel-estimation error. The proof of convergence of the algorithm is derived and supported by simulation results. Performance results show that our algorithm performs well in the situations in which the radio spectrum is limited or good estimations are hard to obtain.

11   Dynamic Resource Allocation

11.1   Introduction

Over the past few decades, wireless communications and networking have witnessed an unprecedented growth and have become pervasive much sooner than anyone could have imagined. In wireless communication systems, two important detrimental effects that decrease network performance are the channel's time-varying nature and CCI. Because of effects such as multipath fading, shadowing, path loss, propagation delay, and noise level, the SINR at a receiver output can fluctuate on the order of tens of decibels. The other major challenge for the system design is the limited available RF spectrum. Channel reuse is a common method used to increase the wireless system capacity by reusing the same channel beyond some distance. However, this introduces CCI that degrades the link quality.

A general strategy to combat these detrimental effects is the dynamic allocation of resources such as transmitted power, modulation rates, channel assignment, and scheduling based on the channel conditions. Power control is one direct approach toward minimizing CCI. The transmit power is constantly adjusted. They are increased if the SINRs at the receivers are low and are decreased if the SINRs are high. Such a process improves the quality of weak links and reduces the unnecessary transmit power. In [2, 92] centralized power control schemes are proposed to balance the carrier-to-interference ratio (CIR) or maximize the minimum CIR in all links. Those algorithms need global information about all link gains and power. The distributed power control algorithms that use only local measurements of SINR are presented in [209, 352].

In adaptive modulation, the system assigns modulation rates with different constellation sizes and spectral efficiencies to different links, according to their channel conditions. In [8, 49, 68, 69, 127, 306, 311], adaptive modulation techniques have been proposed to enhance the spectrum efficiency for wireless channels. The performance approximation and robustness for estimation errors have been investigated. In [134, 225], adaptive coding provides another way for transmission rate control. In [95, 112, 113, 116, 117, 241, 312, 352], many adaptive algorithms are constructed to adaptively control the transmitted power and rate to optimize the system performance.

Scheduling and channel assignment are techniques used to allocate users' transmission over different time and frequency (code) slots. There are trade-offs and practical constraints to allocate these resources. How to efficiently manage these resources has become an important research issue. In [185], the authors present an "opportunistic" transmission scheduling policy for a single-cell TDMA/FDMA system that exploits time-varying channels and maximizes the system performance stochastically. Opportunistic scheduling [245] has attracted a lot of research attention as a special kind of resource allocation over time. The basic idea of "opportunistic scheduling" is to allocate resources to links experiencing good channel conditions while avoiding allocating resources to links experiencing bad channel conditions, thus efficiently utilizing radio resources. This is also referred to as channel-aware scheduling, which explores time/multiuser diversity.

To optimize the system's overall performance, the resource-allocation schemes often allocate most of the radio resources to the users with good channel conditions. However, this allocation can be very unfair, because the users in the same class may pay the same amount of money for their services. An important trade-off exists between system performances and fairness among users. For example, allowing only users with good channel conditions to transmit may result in high throughput, but meanwhile sacrifice the transmissions of other users. In the literature, some definitions of fairness have been considered, such as maximin fairness and proportional fairness [160]. In the first example in this chapter, we define the long-term fairness concept and develop the adaptive algorithms to maintain this fairness.

To satisfy the growing demands for heterogeneous applications of wireless networks, it is critical to deliver flexible, variable-rate services with high spectral efficiencies. Different QoS provisioning over time-varying channel conditions is a key challenge for system design. Resource allocation provides a strategic means to address this challenge. Current wireless systems use SNR as a QoS measure. With the increasing demand for multimedia transmissions, their delay-sensitive QoS cannot be satisfied by the SNR measure itself. So it is necessary to define new QoS measures for multimedia transmission services, and we will discuss the new QoS measure in detail in the second example.

Finally, if the channels can be modeled as a Markov Chain with known transition probabilities, each user can calculate the optimal dynamic strategies to maximize the long-term benefits. In [174], Lambadaris et al. considered the problem of jointly optimal *admission and routing* at a data network node. Another good example is [256], which deals with optimal control of service in *tandem queues*. In [251], the authors consider the problem of *stochastic control of handoffs in a cellular networks* and try to find an optimal policy for the handoff problem. In the last section, a dynamic programming optimization method is introduced to obtain the optimal bit-rate/delay-control policy for packet transmission in wireless networks with fading channels.

11.2   Resource Management with Fairness

In this section, we develop a joint adaptive power-control and rate-allocation scheme by using MQAM adaptive modulation with antenna diversity. The optimization goal is to minimize the overall transmitted power under some constraints: The overall network throughput is not reduced; the time-average throughput of each user is maintained as a constant that is determined by the service for which the user pays. To solve the problem, the problem is heuristically divided into two subproblems. First, users determine the ranges of throughput that they can accept at different times and report these ranges back to the system. An algorithm is developed to ensure fairness at the user level. Second, the system determines the optimal throughput allocation to different users at each time within the acceptable throughput ranges provided by the users. Three adaptive algorithms are developed to solve this subproblem at the system level. The whole scheme can be interpreted as "water filling" each user's throughput in the time domain and allocating network throughput to different users each time, according to their channel conditions. From the simulation results, the proposed scheme reduces the overall transmitted power up to 7 dB and increases the average spectral efficiency up to 1.2 bit/s/Hz, compared with the previous scheme in [250].

For the system model, we assume K cochannel links exist in distinct cells, such as in TDMA or FDMA networks. Each link consists of a mobile unit and its assigned base station. Coherent detection is assumed to be possible so that it is sufficient to model this multiuser system by an equivalent baseband model. Antenna arrays with P elements are used only at the base stations (BSs). Each link is affected by the multipath fading, with the propagation delay far less than one symbol duration. The maximum number of paths is L. For the uplink case, the signal at the pth antenna array element of the ith BS can be expressed as

$$x_i^p(t) = \sum_{k=1}^{K} \sum_{l=1}^{L} \sqrt{\rho_{ki} G_{ki} P_k} \alpha_{ki}^{pl} g_k(t - \tau_{ki}) s_k(t - \tau_{ki}) + n_i^p(t), \qquad (11.1)$$

where ρ_{ki} and G_{ki} are the log-normal shadow fading and the path loss from the kth user to the ith BS, respectively, α_{ki}^{pl} is the lth path fading loss from the kth user to the ith BS's pth antenna, P_k is the transmitted power, $g_k(t)$ is the shaping function, $s_k(t)$ is the message symbol, $n_i^p(t)$ is the ith BS's thermal noise at the pth antenna, and τ_{ki} is the channel propagation delay. Here $\tau_{ii} = 0, \forall i$ (the delay from the mobile to its assigned BS), and $\tau_{ki}, k \neq i$ (the delay from the mobile to other cell's BS) is uniformly distributed within one symbol duration. The channels change slowly and are stable over a frame with hundreds of symbols. The impulse response from the kth mobile to the pth element of the ith base station is defined as $h_{ki}^p = \sum_{l=1}^{L} \alpha_{ki}^{pl} r_{ki}^p$, where r_{ki}^p includes the effects of the transmitter, receiver filter, and shaping function $g_k(t - \tau_{ki})$. We define $n_i^p(n)$ as the sampled noise.

Because of the channel distortions, CCI, and thermal noises, the average receivers' SINR can be very low in most cases. Under this condition, to satisfy the desired BER, only a low modulation rate or even no transmission can be selected. Antenna diversity is an important means to increase the average receiver's SINR. Consequently, MQAM can be applied with different modulation rates for the desired BER. The antenna outputs can be combined by maximal ratio combining (MRC) or selective combining (SC) [148], as shown in Figure 11.1. MRC diversity requires that the individual signals from each branch be compensated in phase and weighted by the square roots of their SINRs and then be summed coherently. If perfect knowledge of the branch amplitudes and phases is assumed, when the noise is spatially white, MRC is the optimal diversity-combining

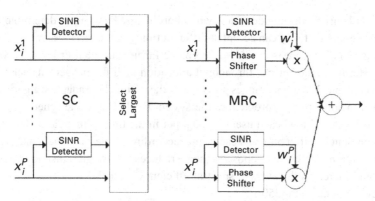

Figure 11.1 Selective combining and maximum ratio combining.

scheme and provides the maximum capacity improvement. The disadvantage of MRC is that it requires all knowledge of the branch parameters. The SC combiner chooses only the branch with the highest SINR. SC is simpler than MRC but yields suboptimal performance. By using the antenna diversity, the ith BS's combiner output can be written as $\mathbf{w}_i^H \mathbf{x}_i$, where $\mathbf{x}_i = [x_i^1 \dots x_i^P]^T$, and $\mathbf{w}_i$ is a $P \times 1$ combiner weight vector given by

$$
\begin{cases}
\text{For MRC:} \ |[\mathbf{w}_i]_j| = \sqrt{\Gamma_i^j}, \\
\text{For SC:} \ [\mathbf{w}_i]_j = \begin{cases} 1, & j\text{th antenna has the largest SINR}, \\ 0, & \text{otherwise} \end{cases}
\end{cases}
$$

where Γ_i^p is the received SINR at the pth antenna element that can be calculated from (11.1). The ith BS's combiner output SINR is given by [112]

$$
\Gamma_i = \frac{P_i \rho_{ii} G_{ii} \left\| \mathbf{w}_i^H \mathbf{h}_{ii} \right\|^2}{\sum_{k \neq i} P_k \rho_{ki} G_{ki} \left\| \mathbf{w}_i^H \mathbf{h}_{ki} \right\|^2 + \mathbf{w}_i^H \mathbf{N}_i \mathbf{w}_i}, \tag{11.2}
$$

where $\mathbf{h}_{ki} = [h_{ki}^1, \dots, h_{ki}^P]^T$, $\mathbf{N}_i = E\{\mathbf{n}_i \mathbf{n}_i^H\}$, and $\mathbf{n}_i = [n_i^1 \dots n_i^P]^T$.

In adaptive modulation, the transmitters and receivers can adaptively select the modulation rates, i.e., throughput, according to the channel conditions. It has been shown that adaptive modulation can greatly increase the spectral efficiency of wireless communications [49, 241]. Here adaptive MQAM modulation is applied. It has been shown that the BER of square MQAM with gray bit mapping as a function of the received SINR Γ and constellation size M is approximately given by [148]

$$
\text{BER}(\Gamma, M) \approx \frac{2}{\log_2 M} \left(1 - \frac{1}{\sqrt{M}} \right) \text{erfc} \left(\sqrt{1.5 \frac{\Gamma}{M-1}} \right), \tag{11.3}
$$

where erfc is the complementary error function. This approximation is good when the SINR Γ is high.

Now the relation between SINR and throughput will be shown. In the ith cell, the ith link between the mobile and its assigned BS uses the modulation with constellation size M_i. Without loss of generality, each user is assumed to have the unit bandwidth. The ith link has throughput $T_i = \log_2(M_i)$. For BER $= 10^{-2}$ and BER $= 10^{-5}$, the required

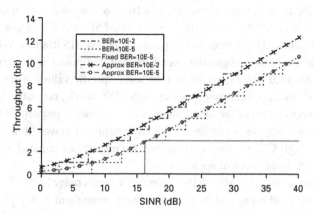

Figure 11.2 BER approximation and BER standard formula for MQAM.

SINRs of different constellation sizes are shown in Figure 11.2. One can see that, for the traditional power control with fixed modulation (8 QAM), the receiver must have a SINR greater than a specific threshold to have any throughput that satisfies BER $= 10^{-5}$. Each user can select a range of different modulation rates. Consequently the targeted receiver's SINR can be chosen within a range.

It is hard for (11.3) to be inverted and differentiated. In [49, 241], the authors introduced BER approximations for different modulation rates as

$$\text{BER}_i \approx c_1 e^{-c_2 \frac{\Gamma_i}{2^{c_3 T_i} - 1}}, \tag{11.4}$$

where $c_1 \approx 0.2$, $c_2 \approx 1.5$, and $c_3 \approx 1$. This approximation is tight when the SINR is high. When (11.4) is rearranged for a specific BER, the ith link's throughput is given by

$$T_i = c_4^i \log_2 \left(1 + c_5^i \Gamma_i\right) \text{ bit/s}, \tag{11.5}$$

where $c_4^i = 1/c_3^i$ and $c_5^i = -c_2^i / \ln(\text{BER}_i / c_1^i)$. In Figure 11.2, the approximation is compared with the expression in (11.3) at BER $= 10^{-2}$ and BER $= 10^{-5}$, respectively. It is shown that (11.5) is a good approximation for throughput versus SINR for a fixed BER.

In reality, the channel-estimation errors can affect the performance of adaptive modulation. The perfect channel estimation is assumed, and it is used in much of the literature. Many analyses for the effects of channel-estimation errors on adaptive modulation can be found in [68, 69, 306].

In a traditional power-control problem [250], the SINR of each user is maintained greater than or equal to some threshold γ_i that can provide adequate link quality. The problem is given by

$$\min_{\gamma_i} \sum_{i=1}^{K} P_i(\mathbf{\Gamma}), \tag{11.6}$$

subject to $\Gamma_i \geq \gamma_i, \ \forall i,$

where $\mathbf{\Gamma} = [\Gamma_1 \cdots \Gamma_K]^T$. In this kind of power control, a fixed and predefined targeted SINR threshold γ_i for the desired modulation rate and BER is assigned to each user. Then the transmitted power is updated to ensure users' targeted SINRs without considering their channel conditions. The system works perfectly in low SINR areas. When the targeted SINRs become high enough, the overall transmitted power will start to increase rapidly. If the targeted SINRs are larger than some specific values, there are no feasible solutions, i.e., the receivers cannot get enough SINR levels, no matter how large the transmitted power is. One of the underlying reasons for such a problem is that the users with bad channel responses require too much transmitted power; thus they introduce unnecessarily high CCI to others. Therefore, having a fixed targeted SINR threshold is not an optimal power control approach.

All links are assumed to apply MQAM with throughput T_i within a range $[T_i^{\min}, T_i^{\max}]$, according to their channel conditions, and the overall network throughput $T = \sum_{i=1}^{K} T_i$ is maintained greater than or equal to a constant R. R is equal to the sum of the fixed targeted throughput in the previous scheme [250] in (11.6). R should be selected such that the system is always feasible. If R is too large, it is likely that the overall network throughput will be larger than the overall system capacity. As a result there will be no solution. Each time, the links with bad channel conditions sacrifice their throughput, i.e., they use lower SINR thresholds, which reduce the unnecessary CCI. The links with good channel conditions use higher SINR thresholds, i.e., more bits per symbol are selected, which increases the network throughput. For each link, the time-average throughput is a constant to ensure fairness, and the throughput is "water filled" at different times. For the whole system at any specific time, the overall network throughput is allocated to different links, according to their channel conditions so as to minimize the overall transmitted power. The value of R is also equal to the sum of all users' time average throughput, so that the sum of users' time-average throughput and the overall network throughput can coincide. This problem can be summarized as

$$\min_{T_i, P_i} \sum_{i=1}^{K} P_i, \tag{11.7}$$

$$\text{subject to} \begin{cases} \text{Feasibility: } (\mathbf{I} - \mathbf{DF})\mathbf{P} \geq \mathbf{u}, \\ \text{Network Performance: } T \geq R, \\ \text{Throughput Range: } T_i^{\min} \leq T_i \leq T_i^{\max}, \\ \text{Fairness: } \lim_{N \to \infty} \frac{\sum_{n=1}^{N} T_i(n)}{N} = \text{const}_i, \end{cases}$$

where $R = \sum_{i=1}^{K} E[T_i(n)]$. Only one type of user is assumed, so $\text{const}_i = \text{const}_j$, $\forall i, j$. The feasibility constraint $(\mathbf{I} - \mathbf{DF})\mathbf{P} \geq \mathbf{u}$ is the matrix expression for the equalities $\Gamma_i \geq \gamma_i$, $\forall i$ [250], where $\mathbf{u} = [u_1, \ldots, u_K]^T$, $u_i = \gamma_i \mathbf{w}_i^H N_i \mathbf{w}_i / (\rho_{ii} G_{ii} \|\mathbf{w}_i^H \mathbf{h}_{ii}\|^2)$, $\mathbf{P} = [P_1, \ldots, P_K]^T$, $\mathbf{D} = \text{diag}\{\gamma_1, \ldots, \gamma_K\}$, and

$$[\mathbf{F}_{ij}] = \begin{cases} 0 & \text{if } j = i \\ \frac{\rho_{ji} G_{ji} \|\mathbf{w}_i^H \mathbf{h}_{ji}\|^2}{\rho_{ii} G_{ii} \|\mathbf{w}_i^H \mathbf{h}_{ii}\|^2} & \text{if } j \neq i \end{cases}.$$

In the problem just defined, the complexity lies in the optimization over time and grows rapidly with the number of users. Next, algorithms are developed to reduce the complexity and distribute the computing efforts to both the system level and the user level.

The difficulties in solving (11.7) lie in the feasibility and fairness constraints. First, in the feasibility constraint, if the users' transmitted power is fixed, the targeted SINR γ_i is linearly constrained. On the other hand, if γ_i is fixed, the constraint is linear for **P**. However, if both SINRs and power are considered, the problem is a bilinear matrix inequality (BMI) problem [205]. The BMI problem is nonconvex and nonlinear. Only limited tools are available in the literature to find the solutions [205]. Second, in the fairness constraint, the throughput is considered at different times. It is very difficult to solve the problem by traditional dynamic programming, because the distributions of the received SINRs and transmitted power are extremely hard to model and calculate. Therefore the problem defined in (11.7) is too difficult to find an analytically optimal solution. A heuristic way is needed to obtain a suboptimal solution with relatively good performances.

If the fairness constraint is not considered, the problem in (11.7) is a pure constrained optimization problem. With the consideration of fairness, the motivation to solve the problem comes from jointly considering the throughput ranges and fairness constraints. First, the users report the ranges of throughput that they can accept. Then the system decides how to allocate the throughput to each user each time, according to these ranges. The acceptable throughput ranges are modified by the users' transmission history. Each time, some users may have more throughput, whereas others have less. Then the users with more throughput will become less aggressive about transmitting and will request smaller throughput ranges in the near future, and vice versa. From the preceding idea, the optimization problem in (11.7) is divided into two subproblems:

1. At the user level, to ensure fairness, the users trace their histories of throughput and report the ranges of throughput that they can accept to the system at the current time.
2. At the system level, for the whole network each time, the system determines the optimal throughput allocation to different users, and this allocation requires the lowest overall transmitted power.

Therefore the overall transmitted power is minimized each time, and fairness is guaranteed. However, the optimal solution for (11.7) is not guaranteed to be achieved. But, from the simulation results later, the significant performance improvements over the traditional system [250] will be shown.

An illustrative example for two users is shown in Figure 11.3. The two axes represent the two users' desired SINRs that are related to their throughput. The provided ranges are the required SINRs for the throughput ranges that the users provide, and these ranges are also restricted by the feasibility constraint. On the dashed curve, the overall network throughput $T = T_1 + T_2$ is a constant. At the system level, the goal is to find what the optimal point each time that requires the minimum overall transmitted power within the range (shown as the polyhedra) and under the overall throughput constraint $T \geq R$. At the user level, the problem is how to change the throughput ranges over different times

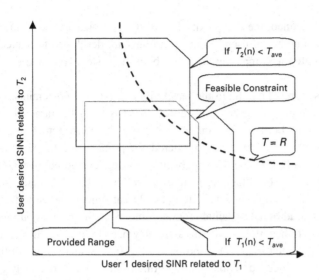

Figure 11.3 Two-user example for problem partition.

to ensure fairness. For example, if user 1 is assigned to have small throughput now, he will be more aggressive about transmitting his data in the future. Consequently, the throughput range will move to the right within the practical range, and user 1 has to be assigned the higher throughput in the future.

For the throughput range at the user level, an adaptive algorithm is developed at the user level to report the acceptable throughput ranges back to the system so as to ensure fairness. The key idea is to adapt the throughput ranges with joint consideration of the fairness constraint. Instead of having a fixed throughput range $[T_i^{\min}, T_i^{\max}]$ for each link, the throughput ranges are adaptively changed by taking into account the links' throughput histories. Assume the ith link can select throughput $T_i^{\min}(n) \le T_i(n) \le T_i^{\max}(n)$ at time n, and the average desired throughput for the ith link is T_i^{ave}. Each time, $T_i^{\min}(n+1)$ and $T_i^{\max}(n+1)$ are modified by the current $T_i(n)$. When $T_i(n)$ is smaller than T_i^{ave}, $T_i^{\min}(n+1)$ and $T_i^{\max}(n+1)$ are increased so that there is a higher probability that the future throughput $T_i(n+1)$ is larger than T_i^{ave}. When $T_i(n)$ is larger than T_i^{ave}, $T_i^{\min}(n+1)$ and $T_i^{\max}(n+1)$ are decreased so that there is a higher probability that $T_i(n+1)$ is smaller than T_i^{ave}. $T_i^{\min}(n+1)$ and $T_i^{\max}(n+1)$ are bounded by $\hat{T}_i^{\min}$ and $\hat{T}_i^{\max}$, which are the practical minimum and maximum throughput boundaries that the ith link can select, respectively. Their values are fixed and predefined by the system. To track the history of T_i, $T_i^{\text{mid}}(n) = T_i^{\text{mid}}(n-1) + \beta[T_i(n) - T_i^{\text{ave}}]$, $0 < \beta < 1$, where β is a constant that depends on how much delay the user can suffer. If the delay constraint is tight, β should be selected as a relatively larger number, so that the throughput range will move quickly to compensate for the user's throughput loss at a specific time. If the user can suffer longer delay, β could be selected as a relatively smaller number, so that the user can wait until the channel becomes better to be compensated back. The selected value of β is also affected by how rapidly the channels change. If the channels change slowly, a smaller β is preferred, so that the user can wait; otherwise, a larger β is selected.

Each time, the throughput window is updated by

$$T_i^{\text{mid}}(n) = T_i^{\text{mid}}(n-1) + \beta \left[T_i(n) - T_i^{\text{ave}} \right],$$
$$T_i^{\text{min}}(n+1) = \min \left\{ \max \left[T_i^{\text{ave}} - T_i^{\text{mid}}(n) + \hat{T}_i^{\text{min}}, \hat{T}_i^{\text{min}} \right], \hat{T}_i^{\text{max}} \right\}, \qquad (11.8)$$
$$T_i^{\text{max}}(n+1) = \max \left\{ \min \left[\hat{T}_i^{\text{max}} - T_i^{\text{mid}}(n) + T_i^{\text{ave}}, \hat{T}_i^{\text{max}} \right], \hat{T}_i^{\text{min}} \right\}.$$

The preceding throughput window may move to the opposite direction of the channel-changing trend. When the channel is bad, the user selects less throughput. But the next time, the user has to select a larger throughput because the throughput window moves to a higher throughput area, even if the channel is still bad. With the consideration of the channel changes, a scheme is developed so that the throughput window follows the channel-changing trend. This problem can be categorized as a dynamic programming problem given by

$$\left[T_i^{\text{min}}(n+1), T_i^{\text{max}}(n+1) \right] = f_n \left\{ \left[T_i^{\text{min}}(n), T_i^{\text{max}}(n) \right], v_n, T_i(n) \right\}, \quad n = 0, \dots, N-1,$$
$$(11.9)$$

where f_n is a function to select the throughput window at time n, and v_n is the control policy that has a different impact on the outcomes of f_n. The problem in (11.9) is extremely difficult to solve, but an intuitive idea can be applied to find a much simpler solution. Because β may not be an integer, the throughput window developed in (11.8) may not be discrete. If the ith user's assigned throughput at the current time n is smaller than the median of all the users' assigned throughput in the adjacent cells, this means that the ith user is possibly still under the bad channel condition. The lower throughput window is assigned to follow the channel condition, by using the floor of the original throughput window. Here the floor is a function that finds the maximum integer immediately less than the real value. On the other hand, if the ith user's throughput is larger than the median of the users' throughput among the adjacent cells, the higher throughput window is assigned to follow the channel condition by using the ceiling of the original throughput window. Here the ceiling is a function that finds the minimum integer immediately greater than the real value.

In addition, when a user is trapped in a bad channel for a long time, instead of assigning him with a very high throughput range, the algorithm should be able to assign this user with lower throughput. By doing so, the user will not cause too much CCI to others, and the system performance can be improved. The history of $T_i^{\text{min}}(n)$ is tracked. If the user detects Z consecutive $T_i^{\text{min}}(n)$ equal to $\hat{T}_i^{\text{max}}$, the user will report the acceptable throughput range as $[\hat{T}_i^{\text{min}}, T_i^{\text{max}}(n)]$, instead of $[T_i^{\text{min}}(n), T_i^{\text{max}}(n)]$. Consequently, the system is able to assign the minimal throughput to the user. The throughput ranges are updated by users to the BS every power-update interval. Because the ranges are discrete and limited by the hardware, the associated overheads to report these ranges are small. In a real system, this information is coded by a powerful error-control code to ensure that it comes through without errors. In each iteration, users' throughput windows are updated by the following algorithm.

Adaptive Algorithm for Each User's Throughput Window

1. **Initialization:** $T_i^{\min}(0) = \hat{T}_i^{\min}$, $T_i^{\max}(0) = \hat{T}_i^{\max}$, $T_i^{\mathrm{mid}}(0) = T_i^{\mathrm{ave}}$

2. **Iteration:**
   $T_i^{\mathrm{mid}}(n) = T_i^{\mathrm{mid}}(n-1) + \beta[T_i(n) - T_i^{\mathrm{ave}}]$;
   if $T_i(n) > \mathrm{median}[T_j(n)]$, $j \in$ all adjacent CCI cells,

   $T_i^{\min}(n+1) = \mathrm{floor}\left(\min\{\max[T_i^{\mathrm{ave}} - T_i^{\mathrm{mid}}(n) + \hat{T}_i^{\min}, \hat{T}_i^{\min}], \hat{T}_i^{\max}\}\right)$;

   $T_i^{\max}(n+1) = \mathrm{floor}\left(\max\{\min[\hat{T}_i^{\max} - T_i^{\mathrm{mid}}(n) + T_i^{\mathrm{ave}}, \hat{T}_i^{\max}], \hat{T}_i^{\min}\}\right)$;

   else

   $T_i^{\min}(n+1) = \mathrm{ceiling}\left(\min\{\max[T_i^{\mathrm{ave}} - T_i^{\mathrm{mid}}(n) + \hat{T}_i^{\min}, \hat{T}_i^{\min}], \hat{T}_i^{\max}\}\right)$;

   $T_i^{\max}(n+1) = \mathrm{ceiling}\left(\max\{\min[\hat{T}_i^{\max} - T_i^{\mathrm{mid}}(n) + T_i^{\mathrm{ave}}, \hat{T}_i^{\max}], \hat{T}_i^{\min}\}\right)$.

3. **Feedback the Acceptable Throughput Ranges to BS:**
   if $T_i^{\min}(n+1) = T_i^{\min}(n) = \cdots = T_i^{\min}(n - Z + 1) = \hat{T}_i^{\max}$,
   report $[\hat{T}_i^{\min}, T_i^{\max}(n)]$;
   else, report $[T_i^{\min}(n), T_i^{\max}(n)]$.

If a user is never trapped in the bad channel for a long period of time, when $T_i(n)$ is continuously less than T_i^{ave} for some time, $T_i^{\min}(n)$ is increased to T_i^{ave}. Then the next $T_i(n+1)$ has to select the throughput equal to or greater than T_i^{ave}; consequently, $T_i^{\mathrm{mid}}(n)$ stops increasing. The same analysis can be applied to $T_i^{\max}(n)$. Because $T_i^{\min}(n)$ and $T_i^{\max}(n)$ are bounded and are linearly modified by $T_i^{\mathrm{mid}}(n)$, $T_i^{\mathrm{mid}}(n)$ is also bounded. If $T_i^{\mathrm{mid}}(n)$ is rearranged and summed over the different times,

$$\frac{\sum_{n=1}^{N} T_i(n)}{N} = T_i^{\mathrm{ave}} + \frac{[T_i^{\mathrm{mid}}(n) - T_i^{\mathrm{ave}}]}{\beta N}. \tag{11.10}$$

The second term on the right-hand side decreases to zero as $N \to \infty$. So $\lim_{N \to \infty} \frac{\sum_{n=1}^{N} T_i(n)}{N} = T_i^{\mathrm{ave}}$, i.e., the system is fair, so that each user's time-average throughput is a constant.

If a user is trapped in the bad channel for a long period of time and detects Z consecutive $T_i^{\min}(n)$ equal to $\hat{T}_i^{\max}$, the user will report the acceptable throughput range as $[\hat{T}_i^{\min}, T_i^{\max}(n)]$. Under this condition, $T_i^{\mathrm{mid}}(n)$ will not be bounded. If the channel becomes better in the future and the system assigns more throughput to this user, $T_i^{\mathrm{mid}}(n)$ will be increased. Consequently, $T_i^{\max}(n)$ will be less than $\hat{T}_i^{\max}$, and the second term on the right-hand side of (11.10) will approach zero asymptotically. If a user is trapped in the bad channel indefinitely, $T_i^{\mathrm{mid}}(n)$ will go to negative infinity, and the fairness constraint cannot be satisfied. In practice, this situation seldom happens. If it does happen, there is no practical meaning to guarantee fairness for this user because this user will cause too much CCI that will reduce the system performance significantly.

For throughput allocation at the system level, the second subproblem will be solved, and three adaptive algorithms will be developed at the system level to allocate throughput to different users each time to generate the minimum overall transmitted power. The first one is a full search algorithm that can guarantee finding the optimal solution each time, but the complexity is very high. The second one is a fast search algorithm that analyzes which users contribute more to the overall transmitted power. The last one is an adaptive algorithm by assuming that the throughput is continuous and approximated by (11.5). Then the throughput allocation result is projected to the closest discrete value that satisfies all the constraints.

First, we study the full-search algorithm. Because there are only a limited number of discrete throughputs T_i that each user can select and there are only a limited number of users, a full-search method can be applied to find the optimal throughput allocation. The users provide the acceptable throughput ranges to the system. The system calculates the overall transmitted power of all combinations of T_i by the iterative algorithm under the constraints in (11.7). The throughput allocation that generates the lowest overall transmitted power is selected. The adaptive algorithm can find the optimal solution each time, but it has very high complexity. The complexity is increased exponentially with the number of users, which is not acceptable in practice. It can be used as a performance bound. The full-search-adaptive algorithm is given as follows.

Full-Search Adaptive Algorithm for Throughput Allocation

1. **Adaptive Modulation:**
   search all possible $T_i(n)$ for every user subject to the constraints.
   find the combination of $T_i(n)$ that minimizes $\sum_{i=1}^{K} P_i$
   calculated by the iteration.
2. **Iteration:**
   - Initialization: $P_1, \ldots, P_K$ = any positive feasible values
   - Antenna Diversity: $\mathbf{w}_i = \arg\max_{\mathbf{w}_i} \Gamma_i$
   - Power-Allocation Update Iteration:
     γ_i = required SINR for $T_i(n)$ and desired BER;
     $\mathbf{D} = \mathrm{diag}(\gamma_1, \ldots, \gamma_K)$; $\mathbf{P} = \mathbf{DFP} + \mathbf{u}$.
3. **Throughput Range Update:**
   Update $T_i^{\mathrm{mid}}(n)$, $T_i^{\mathrm{min}}(n)$, and $T_i^{\mathrm{max}}(n)$.

To reduce complexity, a fast search algorithm is developed. The system needs to find out which users contribute more to the overall transmitted power. The gradient of overall transmitted power to each user's targeted SINR is derived. If the users with larger gradients can sacrifice their SINRs a little bit, the overall transmitted power will be reduced significantly.

In the Perron–Frobenius theorem [85], if the spectrum radius of $\mathbf{DF}$, $\rho(\mathbf{DF})$, i.e., the maximum absolute eigenvalue, is less than 1, the minimum overall transmitted power is achieved when $\Gamma_i = \gamma_i$, $\forall\, i$, and P_{sum} can be written as

$$P_{\mathrm{sum}} = \sum_{i=1}^{K} P_i[T_i(n)] = \mathbf{1}^T(\mathbf{I} - \mathbf{DF})^{-1}\mathbf{u}, \qquad (11.11)$$

where $\mathbf{1} = [1 \ldots 1]^T$. Define $\mathbf{Q} = [\mathbf{I} - \mathbf{DF}]^{-1}$. If $(\mathbf{DF}) \in R^{K \times K}$ and $\rho(\mathbf{DF}) < 1$, then $\mathbf{Q} = \sum_{j=0}^{\infty} (\mathbf{DF})^j$. Because $\mathbf{D} = \mathrm{diag}(\gamma_1, \ldots, \gamma_K)$, and $[\mathbf{F}]_{ij} > 0, \forall i, j$, if γ_j, $j = 1 \ldots K$, $j \neq i$, is fixed, every component in $\mathbf{Q}$ is a function of $(\gamma_i)^j$, $j = 1 \ldots \infty$ with nonnegative coefficients. In vector $\mathbf{u}$, every u_i has the nonnegative coefficients as well.

So $P_{sum} = \sum_{i=1}^{K} P_i = \mathbf{1}^T \mathbf{Q}^{-1} \mathbf{u}$ is also a function of $(\gamma_i)^j$ with nonnegative coefficients. The only situation in which the coefficients are zeros is when the antenna diversity uses a null for the desired mobile user. This hardly happens in practice. Because $\gamma_i > 0, \forall i$, when the other γ_j, $j = 1 \ldots K$, $j \neq i$ is fixed, P_{sum} is a convex and increasing function of γ_i. From (11.5), γ_i is an increasing and convex function of throughput T_i. So P_{sum} is also an increasing and convex function of T_i, when the other T_j, $j = 1 \ldots K$, $j \neq i$ is fixed. Consequently, the overall transmitted power is minimized when the network throughput constraint is equal, i.e., $T = R$. This is because any T_i can be reduced to have smaller overall transmitted power if $T > R$.

Now the gradients of overall transmitted power can be deduced. If the derivatives are taken with respect to γ_i on both sides of (11.11), the ith element of gradient $\mathbf{g} = [g_1 \ldots g_K]^T$ is given by

$$g_i = \frac{\partial P_{sum}}{\partial \gamma_i} = \frac{c_i P_i}{\Gamma_i}, \tag{11.12}$$

where $c_i = \mathbf{1}^T (\mathbf{I} - \mathbf{DF})^{-1} \mathbf{v}_i$, and $[\mathbf{v}_i]_j = 1$, if $j = i$; $[\mathbf{v}_i]_j = 0$, otherwise. The value of c_i reflects how severe the CCI is. When the CCI is large, c_i tells how much the ith user causes CCI to other users. When the CCI is small, $c_i \approx c_j$, $\forall i, j$. Because the adaptive algorithm needs only the direction of the gradients and does not need the amplitudes, the value of c_i can be ignored, i.e., $c_i = 1$, $\forall i$, when the CCI is small. Equation (11.12) is very significant in that it provides a very simple way to find the gradients. In this case, SINRs can be measured at each BS's antenna diversity output, and the feedback channels can be used to get the mobile transmitted power values to calculate the gradients. Consequently the complexity can be reduced greatly.

With the gradients, a greedy algorithm is developed. First, because the network throughput constraint $T = R$ is nonlinear, the overall transmitted power P_{sum} is no longer a convex function of γ_i under this constraint. The gradients of different users are compared. If a user with a larger gradient selects lower throughput, i.e., he requires a lower targeted SINR threshold, the overall transmitted power is greatly reduced. So first the throughput that generates the lowest overall transmitted power is decided for the user with the highest gradient, subject to the constraints. When the throughput of the user with the largest gradient is changed, the throughput of the other users is modified in order from the lower gradient to higher gradient to compensate for the network throughput constraint $T = R$. By doing this, more throughput is allocated to the users with small gradients, and less throughput is assigned to the users with large gradients; Consequently, the overall transmitted power will be reduced significantly. Note that the throughput of the user with the largest gradient may not end up with the lowest throughput $T_i^{min}(n)$, because the increase of the sum of other user's power may be larger than the decrease of this user's power. In the next iteration, the throughput of the user with the largest gradient is fixed, and the system finds the optimal throughput for the user with the second highest gradient, and so on until we find the throughput of the last user in the row. Because every user searches for only a fixed amount of throughput range and reordering is needed, if the gradient is calculated by (11.12) and $c_i \approx c_j$, $\forall i, j$ for simplicity, this suboptimal algorithm has the complexity of only $O(K^2 \log_2 K)$. If the CCI is severe and

$c_i \neq c_j$, then the complexity is $O(K^3 \log_2 K)$. If the user index is rearranged from the largest gradient to the lowest, i.e., $g_1 \geq g_2 \geq \ldots \geq g_K$, and any nonfeasible solution has $P_{\text{sum}} = \infty$, the suboptimal adaptive iterative algorithm is summarized by the following algorithm:

Fast Search Adaptive Algorithm for Throughput Allocation

1. **Initialization:**
   $T_1(0) = T_1^{\text{ave}}, \ldots, T_K(0) = T_1^{\text{ave}},$
   $P_1, \ldots, P_K = $ any feasible positive const.
2. **Adaptive Modulation**
   for $i = 1$ to K
      for $T_i = T_i^{\text{min}}(n)$ to $T_i^{\text{max}}(n)$
        1. Modify from $T_K(n)$ to $T_{i+1}(n)$
     to satisfy the constraint $T = R$ (exhaust T_K first).
        2. Run iteration
           •Antenna Diversity: $\mathbf{w}_i = \arg\max_{\mathbf{w}_i} \Gamma_i$
           •Power-Allocation Update Iteration:
           $\gamma_i = $ required SINR for T_i and desired BER,
           $\mathbf{D} = \text{diag}(\gamma_1 \ldots \gamma_K), \quad \mathbf{P} = \mathbf{DFP} + \mathbf{u}.$
        3.Find T_i that generates the lowest power for the ith user.
     end
   end
3. **Throughput Range Update:**
   Update $T_i^{\text{mid}}(n)$, $T_i^{\text{min}}(n)$, and $T_i^{\text{max}}(n)$.

The algorithm is suboptimal because the optimal throughput for one user may not be optimal for all users. The algorithm may stop at some local minimum points or the boundary points. From the simulation results that will be shown later, the suboptimal algorithm has a relatively good performance.

The feasible constraint in (11.7) is a BMI constraint. Here the approximation of throughput in (11.5) is used, and a projected gradient algorithm [205] is developed to change each user's targeted SINR to find the minimal overall transmitted power. The throughput allocation results are probably not integers. The results are projected to the nearest discrete throughput allocation that satisfies the constraints. Then the preceding two steps are employed again, until the discrete throughput allocations are the same in two consecutive iterations.

First, the projected gradient method will be developed. The throughput is now supposed to be continuous and has the value $\tilde{T}_i$. If each user's targeted SINR is changed by the gradient in (11.12), the overall throughput constraint $\sum_{i=1}^{K} \tilde{T}_i = R$ cannot be satisfied. The gradient needs to be modified such that the overall throughput constraint holds. The plane that is tangent to the curve $\sum_{i=1}^{K} \tilde{T}_i = R$ at point $[\gamma_1, \ldots, \gamma_K]$ needs to be found, where $\gamma_i = (2^{\tilde{T}_i/c_4^i} - 1)/c_5^i$. Without loss of generality, this plane can be moved to the origin. The plane can be expressed as $\sum_{i=1}^{K} k_i x_i = 0$, where $k_i = c_4^i c_5^i/(1 + c_5^i \gamma_i)$. The modified gradient is given by $\mathbf{q} = [q_1 \ldots q_K]^T$. By the definition of projection, vector $\mathbf{q}$ satisfies equation $\|\mathbf{g} - \mathbf{q}\|^2 = \min_{\forall \mathbf{q} \in \text{plane}} \|\mathbf{g} - \mathbf{q}\|^2$. The right-hand side needs to be minimized to get the optimal vector, i.e., the projection $\mathbf{q}$.

The best throughput allocation $\tilde{T}_i$ is obtained from the preceding projected gradient algorithm. $\tilde{T}_i$ needs to be projected to a discrete value. The projection problem can be written as

$$\min_{T_i} \|\tilde{T}_i - T_i\|^2, \tag{11.13}$$

$$\text{subject to} \sum_{i=1}^{K} T_i = R, \ T_i^{\min}(n) \leq T_i \leq T_i^{\max}(n), \ \text{and} \ (\mathbf{I} - \mathbf{DF})\mathbf{P} \geq \mathbf{u},$$

where $\tilde{T}_i$ is projected to the discrete value with the constraint $\sum_{i=1}^{K} T_i = R$. However, the discrete throughput projection may not be feasible or may not be in the ranges. If this is the case, the second closest point needs to be found to see if it satisfies all the constraints. The search is continued until a feasible solution is found. The projected gradient algorithm is given as follows:

Projected Gradient Algorithm

1. **Initialization:**
   $T_1(0) = T_1^{\text{ave}}, \ldots, T_K(0) = T_K^{\text{ave}}$,
   $P_1, \ldots, P_K = $ any feasible positive const.
2. **Iteration:** Stop when T_i is stable.
   - Antenna Diversity: MRC or SC
   - Adaptive Threshold Allocation
     do
     {SINR Range:
     $\gamma_i^{\min} = (2^{T_i^{\min}/c_4^i} - 1)/c_5^i; \ \gamma_i^{\max} = (2^{T_i^{\max}/c_4^i} - 1)/c_5^i$
     Projected Gradient:
     $\mathbf{g} = \nabla P_{\text{sum}}; \ \mathbf{q} = \text{projection}(\mathbf{g}); \ \gamma_i = \gamma_i - \mu \cdot q_i \ \forall i$;
     Within Range:
     if $(\gamma_i > \gamma_i^{\max}) \ \gamma_i = \gamma_i^{\max}$; if $(\gamma_i < \gamma_i^{\min}) \ \gamma_i = \gamma_i^{\min}$}
     while (γ_i not stable, not boundary)
   - Adaptive Modulation: Select $\tilde{T}_i = c_4^i \log_2(1 + c_5^i \gamma_i)$.
   - Throughput Projection:
     Project $\tilde{T}_i$ to the nearest T_i that satisfies the constraints.
   - Power-Update Iteration:
     $\gamma_i = (2^{T_i/c_4^i} - 1)/c_5^i; \ \mathbf{D} = \text{diag}(\gamma_1, \gamma_2 \cdots \gamma_K); \ \mathbf{P} = \mathbf{DFP} + \mathbf{u}$.
3. **Throughput Range Update:**
   Update $T_i^{\text{mid}}(n)$, $T_i^{\min}(n)$, and $T_i^{\max}(n)$.

In the algorithm, μ is a small constant whose value decides the rate of convergence and the variance of the final results. Whether γ_i is stable is decided by comparing the maximum difference of γ_i in two consecutive steps. The algorithm has complexity of $O(K^3)$. However, because two iterations are needed each time, the complexity is higher

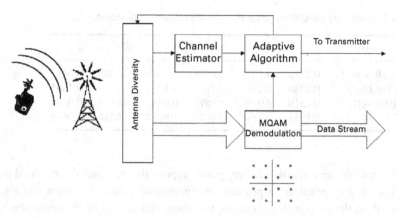

Figure 11.4 Power-control and throughput management system.

than that of the fast search algorithm but still much lower than that of the full-search algorithm, when the number of users is large. From the simulation results, it will be shown that the projected gradient algorithm can find the optimal solution each time.

The algorithm starts from any feasible rate and power allocation. In each iteration, the gradient of the overall transmitted power is calculated and projected onto a plane where the network performance constraint is satisfied. This modified gradient is at least pointing in the direction where the overall transmitted power is increasing. The algorithm modifies the SINR allocation in the opposite direction of this modified gradient, so that the new overall transmitted power is less than or equal to that of the old iteration. When the algorithm finds the SINR allocation solution, this SINR allocation must be feasible, and the transmitted power is updated by fixing the targeted SINRs. This power-update iteration converges to a unique solution [250, 347].

With the adaptive algorithms, a joint power-control and throughput management system is constructed in Figure 11.4. Because of users' multipath fading, shadowing, and random locations in their respective cells, the channel conditions are varying. Therefore accurate techniques for real-time estimations of channel conditions are essential [69, 148, 306]. The fluctuations of channels are assumed to be tracked perfectly by the BSs. This information is sent back to the mobile users via an error-free feedback channel. The time delay in this feedback channel is also assumed to be negligible, compared with the speed of channel and CCI variations. All these assumptions are reasonable in slowly varying channels.

The way the system works and the distribution of computing efforts are shown as follows: At the user level, the users compute and provide the system with their acceptable throughput ranges, according to their transmission histories and current channel conditions. At the system level, in which the BSs have much stronger computing power, the adaptive algorithm module gets the estimation of users' channel responses from the channel-estimation module. Then power control and modulation rates are computed. The power-control and best throughput allocation information is sent back to the mobile users. Then the mobile users, accordingly, adapt their transmission rates and power.

Table 11.1 Normalized transmitted power with respect to numbers of antennas

No. of antennas	2	3	4	5	6	7
MRC (BER = 10^{-2})	0.2722	0.1271	0.0904	0.0598	0.0468	0.0412
SC (BER = 10^{-2})	0.3746	0.2095	0.1677	0.1260	0.1083	0.1065
MRC (BER = 10^{-5})	0.1519	0.0787	0.0572	0.0463	0.0349	0.0264
SC (BER = 10^{-5})	0.1958	0.1248	0.0873	0.0797	0.0716	0.0655

For the mobile device with a battery power supply, the maximum transmitted power is limited. In the optimization problem, the maximum power constraint can also be considered. In the proposed approaches, this constraint can be easily implemented by the full search and fast search algorithms. The algorithms are modified such that only the throughput allocation that satisfies the maximum power constraint will be selected. However, in the projected gradient method, the maximum power constraint will impose another very complex and nonlinear constraint in the proposed adaptive algorithm.

To evaluate the performances of the proposed algorithms, a network with hexagonal cells is simulated. The radius of each cell is 1000 m. Two adjacent cells do not share the same channel, i.e., the reuse factor is 7. There are 50 cells in the networks. One BS is placed at the center of each cell. In each cell, one user is placed randomly with a uniform distribution. In the simulation, the fading is considered as complex Gaussian distributions with three multipaths of equal power. The fading is independent between two resource-allocation intervals. Each BS has a P-element antenna array. Noise power is 10^{-3}, and $Z = 50$. A 3-dB log-normal distribution is considered. The two-slope path-loss model [128] is applied to obtain the average received power as a function of distance. We select the basic path-loss exponent as 2, the additional path-loss exponent as 2, the BS antenna height as 50 m, the mobile antenna height as 2 m, and the carrier frequency as 900 MHz. In Table 11.1, the overall transmitted power of our proposed system is shown with respect to different number of antennas. The value is normalized with the case in which only a single antenna is applied. Two different BER requirements (BER $= 10^{-2}$ and BER $= 10^{-5}$) are shown respectively . The overall transmitted power can have a reduction of about 75% to 95% for MRC compared with the single-antenna case. The performance of SC is consistently worse than that of MRC. Because SC can apply noncoherent modulation, the complexity is much smaller. When the desired BER is decreased, SC performs closer to MRC. With the number of antennas P increasing, from simulations, the decrease of power saturates around $P = 4$. Therefore $P = 4$ is chosen for the rest of the simulations.

In Figures 11.5(a) and 11.5(b), the normalized overall transmitted power as a function of average spectral efficiency (bit/s/Hz) is compared for the fixed scheme [250], fast search scheme, projected gradient algorithm, and optimal full-search scheme with MRC and SC diversity at BER $= 10^{-2}$ and BER $= 10^{-5}$, respectively. We normalize the power with the MRC scheme when the spectral efficiency equals 1. Each user is assumed to have the same desired time-average throughput $T_1^{\mathrm{ave}} = \cdots = T_K^{\mathrm{ave}}$. Define window size $= (T_i^{\mathrm{max}} - T_i^{\mathrm{ave}}) = (T_i^{\mathrm{ave}} - T_i^{\mathrm{min}}) = 2$ bit/s, $\forall i$. Each user is assumed to

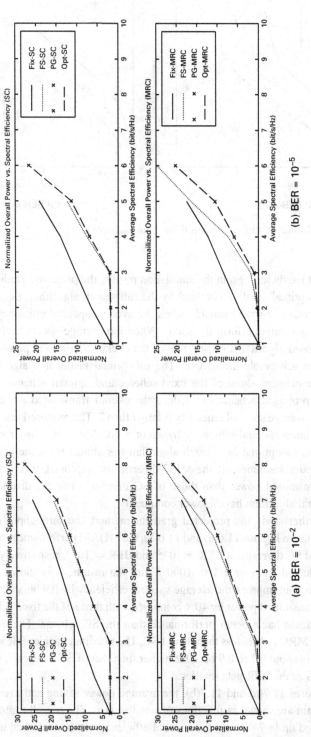

(a) BER = 10^{-2}

(b) BER = 10^{-5}

Figure 11.5 Normalized power (dB) vs. throughput.

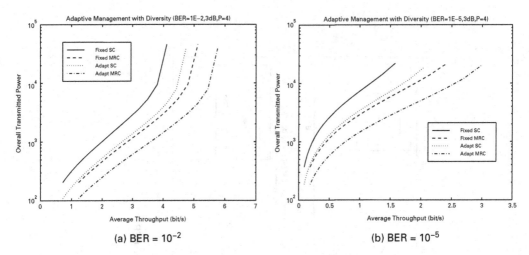

Figure 11.6 MQAM performance with continuous throughput assumption.

have unit bandwidth. From the simulation results, the projected gradient algorithm can find the optimal solution obtained by the full-search algorithm. Because there is only one allocation scheme available when the average spectral efficiency is equal to one, all the algorithms perform the same. When the average spectral efficiency increases, the proposed algorithms greatly reduce the overall transmitted power and increase the maximum achievable throughput. The suboptimal fast search algorithm has the performance between those of the fixed scheme and optimal scheme. The results show that the proposed scheme can reduce the overall transmitted power by about 7 dB when the average spectral efficiency is larger than 2. The proposed scheme also increases the maximum spectral efficiency by about 1 bit/s/Hz. In the lower spectral efficiency range, the suboptimal fast search algorithm has almost the same performance as that of the optimal solution. If the MRC diversity is employed, it reduces about 3–4 dB more transmitted power than those of SC diversity. The SC diversity and proposed suboptimal algorithm have a lower complexity.

To further study the projected gradient method, the throughput is assumed to be continuous. In Figures 11.6(a) and 11.6(b), the MQAM performances are compared with MRC and SC diversity at BER $= 10^{-2}$ and BER $= 10^{-5}$, respectively. The simulations are conducted from time 1 to 1000. From the results, it is shown that the adaptive algorithms can improve the average spectral efficiency by 0.9 bit/s/Hz and decrease the overall transmitted power by 40% compared with those of the fixed schemes. The MRC scheme again has a better performance than the SC scheme. The overall transmitted power of MRC is 40% less than that of SC. The maximum achievable spectral efficiency of MRC is about 0.7 to 0.9 bit/s/Hz higher than that of SC. However, this improvement decreases as the BER gets smaller.

In Figures 11.7(a) and 11.7(b), the average power saving and average spectral efficiency gain are shown as functions of window size. The overall transmitted power can be reduced up to 7 dB, and the spectral efficiency can be increased up to 1.2 bit/s/Hz.

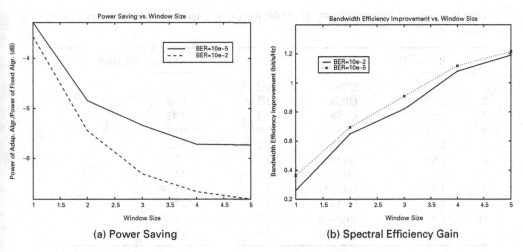

Figure 11.7 Effects of window size.

The power stops decreasing and spectral efficiency increasing speed is reduced, as the window size is growing. This is because of the time-average throughput constraint for each user. The user that gets better throughput at this time must pay back in the future. So there is no need to have a very large window size. Only a limited number of modulation rates are necessary; consequently the system complexity can be simple.

A joint power and throughput optimization framework is proposed to study the performance of adaptive resource allocation in wireless networks. The adaptive power-minimization algorithms are constructed under the fairness constraint by using adaptive modulation with antenna diversity to fully utilize the spectrum, to combat time-varying wireless channels, and to reduce CCI. The proposed scheme can be interpreted as "water filling" each user's throughput in the time domain and allocating the network throughput to different users each time. A joint power and throughput management system is built to adaptively allocate the resources. From the simulation results, the algorithms reduce the total transmitted power of mobile users by up to 7 dB, which is critical in terms of battery life. The spectral efficiency is increased by up to 1.2 bit/s/Hz, which, in turn, increases the network performance.

11.3   Delay-Sensitive Scheduling for Multimedia Transmission

In this section, our goals are to define a new QoS measure for multimedia transmissions and to develop new scheduling schemes, so as to improve system performances and maintain the individual fairness among users. First, we quantify the QoS measure as a user satisfaction factor (USF) for different delay constraints and different applications such as voice, video, and data. The proposed USF is a function of both the number of received bits and the delay-sensitivity profiles. Then, from this QoS measure, we develop four types of scheduling schemes, namely the maxi-min approach, the overall performance approach, the two-step approach, and the

Table 11.2 Required SNRs for different BERs of different adaptive modulation and convolutional coding rates

k	Rate ω_k (W)	Modulation	Coding rate	SNR ρ_k (dB), BER $\leq 10^{-6}$
1	1	QPSK	1/2	4.65
2	1.33	QPSK	2/3	6.49
3	1.5	QPSK	3/4	7.45
4	1.75	QPSK	7/8	9.05
5	2	16QAM	1/2	10.93
6	2.66	16QAM	2/3	12.71
7	3	16QAM	3/4	14.02
8	3.5	16QAM	7/8	15.74
9	4	64QAM	2/3	18.50
10	4.5	64QAM	3/4	19.88
11	5.25	64QAM	7/8	21.94

proportional approach. Simulation results show that these four scheduling schemes achieve better trade-offs between the system performance and individual fairness, compared with the weighted round-robin and the modified proportional fairness scheduling schemes.

We consider a downlink scenario of a single-cell system. The system has K users randomly located within the cell. The transmission bandwidth is W. Slow fading is assumed such that the channel gain is stable within each packet. For simplicity, the channel parameters from different users are assumed perfectly estimated, and the channel information is reliably fed back from mobile users to the BS without any delay. Denote Γ_k as the kth user's SNR: $\Gamma_k = \frac{B_k G_k}{\sigma^2}$, where G_k is the channel gain and B_k is the transmit power for the kth user. The thermal-noise power for different users is assumed to be the same and represented as σ^2. For the downlink system, because of the practical constraints in implementation, such as the limitation of power amplifier and consideration of CCIs to other cells, the transmit power is bounded by $B_{\max}$.

We also assume that, with appropriate channel modulation and coding, the packet loss rate is sufficiently low, meaning that the BER of the channel transmission is kept lower than some desired threshold. Adaptive modulation and adaptive channel coding provide each user with the ability to adjust his/her data transmission rate r_k, according to the channel condition, for achieving the desired BER. We focus on MQAM and convolutional codes as they provide high spectrum efficiency and strong forward error protection, respectively. We select the BER threshold as BER $\leq 10^{-6}$. The required SNRs for BER $= 10^{-6}$ with different modulations and convolutional coding rates using bit interleaved coded modulation (BICM) are listed in Table 11.2, based on the results in [12]. To facilitate discussion, denote the number of combinations in Table 11.2 of different modulation and convolutional coding rates as Q. Define the feasible set of the rate as $\omega = \{\omega_0, \omega_1, \omega_2, \ldots, \omega_Q\}$, and the corresponding set of the required SNR for BER $\leq 10^{-6}$ as $\rho = \{\rho_0, \rho_1, \rho_2, \ldots, \rho_Q\}$, where $\omega_0 = 0$ and $\rho_0 = 0$. Thus all r_k should be selected from the set ω. For the scheduling scheme, the rate should be selected such

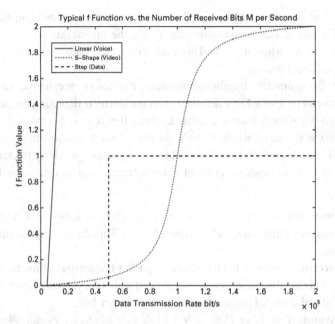

Figure 11.8 Typical f functions vs. received bits M per second.

that the maximal transmission power $B_{\max}$ is able to maintain a BER lower than 10^{-6}. Specifically, the following inequalities should hold if user i's transmit rate ω_j is selected:

$$\rho_j \leq \frac{B_{\max} G_i}{\sigma^2} < \rho_{j+1}, \text{ for } j = 0 \ldots Q - 1; \text{ and } \rho_j \leq \frac{B_{\max} G_i}{\sigma^2}, \text{ for } j = Q. \quad (11.14)$$

We define the USF that represents multimedia users' QoS considering both received bits and delay-sensitivity profile. First, we consider QoS representation when all transmitted bits are received. We define f to quantify QoS as a function of M number of received bits. For different types of services, f can have different definitions. For example, for data transmission, f can be a step function, i.e., only when all information is received can QoS be ensured. For most embedded voice coders, the MSE can be approximated by an exponential function of M. So the corresponding reconstructed peak signal-to-noise ratio (PSNR) can roughly be a linear function of M. For the layered video encoders such as MPEG4-FGS, the base-layer packets should be transmitted first to maintain the basic QoS, and then the enhanced-layer packets are transmitted with the higher-priority packets first. The resulting PSNR is a kind of S shape. Some typical f functions are illustrated in Figure 11.8. Here we normalize the QoS for the different users, so that users' QoS is exactly satisfied when $f = 1$. When $f > 1$, users are oversatisfied, and when $f < 1$, users are undersatisfied. For the different applications, to achieve the same f, the necessary numbers of the received bits M can be different.

For different types of multimedia transmissions, different packets can have different delay-sensitivity profiles. For example, for voice services, the packets should be received with a strict delay. For layered video services, the base layer can be transmitted and decoded earlier than the enhanced layers. We define N as the packet-transmission time with the strictest delay constraint, and n' as the packet-transmission finishing time. The delay-sensitivity profile can be presented by $P(n')$, the probability of finishing the transmission at time n', when $n' \geq N$. This probability represents different delay tolerance levels for different packets.

Here we consider the geometric distribution for delay profiles as an example. Other types of distributions can be applied in a similar way. A parameter α depicting the delay tolerance is assigned when each user is admitted, where $0 \leq \alpha < 1$. Suppose a user finishes the transmission at time n', where $n' \geq N$; the packet is associated with a weight $P(n' = N + i) = (1 - \alpha)\alpha^i$, $i = 0, 1, \dots$. The value of α represents the delay tolerance of the user. Different payload types in terms of delay tolerance can be categorized as follows:

1. **Strict-Delay Constraint:** $\alpha = 0$, $P(n' = N) = 1$, and $P(n') = 0, \forall n' > N$. In this case, the packet must be transmitted before time $(N + 1)$. This fits such applications as voice payloads.
2. **Soft-Delay Constraint:** $0 < \alpha < 1$. The estimated packet-transmission time is $\overline{N} = (N - 1) + \frac{1}{(1-\alpha)}$. This fits the video/image payloads, in which the higher-priority bits can be transmitted and decoded earlier than the lower-priority bits.
3. **No-Delay Constraint:** $\alpha \to 1$, so $P(n' = N + i) \to 0, \forall i \geq 0$. Users can tolerate arbitrary delays. This fits time-insensitive payloads such as e-mails.

In traditional networks, the system assigns the resources to users, according to predefined allocations. There are few feedbacks during transmissions to reflect whether the users really get the desired QoS, even though the wireless channels may fluctuate. Therefore we need to define the USF for QoS measurements such that the system can adapt its resource-allocation strategy under different conditions. Suppose there is an estimator g that estimates the number of transmitted bits at time n', based on the transmission history and channel conditions using the information no later than time n. The most straightforward method is the linear prediction. At the current time n, the resulting prediction of the number of bits received at a future time n' is

$$g(n, n') = \frac{n' \sum_{j=1}^{n-1} T_i(j)}{n - 1}, \ n' \geq N. \tag{11.15}$$

Other types of function g can also be explored, such as a moving window, Kalman filter, etc.

We define USF at time n as

$$\text{USF}_i(n) = \sum_{n'=N}^{\infty} P(n') f[g(n, n')]. \tag{11.16}$$

The physical meaning of USF is the estimated weighted QoS according to the delay-sensitive profile.[1] For example, for the voice services, $\alpha = 0$, then any components of f with $n' > N$ are weighted by zero in (11.16). So the delayed transmission has no impact

---

[1] USF is not the estimation by the distribution of the real packet finishing time but by the delay-sensitive weight instead.

on USF. For the data transmissions with $\alpha \to 1$, USF is always 1. For the soft delay case, USF is affected by both QoS function f and distribution profile of P. Note that this USF defines the short-term QoS, as the delay profile is related to each packet's transmission time. The value of USF represents the user's QoS measurement as follows:

1. $USF(n) > 1$: The user is oversatisfied at the current time n. For the rest of transmission time, he/she may be less aggressive for acquiring resources.
2. $USF(n) = 1$: User's QoS is exactly satisfied. If he/she acquires the resource at the current rate, the final QoS f is equal to 1.
3. $0 \leq USF(n) < 1$: When USF becomes smaller, the user becomes more dissatisfied and has to transmit more aggressively in the rest of packet-transmission time.

From USF, we develop four types of scheduling schemes to improve both individual fairness and system overall performances.

Scheduling Scheme 1: Maximin Approach The first scheduler selects the user with the minimal USF for transmissions so as to improve the minimal performance of the system. The scheduler policy is expressed as

$$\arg\min_i \; w_i[\text{USF}_i(n)] \,,$$

where w_i is the weight value for user i. For example, we can select $w_i > 1$ when user i has no delay constraint and $w_i = 1$ for other applications. So only when other delay-sensitive users are oversatisfied, the user with no delay constraints can have priority for transmission. This approach is fair among users but will have general inferior system performances because the users in the bad channel conditions might consume most of scheduled transmission times.

Scheduling Scheme 2: Overall Performance Approach This type of scheduler tries to maximize the overall USF. Define $\Delta\text{USF}_i(n)$ as the USF change if the scheduler assigns the current time slot to user i. Because each time only one user occupies the channel, we have

$$\arg\max_i \; w_i[\Delta\text{USF}_i(n)].$$

By doing this, the user with the best channel condition has the advantage for transmissions. Consequently, the scheduler can take advantage of multiuser diversity and channel diversity to optimize the system performances. However, the fairness among users is not considered.

Scheduling Scheme 3: Two-Step Approach The philosophy of this approach is to maintain the users' minimal QoS first, and then try to maximize the overall system performances. The approach is the combination of the first two approaches, shown as

$$\arg\min_i \; w_i[\text{USF}_i(n)], \quad \text{if } \text{USF}_i < 1, \exists i;$$
$$\arg\max_i \; w_i[\Delta\text{USF}_i(n)], \text{ otherwise.}$$

This approach can satisfy the basic fairness among users and then optimize the system performances. So it is a trade-off between the first two approaches.

Scheduling Scheme 4: Proportion Approach In [149], proportional fair scheduling is proposed, in which the scheduler selects the user with the largest ratio of current possible maximal rate over the sum of rates in his/her transmission history. By doing this, the scheduler takes into account the current channel condition and maintains the *long-term proportional* fairness. We employ a similar idea. Compared with the second approach, the scheduler weights the user with the inverse of his/her current USF value plus a small positive constant ε. This constant is used to prevent the error of "divided by zero." If the user has a better channel, the rate adaption can transmit more information to improve his/her USF, and the user has the advantage of transmitting. On the other hand, if the USF is small, the user also has the advantage of transmissions. The scheduler can be expressed as

$$\arg\max_{i} \frac{w_i[\Delta \mathrm{USF}_i(n)}{]} \mathrm{USF}_i(n) + \varepsilon.$$

This scheduler provides another trade-off between the performances and fairness.

We conduct the simulations with the following settings to demonstrate the performances of the proposed schemes based on the new QoS measure USF. The channel gain is given by

$$G = \frac{C_0}{r^a[1 + r\lambda_c/(4h_b h_m)]^b} \tag{11.17}$$

where C_0 is a constant that depends on the antenna gain (here $C_0 = 10^{-7}$), r is the distance between the mobile and the BS, a is the basic path-loss exponent (we select two), b is the additional path-loss component (we select one), h_b is the BS antenna height, h_m is the mobile antenna height, and λ_c is the wavelength of the carrier frequency. The mobile antenna height is 2 m, and the base station antenna height is 50 m. The carrier frequency is 1845 MHz. The total bandwidth is 100 KHz. The maximal BS transmission power is 10 W. Shadowing has a standard deviation of 10 dB and shadowing correlation distance is 2 m. The background-noise power is -90 dBm W. All users are uniformly distributed within the range of $[r_0, r_1]$ with $r_0 = 20$ m being the closest distance and $r_1 = 500$ m being the cell radius. Each user has a minimum speed of 2 km/h and a maximum speed of 100 km/h. The period for direction change is 127 steps. The scheduling updates 2000 times per second. In the simulations, we have 100 different settings of users' locations with 60 for each setting. $w_i = 1, \forall i$ and $\varepsilon = 0.001$. There are 13 users in the system, 10 users have voice payloads, 1 user has video payloads, and 2 users have data payloads. Three types of heterogenous payloads are shown in Figure 11.8 and are described in detail as follows:

1. **Type 1 User, Voice Payloads:** To simulate the real-time wireless voice communications, we employ the GSM AMR (adaptive multirate) narrowband speech encoder. This encoder operates with 20 ms per packet with possible encoding rates ranging from 4.75 to 12.2 kbps. So we assume that $\alpha = 0$ and the delay time is 20 ms. We use the linear approximation and assume that $f = 0$ at 4.75 kbps, $f = 1$ at 10 kbps, and $f = 1.421$ at 12.2 kbps.

2. **Type 2 User, Video Payloads:** We consider the embedded video encoder that generates constant rates over time for simplicity. For general variable-bit-rate video encoders, the parameters for function f and M are changed for different video contents and different I, B, and P frames. We assume there are 10 frames per second. The strictest transmission time is 50 ms and delay factor $\alpha = 0.99$. We approximate the QoS function as an S-shaped function $f = \arctan[0.05(M - 5000)] + 0.5\pi$, where $f = 1$ at rate 100 kbps and $f = 2.9442$ at rate 200 kbps. This fits a typical transmission of QCIF (quarter common intermediate format) video with a resolution of 176×144.

3. **Type 3 User, Data Payloads:** We assume the packet length is 10 kbits. Each packet is appended with the CRC check. So we have $f = 1$ if all bits are decoded correctly and $f = 0$ otherwise. The arrival packets are modeled as Poisson arrival with rate equal to 1. We assume the strict delay time is 100 ms and the delay parameter is $\alpha = 1 - 10^{-6}$.

It is worth mentioning that the preceding quantifications of QoS can be generalized. More practical considerations can be taken into account. For example, for voice coders, the talk spurt and silence spurt can have different rate requirements; for video encoders, different scenes and frames may have different parameters; the other QoS measurement can also be considered for the QoS function f.

To compare the performances, we also study the weighted round-robin and proportional fair scheduling schemes to the preceding settings. Because short-term fairness is considered, we modify the proportional fair scheme [160] using the following criteria:

$$\arg\max_i \frac{c_i \, T_i(n)}{\sum_{j=1}^{n-1} T_i(j)}, \qquad (11.18)$$

where $T_i(n)$ is the maximal rate that can be achieved for user i at time n and c_i is the weight factor. The video payloads have 10 times more weight than the other payloads for both round-robin and modified proportional fair scheduling schemes.

In Figures 11.9–11.11, we show the overall transmission proportions, overall throughput, and average USF of the different payloads, respectively. Notice that, for the first type of payload, there are 10 users and $f = 1$ at 10 bps, whereas for the second type of payload, there is one user and $f = 1$ at 100 bps. So if we do not consider the delay requirement, the overall rates of these two types of payloads should be the same. From the figures, we can see that the proposed four schemes give more proportions of time to the first type of users. Consequently, the first types of users have higher throughput and higher USF than those of the second types of users. This is because the proposed scheduling schemes take into consideration the strict short-term delay requirement of the first type of users, whereas the weighted round-robin and modified proportional fairness scheduling schemes do not. In addition, the overall rate of type 3 data payload is 20% of the other two payloads. The weighted round-robin and modified proportional fairness scheduling schemes give about 1/5 of transmission slots to the third type of users compared with the other types. For these types of users, USF is always equal to 1 for all schedulers, because there is almost no delay requirement. The proposed

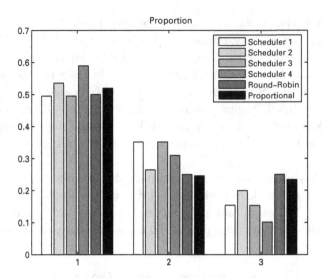

Figure 11.9 Overall transmission proportions of different payloads.

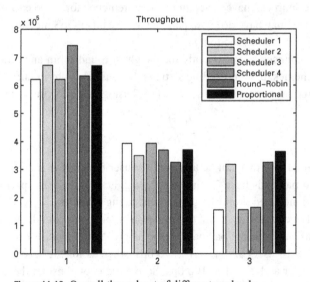

Figure 11.10 Overall throughput of different payloads.

schemes allocate the transmission time only if the other delay-sensitive types of users'
QoS's are satisfied. So better resource allocations are provided with the short-term delay
requirements.

Comparing the proposed four schemes, we notice the following observations. Scheme
1 has the least differences of USF among different payloads. This is because scheme
1 always tries to improve the performance of the user with the least value of USF.
However, this scheme has the lowest throughput. Scheme 2 has the highest overall USF
because there are 10 users of type 1 and 1 user of type 2. The overall throughput is also
high. However, the USF of type 2 is the lowest. This is because scheme 2 tries only to

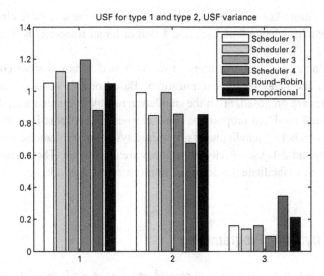

Figure 11.11 Average USFs of different payloads.

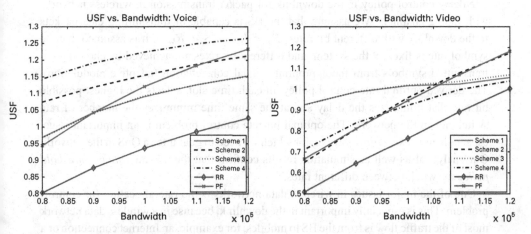

Figure 11.12 USFs for different users.

maximize the overall USF without considering the fairness among users. Schemes 3 and 4 provide certain trade-offs between the system performances and individual fairness. The throughput and USF performances are between scheme 1 and scheme 2. Compared with scheme 3 and scheme 4, scheme 3 finds the trade-off more close to the user fairness whereas scheme 4 finds the trade-off more close to the system performances, under this simulation setting.

In Figure 11.12, we show the USF for the different proposed schemes versus the overall bandwidth. We modify the bandwidth from 80 to 120 kHz so that the system changes from crowded to less crowded. From the figure, we can see that four proposed schemes perform better when the bandwidth is increasing. The gaps between the performances for type 1 users and type 2 users are different. Scheme 1 has the smallest, scheme 2 has the largest, scheme 3 is more fair and close to scheme 1, and scheme 4 is more close to scheme 2. On the whole, scheme 3 and scheme 4 find different trade-offs between scheme 1 and scheme 2.

In summary, we define a new QoS measurement for multimedia transmissions considering heterogeneous and delay-sensitive applications. Based on this measurement, four scheduling schemes are proposed. From the simulation results, compared with the weighted round-robin and modified proportional fair schemes, the proposed schemes find the different trade-offs between individual fairness and system performances, while the heterogenous nature and delay sensitivity of payloads are considered. The schemes can be further generalized to facilitate the design of future wireless networks.

11.4   Rate Control Using Dynamic Programming

In this section, we consider the downlink rate-control problem in a wireless channel. A dynamic programming optimization method is introduced to obtain the optimal bit-rate/delay-control policy in the downlink for packet transmission in wireless networks with fading channels. We assume that the BS is capable of transmitting data packets in the downlink with different bit rates, $R_0 < R_1 < \cdots < R_{M-1}$. It is assumed that the symbol rate is fixed in the system, and different bit rates are achieved by choosing the transmitted symbols from the appropriate signal constellation (adaptive modulation). The derived optimal rate-control policy, in each time slot, selects the highest possible bit rate that minimizes the delay and at the same time minimizes the number of rate switchings in the network. The optimal bit-rate control problem is an important issue, especially in *packet data networks*, for which we need to guarantee a QoS in the network. Our analytical as well as simulation results confirm that there is an *optimal threshold policy* to switch between different rates.

One of the major issues in wireless data networks is the rate-allocation (or control) problem. This is especially important in the downlink, because in a wireless data network most of the traffic flow is from the BS to mobiles, for example, an Internet connection or a multimedia (voice/image/data) connection. In this section, we derive some properties of a class of the optimal rate-control problem by using the theory of dynamic programming (DP). The general nature of the problem considered is as follows. In a wireless network, the BS is capable of transmitting data packets at different rates. The BS transmits the data packets over a wireless channel to mobile users. The received SNR by the mobile users is subject to fluctuation because fading and noise. We assume a finite-state Markov (FSM) model for the wireless channel. The mobile constantly monitors the received SNR. At each measurement instant the mobile observes the state of the channel and determines which state of channel it belongs to. At each decision-making instant, by employing an optimal strategy, the mobile decides whether to send a request to the BS to switch the rate or not. For this purpose there is a feedback channel (assumed to be a BS). Figure 11.13 illustrates the block diagram of a system in which the mobile employs an optimal strategy in choosing the rate in the network.

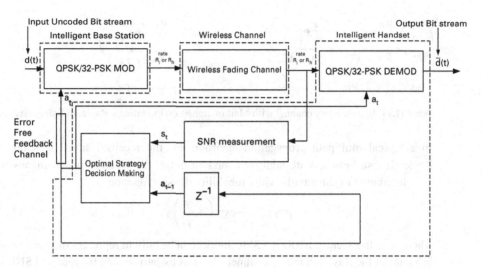

Figure 11.13 Block diagram of a system in which the mobile employs an optimal strategy in choosing the data rate in the network.

The optimal policy that determines the choice of rates (or modulation schemes) jointly minimizes the delay in sending the packets and the number of rate switchings. We show that under certain conditions the optimal strategy has the form of a threshold policy. In a wireless network, shadowing and fading effects result in signal-strength variations in mobile environments. This may cause unnecessary and frequent rate switchings, which is highly undesirable, because it translates to protocol overheads to switch the rate (rate-negotiation phase). An improperly designed rate-control algorithm can result in an unacceptably high level of bouncing (resulting in high signaling costs) and/or a high probability of forced termination. Delaying a rate switching as the signal strength received from the BS starts to deteriorate may result in a lost data-transmission session. A good rate-control algorithm also reduces the occurrence of the involuntary termination of the data transmission in the network.

In the FSM channel (FSMC) model each state corresponds to a specific channel quality that is either noiseless or totally noisy. The idea is to form a FSM model for such wireless channels [325]. Let $S = \{0, 1, \ldots, K - 1\}$ denote a finite set of states. By partitioning the range of the received SNR into a finite number of intervals, FSMC models can be constructed for Rayleigh fading channels [325]. The members of set S correspond to those partitions. Now let $\{S_n\}$, $n = 0, 1, \ldots$ be a *stationary Markov process*. Because a stationary Markov process has the property of *time-invariant transition probabilities*, the transition probability is independent of the time index n and can be written as

$$p_{ss'} \equiv \Pr(S_{n+1} = s' | S_n = s), \quad n = 0, 1, \ldots, \quad s, s' \in \{0, 1, \ldots, K - 1\}. \quad (11.19)$$

If we assume that the transitions happen between only adjacent states, we get

$$p_{ss'} = 0, \ |s - s'| > 1, \ s, s' \in \{0, 1, \ldots, K - 1\}. \quad (11.20)$$

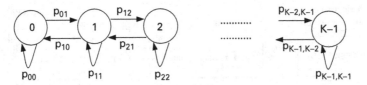

Figure 11.14 K-state noisy channel with Markov transition modeling a Rayleigh fading channel.

In a typical multipath propagation environment, the received signal envelope has Rayleigh distribution. With additive Gaussian noise, the received instantaneous SNR γ is distributed exponentially with probability density function

$$p(\gamma) = \frac{1}{\gamma_0}\exp\left(-\frac{\gamma}{\gamma_0}\right), \ \gamma \geq 0, \tag{11.21}$$

where γ_0 is the average SNR. A FSMC model can be built to represent the time-varying behavior of the Rayleigh fading channel. We start by partitioning the received SNR into a finite number of intervals. Let $\Gamma_0 = 0 < \Gamma_1 < \Gamma_2 \cdots < \Gamma_K = \infty$ be the thresholds of the received SNR. Then the channel is in state k if the received SNR is between Γ_k and Γ_{k+1}. For a packet-transmission system, we assume that a one-step transition in the model corresponds to the channel-state transition after one packet time period T_p. A received packet is said to be in channel state s_k, $k = 0, 1, \ldots, K - 1$, if the SNR values in the packet vary in the range $[\Gamma_k, \Gamma_{k+1})$. In this case, the steady-state probabilities of the channel states are given by

$$\pi_k = \int_{\Gamma_k}^{\Gamma_{k+1}} p(\gamma)d\gamma, \ \ k = 0, 1, \ldots, K - 1. \tag{11.22}$$

In this FSMC model, we allow transitions from a given state to its two adjacent states only. The transition probabilities $p_{ss'}$ in Figure 11.14 can be determined using following equations [325]:

$$p_{k,k+1} = \frac{N(\Gamma_{k+1})T_p}{\pi_k}, \ \ k = 0, 1, \ldots, K - 2, \tag{11.23}$$

and

$$p_{k,k-1} = \frac{N(\Gamma_k)T_p}{\pi_k}, \ \ k = 1, 2, \ldots, K - 1, \tag{11.24}$$

where $T_p = \frac{L_p}{R_s}$ is the packet-transmission time, R_s is the symbol rate in the system, and L_p is the packet size (in our system model both R_s and L_p are fixed for all modulation schemes). In Equations (11.23) and (11.24), $N(\cdot)$ is the level crossing function given by

$$N(\Gamma) = \sqrt{\frac{2\pi\Gamma}{\gamma_0}}f_d\exp\left(-\frac{\Gamma}{\gamma_0}\right), \tag{11.25}$$

where f_d is the maximum Doppler frequency defined as

$$f_d = \frac{v}{\lambda}, \tag{11.26}$$

where v is the mobile's speed and λ is the wavelength. From this model we proceed to obtain an optimal policy for the rate-control problem over Rayleigh fading wireless channels.

Next, we review some of the relevant results from the theory of DP that will be used subsequently to derive the nature of optimal policies for a class of rate-control problems. The stochastic model of the wireless channel is such that the states of the underlying Markov model of the channel evolve according to a time-invariant Markov transition rule independent of past and present actions (chosen rate) taken by the mobile. Let $\{S_t, \ t = 0, 1, \ldots\}$ be a discrete time process. At any given time the state of the channel S_t takes its value from a finite-state space denoted by the set of non-negative integers $\{0, 1, \ldots, K - 1\}$. In our problem this set represents the finite-state space of the underlying Markov model of the channel. At the beginning of the time slot $[t, t + 1)$, for $t = 0, 1, \ldots$, the channel is in state s and the packets are transmitted in the downlink with rate R_i and a decision needs to be taken as to which rate to select for transmitting the packets in the downlink during the time slot $[t, t + 1)$. Let U_t denote the $\{0, 1, \ldots, M - 1\}$-valued random variable that encodes the decision taken at time t, i.e., if $U_t = i$, $i = 0, 1, \ldots, M - 1$, then the rate R_i will be used during the time slot $[t, t + 1)$. We set $I_t \equiv U_{t-1}$, which denotes the bit rate at which the packets are transmitted during the time slot $[t - 1, t)$. Now let us define the aggregate state of the system as (S_t, I_t) that takes values in $\{0, 1, \ldots, K - 1\} \times \{0, 1, \ldots, M - 1\}$. Suppose that for time slot $[t, t + 1)$ the mobile chooses the action (rate) U_t and the aggregate state of the system is (S_t, I_t). Then we incur an instantaneous cost $R(S_t, I_t, U_t)$, which is a bounded *mapping* from the finite space $R : \{0, 1, 2 \ldots, K - 1\} \times \{0, 1, \ldots, M - 1\} \times \{0, 1, \ldots, M - 1\} \mapsto \mathbb{R}$, where $\mathbb{R}$ denotes the set of real numbers. We define a Markov policy, π, as a mapping for choosing the sequence of decisions U_t, $t = 0, 1, \ldots$. Therefore a policy π is a mapping from the aggregate state space to the action space, i.e., $\pi : \{0, 1, 2 \ldots, K - 1\} \times \{0, 1, \ldots, M - 1\} \mapsto \{0, 1, \ldots, M - 1\}$. Given the evolution of the aggregate state of the system $\{S_t, I_t\}_{t=0}^{\infty}$, we are interested in the solution of the following problem. Choose π such that

$$V^{\pi}(s, i) \equiv E_{(s,i)}^{\pi}\left[\sum_{t=0}^{\infty} \beta^t R(S_t, I_t, U_t) \right] \qquad (11.27)$$

is minimized, where (s, i) is the initial state of the system, and $E_{(s,i)}^{\pi}$ denotes the expectation under the policy π, with I_0 being arbitrary, and $0 < \beta \leq 1$ is the discount factor. This problem is called an *infinite-horizon discounted-cost* problem. The preceding cost reflects the fact that, while choosing the rate u_t for time slot $[t, t + 1)$, we would like to take into account the effect of this decision on the future behavior of the system. For the case in which $0 < \beta < 1$, the use of the discount factor is motivated by the fact that a cost to be incurred in the future is less important than one incurred at the present time instant.

It is important to mention that β has a nice practical meaning in the system. A session initiated at time $t = 0$ will last a random number T of time slots. We may interpret $1 - \beta$ as the probability that a session is terminated in a time slot and therefore β is the probability that a session continues in a time slot. Consequently, session-duration random variable T is geometrically distributed with

$$P[T = t + 1] = (1 - \beta)\beta^t. \tag{11.28}$$

Next, we introduce a cost function that captures the desired trade-off between data transmission quality and switching cost, in an appropriate balanced manner for the optimal rate-control (-allocation) problem. To have a reasonable cost-per-stage R, each time the mobile unit switches from one rate to another this should be penalized by a cost associated with rate switching. Let C_s denote the cost of the rate switching. On the other hand a reward (which is a function representing the transmission quality) encourages the mobile unit to switch the rate in order to minimize the delay in the network. We define the transmission delay functions $d_m\{0, 1, \ldots, K - 1\} \to R$, $m = 0, 1, \ldots, M - 1$, as follows

$$d_m(s) \equiv \frac{L_p}{R_m T_s [1 - P_{em}(s)]}, \quad m = 0, 1, \ldots, M - 1, \tag{11.29}$$

where L_p is the packet size in symbols per packet, R_m is the bit rate in bits per second (bps), T_s is the symbol time in seconds (which is fixed in the system), $R_m T_s$, in bps/Hz, represents the number of bits transmitted per symbol (a.k.a. spectral efficiency, for example for 32-ary PSK modulation scheme, $R_m T_s = 5$), and P_{em} is the symbol error rate (SER) for $2^{R_m T_s}$-ary PSK modulation that is a function of channel state (symbol SNR) [26].

The final cost-per-stage function R is defined as follows:

$$R(s, i, u) = \begin{cases} C_s & \text{if } i \neq u, \\ \max[d_i(s) - d_{i-1}(s), d_i(s) - d_{i+1}(s)] & \text{if } 1 \leq i = u \leq M - 2, \\ d_0(s) - d_1(s) & \text{if } i = u = 0, \\ d_{M-1}(s) - d_{M-2}(s) & \text{if } i = u = M - 1. \end{cases} \tag{11.30}$$

For the special case in which $M = 2$, i.e., there are only two admissible rates R_0 and R_1, then the cost function $R(.)$ is simplified to

$$R(s, i, u) = \begin{cases} C_s & \text{if } i \neq u \\ (-1)^i [d_0(s) - d_1(s)] & \text{if } i = u \end{cases}. \tag{11.31}$$

Now the problem at hand is to solve the following infinite-horizon discounted-cost problem

$$V(s, i) \equiv \min_{u=0,1,\cdots,M-1} E_{s,i}^\pi \left[\sum_{t=0}^{\infty} \beta^t R(s, i, u) \right] \tag{11.32}$$

for every (s, i) in $\{0, 1, 2, \ldots, K - 1\} \times \{0, 1, \ldots, M - 1\}$ and policy π. To ensure the existence of the expected infinite-horizon discounted cost, it suffices to have a uniformly bounded cost function $R(S_t, I_t, U_t)$ for all $t \in \{0, 1, \ldots\}$ and $0 < \beta < 1$. In our rate-control problem the state and action spaces are finite, $|R(S_t, I_t, U_t)| < B < \infty$ for $\forall t, \in \{0, 1, \ldots\}$, and with the interpretation of β in a practical system we always have $0 < \beta < 1$. This set of conditions ensures the existence of a solution for our optimal rate-control problem. The policy π satisfying the problem cast in (11.32) is called the optimal policy π^*. We subsequently state a well-known result [26] that yields an implicit equation satisfied by the optimal discounted cost function $V(s, i)$.

THEOREM 22 $V(.)$ *satisfies the optimality equation:*

$$V(s, i) = min_u \left[R(s, i, u) + \beta \sum_{s'} p_{ss'} V(s', u) \right],$$ (11.33)

$$(s, i) \in \{0, 1, \ldots, K - 1\} \times \{0, 1, \ldots, M - 1\},$$

where (s, i) is the initial state of the system, and $p_{ss'}$ is the state transition probability of the FSM model of the wireless channel given by the sets of equations (11.23) and (11.24). In effect, (11.33) provides that the cost incurred by choosing an action u at some time instant is the sum of the instantaneous cost $R(s, i, u)$ and the expected cost for the future $\sum_{s'} p_{ss'} V(s', u)$ multiplied by the given discount factor β. The optimal policy chooses that action u that minimizes this sum.

In what follows, we attempt to find the solution of (11.33) using an iterative method. For this purpose, we define the following quantity:

$$\tilde{V}_{n-1}(s, i) \equiv \sum_{s'} p_{ss'} V_{n-1}(s', i).$$ (11.34)

Then the DP equation is simply

$$V_n(s, i) = \min_{\{u=0,1,\ldots,M-1\}} \left\{ R(s, i, u) + \beta \tilde{V}_{n-1}(s, u). \right\}$$ (11.35)

Equations (11.30), (11.34), and (11.35) are used in our computer simulations to find the solution for (11.33) in general case. This method is called *value iteration* or *successive approximation*. To understand the structure of optimal policy, from now on, we would like to restrict our attention to the mathematically more tractable case. Therefore, without loss of generality, in the following we consider only the case in which $M = 2$ and the set of admissible rates is $u \in \{0, 1\}$ (corresponding to R_0 and R_1). In this case (11.35) can be rewritten as

$$V_n(s, i)$$
$$= \min \left\{ C_s + \beta \tilde{V}_{n-1}(s, i \oplus 1), (-1)^i [d_0(s) - d_1(s)] + \beta \tilde{V}_{n-1}(s, i) \right\},$$ (11.36)

where $\oplus$ denotes modulo two addition. Moreover, the optimal policy π^* is a Markov stationary policy that selects to switch in state (s, i) if and only if

$$C_s + \beta \tilde{V}_{n-1}(s, i \oplus 1) \leq (-1)^i [d_0(s) - d_1(s)] + \beta \tilde{V}_{n-1}(s, i).$$ (11.37)

An important observation regarding the solution of the discounted DP problem given by (11.35) is that it can be interpreted as the fixed point of a well-defined operator T, where $TV = V$. Motivated by the form of DP equation (11.35), we associate R–valued mappings $\tilde{T}\varphi$ and $T_u\varphi$, $u = 0, 1$ defined on $\{0, 1, 2, \ldots, K - 1\} \times \{0, 1\}$ by setting

$$(\tilde{T}\varphi)(s, i) \equiv \sum_{s'} p_{ss'} \varphi(s', i).$$ (11.38)

and

$$(T_u\varphi)(s, i) \equiv R(s, i, u) + \beta(\tilde{T}\varphi)(s, u) \tag{11.39}$$

for $(s, i) \in \{0, 1, 2, \ldots, K - 1\} \times \{0, 1\}$. Next, we introduce the operator T by setting

$$(T\varphi)(s, i) \equiv \min_{u=0,1}(T_u\varphi)(s, i), \quad (s, i) \in \{0, 1, 2, \ldots, K - 1\} \times \{0, 1\} \tag{11.40}$$

for every φ. Now, using the important properties given in [26], we state the following important results.

PROPOSITION 1 *Under the model assumptions [stationary Markov model for the channel, bounded cost-per-stage function $R(.)$, and $0 < \beta < 1$, where $1 - \beta$ is the probability of terminating a session in a time slot], the following hold:*
1. The operator T is a strict contraction mapping.
2. The value function V is the only solution of the fixed point equation

$$\varphi = T\varphi. \tag{11.41}$$

3. Moreover, for every element φ, the recursive scheme

$$\varphi_0 = \varphi, \quad \varphi_{k+1} = T\varphi_k, \quad k = 0, 1, \ldots, \tag{11.42}$$

converges to the value function V in the sense that $\lim_k \|\varphi_k - V\|_2 = 0$, where $\lim_k \varphi_k(s, i) = V(s, i)$ for all $(s, i) \in \{0, 1, 2, \ldots, K - 1\} \times \{0, 1\}$.

Now we use the results of the theorems given in [26] to investigate the structure of the optimal policy. In fact, it turns out that the optimal rate-control policy π^* belongs to the class of threshold policies. A rate-switching policy π is said to be a threshold policy with threshold functions τ_i $i = 0, 1$, if it is a Markov stationary policy such that

$$\pi^*(s, 0) = 1 \quad \text{iff} \quad z(s) \geq \tau_0, \tag{11.43}$$

and

$$\pi^*(s, 1) = 0 \quad \text{iff} \quad z(s) \leq \tau_1, \tag{11.44}$$

where $z(s) \equiv d_0(s) - d_1(s)$, with $z : \{0, 1, 2, \ldots, K - 1\} \to \mathsf{R}$.

PROPOSITION 2 *Under the model assumptions [stationary Markov model for the channel, bounded cost-per-stage function $R(\cdots.)$, and $0 < \beta < 1$, where $1 - \beta$ is the probability of terminating a session in a time slot], the optimal rate control policy π^* is a threshold policy with thresholds $\tau_i^* \in \mathsf{R}$, $i = 0, 1$, which are uniquely determined through the equations*

$$C_s + (-1)^i \beta(\Delta\tilde{T}V)(\tau_i^*) = (-1)^i \tau_i^*, \ \tau_i^* \in \mathsf{R}, \ i = 0, 1. \tag{11.45}$$

Furthermore, $\tau_1^* \leq \tau_0^*$.

Proof 5 Fix (s, i) in $\{0, 1, \ldots, K - 1\} \times \{0, 1\}$, and $z = d_0(s) - d_1(s)$ in R. We begin by rewriting DP equation (11.36) in the following form:

$$V(z, i) = \min\left\{C_s + \beta(\tilde{T}V)(z, i \oplus 1), \ (-1)^i z + \beta(\tilde{T}V)(z, i).\right\} \tag{11.46}$$

The optimal policy π^* is the Markov stationary policy that selects to switch in state (s, i) if and only if

$$C_s + \beta(\tilde{T}V)(z, i \oplus 1) \leq (-1)^i z + \beta(\tilde{T}V)(z, i), \qquad (11.47)$$

or equivalently, if and only if

$$C_s + (-1)^i \beta(\Delta \tilde{T}V)(z) \leq (-1)^i z, \ z \in \mathsf{R}, \qquad (11.48)$$

where

$$(\Delta \tilde{T}V)(z) = (\tilde{T}V)(z, 1) - (\tilde{T}V)(z, 0). \qquad (11.49)$$

For $i = 0$ (respectively $i = 1$) the left-hand side of inequality (11.48) is monotone nonincreasing (resp. nondecreasing) function of z, whereas its right-hand side is strictly increasing (resp. decreasing) function of z. It is now a simple matter to conclude that the switching sets $S_i(z) \equiv \{z \in \mathsf{R} : C_s + (-1)^i \beta(\Delta \tilde{T}V)(z) \leq (-1)^i z\}$, $i = 0, 1$, are nonempty closed and connected sets that are disjointed (owing to the condition $C_s > 0$). In fact, $S_0(z) = [\tau_0^*, \infty)$ with $\tau_0^* = \inf S_0(z)$, and $S_1(z) = (-\infty, \tau_1^*]$ with $\tau_1^* = \sup S_1(z)$, and the optimal policy is of the threshold type. Because $S_0(z)$ and $S_1(z)$ are disjoint sets, we see that $\tau_1^* \leq \tau_0^*$, and this concludes the proof of Proposition 2.   **QED**

Once a rate-control (-allocation) policy (be it optimal or not) has been selected, it is of interest to compute the average *delay* of transmitting the packets over the wireless channel and the expected number of rate switchings that the mobile experiences while the optimal policy is in effect. These two quantities constitute good measures of the effectiveness of a rate-control policy.

We define the average delay D_π of the policy π to be the the mean value of the delay of the selected rate to receive the packets from the base station under the policy π during the packet transmission, namely,

$$D_\pi(s, i) \equiv \mathbf{E}_{s,i}^\pi \left[\sum_{t=0}^\infty \beta^t [I_t d_1(S_t) + (1 - I_t) d_0(S_t)] \right]. \qquad (11.50)$$

On the other hand, the expected number of rate switchings under the policy π is defined by

$$S_\pi(s, i) \equiv \mathbf{E}_{s,i}^\pi \left[\sum_{t=0}^\infty \beta^t \mathbf{1}[I_t \neq U_t] \right], \qquad (11.51)$$

where $\mathbf{1}(\cdot)$ is the indicator function. It is equal to one if the condition $(\cdot)$ is met. Therefore both D_π and S_π can be written as discounted-cost functions. For any Markov stationary policy π, and in particular for any threshold policy, this fact can be exploited for numerical purposes by interpreting D_π and S_π as fixed points for suitably defined contraction mappings. More precisely, to evaluate the average delay, for each Markov stationary policy π, we consider an operator K_π of the form

$$(K_\pi \varphi)(s, i) \equiv (K_{\pi(s,i)} \varphi)(s, i) \qquad (11.52)$$

for every $(s, i) \in \{0, 1, 2, \ldots, K - 1\} \times \{0, 1\}$, where for each $u = 0, 1$, the operator K_u is defined by

$$(K_u \varphi)(s, i) \equiv d_i(s) + \beta(\tilde{T}\varphi)(s, u). \tag{11.53}$$

As in Proposition 1 the operator K_u, $u = 0, 1$, is a contraction mapping and so is K_π. It follows from the Markov property that the average delay D_π is the unique fixed point of K_π, and can be evaluated through the recursion

$$\varphi_0 = 0, \quad \varphi_{k+1} = K_\pi \varphi_k, \quad k = 0, 1, \ldots. \tag{11.54}$$

To compute the expected number of rate switchings, we use the operator K_π^*, which is of the form

$$(K_\pi^* \varphi)(s, i) \equiv (K_{\pi(s,i)}^* \varphi)(s, i) \tag{11.55}$$

for every $(s, i) \in \{0, 1, 2, \ldots, K - 1\} \times \{0, 1\}$, where for each $u = 0, 1$, the operator K_u^* is defined by

$$(K_u^* \varphi)(s, i) \equiv \mathbf{1}[u \neq i] + \beta(\tilde{T}\varphi)(s, u). \tag{11.56}$$

This time, the operators K_u^*, $u = 0, 1$, are contraction mappings, and so is K_π^*. The unique fixed point of K_π^* is S_π, and is obtained through the recursion

$$\varphi_0 = 0, \quad \varphi_{k+1} = K_\pi^* \varphi_k, \quad k = 0, 1, \ldots. \tag{11.57}$$

It is clear that both $D_\pi(s, i)$ and $S_\pi(s, i)$ are functions of channel state $s \in \{0, 1, \ldots, K - 1\}$ and rate $i \in \{0, 1, \ldots, M - 1\}$. It would be useful to calculate average delay and average switching rate over all possible channel states and admissible rates for a fixed C_s. Therefore we have

$$\bar{D}_\pi = \sum_{i'=0}^{M-1} p_{i'} \sum_{s'=0}^{K-1} \pi_{s'} D_\pi(s', i'), \tag{11.58}$$

$$\bar{S}_\pi = \sum_{i'=0}^{M-1} p_{i'} \sum_{s'=0}^{K-1} \pi_{s'} S_\pi(s', i'), \tag{11.59}$$

where π_s is the steady-state probability given by (11.22) and p_i is the probability of selecting rate R_i under the adopted rate-control policy and channel model, i.e., $p_i = Pr[I_t = i]$, $i = 0, 1, \ldots, M - 1$. For the special case in which $M = 2$, we have $p_0 = 1 - p_1$, and the threshold policy given by (11.43) and (11.44) can be rewritten as

$$U_t = I_{t+1} = \begin{cases} 1 & \text{if } Z_t \geq \tau_0 \\ I_t & \text{if } \tau_1 \leq Z_t \leq \tau_0 \\ 0 & \text{if } Z_t \leq \tau_1 \end{cases}. \tag{11.60}$$

Using (11.60) and the i.i.d. assumption on the random variables $\{Z_t, t = 0, 1, \ldots\}$ it can be shown that the sequence of random variables $\{I_t, t = 0, 1, \ldots\}$ form a Markov

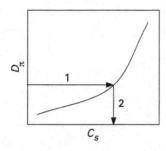 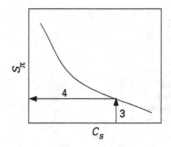

Figure 11.15 A typical graph for D_π and C_π versus switching cost C_s.

Chain on $\{0, 1\}$ with the following transition probabilities:

$$P_{01} = P[I_{t+1} = 1 \mid I_t = 0]$$
$$= P[Z_t \geq \tau_0] \tag{11.61}$$

and

$$P_{10} = P[I_{t+1} = 0 \mid I_t = 1]$$
$$= P[Z_t \leq \tau_1]. \tag{11.62}$$

Using (11.61) and (11.62) we get

$$p_0 = P[I_t = 0] = \frac{P_{10}}{P_{01} + P_{10}}. \tag{11.63}$$

Now (11.63) can be used along with (11.59) to compute $\bar{D}_\pi$ and $\bar{S}_\pi$ for a fixed C_s. Intuitively, as C_s increases, the average delay $\bar{D}_\pi$ increases while the expected number of rate switchings $\bar{S}_\pi$ decreases.

One of the critical parameters on which the optimal rate-control policy clearly depends is the value of the rate-switching cost C_s. Given a rate-switching cost C_s, we can compute an optimal rate-control policy that solves the minimization problem posed in (11.32). From our previous discussion, $\bar{D}_\pi$ is an increasing function of C_s, whereas $\bar{S}_\pi$ is a decreasing function of C_s, similar to the graphs shown in Figure 11.15. As a design procedure, we can start from a desired value of $\bar{D}_\pi$ and use the graph in Figure 11.15 to find the respective value of C_s and from there the respective value of value for the switching $\bar{S}_\pi$. If the resulting value of $\bar{S}_\pi$ is satisfactory, then the rate-control policy is acceptable; otherwise the procedure has to be started over again by choosing a larger $\bar{D}_\pi$. Hence the previously mentioned procedure is summarized by a flowchart in Figure 11.16.

We have studied the rate-control problem in wireless networks and offered a novel method based on DP to obtain an optimal rate-control policy. Now, our simulation results for the solution of the rate-control problem posed in (11.33) are presented. In these simulations, a successive approximation method (a.k.a., *value-iteration* method) is used to solve (11.33). Our simulation results indicate that the optimal strategy for selecting the rates is indeed a *threshold policy*. This corroborates the results of Proposition 2.

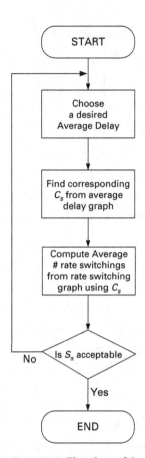

Figure 11.16 Flowchart of the design procedure to select an appropriate switching cost C_s.

The wireless channel is modeled as a FSM chain [325]. The parameters for the simulations are as follows: Symbol rate $R_s = 500$ ks/s, packet size $L_p = 200$ symbols (i.e., $T_p = 0.4$ ms), Doppler frequency $f_d = 10$ Hz. We consider a 15-state Markov model for the fading channel, with average SNR $\gamma_0 = 12$ dB, and SNR thresholds $\Gamma_1 = -2$ dB and $\Gamma_i = \Gamma_{i-1} + \Delta$, $i = 2, 3, \ldots, 14$, where $\Delta = 2$ dB.

First, we consider the optimal rate-control policy in a system with two admissible rates, i.e., $M = 2$. In this case we assume that packets are transmitted in the downlink using either a QPSK modulation or a 16-PSK modulation (i.e., $R_0 = 2R_s$ bps, and $R_1 = 4R_s$ bps, where R_s is the symbol rate). The simulation results for optimal thresholds for two values of C_s are shown in Figure 11.17. These optimal thresholds along with the transmission delay curves, d_0 and d_1, are plotted in the same figure for comparison purposes. The optimal policy $\pi^*(s, i)$ for $C_s = 45$, plotted in Figure 11.17, is illustrated in Table 11.3. The optimal policy $\pi^*(s, i)$, essentially dictates to us the rate to be used if system is in state (s, i). This notion is demonstrated in a matrix form given by Table 11.3.

As we discussed earlier, if there is no cost for switching the rates, i.e., $C_s = 0$ in (11.31), the optimal policy for the rate-control problem is simply to switch to the rate

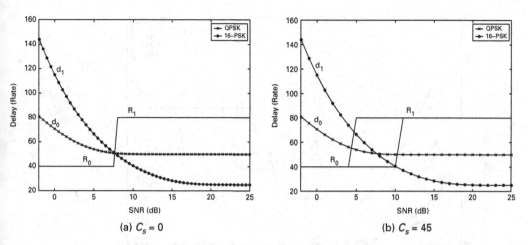

Figure 11.17 Optimal rate-control policy in a system with two admissible rates R_0 and R_1 in a Rayleigh fading channel.

Table 11.3 Optimal policy $\pi^*(s, i)$ in a system with two admissible rates: rate 0, R_0 (QPSK), and rate 1, R_1 (16-ary PSK), for $C_s = 45$

State index k	$\pi^*(s_k, 0)$	$\pi^*(s_k, 1)$
0	0	0
1	0	0
2	0	0
3	0	0
4	0	1
5	0	1
6	0	1
7	1	1
8	1	1
9	1	1
10	1	1
11	1	1
12	1	1
13	1	1
14	1	1

with smaller delay. This means that, in this case, the optimal thresholds are equal ($\tau_1 = \tau_0$) to the intersection of the delay curves d_0 and d_1. Our simulations results confirms this observation. Figure 11.17 illustrates this case, which is obtained for $C_s = 0$ in the cost function given by (11.31). The optimal threshold rate-control policy is also obtained for $C_s = 45$ which is demonstrated in Figure 11.17. It is worth mentioning that as C_s increases, $\tau_0 - \tau_1$ increases as well; in other words, the optimal policy becomes more sluggish. Our simulation results shown in Figure 11.17 clearly support this claim.

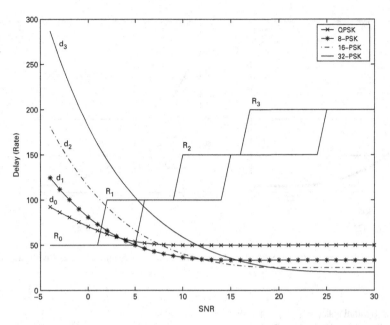

Figure 11.18 Optimal rate-control policy in a multirate system with four admissible rates R_0, R_1, R_2, and R_3 in a Rayleigh fading channel for $C_s = 40$.

We should mention that both optimal policy and convergence rate of the value-iteration method depend on β. As β decreases, the value-iteration method converges faster to the final optimal rate-control policy. A detailed analysis of the value-iteration method and the effect of β on its convergence is provided in [26 (Vol. II)]. The effect of β on the optimal policy is as follows: As β decreases, the probability that the session is terminated in a time slot increases. Because in this case it would be more probable to terminate a session in the next time slot, the optimal policy tends to stay with the current rate for wider range of states (to avoid unnecessary switching costs, as it is more probable to end the session in next slot), i.e., the optimal policy becomes more sluggish. Consequently decreasing β has similar effect on optimal policy as increasing C_s. Because in practical systems β is dictated by the traffic behavior in the system, C_s, which is a parameter in system design, has to be adjusted accordingly. As a result, in our simulations we have fixed $\beta = 0.8$ and only the effect of increasing C_s is studied through simulations.

Next, we consider the optimal rate-control policy in multirate systems. So far we have only considered the optimal rate-control policy for the systems with two admissible rates, R_0 and R_1. The important feature of our proposed optimal rate-control method is that, with a well-defined cost function, it can be generalized to more than two rates. Now we consider the cost function proposed in (11.30) and attempt to find the solution of (11.33). Next, a set of simulations is performed to obtain the optimal rate-control policy for a system with $M = 4$, i.e., there are four admissible rates, R_0, R_1, R_2, and R_3, in the system. In this case, we assume that packets are transmitted in the downlink using one of the following four modulation schemes: QPSK, 8-PSK, 16-PSK, and 32-PSK modulations. The optimal rate-control policy for such a system is illustrated in

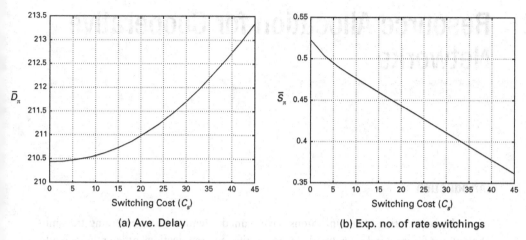

Figure 11.19 Rate-switching cost.

Figure 11.18. As we expect again, the optimal rate-control policy in a multirate system is also a *threshold* policy, as shown in Figure 11.18. In Figure 11.18 the delay curves d_0, d_1, d_2, and d_3, associated with each modulation scheme, are plotted in the same graph along with the optimal rate-control policy (this way we can compare the thresholds against intersection points of delay graphs).

Figure 11.19 illustrates how $\bar{D}_\pi$ and $\bar{S}_\pi$ vary as the switching cost C_s increases. We assess the effectiveness of the proposed method by comparing the average delay $\bar{D}_\pi$ and expected number of rate switchings $\bar{S}_\pi$ for different values of the switching cost C_s. From Figure 11.19(a), $\bar{D}_\pi = 210.5$ when $C_s = 0$, and it is $\bar{D}_\pi = 213.25$ when $C_s = 45$, which represents only a 1.3% increase in average delay. On the other hand, from Figure 11.19(b), $\bar{S}_\pi = 0.235$ for $C_s = 0$, and it is $\bar{S}_\pi = 0.145$ for $C_s = 45$, which represents more than a 38% decrease in rate switchings. Therefore, by sacrificing only a 1% increase in average delay in the system, we can save almost 40% in rate switchings.

We have studied the problem of optimal rate control in wireless networks with Rayleigh fading channels. A stochastic optimization technique based on the DP method is used to obtain the optimal rate-control policy in such networks. Using the results from the theory of DP, it is shown that the optimal rate-control policy is in the form of a *threshold policy* – a property of significance from both the analytical and implementation points of view. Simulation results confirm that the optimal rate-control policy is indeed a threshold policy. These results also demonstrate the effectiveness of our optimal rate-control policy in optimizing the overall delay and number of rate switchings in the network. Simulation results indicate that, by sacrificing only 1% of transmission quality in terms of the average delay, one can achieve almost 40% reduction in rate switchings in the network.

12 Resource Allocation for Cooperative Networks

12.1 Introduction

Recently, cooperative communications have gained attention as an emerging transmit strategy for future wireless networks. Cooperative communications efficiently take advantage of the broadcasting nature of wireless networks. The basic idea is that users or nodes in a wireless network share their information and transmit cooperatively as a virtual antenna array, thus providing diversity that can significantly improve system performance. In cooperative transmission, relays are assigned to help a sender in forwarding its information to its receiver. Thus the receiver gets several copies of the same information via independent channels.

The pioneer work on cooperative transmission can be found, e.g., in [53], where a general information theoretical framework about relaying channels is established. In [272, 273], a CDMA-based two-user cooperative modulation scheme has been proposed. The main idea is to allow each user to retransmit estimates of their partner's received information such that each user's information is transmitted to the receiver at the highest possible rate. This work is extended in [175], where the outage and the ergodic capacity behavior of various cooperative protocols, e.g., decode-and-forward (DF) and amplify-and-forward (AF) cooperative protocols, are analyzed for a three-user case under quasi-static fading channels. In [298], the authors provided SER performance analysis and optimum power allocation for DF cooperative systems in a narrowband Rayleigh fading environment. The work in [138] analyzes the schemes based on the same channel without fading, but with more complicated transmitter cooperative schemes involving dirty paper coding. In [199], a cooperative broadcast strategy was proposed with an objective of maximizing the network lifetime. Recent work in [35] presented theoretical characterizations and analysis for the physical layer of multihop wireless communications channels with different channel models. In the next section, we discuss some of these cooperative communication protocols in detail.

Resource allocation has long been regarded as an effective way to dynamically combat channel fluctuations and reduce CCIs in wireless networks. For example, power control constantly adjusts the transmitted power so as to maintain the received link quality. The merits of the cooperative transmissions in the physical layer have been explored; however, the impact of the cooperative transmissions on the design of the higher layers is not well understood yet. In this chapter, we discuss resource allocation for different layers to optimize the performances over this new paradigm of cooperative communication strategy. Specifically, we divide the discussions into four different layers as follows. The basic problems are stated, and then examples are given in detail.

1. Physical layer
   The major issue here is to optimize the capacity region, minimize the BER, and improve the link quality by power control. An UWB cooperative system is investigated to reduce the BER and improve the coverage area, under cooperative communication. The key is to adjust the source power and relay power to optimize the BER and converge area, according to the source–relay, relay–destination, and source–destination channels.

2. MAC layer
   The major concern in this layer is relay selection and channel allocation, i.e., among all the possible relays and channels that one can improve the source-to-destination link most. Two examples are given. First, the cooperative OFDM networks such as future WLAN and WMAN are studied for the questions of "who helps whom" and "how to cooperate." Second, a distributed relay-selection scheme is proposed to extend the coverage of future cellular networks.

3. Routing layer
   The main problem is route selection. With cooperative communication, the link can be improved by possible relays. So the optimal route selection depends not only on the nodes of the links but also on the relays as well. We propose how to improve the traditional routing schemes by considering the cooperative diversity. Moreover, a new routing table is constructed completely based on the cooperative communication, so that the new cooperative routes can be obtained independently from the traditional routes.

4. Application layer
   For multimedia transmission, the relay can forward some coded and processed information instead of purely relaying the original bits. By doing this, the destination can improve the reconstructed voice/image/video qualities because the natures of multimedia data. We briefly mention these technologies in the Summary at the end of this chapter.

12.2   Cooperative Communication Protocols

In this section, we discuss several cooperative communication protocols in detail. The system block diagram of cooperative communication is illustrated in Figure 12.1, in which there are a source node s, several relays nodes r_i, and a destination node d. The cooperative transmission consists of two phases. In Phase 1, source s broadcasts its information to both destination d and each relay node r_i. The received signals Y_d and Y_{r_i} at destination d and relay r_i can be expressed as

$$Y_d = \sqrt{P_s G_{s,d}} X + n_d, \tag{12.1}$$

and

$$Y_{r_i} = \sqrt{P_s G_{s,r_i}} X + n_{r_i}, \tag{12.2}$$

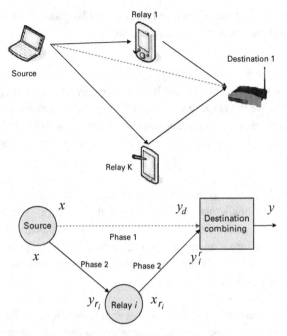

Figure 12.1 Cooperative communication system model.

where P_s represents the transmit power to the destination from the source, X is the transmitted information symbol with unit energy at Phase 1 at the source, $G_{s,d}$ and G_{s,r_i} are the channel gains from s to d and r_i, respectively, and n_d and n_{r_i} are the AWGN. Without loss of generality, we assume that the noise power is the same for all the links, denoted by σ^2. We also assume the channels are stable over each transmission frame.

For *direct transmission*, without the relay nodes' help, the SNR that results from s to d can be expressed by

$$\Gamma_{s,d}^{\mathrm{DT}} = \frac{P_s G_{s,d}}{\sigma^2},\tag{12.3}$$

and the information rate of the direct transmission is

$$R_{s,d} = W \log_2 \left(1 + \frac{\Gamma_{s,d}^{\mathrm{DT}}}{\Gamma}\right),\tag{12.4}$$

where W is the bandwidth for information transmission and Γ is a constant representing the capacity gap.

For the *AF* cooperative transmission, in Phase 2, relay i amplifies Y_{r_i} and forwards it to the destination with transmitted power P_{r_i}. The received signal at the destination is

$$Y_r^i = \sqrt{P_{r_i} G_{r_i,d}} X_{r_i} + n_d',\tag{12.5}$$

where

$$X_{r_j} = \frac{Y_{r_i}}{|Y_{r_i}|}\tag{12.6}$$

is the energy-normalized transmitted signal from the source to the destination at Phase 1, $G_{r_i,d}$ is the channel gain from relay i to the destination, and n'_d is the received noise at Phase 2. Substituting (12.2) into (12.6), we can rewrite (12.5) as

$$Y^i_r = \frac{\sqrt{P_{r_i} G_{r_i,d}}(\sqrt{P_s G_{s,r_i}} X_s + n_{r_i})}{\sqrt{P_s G_{s,r_i} + \sigma^2}} + n'_d. \tag{12.7}$$

Using (12.7), the relayed SNR at the destination for the source, which is helped by relay node i, is given by

$$\Gamma^{AF}_{s,r_i,d} = \frac{P_{r_i} P_s G_{r_i,d} G_{s,r_i}}{\sigma^2(P_{r_i} G_{r_i,d} + P_s G_{s,r_i} + \sigma^2)}. \tag{12.8}$$

Therefore, by (12.4) and (12.8), we have the information rate at the output of MRC as

$$R^{AF}_{s,r_i,d} = \frac{1}{2} W \log_2 \left(1 + \frac{\Gamma^{DT}_{s,d} + \Gamma^{AF}_{s,r_i,d}}{\Gamma} \right). \tag{12.9}$$

If multiple relay nodes are available to help the source and consist of a set L with size N, we have

$$R^{AF}_{s,L,d} = \frac{1}{N+1} W \log_2 \left(1 + \frac{\Gamma^{DT}_{s,d} + \sum_{r_i \in L} \Gamma^{AF}_{s,r_i,d}}{\Gamma} \right). \tag{12.10}$$

In the DF cooperative transmission protocol, the relay decodes the source information in Phase 1 and relays to the destination in Phase 2. The destination combines the direct transmission information and relayed information together. The achievable rate can be calculated by the following maximization:

$$R^{DF}_{s,r_i,d} = \max_{0 \le \rho \le 1} \min\{R_1, R_2\}, \tag{12.11}$$

where

$$R_1 = \log \left[1 + (1 - \rho^2) \frac{P_{s,d} G_{s,r_i}}{\sigma^2} \right] \tag{12.12}$$

and

$$R_2 = \log \left(1 + \frac{P_s G_{s,d}}{\sigma^2} + \frac{P_{r_i} G_{r_i,d}}{\sigma^2} + \frac{2\rho \sqrt{P_s G_{s,d} P_{r_i} G_{r_i,d}}}{\sigma^2} \right). \tag{12.13}$$

In the *estimate-and-forward* (EF) cooperative transmission, the relay, in Phase 2, sends the estimate of the received signal of Phase 1. The destination uses the relay's information as the side information to decode the direct transmission in Phase 1. From [53] and [163], the achievable rate can be written as

$$R^{EF}_{s,r_i,d} = W \log_2 \left(1 + \frac{\Gamma^{DT}_{s,d} + \Gamma^{EF}_{s,r_i,d}}{\Gamma} \right), \tag{12.14}$$

where

$$\Gamma_{s,r_i,d}^{\text{EF}} = \frac{P_s P_{r_i} G_{s,r_i} G_{r_i,d}}{\sigma^2 [P_{r_i} G_{r_i,d} + P_s(G_{s,d} + G_{s,r_i}) + \sigma^2]}. \tag{12.15}$$

There are some other cooperative communication protocols developed in the literature. Here we list several of them.

1. *Coded cooperation [145]*:
   Coded cooperation integrates user cooperation with channel coding. Instead of repeating some form of the received information, the user decodes the partner's transmission and transmits additional parity symbols (e.g. incremental redundancy) according to some overall coding scheme. This framework maintains the same information rate, code rate, bandwidth, and transmit power as a comparable noncooperative system.

2. *Distributed space–time coded cooperation [176]*:
   Space–time coded cooperative diversity protocols for combating multipath fading across multiple protocol layers exploit spatial diversity available among a collection of distributed terminals that relay messages for one another in such a manner that the destination terminal can average the fading. Those terminals can fully decode the transmission by utilizing a space–time code to cooperatively relay to the destination.

3. *Incremental relaying*:
   In this type of cooperation, the destination will broadcast ACK or NACK information after the first stage. The relay retransmits only after receiving NACK. By doing this, the bandwidth efficiency can be greatly improved, because the only bandwidth increase happens when the direct transmission link fails.

4. *Cognitive relaying*:
   The cooperative communication protocol can help cognitive users to reduce the detection time and thus to increase their agility. On the other hand, the relays can detect the spectrum cognitively so as to improve the source-to-destination link.

12.3   Physical-Layer Issue: Power Control

In Figure 12.1, if the relay has been selected, the problem to enhance the link between the source and destination falls into the optimization for the physical layer. The major optimization is power control, where the source and the relay utilize the limited available power. The goal is to improve the performance such as the BER or SNR at the destination. Moreover, the direct transmission in Phase 1 and the relay transmission in Phase 2 are correlated. The capacity region can be further investigated in addition to the power constraint. Some coding technologies such as space–time coding can also be deployed in the relay so as to fully explore the diversity gain of such a cooperative communication system.

In the following subsection, we give an example [283] of how to utilize the cooperative diversity to improve the performance of the physical layer of an UWB system. Because of the limitation on the transmitted power level, any UWB system faces major design challenges in achieving wide coverage while ensuring an adequate system performance.

In this example, an employment of cooperative communications in the UWB is proposed to enhance the performance and the coverage of the UWB by exploiting the broadcasting nature of wireless channels and the cooperation among UWB devices. SER performance analysis and optimum power allocation are provided for cooperative UWB multiband OFDM systems with DF cooperative protocol. To capture the multipath-clustering phenomenon of UWB channels, the SER performance is characterized in terms of cluster- and ray-arrival rates. An optimum power allocation is determined based on two different objectives, namely minimizing the overall transmitted power and maximizing the system coverage. Furthermore, an improved cooperative UWB multiband OFDM scheme is proposed to take advantage of unoccupied subbands. Simulation results are shown to validate the theoretical analysis.

UWB technology shows a great promise for high-speed short-range wireless communications [17]. However, because of the limitation on its transmitted power level, any UWB system faces major design challenges to achieve the desired performance and coverage. To date, limited works have been proposed to improve the coverage of UWB systems. One approach utilizes analog repeaters as used in cellular systems. For example, a pulse-position-modulation UWB repeater is proposed in [47]. Although the analog repeaters are simple, they suffer from noise amplification, which has confined their applications to specific scenarios. Another approach is the employment of multiple-input multiple-output (MIMO) technology in UWB systems. It has been shown that the UWB-MIMO systems can efficiently exploit the available spatial and frequency diversities, and hence greatly improve the UWB performance and coverage range [73, 285]. Nevertheless, it might not be easy to have multiple antennas installed in UWB devices.

The research in [175, 298] has proved the significant potential of cooperative diversity in wireless networks. Current UWB technology, on the other hand, relies on a noncooperative transmission, in which the diversity can be obtained only from MIMO coding or information repetition at the transmitter [18, 285]. Furthermore, many UWB devices are expected to be in home and office environments; most of these devices are not in active mode simultaneously, but they can be utilized as relays to help the active devices. Additionally, because of the TDMA mechanism of the MAC and the network structure of the IEEE 802.15.3a WPAN standard [375], the cooperative protocols can be adopted in UWB WPANs. These facts provide motivation for introducing the concept of cooperative diversity in UWB systems as an alternative approach to improve the UWB performance and coverage without the requirement of additional antennas or network infrastructures.

In this subsection, we propose to enhance the performance of UWB systems with cooperative protocols. The SER performance analysis and optimum power allocation are provided for cooperative UWB multiband OFDM (MB-OFDM) systems employing DF cooperative protocol. To capture the clustering property of UWB channels [76], the performance is characterized in terms of the cluster- and the ray-arrival rates. Moreover, we propose an improved cooperative UWB scheme that is compatible with the current MB-OFDM standard proposal [18]. Both analytical and simulation results show that the proposed cooperative UWB scheme achieves 43% power saving and 85% coverage extension compared with a noncooperative UWB at the same data rate. By allowing

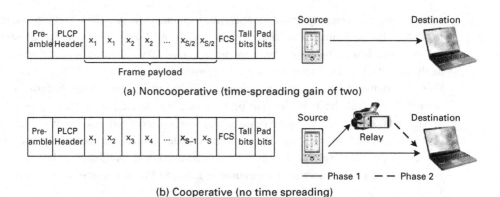

(a) Noncooperative (time-spreading gain of two)

(b) Cooperative (no time spreading)

Figure 12.2 Illustrations of noncooperative and cooperative UWB MB-OFDM systems with the same data rate.

both source and relay to transmit simultaneously, the performance of cooperative UWB is further improved to 52% power saving and 100% range extension.

For the *system model*, we consider a UWB MB-OFDM system [18], in which the available spectrum is divided into several subbands. Within each subband, the data is modulated using OFDM technique. Different bit rates are achieved by using different channel-coding, frequency-spreading, or time-spreading rates. Frequency spreading is obtained by choosing conjugate symmetric inputs to the IFFT (inverse fast Fourier transform), whereas time spreading is achieved by repeating the same data in an OFDM symbol on two different subbands [18]. The receiver combines the data transmitted via different times or frequencies to increase the SNR of the received signal.

As in the IEEE 802.15.3a standard [76], the channel impulse response is based on the Saleh–Valenzuela (S-V) model [264]:

$$h(t) = \sigma^2 \sum_{c=0}^{C} \sum_{l=0}^{L} \alpha(c, l) \delta[t - T(c) - \tau(c, l)], \qquad (12.16)$$

where σ^2 represents total multipath energy, $\alpha(c, l)$ is the gain of the lth multipath component in the cth cluster, $T(c)$ is the delay of the cth cluster, and $\tau(c, l)$ is the delay of the lth path in the cth cluster relative to the cluster-arrival time. The cluster arrivals and the path arrivals within each cluster are modeled as Poisson distribution with rate Λ and rate λ (where $\lambda > \Lambda$), respectively. We assume that $\alpha(c, l)$ are modeled as zero-mean, complex Gaussian random variables with variances

$$\Omega(c, l) = \mathrm{E}\left[|\alpha(c, l)|^2\right] = \Omega(0, 0) \exp\left(-\frac{T(c)}{\Gamma} - \frac{\tau(c, l)}{\gamma}\right), \qquad (12.17)$$

where $\mathrm{E}[\cdot]$ is the expectation operation, $\Omega(0, 0)$ is the mean energy of the first path of the first cluster, Γ is the cluster-decay factor, and γ is the ray-decay factor [76]. The total energy contained in terms $\alpha(c, l)$ is normalized to unity, i.e., $\sum_{c=0}^{C} \sum_{l=0}^{L} \Omega(c, l) = 1$.

In a noncooperative UWB system, each source transmits data directly to its destination. We consider the case in which time-domain spreading with a spreading factor of two is performed, as shown in Figure 12.2(a). In Figure 12.2, $\mathbf{x}_i$ ($1 \le i \le S$) denotes a vector of data symbols to be transmitted in each OFDM symbol, and S represents the number of

OFDM symbols contained in the frame payload. At the destination, the received signal at the nth subcarrier during the kth OFDM symbol duration is

$$y_{s,d}^k(n) = \sqrt{P_k} H_{s,d}^k(n) x(n) + z_{s,d}^k(n), \tag{12.18}$$

where P_k is the transmitted power at the source, $x(n)$ denotes a data symbol to be transmitted at subcarrier n, $H_{s,d}^k(n)$ is the frequency response of the channel from the source to the destination, $z_{s,d}^k(n)$ is additive noise, and no intersymbol interference is assumed. The superscript index k, $k = 1$ and 2, is used to distinguish the signals in two consecutive OFDM symbols. Because time spreading is performed, $x(n)$ is the same in both OFDM symbols. The noise, $z_{s,d}^k(n)$, is modeled as a complex Gaussian random variable with zero mean and variance N_0. From (12.16), the channel frequency response is given by

$$H_{s,d}^k(n) = \sigma_{s,d}^2 \sum_{c=0}^{C} \sum_{l=0}^{L} \alpha_{s,d}^k(c,l) e^{-j2\pi n \Delta f [T_{s,d}(c) + \tau_{s,d}(c,l)]}, \tag{12.19}$$

where $j = \sqrt{-1}$, $\Delta f = 1/T$, and T is the OFDM symbol period. With an ideal band hopping, we assume that the signal transmitted over different frequency bands undergo independent fading, i.e., $H_{s,d}^k(n)$ are independent for different k. When frequency-domain spreading is performed, the same data can be transmitted in more than one subcarrier. For subsequent performance evaluation, we denote Φ_n as a set of subcarriers that carry the data $x(n)$, and $g_F = |\Phi_n|$ as the frequency-spreading gain.

We consider a two-user cooperation over an UWB MB-OFDM system. Each user can act as a source or a relay. The cooperative strategy comprises two phases. In Phase 1, the source sends the data to its destination, and the data are also received by the relay. In Phase 2, the source is silent, while the relay helps forward the source data. With the DF cooperative protocol, the relay decodes the received data and forwards it to the destination. We consider the scenario of no time-domain spreading. In this scenario, the data frame that is transmitted from the source in Phase 1 and from the relay in Phase 2 can be depicted as in Figure 12.2(b). We can see from Figures 12.2(a) and 12.2(b) that, for a fixed frequency-spreading gain, the cooperative UWB scheme without time spreading achieves the same rate as noncooperative UWB scheme with time spreading.

In Phase 1, the received signal at the destination is the same as (12.18) with $k = 1$, and the received signal at the relay is

$$y_{s,r}(n) = \sqrt{P_1} H_{s,r}(n) x(n) + z_{s,r}(n). \tag{12.20}$$

In Phase 2, the relay forwards the decoded symbol with power P_2 to the destination only if the symbol is decoded correctly; otherwise, the relay does not send or remain idle [298]. The received signal at the destination in Phase 2 is

$$y_{r,d}(n) = \sqrt{\tilde{P}_2} H_{r,d}(n) x(n) + z_{r,d}(n), \tag{12.21}$$

where $\tilde{P}_2 = P_2$ if the relay decodes correctly; otherwise $\tilde{P}_2 = 0$. The channel responses $H_{s,r}(n)$ and $H_{r,d}(n)$ are also modeled according to the S-V model with total multipath energy $\sigma_{s,r}^2$ and $\sigma_{r,d}^2$. The noise $z_{s,r}(n)$ and $z_{r,d}(n)$ are complex Gaussian distributed with zero mean and variance N_0. We assume that the channel-state information is known at the receiver, but not at the transmitter. The channel coefficients are assumed to be independent for different transmit–receive links.

Next, we analyze the average SER performance of DF cooperative UWB MB-OFDM systems. We focus on the analysis for UWB systems with MPSK signals as used in [18]. The analysis for MQAM is similar, and we omit it here because of space limitation.

With the knowledge of channel-state information, the destination coherently combines the received signals from the source and the relay. Assume that each transmitted symbol has unit energy; then the instantaneous SNR of the MRC output can be written as [280]

$$\eta = \frac{P_1}{N_0} \sum_{n \in \Phi_n} |H_{s,d}(n)|^2 + \frac{\tilde{P}_2}{N_0} \sum_{n \in \Phi_n} |H_{r,d}(n)|^2. \tag{12.22}$$

The conditional SER in case of MPSK signals is given by [280]

$$P_{e|\{H\}} = \Psi(\eta) = \frac{1}{\pi} \int_0^{f_M} \exp\left(-\frac{b\eta}{\sin^2 \theta}\right) d\theta, \tag{12.23}$$

where $b = \sin^2(\pi/M)$. From (12.20) and (12.23), the instantaneous SNR at the MRC output of the relay is $\eta_{s,r} = \frac{P_1}{N_0} \sum_{n \in \Phi_n} |H_{s,r}(n)|^2$, and the conditional probability of incorrect decoding at the relay is $\Psi(\eta_{s,r})$. Taking into account the two possible cases of $\tilde{P}_2$, the conditional SER in (12.23) can be reexpressed as

$$P_{e|\{H\}} = \Psi(\eta)|_{\tilde{P}_2=0} \Psi(\eta_{s,r}) + \Psi(\eta)|_{\tilde{P}_2=P_2} \left[1 - \Psi(\eta_{s,r})\right]. \tag{12.24}$$

We substitute (12.22) into (12.24) and average over the channel realizations, resulting in the average SER

$$P_e = \frac{1}{\pi} \int_0^{f_M} \mathcal{M}_{\eta_{s,d}}(b_\theta) \mathcal{M}_{\eta_{r,d}}(b_\theta) d\theta \left[1 - \frac{1}{\pi} \int_0^{f_M} \mathcal{M}_{\eta_{s,r}}(b_\theta) d\theta\right]$$
$$+ \frac{1}{\pi^2} \int_0^{f_M} \mathcal{M}_{\eta_{s,d}}(b_\theta) d\theta \int_0^{f_M} \mathcal{M}_{\eta_{s,r}}(b_\theta) d\theta, \tag{12.25}$$

where $f_M = \pi - \pi/M$, $b_\theta = \frac{b}{\sin^2 \theta}$,

$$\eta_{s,d} = \frac{P_1}{N_0} \sum_{n \in \Phi_n} |H_{s,d}(n)|^2, \tag{12.26}$$

$$\eta_{r,d} = \frac{P_2}{N_0} \sum_{n \in \Phi_n} |H_{r,d}(n)|^2, \tag{12.27}$$

and $\mathcal{M}_\eta(s) = \mathrm{E}\left[\exp(-s\eta)\right]$ is the moment-generating function (MGF) of η [280]. Observe that the MGFs of $\eta_{s,d}$, $\eta_{s,r}$ and $\eta_{r,d}$, are in terms of the multipath coefficients whose amplitudes are Rayleigh distributed, as well as the multipath delays that are based on the Poisson process.

In general, it is difficult, if not impossible, to obtain closed-form formulations of the MGFs in (12.25). In this case, we exploit an approach in [284] that approximates $\mathcal{M}_{\eta_{x,y}}(s)$ as

$$\mathcal{M}_{\eta_{x,y}}(s) \approx \prod_{n=1}^{g_F} \left[1 + \frac{s P_x \sigma_{x,y}^2 \beta_n(\mathbf{R}_{x,y})}{N_0}\right]^{-1}, \tag{12.28}$$

where $P_x = P_1$ if x is the source and $P_x = P_2$ if x is the relay. In (12.28), $\beta_n(\mathbf{R}_{x,y})$ denotes the eigenvalues of a matrix $\mathbf{R}_{x,y}$, which is a correlation matrix whose each diagonal component is one and the (i, j)th $(i \neq j)$ component is given by

$$\mathbf{R}_{x,y}(i, j) = \Omega_{x,y}(0, 0)Q_{i,j}\left(\Lambda_{x,y}, \Gamma_{x,y}^{-1}\right)Q_{x,y}\left(\lambda_{x,y}, \gamma_{x,y}^{-1}\right), \qquad (12.29)$$

where $Q_{i,j}(a, b) = (a + b + \mathbf{j}2\pi(n_i - n_j)\Delta f)/(b + \mathbf{j}2\pi(n_p - n_q)\Delta f)$, in which n_i denotes the ith element in the set Φ_n. Note that the MGF in (12.28) is exact if $g_F = 1$ ($\Phi_n = \{n\}$). By substituting (12.28) into (12.25), we get the average SER performance:

$$P_e \approx F[U_{s,d}(\theta)U_{r,d}(\theta)]\{1 - F[U_{s,d}(\theta)]\} + F[U_{s,d}(\theta)]F[U_{s,r}(\theta)], \qquad (12.30)$$

where $U_{x,y}(\theta) = \prod_{n=1}^{g_F}\left(1 + \frac{v_{x,y}(n)}{\sin^2\theta}\right)$, $F[x(\theta)] = \frac{1}{\pi}\int_0^{f_M}\frac{1}{x(\theta)}d\theta$, and $v_{x,y}(n) = bP_x\sigma_{x,y}^2\beta_n(\mathbf{R}_{x,y})/N_0$.

To gain more insight into the cooperative UWB performance, we also provide approximate SER formulations that involve no integration as follows. By removing the negative term in (12.30) and bounding $1 + v_{x,y}(n)/\sin^2\theta$ with $[1 + v_{x,y}(n)]/\sin^2\theta$, we get

$$P_e \approx \prod_{n=1}^{g_F}[1 + v_{s,d}(n)]\{A_{2g_F}[1 + v_{r,d}(n)] + A_{g_F}^2[1 + v_{s,r}(n)]\}, \qquad (12.31)$$

where $A_i = \frac{1}{\pi}\int_0^{f_M}\sin^{2i}\theta d\theta$. If all channel links are available, i.e., $\sigma_{s,d}^2 \neq 0$, $\sigma_{s,r}^2 \neq 0$, and $\sigma_{r,d}^2 \neq 0$, the SER of cooperative UWB scheme can be further approximated by ignoring all 1's in (12.31) as

$$P_e \approx \prod_{n=1}^{g_F}v_{s,d}(n)\left[A_{2g_F}v_{r,d}(n) + A_{g_F}^2v_{s,r}(n)\right]. \qquad (12.32)$$

In Figure 12.3, we compare the preceding SER approximations with SER simulation curves of the cooperative UWB system with frequency-spreading gains $g_F = 1$ and 2. The simulated MB-OFDM system has $N = 128$ subcarriers, the subband bandwidth is 528 MHz, and the channel-model (CM) parameters follow those for CM 1 [76]. For fair comparison, we plot average SER curves as functions of P/N_0. Clearly, the theoretical SER (12.30) closely matches the simulation curves. SER approximations (12.31) are close to the simulation curves for the entire SNR range, whereas SER approximations (12.32) are loose at low SNR but tight at high SNR.

We compare here the performance of cooperative and noncooperative UWB MB-OFDM systems with the same data rate. Because of space limitation, we focus on the case of frequency-spreading gains $g_F = 1$ and 2. Assuming equal power allocation ($P_1 = P_2 = P/2$ as in [18]), the performance of the noncooperative UWB system with a time-spreading gain of two can be evaluated as $P_e = (G_{NC}P/N_0)^{-2g_F}$, i.e., the diversity gain is twice the frequency-spreading gain. The coding gain is $G_{NC} = b\sigma_{s,d}^2/(2\sqrt{A_2})$ if

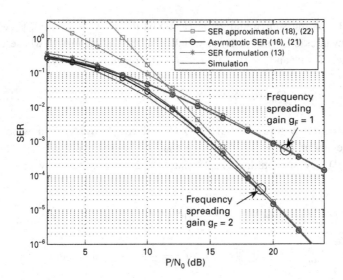

Figure 12.3 Comparison of SER formulations and simulation results for the DF cooperative UWB system: $\sigma_{s,d}^2 = \sigma_{s,r}^2 = \sigma_{r,d}^2 = 1$, $P_1 = P_2 = P/2$.

$g_F = 1$ and $G_{NC} = b\sigma_{s,d}^2\sqrt{1 - B_{s,d}^2/(2A_4^{\frac{1}{4}})}$ if $g_F = 2$. Here, $B_{x,y}$ is related to the CM parameters as

$$B_{x,y} = \Omega_{x,y}(0,0)\frac{\left[\left(\Lambda_{x,y} + \Gamma_{x,y}^{-1}\right)^2 + q\right]^{\frac{1}{2}}\left[\left(\lambda_{x,y} + \gamma_{x,y}^{-1}\right)^2 + q\right]^{\frac{1}{2}}}{\left[\left(\Gamma_{x,y}^{-1}\right)^2 + q\right]^{\frac{1}{2}}\left[\left(\gamma_{x,y}^{-1}\right)^2 + q\right]^{\frac{1}{2}}}, \tag{12.33}$$

in which $q = (2\pi\mu\Delta f)^2$ and μ denotes the subcarrier separation. For the cooperative scheme, denote $r = P_1/P$ as the power ratio of the transmitted power P_1 at the source over the total power P. According to (12.32), the approximate SER of DF cooperative UWB system can be expressed as

$$P_e = (G_{DF}P/N_0)^{-2g_F}, \tag{12.34}$$

i.e., the cooperative UWB scheme achieves the same diversity gain as the non-cooperative scheme. The factor G_{DF} represents cooperative gain which can be determined as $G_{DF} = b\sigma_{s,d}\sigma_{s,r}\sigma_{r,d}r/[A_1^2\sigma_{r,d}^2 + A_2\sigma_{s,r}^2r/(1-r)]^{1/2}$ if $g_F = 1$ and $G_{DF} = b\sigma_{s,d}\sigma_{s,r}\sigma_{r,d}r[(1-B_{s,d}^2)(1-B_{s,r}^2)(1-B_{r,d}^2)]^{\frac{1}{4}}/[A_2^2\sigma_{r,d}^2(1-B_{r,d}^2) + A_4\sigma_{s,r}^2(1-B_{s,r}^2)r^2/(1-r)^2]^{\frac{1}{4}}$, if $g_F = 2$. Because both noncooperative and cooperative UWB systems achieve the same diversity order, it is interesting to compare the coding gain and the cooperative gain. The ratio $\xi = G_{DF}/G_{NC}$ in case of $g_F = 1$ and 2 are

$$\xi = V\left(A_1^2/A_2\sigma_{r,d}^2 + r/(1-r)\sigma_{s,r}^2\right)^{-\frac{1}{2}}, \tag{12.35}$$

$$\xi = V\left(\frac{A_2^2(1-B_{s,d}^2)\sigma_{r,d}^2}{A_4(1-B_{s,r}^2)} + \frac{r^2(1-B_{s,d}^2)\sigma_{s,r}^2}{(1-r)^2(1-B_{r,d}^2)}\right)^{-\frac{1}{4}}, \tag{12.36}$$

respectively, where $V = 2r\sigma_{s,r}\sigma_{r,d}/\sigma_{s,d}$.

Next, we provide an optimum power allocation for the cooperative UWB MB-OFDM system with two different objectives, namely minimizing overall transmitted power and maximizing the coverage.

We define $\mathbf{P} = [P_1 \ P_2]^T$ as a power-allocation vector. Our objective is to minimize the overall transmitted power under the constraint on the SER performance and the transmitted power level. The optimization problem is formulated as

$$\min_{\mathbf{P}} \ P = \sum_i P_i, \tag{12.37}$$

$$\text{s.t.} \ \begin{cases} \text{performance: } P_e \leq \varepsilon \\ \text{power: } P_i \leq \bar{P}, \ \forall i \end{cases},$$

where ε denotes the required SER and $\bar{P}$ is the maximum transmitted power for each subcarrier. The first constraint in (12.37) is to ensure the performance requirement. The second constraint is related to the limitation on the transmitted power level.

Consider first the formulated problem in (12.37) without the maximum power constraint. Using the SER in (12.32), the optimum power allocation can be determined, after some manipulations, as

$$P_1 = rP \text{ and } P_2 = (1-r)P, \tag{12.38}$$

where, in the case of $g_F = 1$:

$$P = N_0/(V\sqrt{\varepsilon}) \left(A_2 r/(1-r)\sigma_{s,r}^2 + A_1^2\sigma_{r,d}^2 \right)^{1/2}, \tag{12.39}$$

$$r = (\sigma_{s,r} + K)(3\sigma_{s,r} + K)^{-1}, \tag{12.40}$$

in which $K = \sqrt{\sigma_{s,r}^2 + (8A_1^2/A_2)\sigma_{r,d}^2}$, and in the case of $g_F = 2$:

$$P = \frac{N_0}{V} \left[\frac{A_4 r^2 \sigma_{s,r}^4 (1 - B_{s,r}^2) + A_2^2 (1-r)^2 \sigma_{r,d}^4 (1 - B_{r,d}^2)}{\varepsilon (1-r)^2 (1 - B_{s,d}^2)(1 - B_{s,r}^2)(1 - B_{r,d}^2)} \right]^{1/4}, \tag{12.41}$$

$$r = \frac{4^{\frac{1}{3}}c^2 + 2(c_{s,r} + 3c_{r,d})c + 4^{\frac{2}{3}} \left(c_{s,r}^2 - 12c_{s,r}c_{r,d} \right)}{6(2c_{s,r} + c_{r,d})c}, \tag{12.42}$$

in which $c = \left[3(2c_{s,r} + c_{r,d})(12c_{s,r}c_{r,d} + 81c_{r,d}^2)^{1/2} + 2c_{s,r}^3 + 72c_{s,r}c_{r,d} - 27c_{r,d}^2 \right]^{1/3}$. The results in (12.40) and (12.42) reveal that the asymptotic power allocation of cooperative UWB systems depends on only the quality of the source–relay link and relay–destination link, but not on the source–destination link.

In the case of joint encoding across subcarriers, the optimum power allocation also depends on the multipath clustering property of UWB channels through parameters $B_{x,y}$. Table 12.1 shows that the optimum power allocation that is obtained from (12.38) and that is based on SER in (12.31) agrees with the results that are obtained via exhaustive search to minimize (12.30) for all considered scenarios. According to the SER expressions in (12.31), the ratio between the power of cooperative and noncooperative UWB systems

Table 12.1 Comparisons of optimum power allocation obtained via exhaustive search and analytical results: $\sigma_{s,d}^2 = 1$, CM 1, $\varepsilon = 5 \times 10^{-2}$

$\sigma_{s,r}^2$	$\sigma_{r,d}^2$	g_F	Search	From (12.31)	From (12.38)
10	1	1	0.5321	0.5356	0.5247
10	1	2	0.5072	0.5095	0.5023
1	10	1	0.7873	0.7772	0.7968
1	10	2	0.8082	0.7882	0.8316

with the same spreading gain is

$$\frac{P_{\text{DF}}}{P_{\text{NC}}} = \frac{N_0 P_e^{-1/(2g_F)} G_{\text{DF}}^{-1}}{N_0 P_e^{-1/(2g_F)} G_{\text{NC}}^{-1}} = \frac{G_{\text{NC}}}{G_{\text{DF}}} = \frac{1}{\xi}. \tag{12.43}$$

Substituting power allocation in (12.40), (12.42) into (12.43), we can show that if the source–relay link is of high quality (e.g., $\sigma_{s,d}^2 = \sigma_{r,d}^2 = 1, \sigma_{s,r}^2 = 10$), then the cooperative scheme yields about 50% power saving compared with the noncooperative scheme with the same rate; if the relay–destination link is of high quality (e.g. $\sigma_{s,d}^2 = \sigma_{s,r}^2 = 1$, $\sigma_{r,d}^2 = 10$), the power saving of 80% can be achieved.

With the maximum power limitation, it is difficult to obtain a closed-form solution to the problem in (12.37). In this case, we provide a solution as follows. Let P_1 and P_2 be the transmitted power that we obtain by solving (12.37) without the maximum power constraint, and let $\hat{P}_1$ and $\hat{P}_2$ denote our solution.

- If $\min\{P_1, P_2\} > \bar{P}$, then no feasible solution to (12.37).
- Else if $\max\{P_1, P_2\} \le \bar{P}$, then $\hat{P}_1 = P_1$ and $\hat{P}_2 = P_2$;
- Otherwise, (i) Let $j = \arg\max_i\{P_1, P_2\}$ and $j' = \arg\min_i\{P_1, P_2\}$. (ii) Set $P_j = \bar{P}$ and find $P_{j'}$ such that the desired SER performance is satisfied, i.e., $P_{j'}$ is obtained by solving $P_e - \varepsilon = 0$ where P_e is according to (12.31) or (12.32) with P_j replaced by $\bar{P}$. (iii) If $P_{j'} \le \bar{P}$, then $\hat{P}_j = \bar{P}$ and $\hat{P}_{j'} = P_{j'}$; otherwise, no feasible solution to (12.37).

We determine the optimum power allocation and the relay location so as to maximize the distance between the source and the destination with constraint on the error performance. The geometry on the channel-link qualities is taken into account by assuming that the total multipath energy $\sigma_{x,y}^2$ is modeled as

$$\sigma_{x,y}^2 = \kappa D_{x,y}^{-\nu}, \tag{12.44}$$

where κ is a constant whose value depends on the propagation environment, ν is the propagation-loss factor, and $D_{x,y}$ represents the distance between node x and node y. Given a fixed total transmitted power P, we aim to jointly determine power allocation $r = P_1/P$ and relay location to maximize the distance $D_{s,d}$. From the SER performance obtained in (12.31), we can see that the performance of the cooperative UWB system is related not only to the power allocation but also to the location of the nodes. Obviously, the optimum relay location must be on the line joining the source and the destination. In

Table 12.2 Power allocation, relay location, and maximum coverage of cooperative UWB MB-OFDM systems

P/N_0 (dB)	Exhaustive search			Analytical solution		
	r	$D_{s,r}$	$D_{s,d}$	r	$D_{s,r}$	$D_{s,d}$
25	0.86	13.00	14.06	0.88	13.74	14.87
35	0.55	15.53	33.82	0.58	15.12	33.98
25	0.89	17.11	19.14	0.88	17.31	19.79
35	0.52	13.21	43.87	0.54	13.27	43.92

this case, the distance $D_{s,d}$ can be written as a summation of the distance of the source–relay link and relay–destination link, i.e., $D_{s,d} = D_{s,r} + D_{r,d}$. Now we formulate an optimization problem as follows:

$$\max_{r, D_{s,r}, D_{r,d}} \quad D_{s,r} + D_{r,d}, \qquad (12.45)$$

$$\text{s.t.} \begin{cases} \text{Performance: } P_e \leq \varepsilon \\ \text{Power: } rP \leq \bar{P}, \ (1-r)P \leq \bar{P}, \ 0 < r < 1 \end{cases}.$$

With the SER formulations derived in (12.31), the Lagrangian multiplier method can be applied to solve (12.45) without a maximum power constraint, and a solution under the maximum power constraint can be obtained by a similar solution discussed previously.

Table 12.2 shows the optimum power allocation and the distances obtained via exhaustive search [using the SER in (12.30)] and that from analytical solutions [using SER in (12.31)]. Clearly both results match closely. Interestingly, when P/N_0 is small, the maximum coverage is achieved when the relay is located far away from the source, and almost all of the transmitted power P is allocated at the source. On the other hand, when P/N_0 is high ($P/N_0 > 30$ dB), the optimum relay location is about the midpoint between source and destination, and the power should be equally allocated at the source and the relay. The reason is that, at small SNR, the transmitted power is not large enough for the cooperative system to achieve the performance of diversity of the order of two; hence the forwarding role of the relay is less important and almost all of the power should be used at the source. At high enough SNR, diversity of the order of two can be achieved, so the relay should be in the middle to balance the channel quality of the source–relay link and the relay–destination link.

The current multiband standard proposal [18] allows several UWB devices to transmit at the same time using different subbands. However, in a short-range scenario, the number of UWB devices that simultaneously transmit their information tends to be smaller than the number of available subbands. Here we propose an improved cooperative UWB strategy that makes use of the unoccupied subbands as follows. The time-domain spreading with a spreading factor of two is performed at the source. The improved cooperative UWB scheme comprises two phases, each corresponding to one OFDM symbol period. In Phase 1, the source broadcasts its information using one subband. In Phase 2, the source repeats the information using another subband to gain the diversity from time

Figure 12.4 Illustration of an improved cooperative UWB MB-OFDM.

spreading, while the relay forwards the source information using an unoccupied subband. The destination combines the received signals from the source directly in Phase 1 and Phase 2, and the signal from the relay in Phase 2. Figure 12.4 illustrates an example of the improved cooperative UWB system. In Figure 12.4, the source and the relay are denoted respectively by S and R. It is worth noting that the improved cooperative UWB scheme is compatible with the current multiband standard proposal [18] that allows multiuser transmission using different subbands. In addition, the proposed scheme yields the same data rate as the noncooperative scheme with the same spreading gain.

Let P_1 and P_2 denote the transmitted power at the source in Phase 1 and Phase 2, and let P_3 denote the power at the relay. Following the same procedures as in (12.31), the SER of the improved cooperative UWB scheme can be approximated as

$$P_e \approx F\left[U_{s,d}^2(\theta)U_{r,d}(\theta)\right](1 - F[U_{s,d}(\theta)]) + F\left[U_{s,d}^2(\theta)\right]F[U_{s,r}(\theta)], \quad (12.46)$$

in which the asymptotic performance can be determined as

$$P_e = (G_I P/N_0)^{-3g_F}, \quad (12.47)$$

where $G_I = W/[A_1 A_2 \sigma_{r,d}^2 + A_3 \sigma_{s,r}^2 r_1/r_3]^{\frac{1}{3}}$ if $g_F = 1$, and $G_I = WZ/[A_2 A_4 \sigma_{r,d}^4 (1 - B_{r,d}^2) + A_6 \sigma_{s,r}^4 (1 - B_{s,r}^2) r_1^2/r_3^2]^{\frac{1}{6}}$ if $g_F = 2$, in which $W = b[\sigma_{s,d}^4 \sigma_{s,r}^2 \sigma_{r,d}^2 r_1^2 r_2]^{\frac{1}{3}}$ and $Z = [(1 - B_{s,d}^2)^2 (1 - B_{s,r}^2)(1 - B_{r,d}^2)]^{\frac{1}{6}}$. The improved cooperative UWB system provides an overall performance of diversity order $3g_F$. Based on the preceding SER formulations, the optimum power allocation can be determined.

We perform computer simulations to compare the performance of the proposed cooperative UWB schemes and to validate the derived theoretical results. In all simulations, UWB MB-OFDM system has 128 subcarriers, the signal is based on QPSK, and the subband bandwidth of 528 MHz. The propagation-loss factor is $\nu = 2$. The source is located at position $(0, 0)$.

In Figure 12.5, we compare the average SER performances of UWB systems with different cooperative strategies. The locations of the relay and the destination are (1 m, 0) and (2 m, 0). All channel links are modeled by CM 1. Equal power allocation is used. To compare fairly, we present the SER curves as functions of P/N_0. We can see from Figure 12.5 that both noncooperative and cooperative UWB systems achieve an overall performance of diversity of the order of $2g_F$. In the case of $g_F = 1$, the cooperative UWB scheme outperforms the noncooperative UWB by 2 dB at a SER of 10^{-3}. This agrees with the analysis in (12.35) that shows that the performance gain of the DF cooperative UWB compared with the noncooperative UWB is $\xi = [(1 + A_1^2/A_2)\sigma_{s,d}^2]^{1/2} = 1.59$.

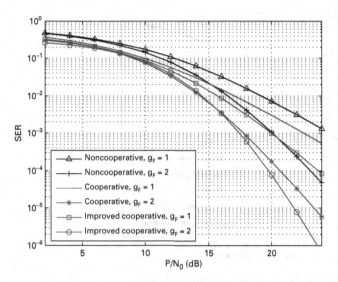

Figure 12.5 SER performance vs. P/N_0.

In the case of $g_F = 2$, the performance of cooperative scheme is about 2.5 dB better than that of noncooperative scheme. This also corresponds to the analysis in (12.36) in which the performance gain is $\xi = [(1 + A_2^2/A_4)\sigma_{s,d}^2]^{1/4} = 1.81$. Additionally, Figure 12.5 illustrates that the improved cooperative UWB scheme provides the performance of diversity of the order of $3g_F$ and yields about a 2-dB performance improvement over the cooperative UWB scheme.

Figures 12.6 and 12.7 compare the total transmitted power of noncooperative and cooperative systems to achieve the same range. The performance requirement is $P_e \leq 5 \times 10^{-2}$. The relay is located in the middle between the source and the destination ($D_{s,d} = D_{s,r}/2$). All channel links are modeled by CM 4. In Figure 12.6, we consider the case of no limitation on the transmitted power level. By increasing the frequency-spreading gain, the overall transmitted power can be reduced by 60%. With the same g_F, the cooperative scheme achieves 43% power saving compared with the noncooperative scheme. This is consistent with the analytical results in (12.43), in which the power ratio of cooperative and noncooperative schemes can be calculated as $P_{DF}/P_{NC} = 0.59$ when $g_F = 1$ and $P_{DF}/P_{NC} = 0.54$ when $g_F = 2$. Figure 12.6 also shows that using the improved cooperative UWB scheme can achieve up to 52% power saving compared with the noncooperative scheme. In Figure 12.7, the maximum power constraint is taken into account, and the power is allocated according to the proposed suboptimal solution. The power limitation is set at $P_i/N_0 \leq 19$ dB. The tendencies observed in Figure 12.7 are similar to those in Figure 12.6. The improved cooperative scheme saves about 50% overall transmitted power in the case of $g_F = 1$ and saves about 20% in case of $g_F = 2$.

Next, we study the coverage of the UWB system under different cooperative strategies. All channel links are based on CM 4. The SER performance requirement is fixed at 5×10^{-2}. In Figure 12.8, we depict the coverage as a function of P/N_0. The transmitted power level is limited by $P_i/N_0 \leq 19$ dB. For the cooperative scheme, the relay location

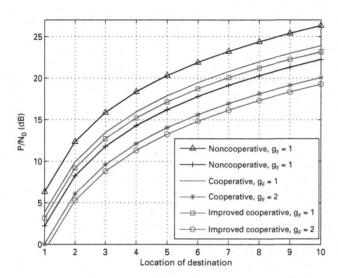

Figure 12.6 P/N_0 vs. destination location (no power limitation).

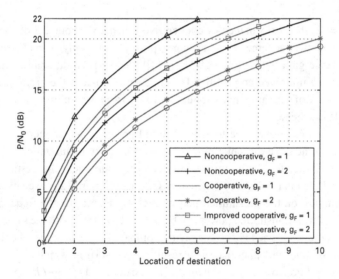

Figure 12.7 P/N_0 vs. destination location (with power limitation).

and the power allocation are designed such that the distance $D_{s,d}$ is maximized. With the same P/N_0 and the same transmission data rate, the coverage of UWB system can be increased by 85% using the cooperative scheme, and it can be increased by 100% using the improved cooperative scheme.

On the whole, we propose to enhance the performance of UWB systems by employing cooperative diversity. We analyze the SER performance and provide optimum power allocation of cooperative UWB MB-OFDM systems with the DF protocol. Both non-cooperative and cooperative schemes achieve the same diversity of the order of twice the frequency-spreading gain for every channel environment. The cooperative gain, on

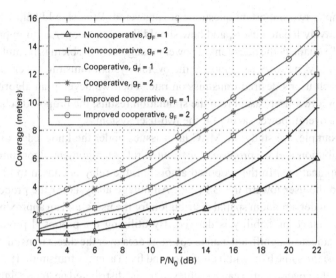

Figure 12.8 $D_{s,d}$ vs. P/N_0 (with power limitation).

the other hand, depends on the clustering property of UWB channels. By taking advantage of the relay location and properly allocating the transmitted power, the cooperative gain can be improved such that the cooperative UWB achieves superior performance to the noncooperative scheme with the same data rate. It turns out that, at low SNR, the coverage is maximized if the relay is located farthest away from the source, and almost all of the transmitted power is allocated at the source; at high SNR, the coverage is maximized if the relay is located in the midpoint between source and destination, and equal power allocation is used. We also propose to further improve the cooperative UWB scheme by allowing the source and the relay nodes to simultaneously retransmit the information. Simulation results confirm the theoretical analysis that the cooperative UWB scheme can achieve 43% power saving and 85% coverage extension compared with the noncooperative scheme, whereas the improved cooperative UWB scheme can achieve 52% power saving and 100% coverage extension.

12.4   MAC Layer Issue: Relay Selection and Channel Allocation

In most of the previous works, the cooperating relays are just assumed to exist and are already coupled with the source nodes in the network. These works also assumed fixed channel variances between all of the nodes in the network, which implies a fixed network topology. If the random users' spatial distribution and the associated propagation path losses between different nodes in the network are taken into consideration, then these assumptions, in general, are no longer valid. In Figure 12.1, we can see that the source–destination pair can select different relays for cooperation and the performance can be improved differently. We will show two examples in this section for relay-selection problems.

In the first example, two important questions are answered: Who should help whom among the distributively located users, and how should the users cooperate to improve the performance? To quantify the questions, a power management problem is formulated over a multiuser OFDM network to minimize the overall system transmit power under the constraint of each user's desired transmission rate. Then we develop an algorithm to find solutions for a two-user case. From the simulation results, the proposed scheme achieves up to 50% overall power saving for the two-user system.

In the second example, we study the MAC layer issues under the knowledge of the users' spatial distribution and we propose a distributed relay-assignment algorithm for cooperative communications. In the proposed algorithm, the relay is chosen to be the nearest neighbor to the user toward the BS (AP). An outage analysis for the proposed scheme is provided under a random spatial distribution for the users, and an approximate expression for the outage probability is derived. By utilizing the proposed protocol, simulation results indicate a significant gain in coverage area over the direct transmission scheme under almost the same bandwidth efficiency and fixed average transmitted power. A 350% increase in the coverage area can be achieved by the distributed nearest-neighbor protocols. This coverage increase can also be translated to energy efficiency over direct transmission when fixing the total coverage area.

12.4.1   OFDM Networks

OFDM is a mature technique used to mitigate the problems of frequency selectivity and intersymbol interferences. The optimization of subcarrier assignment for different users offers substantial gains to the system performances [335, 349, 350]. In addition, the fact that each user can assign the transmission over different subcarriers gives an opportunity for cooperative transmission among users.

Most of existing research concentrates on improving the peer-to-peer link quality. However, there are many questions for multiple-user resource allocation over cooperative transmission that remain unanswered. The most important ones are "who helps whom" and "how to cooperate." We concentrate on solving these two major problems in cellular networks or wireless local area networks.

For an uplink multiuser OFDM system, we suppose that there are N subcarriers and K users in the network. We represent T_i as a transmission rate of the ith user and the rate is split onto N subcarriers in which r_i^n denotes the transmission rate at the nth subcarrier with the corresponding transmit power P_i^n. From information theory [52], we have

$$r_i^n = W \log_2 \left(1 + \frac{P_i^n G_i^n}{\sigma^2 \Gamma} \right), \tag{12.48}$$

where Γ is a constant for the capacity gap, G_i^n is the channel gain, and σ^2 is the thermal-noise power. Without lost of generality, we assume that the noise power is the same for all subcarriers and all users.

In the current OFDM system such as the IEEE 802.11a/g standard [376], the MAC layer provides two different wireless access mechanisms for wireless medium sharing, namely, the distributed coordination function (DCF) and the point coordination function (PCF). The DCF achieves automatic medium sharing among users by using CSMA/CA. The PCF, however, is a centralized control mechanism. In both mechanisms, TDMA technology is utilized for all users to share the channel, i.e., at each time, only one user occupies all the bandwidth.

The goal is to minimize the overall power consumption, under the constraint on each user's minimal rate. If there is no cooperation among users, because of the TDMA utilization, the overall power-minimization problem is the same as minimizing each user's power independently. We define $\mathbf{P}_i = [P_i^1, \ldots, P_i^N]$ as a power assignment vector, the ith user's power-minimization problem can be expressed as

$$\min_{\mathbf{P}_i} \sum_{n=1}^{N} P_i^n, \tag{12.49}$$

$$\text{s.t.} \sum_{n=1}^{N} r_i^n = T_i. $$

From (12.48), the preceding constrained optimization can be solved by the traditional water-filling method. By representing

$$I_i^n = \frac{\Gamma \sigma^2}{G_i^n}, \tag{12.50}$$

the optimal solution of the water-filling method is given by

$$P_i^n = \left(\mu_i - I_i^n\right)^+ \text{ and } r_i^n = W \log_2\left(1 + \frac{P_i^n}{I_i^n}\right), \tag{12.51}$$

where $y^+ = \max(y, 0)$ and μ_i is solved by bisection search of

$$\sum_{n=1}^{N} W \log_2\left(1 + \frac{(\mu_i - I_i^n)^+}{I_i^n}\right) = T_i. \tag{12.52}$$

Note that the solution in (12.51) is based on the assumption that all users do not cooperate with each other. Because of the broadcasting nature of the wireless channels, not only the BS but also other users are able to receive the transmitted information. Therefore, if other users can cooperate and help relay the information, multiple-node diversity can be explored and the system performance can be significantly improved. One fundamental question to answer is how to group users for cooperation, i.e., "who helps whom?." Next, we construct the cooperative transmission framework and then formulate the cooperative problem as an assignment problem.

In the proposed cooperative transmission framework, we adopt a similar concept to the AF scheme in [175]. The difference is that our framework does not require a stage dedicated to relay or transmission. At one time period such as a frame or a time slot, a user transmits data and all other users including BS can listen. In the next time period, another user tries to transmit his/her own data, while he/she can help others to transmit if his/her location and channels are better. One example is shown in Figure 12.9 in which user i relays user j's data to the BS. At time one, user j transmits data, while all other

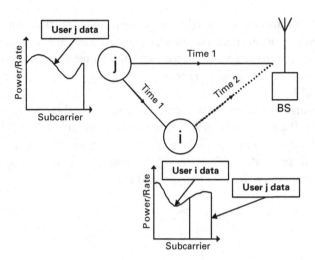

Figure 12.9 Cooperative transmission scheme example.

users including the BS can listen. In the next time period, user i tries to transmit his/her own data, while he/she can help user j to transmit at the same time if his/her location and channels are good. User i can allocate some of his/her subcarriers to relay user j's data, so as to reduce user j's transmission power. In doing so, user i has to transmit his/her own data in the rest of his/her own subcarriers. Consequently, the power for user i will be increased to maintain his/her own data transmission. So there are trade-offs on whether or not to help others. From the system optimization point of view, the overall power of both user i and user j can be minimized by selecting the proper number of subcarriers for cooperation, i.e., the question of "how to cooperate." Moreover, because of the users' different locations and channel conditions, some users are more effective to help others' transmissions. Hence it is essential to find the optimal cooperative groups, i.e., the question of "who helps whom." First, we try to formulate the cooperative transmission problem as an assignment problem.

We define $\mathbf{A}_{KN \times KN}$ in Figure 12.10 as an assignment matrix whose element $\mathbf{A}_{u,v} \in \{0, 1\}$, where $u = 1, \ldots, KN$ and $v = 1, \ldots, KN$. For notation convenience, we denote $(a, n) = (a - 1)N + n$. We use (i, n) to represent the helping user i at subcarrier n, and (j, n') as the helped user j at subcarrier n'. Note that the user index i and $j \in 1, \ldots, K$. The value of each element of $\mathbf{A}$ has the following interpretations:

1. $\mathbf{A}_{(i,n),(i,n)} = 1$ means the ith user's nth subcarrier transmits its own information to the BS.
2. $\mathbf{A}_{(i,n),(j,n')} = 1$, for $i \neq j$ means the ith user's nth subcarrier relays transmit information from the jth user's n' th subcarrier .

We can observe that $\sum_{v=1}^{KN} \mathbf{A}_{u,v} = 1, \forall u = 1, \ldots, KN$, i.e., each subcarrier contains only information from one user at a time. Note that, in case of $\mathbf{A} = \mathbf{I}_{KN \times KN}$, the solution of the proposed scheme is the same as the one using the water-filling method. We also show as an example in Figure 12.10 in which user 2 uses its subcarrier 1 to relay the data

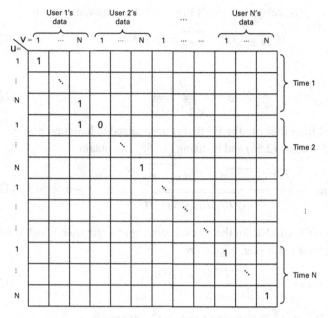

Figure 12.10 Assignment matrix.

for user 1 at the Nth subcarrier, i.e., $\mathbf{A}_{(2,1),(1,N)} = 1$. As shown in Figure 12.10, each set of N rows represents data transmitted at a specific time and each set of N columns represents whose data are being transmitted at that time.

We define $\mathbf{P}_{K \times N} = [\mathbf{P}_1', \ldots, \mathbf{P}_K']'$ as the power-allocation matrix and $\mathbf{G}_{KN \times KN}$ as the channel-gain matrix whose its elements obey the following conditions:

1. $G_{(j,n'),(i,n)}$, for $i \neq j$, denotes the channel gain from the jth user at the n' subcarrier to the ith user at the nth subcarrier.
2. $G_{(i,n),(i,n)}$ represents the channel gain from the ith user at the nth subcarrier to the BS.
3. To prevent $\mathbf{A}_{(i,n),(i,n')} = 1$, for $n \neq n'$, we define $G_{(i,n),(i,n')} = 0$, $\forall n \neq n'$, and $\forall i$.

For the AF cooperative scheme, the receiver at the BS combines the directly received signal and the relayed signal together by MRC. In what follows, we derive r_i^n in (12.48) that incorporates cooperative transmission. From Figure 12.9, we express the SNR that results from the direct transmission from the jth user at the (n')th subcarrier to the BS by

$$\Gamma_j^{n'}(d) = \frac{\mathbf{A}_{(j,n'),(j,n')} P_j^{n'} G_{(j,n'),(j,n')}}{\sigma^2}. \tag{12.53}$$

Next, we consider the SNR at the BS that results from user i relaying user j's information to the BS. Assuming that $X_{j,i}$ is the transmit signal from user j to user i, the received signal at user i is

$$R_{j,i} = \sqrt{P_j^{n'} G_{(j,n')(i,n)}} X_{j,i} + \sigma. \tag{12.54}$$

User i amplifies $R_{j,i}$ and relays it to the BS in which the received signal is

$$R_{i,\text{BS}} = \sqrt{P_i^n G_{(i,n)(i,n)}} X_{i,\text{BS}} + \sigma, \tag{12.55}$$

where

$$X_{i,\text{BS}} = \frac{R_{j,i}}{\sqrt{E|R_{j,i}|^2}} \tag{12.56}$$

is the transmit signal from user i to the BS that is normalized to have unit power.

Substituting (12.54) into (12.56) and by using (12.55), we obtain

$$R_{i,\text{BS}} = \frac{\sqrt{P_i^n G_{(i,n)(i,n)}}\left(\sqrt{P_j^{n'} G_{(j,n')(i,n)}} X_{j,i} + \sigma\right)}{\sqrt{P_j^{n'} G_{(j,n')(i,n)} + \sigma^2}} + \sigma. \tag{12.57}$$

Using (12.57), the relayed SNR for the n'th subcarrier of the jth user, which is helped by the nth subcarrier of the ith user, is given by

$$\Gamma_j^{n'}(r) = \frac{P_i^n P_j^{n'} G_{(i,n)(i,n)} G_{(j,n')(i,n)}}{\sigma^2 \left[P_i^n G_{(i,n)(i,n)} + P_j^{n'} G_{(j,n')(i,n)} + \sigma^2\right]}. \tag{12.58}$$

Therefore, by (12.53) and (12.58), we can rewrite (12.48) as

$$r_j^{n'} = W \log_2\left[1 + \frac{\Gamma_j^{n'}(d) + \Gamma_j^{n'}(r)}{\Gamma}\right]. \tag{12.59}$$

The problem in this case can be considered as the cooperative group problem. Specifically, we determine the assignment matrix $\mathbf{A}$ and the corresponding power-allocation matrix $\mathbf{P}$ that minimize the overall power and satisfy all the constraints. The optimization problem can be formulated as

$$\min_{\mathbf{A},\mathbf{P}} \sum_{i=1}^{K} \sum_{n=1}^{N} P_i^n, \tag{12.60}$$

$$\text{s.t.} \begin{cases} \text{Transmission rate: } \sum_{n=1}^{N} r_i^n = T_i, \forall i; \\ \text{Assignment: } \sum_{v=1}^{KN} \mathbf{A}_{uv} = 1, \forall u = 1, \ldots, KN; \\ \text{Positive power: } P_i^n \geq 0, \forall i, n. \end{cases}$$

In the case of $\mathbf{A} = \mathbf{I}_{KN \times KN}$, the problem in (12.60) is exactly the same as (12.49), and the water-filling method can be used to find the optimal solution.

Note that the problem in (12.60) can be viewed as a generalized assignment problem, which is an NP-hard problem [201]. To solve (12.60), we divide the problem into two subproblems. The first subproblem is to find the optimal $\mathbf{P}$ for a fixed $\mathbf{A}$. Then, in the second subproblem, we try to find the optimal $\mathbf{A}$.

Next, we provide algorithms to solve the problem in (12.60). We first solve the problem of finding optimal power allocation $\mathbf{P}$ with fixed assignment matrix $\mathbf{A}$. After that we try to find the best $\mathbf{A}$.

In this case, we assume that the assignment matrix $\mathbf{A}$ is known. We show the characteristic of the solution by the following theorem.

THEOREM 23 *For a fixed* **A**, *there is only one local optimum that is also the global optimum for (12.60).*

Proof 6 All the users are divided into two groups. The first group of users does not cooperate with others, i.e., $A_{(i,n),(j,n')} = 0$, $\forall i \neq j$ or $n \neq n'$. Therefore, the problem can be considered in the same way as the single-user case, and the water-filling method can be used to find the only local optimum that is also the global optimum for these kinds of users.

The second group of users does cooperate with other users, i.e., $\exists A_{(i,n),(j,n')} = 1$, $\forall i \neq j$ or $n \neq n'$. For a fixed **A**, the optimization problem for this kind of user can be expressed as

$$\min_{\mathbf{P}} \sum_{i=1}^{K} \sum_{n=1}^{N} P_i^n, \tag{12.61}$$

$$\text{s.t.} \begin{cases} P_i^n \geq 0, \forall i, n; \\ \sum_{n=1}^{N} r_i^n = T_i, \forall i. \end{cases}$$

Observe that the optimization goal is linear, and the first constraint is linear as well. We express the second constraint by use of (12.48) as

$$\prod_{n=1}^{N} \left[1 + \frac{\Gamma_i^n(d) + \Gamma_i^n(r)}{\Gamma} \right] = 2^{\frac{T_i}{W}}. \tag{12.62}$$

For a fixed **A** and using (12.53) and (12.58), all power components (P_i^n or $P_j^{n'}$) have a quadrature form, i.e., the expression in the preceding equation is a polynomial function with the maximum order of two. Moreover, all of the coefficients are positive; hence, the constraint is convex. From [34], we know that the only local optimum for this kind of convex optimization problem is also the global optimum. Because the transmission of information of all users is divided into different time slot in TDM, the optimal solutions for the preceding two groups of users are also the optimal solution for (12.60). **QED**

With the fixed **A**, any nonlinear or convex optimization methods [34] can be used to solve (12.61). For example, we can use the barrier method to covert the constrained optimization problem in (12.61) to an unconstrained optimization problem and solve it iteratively. Within each iteration, we can use the Newton method to solve the unconstrained optimization problem. Here, we use the MATLAB FMINCON function to solve the optimization problem in (12.61).

Next, we show a method to find the optimal assignment matrix **A**. Because any element of **A** has a value of either 0 or 1 and the dimension of **A** is $KN \times KN$, we can use full search to find the optimal solution. For any specific **A**, we calculate the overall transmit power and select the one that generates the minimal power. However, the computational complexity is extremely high, especially when a large number of subcarriers are utilized and there are a substantial number of users in the OFDM systems. Some simplified algorithms such as branch-and-bound [201] can be applied to reduce the complexity.

Table 12.3 Two-User Searching Algorithm for **A**

Initialization: $\mathbf{A} = I_{2N \times 2N}$, and Calculate (12.61).

Iteration:
  Hypotheses:
    If user 1's subcarrier n helps user 2's subcarrier n':
      $[\mathbf{A}]_{n,n} = 0$ and $[\mathbf{A}]_{n,n'+N} = 1$;
    If user 2's subcarrier n' helps user 1's subcarrier n:
      $[\mathbf{A}]_{n'+N,n'+N} = 0$ and $[\mathbf{A}]_{n+N,n'} = 1$;
    Solve (12.61).
  Among all hypotheses, find the maximal power reduction.
End when no power reduction, return **A** and **P**.

We propose a two-user greedy algorithm to find the optimal **A** as given in Table 12.3. The basic idea is to modify **A** for one subcarrier per time. Initially, **A** is assigned as an identity matrix, which is basically the water-filling scheme. Then, among N subcarriers of users, we make N hypotheses that the nth subcarrier is assigned to help the other user and the remaining subcarriers are unchanged. Among all of these hypotheses, the algorithm selects the one that maximally reduces the overall power and the remaining $N - 1$ subcarriers make another hypothesis. This process is stopped when the power cannot be further reduced. Note that in this case the searching complexity of the proposed algorithm is N^3 and the algorithm is suboptimal because of the greedy local search. It is worth mentioning that the proposed algorithm does not need to be applied in real time. The BS can calculate the conditions for different types of cooperation off-line and then apply the corresponding cooperation according to different conditions on-line.

To evaluate the performances of the proposed scheme and analyze the questions like "who helps whom" and "how to cooperate," we set simulation parameters as follows: There were a total of $K = 2$ users in the OFDM network. A BS was located at coordinate $(0,0)$, user 1 was fixed at coordinate $(10 \text{ m},0)$, and user 2 was randomly located within the range of $[-40 \text{ m}, 40 \text{ m}]$ in both the x axis and the y axis. The propagation loss factor was set to 3. The noise level was $\sigma^2 = 0.01$ and we selected $\Gamma = 1$. An OFDM modulator for each user utilized $N = 32$ subcarriers and each subcarrier occupied a bandwidth of $W = 1$.

In Figure 12.11, we show a comparison of the overall power (in dB) of the water-filling scheme, user 1 helps user 2 (1-H-2) scheme, and user 2 helps user 1 (2-H-1) scheme. In this simulation, user 2 moved from location $(-40 \text{ m}, 0)$ to $(40 \text{ m}, 0)$. The transmission rate for each user was $T_i = 2$ nW and half of the subcarriers were used for helping others. We observe that when user 2 was located close to the BS, the 2-H-1 scheme can reduce the overall power up to 50%. The reason is that user 1 can use user 2 as a relay node to transmit user 1's information such that user 1's power can be reduced, while user 2 was so close to the BS that, even with only half of subcarriers to carry his/her own information, the power increase for user 2 is still inferior to the power reduction for user 1. On the other hand, when user 2 was located far away from the BS, even in a direction opposite to user 1 such as $(-40 \text{ m},0)$, the 1-H-2 scheme can reduce the overall transmit

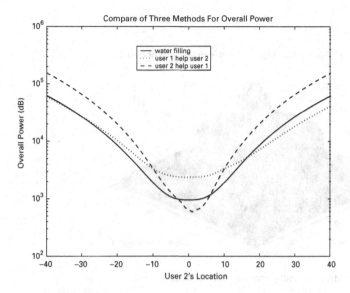

Figure 12.11 Comparison of overall power of three different methods.

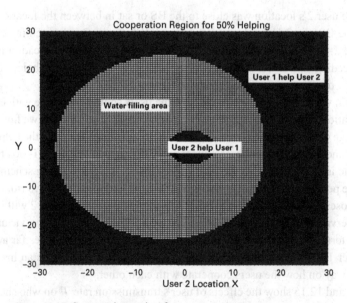

Figure 12.12 Cooperative region for two-user system.

power. This can be explained by the same reason as previously given. In the extreme case, when user 2 was located very far away from the BS compared with the distance of user 1 to the BS, both user 1 and the BS can be considered as multiple sinks for user 2's signal. The phenomenon can be viewed as the so-called "virtual multiple antenna diversity."

In Figure 12.12, we show the region where different schemes should be applied based on user 2's location. With the same simulation setups as in the previous case, we can

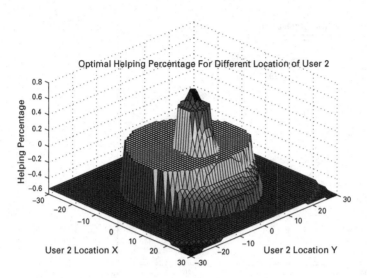

Figure 12.13 Cooperative percentage for two-user system.

see that when the user 2's location was close to the BS or sat in between the location of the user 1 and the BS, the 2-H-1 scheme is preferred. When user 2 is located far away from the BS, the 1-H-2 scheme dominates. In the case in which user 2's location is in between the preceding two cases, the water-filling scheme is the optimum choice. This figure answers the question of "who helps whom."

In Figure 12.13, we answer the question of how the users should cooperate with each other. The simulation setups are the same as in the previous case, except that we find the optimal percentage of subcarriers that were used for helping others. We show the helping percentage as a function of different user 2's locations. When the percentage is positive, the 2-H-1 scheme is activated; when the percentage is zero, the water-filling scheme is chosen; when the percentage is negative, the 1-H-2 scheme is applied. Observe from the figure that the closer user 2's location is to the BS, the larger percentage user 2 will help user 1. This observation follows from the fact that user 2's own transmission virtually costs nothing in terms of power usage. On the other hand, when user 2 moves far away from the BS, user 1 helps user 2 more and more. Therefore the figure gives an insight into the optimal way on how the users cooperate with each other.

Figures 12.14 and 12.15 show the effects of user's transmission rate T_i on who should help whom and how users should cooperate. We modified T_i from 0.5 nW to 10 nW for all users. The user 2 moved from $(-30 \text{ m}, 0)$ to $(30 \text{ m}, 0)$. As we can see in Figure 12.14, the helping regions are changed with increases of the transmission rate. The water-filling region increases to it maximum around the rate equal to 3.4 nW; then the water-filling region is reduced until all regions are occupied by 1-H-2 or 2-H-1 regions. The reason for the changes of the water-filling region is that when the power grows exponentially with an increase in the transmission rate, a user can help with a small percentage and then can reduce a lot of power in the proposed OFDM cooperative network. In Figure 12.15, however, we observe that the helping percentages are shrinking as the users' rates

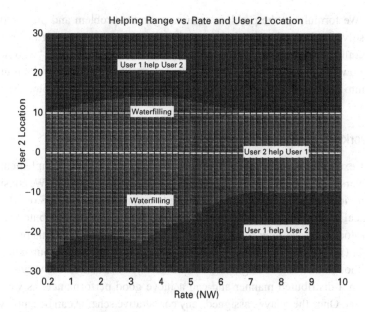

Figure 12.14 Cooperative region vs. user's rate.

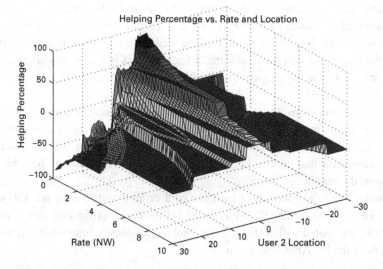

Figure 12.15 Helping percentage vs. user's rate.

keep increasing. This is because the users need more subcarriers to carry their own data, so the number of subcarriers to help others is reduced. Note also from the figure that, when the transmission rate is high, the original water-filling area becomes the 1-H-2 or 2-H-1 area. But the percentage of helping is very small.

On the whole, we constructed a power-control cooperative transmission framework over multiuser OFDM networks. We are able to improve the system performance by exploring the broadcasting nature of the wireless channels and the possibility of cooperation among users. We formulated the problem as an assignment problem and proposed an algorithm to solve it. From the simulation results, the proposed scheme can save up to 50% of overall transmit power. The major contributions are to analyze "who helps whom" and "how to cooperate" for the distributed users in the network, which gives some insight into the design of future wireless cooperative transmission protocols.

12.4.2   Cellular Networks

In the second example, we address the relay-assignment problem for implementing cooperative diversity protocols to extend coverage area in wireless networks. We consider an uplink scenario in which a set of users is trying to communicate to a BS or AP and propose practical algorithms for the relay assignment. We propose a distributed relay-assignment protocol that we refer to as the nearest-neighbor protocol. In this protocol, the helping user (relay) is chosen such as to be the nearest neighbor to the source toward the BS/AP. Although this choice might not be optimal in all scenarios, it is very simple to implement in a distributed manner and can achieve good performance, as we will demonstrate later. Once the relay is assigned, any cooperative scheme can be employed. We consider a modified version of the incremental relaying protocol proposed in [175]. In our modified scheme, if a user's packet is not captured by the BS/AP, the BS/AP is going to feed back a bit, indicating the transmission failure. In this case, if the assigned relay has received the source's packet correctly, it will forward this packet to the BS/AP. Moreover, we do not assume the storage of the analog signal of the first transmitted packet to the BS/AP. As will be demonstrated later, the loss in bandwidth efficiency is negligible in incremental relaying compared with that of direct transmission for practical ranges of the SNR [175].

Furthermore, simulations are carried out to validate the theoretical results derived for the described protocol. We consider the application of the proposed protocol in coverage-area extension in wireless networks. We consider an indoor WLAN scenario, and simulations show that up to a 350% increase in the coverage area can be achieved by the proposed protocol.

We consider a wireless network, which can be a cellular system or a WLAN, with a circular cell of radius ρ. The BS/AP is located at the center of the cell, and N users are uniformly distributed within the cell. The probability density function of the user's distance r from the BS/AP is thus given by

$$q(r) = \frac{2r}{\rho^2}, \qquad 0 \le r \le \rho, \tag{12.63}$$

and the user's angle is uniformly distributed between $[0, 2\pi)$. The wireless link between any two nodes in the network is subject to narrowband Rayleigh fading, propagation path loss, and AWGN. The channel fades for different links are assumed to be statistically mutually independent. This is a reasonable assumption as the nodes are usually well spatially separated. For medium access, the nodes are assumed to transmit over orthogonal channels, and thus no mutual interference is considered in the signal model. All nodes in the network are assumed to be equipped with single-element antennas, and transmission at all nodes is constrained to the half-duplex mode, i.e., any terminal cannot transmit and receive simultaneously [175].

In the direct transmission scheme, which is employed in current wireless networks, each user transmits his signal directly to the BS/AP. The signal received at the destination d (BS/AP) from source user s can be modeled as

$$y_{sd} = \sqrt{P_{TD}Kr_{sd}^{-\eta}}h_{sd}x + n_{sd};$$ (12.64)

where P_{TD} is the transmitted signal power in the direct transmission mode, x is the transmitted data with unit power, h_{ij} is the channel fading gain between two terminals i and j, and i, j are any two terminals in the network. The channel fade of any link is modeled as a zero-mean circularly symmetric complex Gaussian random variable with unit variance. In (12.64), K is a constant that depends on the antennas design, η is the path-loss exponent, and r_{sd} is the distance between the two terminals. K, η, and P_{TD} are assumed to be the same for all users. The term n_{sd} in (12.64) denotes additive noise. All the noise components are modeled as AWGN with variance N_0.

We characterize the system performance in terms of outage probability. Outage is defined as the event that the received SNR falls below a certain threshold γ; hence, the probability of outage P_O is defined as

$$\mathcal{P}_0 = \mathcal{P}[\text{SNR}(r) \leq \gamma].$$ (12.65)

The SNR threshold γ is determined according to the application and the transmitter/receiver structure. If the received SNR is higher than the threshold γ, the receiver is assumed to be able to decode the received message with negligible probability of error. If an outage occurs, the packet is considered lost. The main drawback of direct transmission is that the BS/AP receives only one copy of the message from the source, which makes the communication susceptible to failure because of fading. On the other hand, when cooperative diversity is employed, the BS/AP can receive more than one copy of the message. Cooperative transmission, in general, comprises two stages: In the first stage the source transmits and both the relay and the destination receive, and in the second phase the relay, if necessary, forwards to the destination.

In this example, we adopt for the cooperating protocol a modified version of the incremental relaying protocol in [175]. In this modified protocol, if a user's packet is lost, the BS/AP broadcasts NACK so that the relay assigned to this user can retransmit this packet again. The relay will transmit the packet only if it is capable of capturing the packet, i.e., if the received SNR at the relay is above the threshold. In practice, this can be implemented by utilizing a CRC code in the transmitted packet. This is the first difference between the modified and original incremental relaying protocol in [175] that employs AF at the relay. The signal received from the source to the destination d and the relay[1] in the first stage can be modeled as

$$y_{sd} = \sqrt{P_{TC}Kr_{sd}^{-\eta}}h_{sd}x + n_{sd},$$ (12.66)

$$y_{sl} = \sqrt{P_{TC}Kr_{sl}^{-\eta}}h_{sl}x + n_{sl},$$ (12.67)

[1] We denote the relay by l so as not to confuse it with r that denotes distance.

where P_{TC} is the transmission power in the cooperative mode and will be determined rigorously later to ensure the same average transmitted power in both the direct and cooperative scenarios. If the SNR of the signal received at the destination from the source falls below the threshold γ, the destination asks for a second copy from the relay. Then if the relay was able to receive the packet from the source correctly, it forwards it to the destination

$$y_{ld} = \sqrt{P_{TC} K r_{ld}^{-\eta}} h_{ld} x + n_{ld}. \tag{12.68}$$

A second difference between our modified protocol and the conventional incremental relaying in [175] that, in the case of packet failure in the first transmission from the source, the BS/AP does not store this packet to combine it later with the packet received from the relay. Storing the packet from the first transmission was assumed in most of the previous works on cooperative diversity, as it enhances the received SNR by applying a maximal ratio combiner, for example. However, a crucial implication of this assumption is that the destination has to store an analog form of the signal, which is not practical. This could be practically solved, for example, by storing a quantized version of the signal, and the quantization noise should then be taken into account in the analysis. Note that the relay-assignment algorithm can be applied with *any* cooperative scheme, not only the modified incremental relaying that we already discussed.

To construct a distributed relay-assignment protocol, we assume that each user can know his distance to the BS/AP through, for example, calculating the average received power. According to this protocol, the assigned relay is chosen to be the nearest neighbor to the user toward the BS/AP. This can be done by a simple distributed protocol in which each relay sends out a "Hello" message searching for his nearest neighbors. This can be done using a time-of-arrival (TOA) estimation; for example, see [269] and [300]. The user selects the nearest-neighbor node with a distance closer to the BS/AP than the user himself. For example, as illustrated in Figure 12.16, a user at a distance r_{sd} from the BS/AP will choose his nearest neighbor inside the circle of radius r_{sd} with the BS/AP as its center. In this figure, the nearest neighbor is at a distance r_{sl} from the source. Each user will be assigned a relay to help him, and this nearest-neighbor discovery algorithm can be run periodically according to the mobility of the users and how often they change their locations.

First, we derive an outage probability expression for the direct transmission scheme. As discussed before, the outage is defined as the event that the received SNR is lower than a predefined threshold which we denote by γ. From the received signal model in (12.64), the received SNR from a user at a distance r_{sd} from the BS/AP is given by

$$\text{SNR}(r_{sd}) = \frac{|h_{sd}|^2 K r_{sd}^{-\eta} P_{TD}}{N_0}, \tag{12.69}$$

where $|h_{sd}|^2$ is the magnitude square of the channel fade and follows an exponential distribution with unit mean. Hence the outage probability for the direct transmission mode P_{OD} conditioned on the user's distance can be calculated as

$$\mathcal{P}_{OD}(r_{sd}) = \mathcal{P}[\text{SNR}(r_{sd}) \leq \gamma] = 1 - \exp\left(-\frac{N_0 \gamma r_{sd}^{\eta}}{K P_{TD}}\right). \tag{12.70}$$

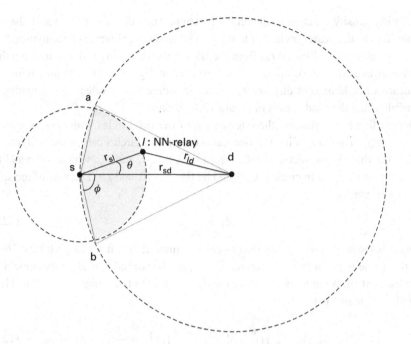

Figure 12.16 Illustrating cooperation under nearest-neighbor protocol: The nearest neighbor is located inside the circle of radius r_{sd} at a distance r_{sl} from the source. Therefore the shaded area should be empty of any users.

To find the average outage probability over the cell, we need to average over the user distribution in (12.63). The average outage probability is thus given by

$$\mathcal{P}_{OD} = \int_0^\rho \mathcal{P}_{OD}(r_{sd})q(r_{sd})dr_{sd} = 1 - \frac{2}{\eta\rho^2}\left(\frac{KP_{TD}}{N_0\gamma}\right)^{\frac{2}{\eta}}\Gamma\left(\frac{2}{\eta}, \frac{N_0\gamma\rho^\eta}{KP_{TD}}\right), \quad (12.71)$$

where $\Gamma(\cdot, \cdot)$ is the incomplete gamma function, and it is defined as [90]

$$\Gamma(a, x) = \int_0^x \exp^{-t} t^{a-1} dt. \quad (12.72)$$

Next, we analyze the outage probability for the nearest-neighbor protocol. We can show that the outage probability expression, which we refer to as $\mathcal{P}_{ONN}$, for given source–relay–destination locations is given by

$$\mathcal{P}_{ONN}(r_{sd}, r_{sl}, r_{ld}) = \left[1 - \exp\left(-\frac{N_0\gamma r_{sd}^\eta}{KP_{TC}}\right)\right]$$
$$\times \left\{1 - \exp\left[-\frac{N_0\gamma\left(r_{sl}^\eta + r_{ld}^\eta\right)}{KP_{TC}}\right]\right\}. \quad (12.73)$$

We omit the derivation because of space limitations. To find the total probability, we need to average over all possible locations of the user and the relay. The user's location distribution with respect to the BS/AP is still given as in the direct transmission case (12.63). The relay's location distribution, however, is not uniform. Next, we calculate the probability density function of the relay's location. According to our protocol, the relay is chosen to be the nearest neighbor to the user which is at a closer distance to the BS/AP, i.e., if the user is at a distance r_{sd} from the BS/AP then the relay is the nearest neighbor to the user in a circle of distance r_{sd}, as illustrated in Figure 12.16. The probability that the nearest neighbor is at distance r_{sl} from the source is equivalent to calculating the probability that the shaded area in Figure 12.16 is empty.

Denote this area, which is the intersection of the two circles with centers s and d, by $A(r_{sd}, r_{sl})$. The area of intersection between the two circles can be divided into two parts: A_1, which is the sector $<asb>$ from the circle s, and A_2, which is the addition of the two small areas in circle d enclosed by the arcs $\widehat{as}$ and $\widehat{sb}$. The area of the sector $<asb>$ is given by

$$A_1 = \phi r_{sl}^2, \tag{12.74}$$

where ϕ is the angle $\angle dsb$. From the isosceles triangle $\triangle dsb$, it is straightforward to see that this angle is given by $\phi = \arccos[r_{sl}/(2r_{sd})]$. The second part A_2 from circle d can be calculated from the total sector area $<dsa>$ less the triangular area $\triangle dsa$. Hence area A_2 can be given as

$$A_2 = 2\left[(\frac{\pi}{2} - \phi)r_{sd}^2 - \frac{r_{sl}}{2}\sqrt{r_{sd}^2 - \frac{r_{sl}^2}{4}}\right]. \tag{12.75}$$

Adding the two areas together, we get the total area expression as follows:

$$A(r_{sd}, r_{sl}) = r_{sl}^2 \arccos\left(\frac{r_{sl}}{2r_{sd}}\right) + \pi r_{sd}^2$$

$$- 2r_{sd}^2 \arccos\left(\frac{r_{sl}}{2r_{sd}}\right) - r_{sl}\sqrt{r_{sd}^2 + \frac{r_{sl}^2}{4}}. \tag{12.76}$$

Let the probability density function (PDF) of the nearest neighbor be denoted by the function $\mathcal{P}_{r_n}$, where r_n is a random variable denoting the nearest-neighbor distance. Because we have N users uniformly distributed in a circular area of radius ρ, the probability of finding no users in an area $A(r_{sd}, r_{sl})$ is given by

$$\mathcal{P}_{r_n}(r_n \geq r_{sl}) = \left[1 - \frac{A(r_{sd}, r_{sl})}{\pi\rho^2}\right]^N. \tag{12.77}$$

Hence, the PDF of r_n can be calculated as

$$\mathcal{P}_{r_n}(r_{sl}) = \frac{\partial \mathcal{P}_{r_n}(r_n \leq r_{sl})}{\partial r_{sl}}$$

$$= \frac{N}{\pi\rho^2}\left(1 - \frac{A(r_{sd}, r_{sl})}{\pi\rho^2}\right)^{N-1} \frac{\partial A(r_{sd}, r_{sl})}{\partial r_{sl}}. \tag{12.78}$$

To find the outage probability as a function of the source distance r_{sd}, we need to average over all possible relay locations that is specified by the pair of distances (r_{sl}, r_{ld}). Because a relay at a distance r_{sl} from the source is uniformly distributed over the angle $\theta = \angle dsl$ as in Figure 12.16, it is much easier to determine the location of the relay in polar form with the source s being the origin. In this case the angle θ takes values between $-\arccos(\frac{r_{sl}}{2r_{sd}}) \leq \theta \leq \arccos(\frac{r_{sl}}{2r_{sd}})$. Hence we can write the conditional outage probability in (12.73) in terms of θ instead of r_{ld} by substituting

$$r_{ld} = \sqrt{r_{sd}^2 + r_{sl}^2 - 2r_{sd}r_{sl}\cos(\theta)}. \tag{12.79}$$

The value of r_{sl} can take values between 0 and $2r_{sd}$. Now, we average the outage probability expression in (12.73) over all possible relay locations

$$\mathcal{P}_{ONN}(r_{sd}) = \int_0^{2r_{sd}} \frac{1}{2\arccos\left(\frac{r_{sl}}{2r_{sd}}\right)} \int_{-\arccos(\frac{r_{sl}}{2r_{sd}})}^{\arccos(\frac{r_{sl}}{2r_{sd}})}$$
$$\mathcal{P}_{ONN}(r_{sd}, r_{sl}, \theta)\mathcal{P}_{rn}(r_{sl})d\theta dr_{sl}^2, \tag{12.80}$$

where $\mathcal{P}_{ONN}(r_{sd}, r_{sl}, \theta)$ is defined as the conditional outage probability in (12.73) after substituting for r_{ld} as a function of θ as in (12.79). The term $\mathcal{P}_{rn}(r_{sl})$ in (12.80) is defined in (12.78). To find the unconditional outage probability of the cell we average over all possible source locations:

$$\mathcal{P}_{ONN} = \int_0^\rho \frac{2r_{sd}}{\rho^2} \mathcal{P}_{ONN}(r_{sd})dr_{sd}. \tag{12.81}$$

The outage probability expression in (12.81) can be calculated only numerically. In the following discussion, we derive an approximate expression for the outage probability under the following two assumptions. Because the relay is chosen to be the nearest neighbor to the source, the SNR received at the relay from the source is rarely below the threshold γ; hence, we assume that the event of the relay being in outage is negligible. The second assumption is that the nearest neighbor always lies on the intersection of the two circles, as points a or b in Figure 12.16. This second assumption is a kind of worst-case scenario, because a relay at distance r_{sl} from the source can be anywhere on the arc $\overparen{bla}$, and a worst-case scenario is to be at points a or b because these are the furthest points from the BS/AP on the arc $\overparen{bla}$. This simplifies the outage calculation as the conditional outage probability (12.73) is now a function of only the source distance r_{sd} as follows:

$$\mathcal{P}_{ONN}(r_{sd}) \simeq \left[1 - \exp(-\frac{N_0\gamma r_{sd}^\eta}{KP_{TC}})\right]^2 \tag{12.82}$$

Substituting (12.82) into (12.81), and using the definition of the incomplete gamma function in (12.72), we get

$$\mathcal{P}_{ONN} \simeq 1 - \frac{4}{\eta\rho^2}\left(\frac{KP_{TC}}{N_0\gamma}\right)^{\frac{2}{\eta}}\Gamma\left(\frac{2}{\eta}, \frac{N_0\gamma\rho^\eta}{KP_{TC}}\right)$$
$$+ \frac{2}{\eta\rho^2}\left(\frac{KP_{TD}}{2N_0\gamma}\right)^{\frac{2}{\eta}}\Gamma\left(\frac{2}{\eta}, \frac{2N_0\gamma\rho^\eta}{KP_{TD}}\right), \tag{12.83}$$

This approximation is tight, as will be shown by computer simulations later.

We performed some computer simulations to compare the performance of the proposed relay-assignment protocol and direct transmission and to validate the theoretical results we derived. In all of our simulations, we compared the outage performance of two different transmission schemes: direct transmission and the nearest-neighbor protocol. Along with the simulation curves, we also plotted the theoretical outage performance that we derived for the two schemes. In all of the simulations, the channel between any two nodes (either a user and the BS/AP or two users) is modeled as a random Rayleigh fading channel with unit variance.

For fairness in comparison between the proposed cooperative schemes and the direct transmission scheme, the average transmitter power is kept fixed in both cases, and this is done as follows. Because a packet is either transmitted once or twice in the cooperative protocol, the average transmitted power in the cooperative case can be calculated as

$$E(\text{TX power}) = P_{TC}\mathcal{P} \text{ (source only transmits)}$$
$$+2P_{TC}\mathcal{P} \text{ (source and relay transmit)}. \tag{12.84}$$

The event that only the source transmits is the union of the events that the $s - d$ link is not in outage, or both the $s - d$ and $s - l$ links are in outage. Hence, the probability of this event can be given by

$$\mathcal{P}(\text{Source only transmits}) = 1 - \mathcal{P}_{OD}^{sd}(P_{TC}) + \mathcal{P}_{OD}^{sd}(P_{TC})\mathcal{P}_{OD}^{sl}(P_{TC}), \tag{12.85}$$

where $\mathcal{P}_{OD}^{sd}(P_{TC})$ denotes the outage probability of the direct transmission between the source and the destination when the source is using transmitting power P_{TC} in the cooperative mode, and $\mathcal{P}_{OD}^{sl}(P_{TC})$ denotes the corresponding probability for the $s - l$ link. The events in which both the source transmits and the relay transmits are just the complements of the previous event is the event in which the $s - d$ link is in outage and the $s - l$ link is not. It is thus given by

$$\mathcal{P}(\text{source and relay transmit}) = \mathcal{P}_{OD}^{sd}(P_{TC})[1 - \mathcal{P}_{OD}^{sl}(P_{TC})]. \tag{12.86}$$

Substituting (12.85) and (12.86) into (12.84), the average transmitted power in the cooperative mode can be given by

$$E(\text{TX power}) = P_{TC}[1 + \mathcal{P}_{OD}^{sd}(P_{TC}) - \mathcal{P}_{OD}^{sd}(P_{TC})\mathcal{P}_{OD}^{sl}(P_{TC})]. \tag{12.87}$$

The power used in transmitting in the direct scheme P_{TD} should be set equal to the quantity in (12.87) in order to have the same average transmitted power. One can see that $P_{TD} \geq P_{TC}$ as expected. In our simulations, we set $P_{TD} = P_{TC}\left[1 + \mathcal{P}_{OD}^{sd}(P_{TC})\right]$ which is in favor of the direct transmission.

A remark on the bandwidth efficiency is now in order. Note that for the cooperative scheme we either utilize the same resources as the direct transmission mode or twice these resources with the same probabilities defined in (12.85) and (12.86). This means that the relation between the bandwidth efficiency of the direct and cooperative transmissions is also governed in the same manner as for the power in (12.87). For practical outage performance in the range of 0.01, the loss in the bandwidth efficiency is thus negligible and we do not take it into account in our simulations.

We consider a WLAN scenario for our simulations. The cell radius is taken to be between 10 and 100 m. The AWGN has variance $N_0 = -70$ dBm, and the path-loss exponent is set to $\eta = 2.6$. The number of users in the cell attached to the AP is taken to

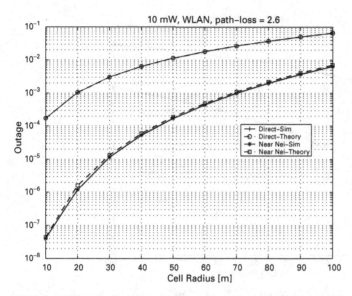

Figure 12.17 Average outage probability versus the cell radius in m for direct and cooperative transmissions. Simulation curves are drawn as solid and theoretical curves as dotted. The average transmitter power is fixed at 10 mW, and the path loss is set to 2.6.

be $N = 10$. The SNR threshold γ is taken to be 20 dB, which is higher than that for the cellular system, because the information transmitted over a WLAN is usually data, which needs a higher quality than the voice signals usually transmitted over cellular systems.

Next, we discuss the simulation results for the WLAN that are demonstrated in Figures 12.17–12.20. Figures 12.17 and 12.18 depict the results for fixed average transmitted power of 10 and 30 mW, respectively. For the 10-mW case, if we require the outage performance to be around 0.001, then the maximum cell size achieved by the direct transmission case is about 20 m. The nearest-neighbor protocol can achieve about 70 m. Hence the cooperative scheme can increase the cell size by about 350% in this case. The theoretical curves still match our expectations, which validates our analysis. Figure 12.18 depicts the results for the 30-mW case. At 0.001 outage performance, the cell radius can increase to 30 m by direct transmission, whereas the nearest-neighbor protocol can extend the cell radius to more that 100 m.

Furthermore, we study the gains that can be achieved from the proposed protocols in terms of energy efficiency. Figures 12.19 and 12.20 depict the results for cell radii of 50 and 100 m, respectively. The average transmitted power is changed from 5 mW (7 dBm) to 30 mW (14.7 dBm). For both the 50- and 100-m cases, it can be seen from the figures that the direct transmission scheme can no longer achieve the 0.001 outage performance for the whole simulated power range. For the 50-m case in Figure 12.19, the nearest neighbor can achieve this performance at less than 5 mW (7 dBm). For the 100-m case in Figure 12.20, the nearest neighbor requires around 14 dBm.

We propose a distributed nearest-neighbor protocol for relay assignments in cooperative communications. In the proposed protocol the relay is selected to be the nearest neighbor to the user toward the BS/AP. Outage performance analysis is provided for the

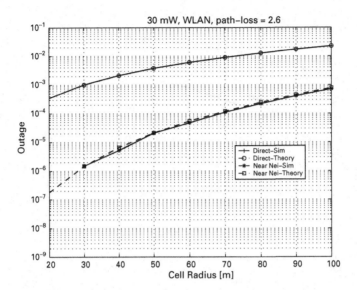

Figure 12.18 Average outage probability versus the cell radius in m for direct and cooperative transmissions. Simulation curves are drawn as solid and theoretical curves as dotted. The average transmitter power is fixed at 30 mW, and the path loss is set to 2.6.

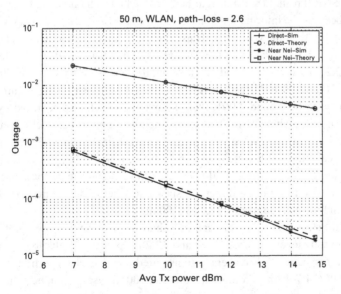

Figure 12.19 Average outage probability versus the transmitted power in dBm for direct and cooperative transmissions. Simulation curves are drawn as solid and theoretical curves as dotted. The cell radius is fixed at 50 m, and the path loss is set to 2.6.

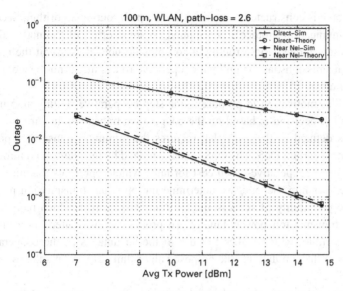

Figure 12.20 Average outage probability versus the transmitted power in dBm for direct and cooperative transmissions. Simulation curves are drawn as solid and theoretical curves as dotted. The cell radius is fixed at 100 m, and path loss is set to 2.6.

proposed protocol. We show that the bandwidth efficiency loss that is due to cooperation is negligible for practical operation conditions: For an outage probability of 0.01 the loss in the bandwidth efficiency is approximately 0.01. Moreover, simulation results are carried for a WLAN scenario. Under the same average transmitted power, simulation results reveal an increase in the coverage area up to 350%. Our theoretical calculations match the simulation results.

12.5   Network Layer Issue: Cooperative Routing

In this section, we study the impact of a network layer by cooperative communication. In a network layer, routing algorithms select the multihop links between the source and the destination with a minimal cost such as the overall power. In MANETs, nodes spend most of their power in communication, either sending their own data or relaying other nodes' data [70]. Therefore designing power-efficient routing algorithms is one of the major concerns in wireless ad hoc networks. Furthermore, communication power can be reduced by jointly considering other layers' protocols, which make use of the special features of the wireless medium. Because ad hoc networks are considered, the algorithms need to be implemented in a distributed way.

The cooperative routing problem has been recently considered in the literature [162, 180, 195, 278, 346]. Most of the current cooperative-based routing algorithms, such as the cooperation along the minimum-energy noncooperative (CAN) path [162], progressive cooperative (PC) [162], and cooperative routing along truncated nonco-operative route (CTNCR) [346], are implemented in two consecutive steps. First, a noncooperative route is constructed using any shortest-path routing algorithm. Second, cooperative communication protocols are applied on some or all of the nodes along the established route. Indeed, these routing algorithms do not fully exploit the merits of cooperative transmissions, as the optimal cooperative routes might not be along the noncooperative routes.

This fact provides motivation to propose a one-step cooperative routing algorithm, in which the routing decision is based on the cooperative transmissions directly. The optimum route is defined as the route that requires the minimum transmitted power while guaranteeing a certain QoS. The QoS is characterized by the end-to-end throughput. First, we derive a cooperative-based cost formula, which represents the minimum transmitted power per hop. Second, we determine the optimum transmission parameters, such as the probability of success and the transmission rate, which require the least amount of transmission power. Finally, we propose a distributed cooperative-based routing algorithm, which makes full use of the advantages of the cooperative transmissions. From the simulations, the proposed algorithm outperforms the two-step algorithms by 9.3%.

Next, we describe the network model and formulate the minimum-power routing problem. We consider a graph $G(N, E)$ with N nodes, E edges, and weight $d_{i,j}^{\alpha}$ for each link $(i, j) \in E$. Given any source–destination pair $(S, D) \in \{1, \ldots, N\}$, the goal is to find the $S - D$ route that minimizes the total transmitted power along this route subject to having a specific throughput. For a given source–destination pair, denote Ω as the set of all possible routes, where any route is defined by its hops. For a route $\omega \in \Omega$, denote ω_i as the ith hop of this route. Thus the problem can be formulated as

$$\min_{\omega \in \Omega} \sum_{\omega_i \in \omega} P_{\omega_i} \quad \text{s.t.} \quad \eta_\omega \geq \eta_0 , \quad (12.88)$$

where P_{ω_i} denotes the transmitted power over the ith hop, η_ω is the end-to-end throughput, and η_0 represents the minimum desired value of the end-to-end throughput. Let η_{ω_i} denote the throughput of the ith hop, which is defined as the number of successfully transmitted bits per second per hertz (b/s/Hz) of a given hop. Furthermore, the end-to-end throughput of a certain route ω is defined as the minimum of the throughput values of the hops constituting this route, i.e.,

$$\eta_\omega = \min_{\omega_i \in \omega} \eta_{\omega_i} . \quad (12.89)$$

It has been proven in [180] that the minimum-energy cooperative path (MECP) routing problem, i.e., to find the minimum-energy route using cooperative radio transmission, is *NP-complete*. This is due to the fact that the optimal path could be a combination of cooperative transmissions and broadcast transmissions.

To propose a routing algorithm with polynomial complexity, we utilize two types of building blocks, as shown in Figure 12.21. The first type is the traditional direct transmission (DT) building block and the second type is the cooperative transmission (CT) block. In Figure 12.21, The DT block is represented by the link (i, j), where node i is the sender and node j is the receiver over this link. The CT building block is represented by the links (x, y), (x, z), and (y, z), where node x is the sender, node y is a

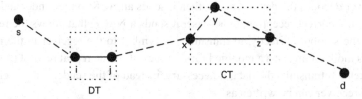

Figure 12.21 Cooperative transmission (CT) and direct transmission (DT) modes as building blocks for any route.

relay, and node z is the receiver. Because of the spatial diversity, the CT block requires less transmitted power than the DT block to achieve the same QoS. The route can be considered as the cascade of any number of these two types of blocks, and the route cost is the sum of the transmitted power along the route. Therefore the minimization problem in (12.88) can be solved by applying any distributed shortest path routing algorithm, such as the Bellman–Ford algorithm [28].

Next, we describe the system model of the DT and CT blocks. Let $h_{u,v}$, $d_{u,v}$, and $n_{u,v}$ represent the channel coefficient, length, and additive noise of the hop (u, v), respectively. We assume $h_{u,v}$ is a complex Gaussian random variable with zero mean and unit variance. For the DT between node i and node j, the received symbol can be modeled as

$$r_{i,j}^{D} = \sqrt{P^D \, d_{i,j}^{-\alpha}} \; h_{i,j} \; s + n_{i,j}, \tag{12.90}$$

where P^D is the transmitted power in the DT mode, α is the path-loss exponent, and s is the transmitted symbol.

For the CT, we consider a modified version of the DF incremental relaying cooperative scheme, proposed in [175]. The transmission scheme for sender x, relay y, and receiver z, can be described as follows. The sender sends its data in the current time slot. Because of the broadcast nature of the wireless medium, both the receiver and the relay receive noisy versions of the transmitted data. The received symbols at the receiver and the relay can be modeled as

$$r_{x,z}^{C} = \sqrt{P^C \, d_{x,z}^{-\alpha}} \; h_{x,z} \; s + n_{x,z}, \tag{12.91}$$

$$r_{x,y}^{C} = \sqrt{P^C \, d_{x,y}^{-\alpha}} \; h_{x,y} \; s + n_{x,y}, \tag{12.92}$$

where P^C is the transmitted power in the CT mode.

Once the data are received, the receiver and the relay decode them. We assume that the relay or the receiver decides that the received symbol is correctly received if the received SNR is greater than a certain threshold, which depends on the transmitter and the receiver structures. Such a system suffers from error propagation, but its effect can be negligible. The rationale behind this is that, when the relays operate in a high SNR regime, the dominant source of error is the channel being in outage, i.e., deep fade, which corresponds to the SNR falling below some threshold. This result has been proven in [367].

If the receiver decodes the data correctly, then it sends an ACK to the sender and the relay to confirm a correct reception. Otherwise, it sends a NACK that allows the relay, if it received the symbol correctly, to transmit this symbol to the receiver in the next time slot. This model represents a modified form of the automatic repeat request (ARQ) in which the relay retransmits the data, if necessary, instead of the sender. The received symbol at the receiver can be written as

$$r_{y,z}^C = \sqrt{P^C \, d_{y,z}^{-\alpha}} \, h_{y,z} \, s + n_{y,z}. \tag{12.93}$$

In general, the relay can transmit with a power that is different from the sender power P^C. However, this complicates the problem of finding the minimum-power formula. For simplicity, we consider that both the sender and the relay send their data employing the same power P^C. Flat quasi-static fading channels are considered; hence, the channel coefficients are assumed to be constant during a complete frame and may vary from one frame to another. Finally, the noise terms are modeled as zero-mean, complex Gaussian random variables with equal variance N_0.

Next, we derive the required power for the two transmission modes in order to achieve certain throughput. Because the throughput is a continuous monotonously increasing function of the transmission power, the optimization problem in (12.88) has the minimum when $\eta_\omega = \eta_0, \forall \omega \in \Omega$. Because the end-to-end throughput $\eta_\omega = \min_{\omega_i \in \omega} \eta_{\omega_i}$, then the optimum power allocation, which achieves a desired throughput η_0 along the route ω, forces the throughput at all the hops η_{ω_i} to be equal to the desired one, i.e.,

$$\eta_{\omega_i} = \eta_0 , \quad \forall \, \omega_i \in \omega. \tag{12.94}$$

The throughput of link ω_i can be calculated as

$$\eta_{\omega_i} = p_{\omega_i}^S \cdot R_{\omega_i}, \tag{12.95}$$

where $p_{\omega_i}^S$ and R_{ω_i} denote the per-link probability of success and transmission rate, respectively. We consider that the desired throughput can be factorized as $\eta_0 = p^S R$. In the following, we calculate the transmitted power in order to achieve the p^S and R for both the DT and the CT modes.

For the DT in (12.90), mutual information between sender i and receiver j can be given by

$$I_{i,j} = \log \left(1 + \frac{P^D \, d_{i,j}^{-\alpha} \, |h_{i,j}|^2}{N_0} \right). \tag{12.96}$$

Without loss of generality, we have assumed unit bandwidth in (12.96). The outage probability is defined as the probability that the mutual information is less than the transmission rate R_D. The channel gain $|h_{u,v}|^2$ between any two nodes u and v is Rayleigh distributed with parameter one [239]. Therefore the outage probability of the link (i, j) can be calculated as

$$p_{i,j}^O = \Pr(I_{i,j} \leq R_D) = 1 - \exp \left[-\frac{(2^{R_D} - 1) \, N_0 \, d_{i,j}^\alpha}{P^D} \right]. \tag{12.97}$$

If an outage occurs, the data are considered lost. The probability of success is calculated as $p_{ij}^S = 1 - p_{ij}^O$. Thus, to achieve the desired p_D^S and R_D for the DT mode, the required transmitted power by the sender is

$$P^D = \frac{(2^{R_D} - 1)\, N_0\, d_{i,j}^\alpha}{-\log(p_D^S)}. \tag{12.98}$$

For the CT mode, the total outage probability is given by

$$p_{x,y,z}^O = \Pr(I_{x,z} \leq R^C) \cdot \Pr(I_{x,y} \leq R^C) \tag{12.99}$$
$$+ \Pr(I_{x,z} \leq R^C)\big[1 - \Pr(I_{x,y} \leq R^C)\big]\Pr(I_{y,z} \leq R^C),$$

where R^C denotes the transmission rate for each time slot. In (12.99), the first term corresponds to the event when both the sender–receiver and the sender–relay channels are in outage, and the second term corresponds to the event when both the sender–receiver and relay–receiver channels are in outage but the sender–relay is not. Consequently, the probability of success of the CT mode can be calculated as

$$p_C^S = \exp\left(-g\, d_{x,z}^\alpha\right) + \exp\left[-g\,(d_{x,y}^\alpha + d_{y,z}^\alpha)\right]$$
$$- \exp\left[-g\,(d_{x,y}^\alpha + d_{y,z}^\alpha + d_{x,z}^\alpha)\right], \tag{12.100}$$

where

$$g = \frac{(2^{R^C} - 1)\, N_0}{P^C}. \tag{12.101}$$

In (12.99) and (12.100), we assume that the receiver decodes the signals received from the relay either at the first time slot or at the second time slot, instead of combining the received signals together. In general, MRC [36] at the receiver will give a better result. However, it requires the receiver to store an analog version of the received data from the sender, which is not practical. The probability that the source transmits only, denoted by $\Pr(\phi)$, is calculated as

$$\Pr(\phi) = 1 - \Pr(I_{x,z} \leq R^C) + \Pr(I_{x,z} \leq R^C)\Pr(I_{x,y} \leq R^C)$$
$$= 1 - \exp\left(-g\, d_{x,y}^\alpha\right) + \exp\left[-g\,(d_{x,y}^\alpha + d_{x,z}^\alpha)\right], \tag{12.102}$$

where the term $[1 - \Pr(I_{x,z} \leq R^C)]$ corresponds to the event when the sender–receiver channel is not in outage, whereas the other term corresponds to the event when both the sender–receiver and the sender–relay channels are in outage. Therefore the probability that the relay cooperates with the source is calculated as $\overline{\Pr(\phi)} = 1 - \Pr(\phi)$. The average transmission rate of the cooperative transmission mode can be calculated as

$$R_{\text{avg}}^C = R^C \Pr(\phi) + \frac{R^C}{2}\overline{\Pr(\phi)} = \frac{R^C}{2}\left[1 + \Pr(\phi)\right], \tag{12.103}$$

where R^C corresponds to the transmission rate if the sender is sending alone in one time slot and $R^C/2$ corresponds to the transmission rate if the relay cooperates with the sender in the consecutive time slot.

By approximating the exponential functions in (12.100) as $\exp(-x) \approx 1 - x + x^2/2$, we obtain

$$g \approx \sqrt{\frac{1 - p_C^S}{d_{eq}}} \,, \tag{12.104}$$

where $d_{eq} = d_{x,z}^\alpha (d_{x,y}^\alpha + d_{y,z}^\alpha)$. Thus R^C can be obtained using (12.103) as

$$R^C = \frac{2\, R_{avg}^C}{1 + \Pr(\phi)}$$

$$\approx \frac{2\, R_{avg}^C}{2 - \exp\left(-\sqrt{\frac{1-p_C^S}{d_{eq}}}\, d_{x,y}^\alpha\right) + \exp\left[-\sqrt{\frac{1-p_C^S}{d_{eq}}}\,(d_{x,y}^\alpha + d_{x,z}^\alpha)\right]} \,, \tag{12.105}$$

where we substituted (12.104) in (12.102). In addition, the required power per link can be calculated using (12.101) and (12.104) as

$$P^C \approx (2^{R^C} - 1)\, N_0\, \sqrt{\frac{d_{eq}}{1 - p_C^S}} \,. \tag{12.106}$$

Finally, the average transmitted power of the cooperative transmission can be calculated as

$$P_{avg}^C = P^C \Pr(\phi) + 2\, P^C \overline{\Pr(\phi)} = P^C\left[2 - \Pr(\phi)\right] \,, \tag{12.107}$$

where $\Pr(\phi)$ and P^C are given in (12.102) and (12.106), respectively.

Futhermore, we derive the closed-form expressions to approximate the optimum probability of success and transmission rate that require the least transmitted power per link, given that their product is equal to the desired throughput.

First, we study the case of the DT as follows. By substituting $R_D = \eta_0/p_D^S$ in (12.98), the power of the DT mode can be written as

$$P^D = \frac{(2^{\eta_0/p_D^S} - 1)\, N_0\, d_{i,j}^\alpha}{-\log(p_D^S)} \,. \tag{12.108}$$

We use the approximation $\log(p_D^S) \approx \frac{p_D^S - 1}{p_D^S}$, which is the first term in the logarithm series representation in [90], 1.512. Therefore, the optimization problem can be formulated as

$$\min_{p_D^S} P^D \approx \frac{(2^{\eta_0/p_D^S} - 1)\, p_D^S\, N_0\, d_{i,j}^\alpha}{1 - p_D^S} \,. \tag{12.109}$$

It can be shown in (12.109), through the second derivative, that the objective function P^D is convex in p_D^S, where $0 \leq p_D^S \leq 1$. Therefore the optimal solution p_D^{S*} can be obtained by solving the equation $\nabla P^D(p_D^{S*}) = 0$ and it is given by

$$p_D^{S*} = \frac{\eta_0\, \log(2)}{1 + \eta_0 \log(2) + \mathbf{L}\left\{-\exp\left[-1 - \eta_0 \log(2)\right]\right\}} \,, \tag{12.110}$$

where $\mathbf{L}(.)$ is the Lambert W function, and it is the inverse function of $I(W) = W\exp(W)$.

Second, we consider the CT. It is difficult to get a closed-form expression for the optimum probability of success, which minimizes the required transmitted power (12.107). Hence we propose a heuristic algorithm, which considers that the probability of success required on each link of the CT block is equal to the obtained DT probability of success (12.110). Thus CT probability of success is calculated using (12.99) as

$$p_C^{S*} = 1 - \left(1 - p_D^{S*}\right)^2 \left(1 + p_D^{S*}\right). \tag{12.111}$$

We will show later in the simulations that the analytical results give very good approximations to the optimum ones, which are obtained through an exhaustive search along the probability of success range.

We describe two cooperative-based routing algorithms, which require polynomial complexity to find the minimum-power route. We assume that each node broadcasts periodically HELLO packet to its neighbors to update the topology information. In addition, we consider a simple MAC protocol, which is the conventional TDMA scheme with equal time slots.

First, we describe the proposed MPCR (minimum power cooperative routing) algorithm for a wireless network of N nodes. The MPCR algorithm can be distributively implemented by the Bellman–Ford shortest-path algorithm [28]. The derived optimal power for DT and CT is utilized to construct the minimum-power route.

In the Bellman–Ford shortest-path algorithm, each node $i \in \{1, \dots, N\}$ executes the iteration $D_i = \min_{j \in N(i)} (d_{i,j}^\alpha + D_j)$, where $N(i)$ denotes the set of neighboring nodes of node i and D_j represents the latest estimate of the shortest path from node j to the destination [28], which is included in the HELLO packet. Therefore the MPCR algorithm is implemented by letting each node calculate the costs of its outgoing links then apply the Bellman–Ford algorithm. Next, we explain how each node calculates the cost of its outgoing links in detail.

Consider node x, which may be the source of any intermediate node, as the sender. It calculates the cost of the (x, z)th hop, where $z \in N(x)$ is the receiver, as follows. First, for every other node $y \in N(x)$, $y \neq z$, the sender x calculates the cost of the CT in (12.107) employing node y as a relay. Given the desired throughput, the optimum probability of success is obtained by (12.111). Second, the cost of the (x, z)th hop is the minimum cost among all the costs obtained in the first step. If the minimum cost corresponds to a certain relay $y*$, node x employs this relay to help the transmission. Otherwise, it uses the DT. The computational complexity of calculating the costs at each node is $O(N^2)$ as it requires two nested loops, and each has the maximum length of N to calculate all the possible cooperative transmission blocks.

Second, we present the cooperation along the shortest noncooperative path (CASNCP) algorithm to be considered for comparison. It is similar to the heuristic algorithms proposed in [162] and [346]. However, it is implemented in a different way using the proposed cooperative-based link-cost formula.

First, the shortest noncooperative path (SNCP) algorithm is implemented using the distributed Bellman–Ford algorithm, to choose the conventional shortest-path route, denoted by ω_S. Second, for each three consecutive nodes on ω_S, the first, second, and third nodes behave as the sender, relay, and receiver, respectively. The sender calculates

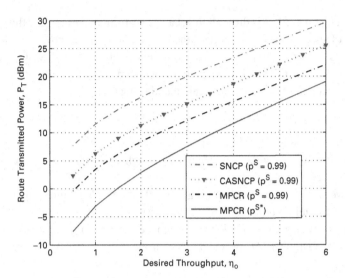

Figure 12.22 Required transmitted power per route versus the desired throughput for $N = 100$ nodes and $\alpha = 4$ in a 200 m $\times$ 200 m grid.

the link cost of this cooperative transmission block using (12.107), and chooses the transmission mode, which requires less power. The maximum number of nodes in the shortest-path route ω_S is N, because the Bellman–Ford algorithm has a computational complexity of $O(N^3)$ [28]. Thus the total computational complexity of implementing the MPCR algorithm or CASNCP algorithm with the Bellman–Ford algorithm is $O(N^3)$.

Finally, we present some computer simulations to illustrate the power savings of our proposed MPCR algorithm. We consider a 200 m $\times$ 200 m grid, where N nodes are uniformly distributed. The AWGN has variance $N_0 = -70$ dBm, and the path loss is $\alpha = 4$. Given a certain network topology, we randomly choose a source–destination pair and apply the various routing algorithms to choose the corresponding route. For each algorithm, we calculate the total transmitted power per route. Finally, these quantities are averaged over 1000 different network topologies.

Figure 12.22 depicts the transmitted power per route for a network of $N = 100$ nodes. It shows the required transmitted power of the MPCR algorithm versus the desired throughout, using the optimum probabilities of success. In addition, it shows the required transmitted power of the different routing algorithms calculated at fixed probability of success $p^S = p_D^S = p_C^S = 0.99$. At $\eta_0 = 3$ b/s/Hz the required transmitted power of the MPCR algorithm, calculated at the optimum probability of success, is less than the one calculated at $p^S = 0.99$ by 4.6 dBm. This clarifies the importance of using our proposed optimum transmission parameters rather than caring only about transmitting with high probability of success.

For the $p^S = 0.99$ curves, it is shown that the SNCP algorithm, which applies the Bellman–Ford shortest-path algorithm, requires the highest transmitted power per route. Applying the CT mode on each three consecutive nodes in the SNCP route results in a

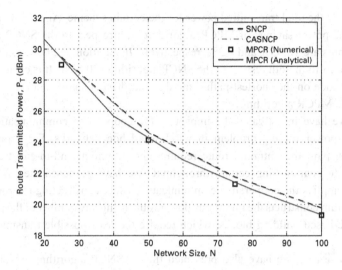

Figure 12.23 Required transmitted power per route versus the number of nodes for $\eta_0 = 6$ b/s/Hz and $\alpha = 4$ in a 200 m $\times$ 200 m grid.

reduction in the required transmitted power as shown in the CASNCP algorithm's curve. Finally, the MPCR algorithm requires the least transmitted power.

One of the major results is that the MPCR algorithm requires less transmitted power than the CASNCP algorithm. Intuitively, this result is because the MPCR applies the cooperative-based link-cost formula to construct the minimum-power route. On the other hand, the CASNCP algorithm first constructs the shortest-path route, then it applies the CT protocol on the established route. Therefore the CASNCP algorithm is limited to apply the cooperative-communication protocol on certain number of nodes, while the MPCR algorithm can consider any node in the network to be in the CT blocks, which constitute the route. Thus the MPCR algorithm reduces the required transmitted power more than the CASNCP algorithm.

Figure 12.23 depicts the required transmitted power per route by the different routing algorithms for different number of nodes at $\eta_0 = 6$ b/s/Hz. It is shown that the numerical curve of the MPCR algorithm, which is obtained through exhaustive search, almost coincides with the analytical curve. In addition, it is shown that the required transmitted power by any routing algorithm decreases with the number of nodes. Intuitively, the higher the number of nodes in a fixed area and the closer the nodes are to each other, the lower the required power between these nodes, which results in a lower required end-to-end transmitted power.

We also calculate the power-saving ratio as a measure of the improvement of the MPCR algorithm. The power saving of a cooperative-based routing scheme with respect to a noncooperative scheme can be calculated as

$$\text{P. S.} = \frac{P_T(\text{noncooperative}) - P_T(\text{cooperative})}{P_T(\text{noncooperative})} \%, \qquad (12.112)$$

where $P_T(.)$ denotes the total transmitted power for the certain scheme. At $N = 80$ and $\eta_0 = 6$ b/s/Hz, the power savings of MPCR algorithm with respect to the SNCP and CASNCP algorithms are 10.05% and 9.3%, respectively. In addition, the power saving of the CASNCP algorithm with respect to the SNCP algorithm is 0.83%. It is clear that applying the CT mode on the shortest-path route does not have a significant effect, as compared with the MPCR algorithm.

In summary, we have investigated the impact of the cooperative communications on the minimum-power routing problem in wireless ad hoc networks. For a given source–destination pair, the optimum route requires the minimum end-to-end transmitted power while guaranteeing certain throughput. We have proposed the MPCR algorithm, which applies the cooperative communication while constructing the route. The MPCR algorithm constructs the minimum-power route using any number of the proposed cooperative-based building blocks, which require the least possible transmitted power.

For comparison issues, we have also presented the CASNCP algorithm, which is similar to most of the existing cooperative routing algorithms. The CASNCP algorithm first constructs the conventional shortest-path route then applies a cooperative-communication protocol upon the established route. From the simulation results, the power savings of the MPCR algorithm with respect to the shortest-path and CASNCP algorithms are 10.05% and 9.3%, respectively.

12.6   Summary on Cooperative Cross-Layer Optimization

Overall, cooperative communication has gained much attention as an emerging transmit paradigm for future wireless networks by efficiently taking advantage of the broadcasting nature of wireless networks, as well as exploiting the inherent spatial and multiuser diversities by treating the cooperative relays as virtual antenna arrays. Most works in the literature focus on how to improve link quality in the physical layer, whereas the resource-allocation and networking issues in the cross-layer point of view are less studied. In this chapter, we discussed resource allocation among sources and relays to optimize the system performances in multiple layers.

The cross-layer impact of cooperative communication lies in the fact that there is a new degree of freedom for the traditional resource-allocation problems. For example, for power control, the relay's power can determine the performances at the destination. With different power controls, the rate adaptation can be adjusted to fully utilize the spectrum. With a limited spectrum and multiple users, relay section and channel allocation investigate the problems for multiple access and spectrum access. Moreover, in the network layer, the routing metrics can be significantly different from the traditional ones under the cooperative transmission, as the routes with the cooperative users can greatly improve the route performances.

In addition to the physical, MAC, and network layers, the impacts of cooperative communication can also be seen in application layers. In [100], the authors considered, in a capacity-achieving case, AF cooperation combined with a refinable single-description coding in which only the base layer is transmitted using cooperation. In [171], the best strategy is investigated from the viewpoint of a resource-allocation protocol to match source coding with cooperation diversity for conversational multimedia communications by studying the distortion performance for different schemes. The results show that the best performance is obtained when all layers of a layered-coded source are sent with user cooperation (using DF in most cases) if the source–destination channel is bad and with no user cooperation if the source–destination channel is good. The results also show that the gains from cooperative diversity outweigh the loss that is due to the sacrifice in overall bandwidth and that cooperation performance is sensitive to the proportion of communication capacity allocated for cooperation.

13 Game-Theoretic Approaches for Resource Allocation

13.1 Introduction

In recent years, wireless networks, especially ad hoc networks that consist of a collection of radio transceivers without requiring centralized administration or a prearranged fixed network infrastructure, have been studied intensively. Considering the application scenarios in which the users are "selfish" and act noncooperatively to maximize their own interests, the performances of such networks will deteriorate dramatically because of the inefficient competition for the wireless resources among selfish users. The greediness of selfish users and the distributed network structure challenge the feasibility of the conventional approaches and require novel techniques for distributed and efficient networking. Thus ensuring cooperation among selfish users becomes an important issue for designing wireless networks.

To ensure the cooperation and study the behaviors of selfish users, game theory is a successful economy tool, which studies the mathematical models of conflict and cooperation between intelligent and rational decision makers. In the literature, different types of game approaches have been introduced to several areas of wireless communications. One of the most important is the pricing anarchy, in which a price is taxed for the resource usage so that cooperative behaviors can be enforced. Noncooperation game theory was studied in [268] for power-control problems, in which the pricing technique was used to achieve Pareto optimality. In [369], resource allocation was studied for a forward link two-cell CDMA voice network with multiple service classes. Noncooperative game theory has also been studied for self-organizing mobile ad hoc wireless networks (MANET). In [38, 368], reputation-based game approaches were proposed to encourage packet forwarding among users. In [206], the authors proposed a noncooperative approach for encouraging collaboration in MANETs. In regard to the cooperation game theory, a framework was described in [343] for bandwidth allocation of elastic services in high-speed networks. In [108], a cooperative game approach called the Nash bargaining solution (NBS) was studied in the scenario of power, rate, and subchannel allocation for single-cell OFDMA systems, so as to achieve fair and efficient performances. In [109], an optimal repeated game framework was proposed to enforce cooperation by threat of future punishment. The framework is applied to the rate-control problem in multiple-access networks with greedy users. In [105, 106], the game-theory approach was proposed to combine with the idea in [297] for resource-allocation problems in multicell OFDMA systems.

In this chapter, we plan to demonstrate several different philosophies using game-theory approaches. We employ the commonly accepted principles of economy analysis, particularly game theory, to tackle the dynamic resource-allocation problem in wireless networks by motivating self-interested users to adopt a social behavior by sharing resources efficiently and thus to improve the overall system quality. It also provides autonomy for users to make their own decisions given their own channel conditions, and flexibility for accommodating multiuser diversity. Particularly, we develop novel gaming approaches to accomplish the aforementioned goals with the following aspects:

1. Noncooperative static games

   We provide three examples for noncooperative games. Traditionally, pricing anarchy is investigated in this approach. A price is defined to control the usage of wireless radio resources, which is similar to tax in the real world. Compared with the schemes in the literature, our schemes have their unique merits in that there is no need to compute the complicated price itself and the utility can be more sophisticated to represent the application's real benefit.

   First, the noncooperative power-control game and through game are proposed for nonlinear distributed resource allocation. For the traditional schemes with one game and one utility function, nonlinearity can result in less efficient equilibria. By using multiple related games and multiple utility functions for power-control and rate-adaptation problems, each game is linear. Consequently, better optimality can be achieved.

   Second, we propose a referee-based approach, in which a virtual referee monitors the network performances. If the performance is good, the referee will do nothing. Otherwise, the referee will mediate the game rules so that the performance can be enhanced. We propose noncooperative game schemes to conduct channel assignment, adaptive modulation, and power control for multicell OFDMA networks.

   Third, we improve the wireless network performances by using cooperative transmission. Cooperative communications have been shown to be able to greatly improve system performances by exploiting the broadcasting nature of wireless channels and cooperation among users. Most existing works concentrate on improving the peer-to-peer link quality. We focus on resource allocation among multiple users such that the system performances can be improved. Here a two-level game is proposed for the sources and relays to achieve optimal resource allocation in a distributed way.

2. Noncooperative repeated games

   The design goal is to punish any deviating greedy users that do not cooperate. For the two examples we will show, we utilize the repeated game theory and machine-learning theory to accomplish such a goal. In particular, we employ this approach for distributed rate control of multiple-access networks and packet-forwarding networks, respectively. In the former case, selfish users would transmit noncooperatively with a high rate. This causes a high collision probability in the multiple-access protocols, which results in low system performances. A Cartel Maintenance repeated game framework is proposed to enforce cooperation among selfish users. The proposed framework designs a closed-form trigger–punishment game rule to encourage the users to follow the cooperative strategy, whereas in the latter case, autonomous users would be reluctant to forward others' packets because of their greediness. However, such greediness would deteriorate both the system efficiency and users' performances. Hence it is a crucial goal to design a distributed mechanism for enforcing cooperation among greedy users for packet forwarding. In second example, users achieve better cooperation by learning others' behaviors in a distributive way. We proposed a distributed self-learning repeated game scheme to achieve the preceding goal.

3. Games with cooperation

In this type of approach, the design goal is to give the autonomous users mutual benefits for cooperation, such that the users can distributively converge to the co-operative operating point. In the first example, we introduce new approaches based on cooperative game theory. The incentive and mutual benefits are obtained through a bargaining process. In particular, such an approach is employed to allocate channel, rate, and power for multiple-users/multiple-channels within a cluster of wireless networks. The approach considers a new fairness criterion, which is a generalized proportional fairness based on NBS, and coalitions. In the second example, we study correlated equilibrium, which can bring mutual benefits to all users. However, because of the available limited information, users can behave noncooperatively. A nonregret learning scheme can achieve the set of correlated equilibria. We employ such an idea to the cognitive radio for distributed opportunistic spectrum access.

The uniqueness and significance of the gaming approaches are primarily targeted at employing the assorted game-theory tools to develop a self-organized framework for dynamic resource allocation. The ideas can significantly improve and advance the models and perspectives for the design and analysis of distributive resource allocation over wireless networks from the game-theory point of view.

13.2   Noncooperative Static Games

In this section, we study three distributed approaches using noncooperative games. In the first approach, we investigate the correlation between power control and rate adaption, and then propose games of multiple dimensions. In the second example, we propose a referee-based approach for multiple-cell OFDMA networks. The idea is to intervene and change the game rule whenever it is necessary, so that the distributed users can play the game and converge to the desired equilibria. Finally, we use the Stackelberg game to analyze the relation between the source and relay nodes in the cooperative transmission networks.

13.2.1   Noncooperative Power Control and Throughput Games

In this example [115], we study the game for multiple users with power-control and rate-adaptation abilities. As we have mentioned, an individual's greediness for radio resources can greatly reduce the efficiency. To achieve better system performances, our primary concern is to design the utility functions and the rules of the games. One of the goals is to motivate individual users to adopt a social behavior and enhance the system performance by sharing the resources. Consequently, we can make the distributed self-optimizing decisions compatible with the demand for a higher overall system performance. In doing so, we link both power control and adaptive modulation by designing games at both the user level and the system level. A noncooperative power-control game (NCPCG) is designed at the user level. At the system level, the optimization goal is to maximize the overall system throughput under the maximal transmitted power constraint. A noncooperative throughput game (NCTG) is designed. There may be multiple Nash equilibria in this game. A distributed algorithm is constructed to achieve the better Nash equilibrium by employing a proposed game rule and an initialization method. An optimal but complex centralized algorithm that achieves the optimal system performance is developed as a performance upper bound. From simulations, the proposed games are optimal for the power at the user level and can be optimal or near optimal for network throughput at the system level.

For the system model, we consider K cochannel uplinks that may exist in distinct cells of wireless networks. Each link consists of a mobile and its assigned base station (BS). We assume the average transmitted power for different modulation constellations is normalized. Define N_i as noise level. The ith user's SINR is

$$\Gamma_i = \frac{P_i h_{ii}}{\sum_{k \neq i} P_k h_{ki} + N_i},\tag{13.1}$$

where P_k is the kth user's transmitted power and h_{ki} is the channel gain from the kth user to the ith BS. Adaptive modulation provides the links with abilities to match the effective bit rates (throughput), according to interference and channel conditions. MQAM is a modulation method that has high spectrum efficiency. We assume each user has a unit bandwidth. In [241], for a desired throughput T_i of MQAM, the ith user's BER can be approximated as

$$\text{BER}_i \approx c_1 e^{-c_2 \frac{\Gamma_i}{2^{T_i}-1}},\tag{13.2}$$

where $c_1 \approx 0.2$ and $c_2 \approx 1.5$ when BER_i is small. For a specific desired BER_i, the ith link's required SINR for the desired throughput T_i can be expressed as $\gamma_i(T_i) = \frac{2^{T_i}-1}{c_3}$, where $c_3^i = -\frac{c_2}{\ln(\text{BER}_i/c_1^i)}$.

If a user's throughput is too large, CCI is severe and it is possible that there exists no feasible power allocation for the desired throughput and BER. To prevent the system from not being feasible, we need to analyze the feasibility condition. First, we use the targeted SINR γ_i and require that the received SINR Γ_i be larger than or equal to this targeted SINR, i.e., $\Gamma_i \geq \gamma_i, \forall i$, in order to ensure the desired BER for the throughput T_i. Rewriting these inequalities in a matrix form, we have

$$(\mathbf{I} - \mathbf{DF})\mathbf{P} \geq \mathbf{Du},\tag{13.3}$$

where $\mathbf{I}$ is an identity matrix, $\mathbf{u} = [u_1, \ldots, u_K]^T$ with $u_i = N_i/(|h_{ii}|^2)$, $\mathbf{D} = \text{diag}\{\gamma_1, \ldots, \gamma_K\}$, and

$$[\mathbf{F}_{ij}] = \begin{cases} 0, & \text{if } j = i \\ \frac{|h_{ji}|^2}{|h_{ii}|^2}, & \text{if } j \neq i \end{cases}. \tag{13.4}$$

The preceding inequality is a bilinear matrix inequality [205], i.e., the power vector is linearly constrained if the targeted SINR vector (throughput) is fixed, and vice versa. Because linearity can achieve global optimum, this motivates us to employ the two-game approach developed later. By the Perron–Frobenius theorem [355], there exists a feasible solution with positive power and rate allocation only if the maximum eigenvalue of $\mathbf{DF}$ [spectrum radius $\rho(\mathbf{DF})$] is inside the unit circle.

Next, we construct games for distributive resource allocation. In wireless communication networks, because of the bandwidth limitation, it is impractical for the mobile users to communicate, and thus cooperate with each other, so as to optimally utilize the wireless resources. Each individual mobile user tries to maximize its performance, based only on its perceived self-interest. However, self-interest will cause the system to be balanced in some undesired and nonoptimal equilibria. We design the game rules for the users' competitions such that the system will be balanced in the desired optimal and efficient resource allocation. Because power and throughput are bilinearly constrained, it is natural to divide the optimization efforts into the system level and the user level. We define value function v_i as the connection between two levels. The goals for both levels are given as follows:

1. **User Level:** The goal is to define a utility function u_i and then each user can compete with other users in a NCPCG to maximize its utility function. There are some practical constraints such as the maximum transmitted power $P_{\max}$. The proposed NCPCG is formulated as

$$\max_{P_i \leq P_{\max}} u_i(P_i, \mathbf{P}_{-i}, v_i), \tag{13.5}$$

where $\mathbf{P}_{-i} = [P_1 \ldots P_{i-1} P_{i+1} \ldots P_K]^T$, and v_i is the assigned value function that is related to throughput T_i. At the user level, the transmitted power P_i is optimized by the proposed NCPCG, and v_i is assigned by the BS. When the throughput T_i is equal to 0, no transmitted power is needed and v_i should be zero. Otherwise, we define the value function as a function of the desired throughput as

$$v_i = \begin{cases} \ln\left(\frac{2^{T_i}-1}{c_3^i}\right) + 1, & \text{if } T_i > 0 \\ v_i = 0, & \text{if } T_i = 0 \end{cases}, \tag{13.6}$$

where v_i is a function of only throughput T_i and c_3^i. c_3^i is related to the desired BER and is usually predefined and fixed. When the CCI is high, from system optimization point of view, the cost for the desired v_i should be increased to reduce CCI. We represent this cost as $\ln \Gamma_i$, where Γ_i reflects the severity of the CCI and can be fed back from the BS to the mobile. We define the utility function as

$$u_i = P_i(v_i - \ln \Gamma_i). \tag{13.7}$$

NCPCG is played iteratively until convergence. It can be shown that u_i satisfies the three requirements of the "standard function" in [347] for iterative functions. Consequently, the NCPCG converges to the unique optimum that achieves the minimal SNR for the desired BER.

2. **System Level:** The goal is to assign a user its value function v_i by a NCTG, such that the overall system throughput $\sum_{i=1}^{K} T_i$ is maximized under the constraint $P_i \leq P_{\max}$, $\forall i$. When the system is balanced, T_i and P_i are functions of $\mathbf{v}$, where $\mathbf{v} = [v_1 \ldots v_K]^T$. The overall network throughput is optimized by the NCTG, and the corresponding v_i, $\forall i$, are assigned to the users for the NCPCG. The problem can be formulated as

$$\max_{\mathbf{v}} \sum T_i(\mathbf{v}), \tag{13.8}$$

$$\text{subject to } P_i(\mathbf{v}) < P_{\max}, \forall i.$$

At the system level, the NCTG is constructed for the users to compete distributively while the system maintains feasibility. We define Λ as an indication function for system feasibility. When the BS detects that all required transmitted powers for the desired BER and throughput are less than or equal to $P_{\max}$, Λ equals 1; otherwise it equals 0. Because the users compete with each other for the throughput, we define each user's utility function $\bar{u}_i$ for NCTG as a product of its throughput T_i and Λ, i.e.,

$$\textbf{(NCTG)} \max_{T_i} \bar{u}_i = T_i \Lambda. \tag{13.9}$$

The game starts from any feasible initial values and is balanced when no user can increase its throughput. The existence of Nash equilibrium can be shown by a similar proof in [268]. However, there might be multiple Nash equilibria, which will be shown in the simulations later. If the users with bad channels get high throughput, they will produce large CCI to other users. Consequently, the system overall throughput will be reduced. So the way to initialize the proposed game and the way to design the game rule for each user to compete its throughput play a critical role on finding the better optimum. The idea for initialization comes from the following theorem:

THEOREM 24 *Define maximal achievable SINR as $\hat{\Gamma}_i$ when $P_i = P_{\max}$, $\forall i$. Then the value $v_i = 1 + \ln\left(\frac{2^{\lfloor \log_2(1+c_3^i \hat{\Gamma}_i) \rfloor} - 1}{c_3^i}\right)$ is feasible for both games, where $\lfloor \rfloor$ is a floor function for the maximal integer that is smaller.*

Proof: Define $\hat{\gamma}_i = e^{v_i - 1} = (2^{\lfloor \log_2(1+c_3^i \hat{\Gamma}_i) \rfloor} - 1)/c_3^i$. Because $\log_2$ is an increasing function, $\hat{\Gamma}_i \geq \hat{\gamma}_i$, $\forall i$. Because $\mathbf{D}, \mathbf{F} \in \mathfrak{R}^{K \times K}$ and $|\rho(\mathbf{DF})| < 1$, we have $\mathbf{P}' = \sum_{j=0}^{\infty} (\mathbf{DF})^j \mathbf{Du}$. Because any component in $\mathbf{D}$, $\mathbf{F}$, and $\mathbf{u}$ is nonnegative, all components in $\mathbf{P}'$ are nondecreasing functions of γ_i. When we select the targeted SINR $\hat{\gamma}_i \leq \hat{\Gamma}_i$, any component of the power vector must be smaller than or equal to $P_{\max}$. So the value functions satisfy the maximum power constraint and the system must be feasible. **QED**

First every user transmits the maximal power. The BS detects the received SINR. Using the preceding theorem, the BS decides what the largest achievable throughput is

Table 13.1 Distributed system algorithm

Initial: $P_i = P_{max}$, $\forall i$, calculate $\mathbf{v}_i$ in Theorem 24 and send back to mobiles.

Iterations: 1. Wait until NCPCG converges

2a. Power-Increase Criteria at Users:
 If H_i condition is satisfied, send throughput increase request to the BS.

2b. Feasibility Detection at the BS:
 Increase throughput for requesting user with highest H_i, detect if feasible.

2c. Feedback to Users:
 If system not feasible: reduce throughput to original value;
 else if no more request: Wait.

---

and sends back the corresponding v_i. The system is sure to be feasible but not necessarily optimal. By doing this, the users with good channels will get higher throughput. Each user will then play the NCPCG until convergence. The convergence is achieved when each user detects that the NCPCG utility is stable, because this means that the interfering users' power and utilities are also stable. After NCPCG convergence, the users decide if they can increase their throughput while the system is still feasible. We need to find the criteria for the users to decide when to send the requests to the BS for throughput increasing. We define $\gamma_i(T_i)$ as the required SINR for the desired throughput T_i. When $T_i > 0$, if we assume the interferences, noise, and channel gains are fixed, the required power for throughput $T_i + 1$ will be $P_i[2^{(T_i+1)} - 1]/(2^{T_i} - 1)$, where P_i is the current power. We compare this desired power with βP_{max}, where β is a constant and $0 \le \beta \le 1$. If the desired power is larger than βP_{max}, the user can send a request to the BS to increase its throughput by one. When this user increases its power, it causes interference to the others. Others have to increase their power, which causes interference to this user as well. So this user has to increase its power more. β is such a factor that takes into consideration this "mutual-interfering" effect. When $T_i = 0$, all received power is the interference-plus-noise power, which is defined as I_i. The channel gain h_{ii} can be estimated during the initialization when this user transmits the maximal power. This user can calculate the estimated received SINR$= \frac{\beta P_{max} h_{ii}}{I_i \gamma_i(1)}$ by transmitting power βP_{max}. If the received SINR is larger than $\gamma_i(1)$, the user will send the throughput increase request. Let us define H_i as the throughput request factor; the criterion for the users to request their throughput to increase by one is when $H_i > 1$, where

$$
H_i = \begin{cases} \frac{\beta P_{max}(2^{T_i} - 1)}{P_i(2^{(T_i+1)} - 1)}, & \text{if } T_i > 0 \\ \frac{\beta P_{max} h_{ii}}{I_i \gamma_i(1)}, & \text{if } T_i = 0 \end{cases}. \tag{13.10}
$$

The preceding game rule for the NCTG and the distributed adaptive algorithm are summarized in Table 13.1.

After initialization, two games are played iteratively. First, users play the NCPCG. After convergence, users play the NCTG, calculate H_i, and send requests if necessary. If the request is granted, the selected user will increase the throughput, and then the

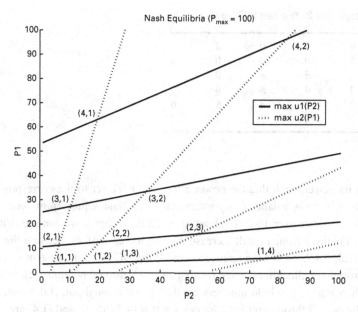

Figure 13.1 Nash equilibria of the NCPCG.

NCPCG is played again. Both games are distributive because only local information is necessary to play. It is worth mentioning that, to determine the largest H_i, adjacent BSs should exchange the values of H_i, which can be implemented with limited signaling.

Next, we construct a centralized scheme as a performance upper bound. The distributed algorithm may not be optimal. First, there is a probability that the users do not send requests, whereas the system might be feasible if users sent requests. Second, there exist Nash equilibria that are not global optimum. To understand the performance loss, we need to find the optimal solution as a performance bound. The most straightforward idea is to let the system centrally decide how to allocate throughput to users with the assumption that all channel responses are known. The idea is not implementable because the channel conditions from the users to the BS in other cells are hard to obtain. The problem becomes a constrained optimization: to maximize overall throughput under the maximum power and maximum eigenvalue constraints, which can be solved by standard nonlinear programming:

$$\max_{v_i} \sum_{i=1}^{K} T_i(\mathbf{P}), \tag{13.11}$$

$$\text{s.t. } |\rho(\mathbf{DF})| < 1, \ P_i \leq P_{\max}, \ \forall \, i.$$

We evaluate the performances of the proposed algorithms by two simulation setups. First, we consider a two-user case. Here we assume that $h_{11} = h_{22} = 1$, $h_{21} = 0.01$, $h_{12} = 0.07$, $N_1 = N_2 = 1$, BER $= 10^{-3}$, and $P_{\max} = 100$. In Figure 13.1, we show the Nash equilibria of the NCPCG when the different throughput allocations are given. On any solid line, u_1 gets the maxima. On any dotted line, u_2 has the optima. Starting from any feasible power allocation, each user tries to maximize its utility function

Table 13.2 Strategic form for two-user NCTG example

5	0	0	0	0	0
4	**5**	0	0	0	0
3	4	**5**	0	0	0
2	3	4	5	**6**	0
1	2	3	4	5	0
(u_1, u_2)	1	2	3	4	5

by controlling its power, such that the power allocation is closer to the corresponding lines. When the system is balanced, any intersection is a Nash equilibrium, where we denote the throughput as user 1's throughput, user 2's throughput. We can see that the maxima for u_1 obtained from P_1 will increase with an increase of P_2, because the CCI increases. In Table 13.2, we list the strategic form of the NCTG at the system level for all nonzero throughput allocations. Each row lists user 1's throughput and each column lists user 2's throughput. The bold numbers are the overall throughput. If the system is not feasible, the overall throughput is 0. We can see that (4,2), (2,3), and (1,4) are Nash equilibria, because no user can improve its throughput alone. However, (2,3) and (1,4) are not desired Nash equilibria for the optimal overall network throughput. The proposed distributed algorithm in Table 13.1 will be initialized at (3,2). If β is properly selected (in this case, $\beta > 0.32$), because $H_1 > H_2$, the algorithm will increase user 1's throughput first and converge to the optimal Nash equilibrium (4,2). So from this example we can see that we can achieve both power optimum and throughput optimum by playing the NCPCG at the user level and the NCTG at the system level.

We set up another simulation to test the proposed algorithms. A network is constructed with one cell at the center and the other six at the degrees of [0, 60, 120, 180, 240, 300], respectively. One BS is located at the cell's center and one user is randomly located within each cell. The cell radius is $r = 1000$ m, the minimal distance between the user and the BS is $r' = 50$ m, and the distance between centers of two adjacent cells is $R = r R_u$ m, where R_u is the reuse factor. $P_{max} = 2$ W, $\eta = 3.5$, $\alpha = 10^{-3}$, BER $= 10^{-3}$, and $N_i = 10^{-11}$ W. The simulations run 10^5 times.

Figure 13.2 compares overall network throughput versus R_u with different β. When R_u is small, CCI is severe. After initialization, by sending the maximal transmitted power, most users get a throughput of zero. The overall network throughput is increased by the users' throughput-increasing requests. If β is large enough, the proposed games can achieve optimal system performance when R_u is small enough. The overall network throughput is minimal when $R_u \approx 0.25$, because different BSs and users are mixed together and CCI is the most severe under this condition. When R_u is large, CCI is minor. After the initialization, most users get the desired throughput. The overall network throughput is refined when β is large. The optimal system performance can be achieved when both R_u and β are large enough. When R_u is in the middle range, the proposed games may fall into some local minima and produce suboptimal solutions, even when $\beta = 1$. The overall throughput can be improved by increasing β.

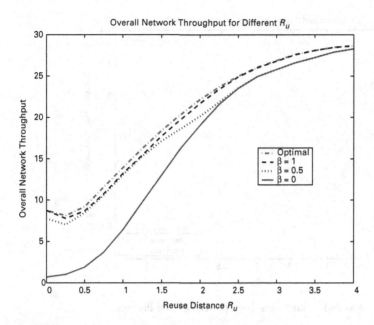

Figure 13.2 System throughput vs. R_u.

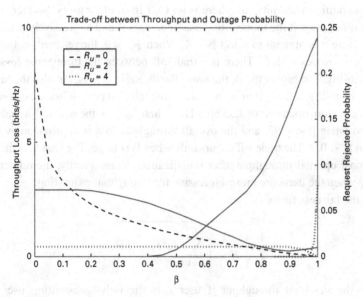

Figure 13.3 Trade-off between rejected and throughput loss for β.

However, the overall throughput improvement by increasing β is at the expense of possible high-request-rejection probability defined as the ratio of the number of rejected requests over the total number of requests. Figure 13.3 shows the throughput loss compared with the optimal solution and rejected probability versus β for different R_u. When $R_u = 0$, the rejected probability is always zero, and the throughput loss is

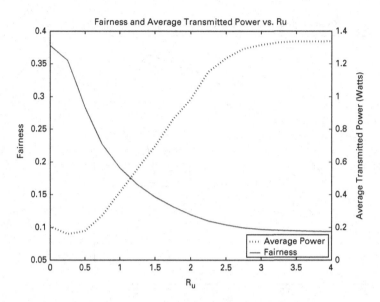

Figure 13.4 Fairness and average transmitted power vs. reuse distance.

monotonically decreasing with β. The optimal solution is that only the user with the best channel condition transmits, and there is no CCI from other users. So there is no penalty from other users if the transmitting user increases β and aggressively sends the request. Therefore it is optimal to select β = 1. When $R_u = 2$, the rejected probability monotonically increases with β. There is a trade-off between the throughput loss and rejected probability. The higher the β, the lower the throughput loss, but also the higher the rejected probability. If the system wants a very low rejected probability, we can select β = 0.4 with a performance loss of 2.35 bit/s/Hz. When $R_u = 4$, the rejected probability is almost zero when β < 0.97, and the overall throughput loss is approximately 0.44 bit/s/Hz when β < 0.9. The trade-off occurs only when β is large. The reason is that the users get almost optimal throughput after initialization. Consequently, the refinement happens only when the users are more aggressive for throughput requesting.

We define the fairness factor as

$$\varrho = \frac{1}{\bar{T}}\sqrt{\frac{1}{K-1}\sum_{i=1}^{K}(T_i/\hat{T}_i - \bar{T})^2}, \tag{13.12}$$

where $\hat{T}_i$ is the maximal throughput if user i is the only transmitting user and $\bar{T} = \text{average}(T_i/\hat{T}_i)$. The physical meaning of ϱ is the normalized variance of users' throughput compared with that of the single-user case. The higher the ϱ, the more unfair among users, i.e., the users' throughput is more affected by CCI. ϱ is one of the possible definitions to measure fairness. Figure 13.4 shows fairness and average transmitted power versus R_u with β = 1. When R_u is small and CCI is severe, ϱ is large and the users with the better channel conditions occupy most of the resources. The average transmitted power is also low, because most users cannot transmit. When R_u becomes

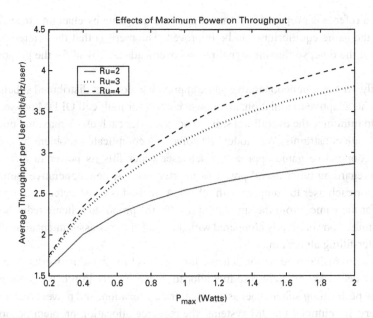

Figure 13.5 Average throughput per user vs. P_{max}.

large and CCI is reduced, the users with worse channel conditions can compete for their transmissions, whereas the users with better channel conditions are not so dominant. Consequently, ϱ is reduced and users transmit more fairly, as in the single-user case. The average transmitted power is increased and saturated with increasing R_u, because most users can transmit according to their own channel conditions, regardless of the low CCI from others.

Figure 13.5 shows the average throughput per user versus P_{max} for different R_u with $\beta = 1$. We can see that the average throughput increases more slowly when P_{max} is large, because the CCI is increasing especially when R_u is small. When R_u is decreasing, the point where average throughput per user saturates moves to the lower P_{max}. There is no need for higher P_{max} if the performance curve is saturated already. So when R_u is decreasing, we can reduce P_{max} accordingly.

Overall, to achieve better overall network throughput, we construct NCPCG and NCTG at the user level and at the system level, respectively. From the simulations, the proposed games converge to near-optimal solutions compared with the optimal solutions obtained from the centralized scheme. The proposed two-game approach explores the bilinear matrix inequality (BMI) nature of the resource allocation to avoid local optima and consequently the approach has high performances in a distributed implementation.

13.2.2   Referee-Based Approach

In this subsection, we propose a new type of approach based on refereeing. The basic idea is that the distributed system can converge to the desired equilibrium most of time. But, occasionally, the system might be balanced in a certain outcome with low efficiency.

In this case, a referee is proposed and tries to mediate the game by changing the rules. As a result, the game equilibrium can be improved. The merit is that the referee does not process all the time, so that the signaling and overheads are trivial for the proposed approach.

Specifically, we use a noncooperative game approach to conduct distributed subchannel assignment, adaptive modulation, and power control for multicell OFDM networks. The goal is to minimize the overall transmitted power under each user's maximal power and minimal rate constraints. We model and solve this complicated problem by a distributed noncooperative game approach: Each user water fills its power to different subchannels regarding other users' power as interferences. A noncooperative game is constructed for each user to compete with others. A method is constructed as a mediator (judge) for the game. From the simulation results, the proposed scheme reduces the overall transmitted power greatly compared with the fixed channel-assignment algorithm and pure water-filling algorithm.

OFDM is a promising modulation scheme for wireless broadband networks. In multiuser OFDM networks, efficient resource allocation can greatly improve system performances by performing subchannel assignment, rate allocation, and power control for different users. In multicell OFDM systems, the resource-allocation problem becomes more complicated, even if the assignment of subchannels to users is predetermined, because users in different cells reuse the same subchannels and cause interferences to each other. If the number of cochannel users is relatively large, the interference seen by a user in a subchannel can be approximated by a Gaussian random variable applying the central-limit theorem. In this case, the water-filling algorithm provides a good solution. When the channel assignment is fixed, many iterative water-filling methods are proposed in [50, 252, 314, 319, 350] to maximize the rate with power constraints. However, if the subchannel assignment is not predetermined, all possible combinations of cochannel users should be checked to determine the optimal resource allocation. In [297], the authors present heuristic distributed algorithms, which are based on iterative water filling with removing subchannels of low SINRs.

In the multicell systems, the mobile users do not have knowledge of other users' conditions and cannot cooperate with each other. They act selfishly to maximize their own performances in a distributed way. Such a fact motivates us to adopt game theory [80]. Resource allocation can be modeled as a noncooperative game that deals largely with how rationally and intelligently individuals interact with each other in an effort to achieve their own goals. In the resource-allocation game, each mobile user is self-interested and attempts to optimize his utility function, which represents the user's performance and controls the outcomes of the game.

In this example, we want to minimize the overall transmitted power under each user's maximal power and minimal rate constraints. By the noncooperative game-theory approach, we find the following facts: If the CCIs are small, users can share the subchannels for transmission. In this case, by carefully designing the utility function, the noncooperative game (for each user to compete for the resources) will be balanced in an optimal and unique Nash equilibrium point (NEP). If the CCIs are severe for some subchannels, the NEP may not be optimal and there might be multiple NEPs. To deal with this situation, some users with bad channels or large interferences to others must be kicked out from using these subchannels, so that the rest of the users can make good use of the corresponding subchannels. We design the utility function for each user, define the criterion as a game rule to kick out users, and develop an adaptive algorithm for resource allocation. From the simulation results, we can see that the proposed scheme can reduce the overall transmitted power greatly compared with the fixed channel-assignment algorithm and pure water-filling algorithm.

The K cochannel cells are taken into consideration that may exist in distinct cells of OFDM networks. Each cell consists of a mobile user and its assigned BS. Assume that coherent detection is possible so that it is sufficient to model this multiuser system by an equivalent baseband model. We also assume that the different links among cells are synchronized.[1] The total number of OFDM subchannels is L. For the uplink case, the sampled signal on the lth subchannel of the ith user can be expressed as

$$x_i^l(n) = \sum_{k=1}^{K} \sqrt{P_k^l G_{ki}^l} s_k^l(n) + n_i^l(n), \tag{13.13}$$

where P_k^l and G_{ki}^l are the transmitted power and propagation loss from the kth user to the ith BS in the lth subchannel, respectively, s_k^l is the message symbol from the kth user to the ith BS at time n, and $n_i^l(n)$ is the sampled thermal noise. We assume that the channels change slowly. Without loss of generality, we assume $N_i^l = E(\|n_i^l\|^2) = N_0$. The ith user's SINR at subchannel l can be expressed as

$$\Gamma_i^l = \frac{P_i^l G_{ii}^l}{\sum_{k \neq i} P_k^l G_{ki}^l + N_0}. \tag{13.14}$$

Rate adaptation such as adaptive modulation provides each subchannel with the ability to match the effective bit rates, according to the interference and channel conditions. MQAM is a modulation method with high spectrum efficiency. In [49], for a desired rate r_i^l of MQAM, the BER of the lth subchannel of the ith user can be approximated as a function of the received SINR Γ_i^l by

$$\text{BER}_i^l \approx c_1 e^{-c_2 \frac{\Gamma_i^l}{2^{r_i^l}-1}}, \tag{13.15}$$

where $c_1 \approx 0.2$ and $c_2 \approx 1.5$ with small BER_i^l. Rearranging (13.15), for a specific desired BER_i^l, the ith user's transmission rate of the lth subchannel for the SINR Γ_i^l and the desired BER_i^l can be expressed as[2]

$$r_i^l = W \log_2(1 + c_3^i \Gamma_i^l), \tag{13.16}$$

where W is the bandwidth and $c_3^i = -\frac{c_2}{\ln(\text{BER}_i^l/c_1)}$. For simplicity, we assume all the subchannels and users have the same BER requirement, i.e., $\text{BER}_i^l = \text{BER}, \forall i, l$.

---

[1] This assumption can be relaxed using the same approach.

[2] Here the rate of MQAM is assumed to be continuous. Discrete MQAM can be applied in a similar way for the approach developed later.

Each user requires the rate R_i and distributes its rate into L subchannels, i.e., $\sum_{l=1}^{L} r_i^l \geq R_i$. We define rate-allocation matrix $[\mathbf{r}]_{il} = r_i^l$. Each user's transmitted power is bounded by $P_{\max}$. Define the $K \times L$ channel assignment matrix $\mathbf{A}$ with $[\mathbf{A}]_{il} = 1$, if $r_i^l > 0$; $[\mathbf{A}]_{il} = 0$, otherwise. Therefore our objective is to minimize the overall transmitted power under the minimal rate and maximal power constraints, i.e.,

$$\min_{\mathbf{A},\mathbf{r}} f(\mathbf{r}) = \sum_{i=1}^{K} \sum_{l=1}^{L} P_i^l, \tag{13.17}$$

$$\text{s.t.} \begin{cases} \sum_{l=1}^{L} r_i^l - R_i \geq 0, \ \forall i, \\ \sum_{l=1}^{L} P_i^l - P_{\max} \leq 0, \ \forall i, \\ r_i^l, P_i^l \geq 0, \ \forall i, l. \end{cases}$$

Because power is a continuously increasing function of rate, the optimum occurs when $\sum_{l=1}^{L} r_i^l = R_i$. The problem in (13.17) is very difficult to solve by centralized constrained nonlinear integer optimization, because the complexity and communication overhead grow fast as the number of users increases. This motivates us to develop a distributed algorithm with limited controls by using game-theory approach.

Next, our focus is to solve (13.17) by noncooperative game theory. First, we analyze the system's feasible region. Then we construct the game. A two-user two-subchannel example is given to show insight. The properties of the the NEP are analyzed. Finally, an iterative algorithm for multiple users with a game mediator is developed.

To ensure the desired BER, for every subchannel, every user should have a SINR of no less than the required SINR γ_i^l, i.e., $\Gamma_i^l \geq \gamma_i^l$, $\forall i, l$. Rewriting these inequalities in a matrix form, we have

$$(\mathbf{I} - \mathbf{D}^l \mathbf{F}^l) \mathbf{P}^l \geq \mathbf{v}^l, \ \forall l, \tag{13.18}$$

where $\mathbf{I}$ is a $K \times K$ identity matrix, $\mathbf{v}^l = [v_1^l, \ldots, v_K^l]'$ with $v_i^l = N_0 \gamma_i^l / G_{ii}$, $\mathbf{D}^l = \text{diag}\{\gamma_1^l, \ldots, \gamma_K^l\}$, and

$$[\mathbf{F}_{ij}^l] = \begin{cases} 0 & \text{if } j = i \\ \dfrac{G_{ji}^l}{G_{ii}^l} & \text{if } j \neq i \end{cases}.$$

By the Perron–Frobenius theorem, there exists a positive power allocation if and only if the maximum eigenvalue of $\mathbf{D}^l \mathbf{F}^l$, i.e., spectrum radius $\rho(\mathbf{D}^l \mathbf{F}^l)$, is inside the unit circle. When $|\rho(\mathbf{D}^l \mathbf{F}^l)| < 1$, the optimal power solution is

$$\mathbf{P}^l = \begin{cases} (\mathbf{I} - \mathbf{D}^l \mathbf{F}^l)^{-1} \mathbf{v}^l, & |\rho(\mathbf{D}^l \mathbf{F}^l)| < 1 \\ +\infty, & \text{otherwise} \end{cases}. \tag{13.19}$$

The system feasibility region Ω is defined as the supporting domain where there exist solutions and the power constraint in (13.17) is satisfied. The condition for (13.19) to have finite solutions is a necessary condition for the existence of feasible Ω and convergence of the algorithm proposed later.

Each user wants to minimize its transmitted power by allocating its rate into the different subchannels in a distributed way, regardless of other users. Defining $\mathbf{r}_i = [r_i^1 \dots r_i^L]^T$, the noncooperative game can be written as

$$\text{Game: } \arg\min_{\mathbf{r}_i \in \Omega} u_i = \sum_{l=1}^{L} P_i^l, \text{ s.t. } \sum_{l=1}^{L} r_i^l = R_i, \qquad (13.20)$$

where u_i is the utility function defined as the ith user's transmit power. If the interferences from others are fixed, it is a water-filling problem. Defining

$$I_i^l = \frac{\sum_{k \neq i} P_k^l G_{ki}^l + N_0}{c_3^i G_{ii}^l}, \qquad (13.21)$$

the solution is

$$P_i^l = (\mu_i - I_i^l)^+ \text{ and } r_i^l = \log_2\left(1 + \frac{P_i^l}{I_i^l}\right), \qquad (13.22)$$

where $y^+ = \max(y, 0)$. μ_i is solved by a bisection search of

$$\sum_{l=1}^{L} \log_2\left[1 + \frac{(\mu_i - I_i^l)^+}{I_i^l}\right] = R_i. \qquad (13.23)$$

However, the interferences from other users do change. Based on game theory [80], the system will be balanced in a Nash equilibrium defined as follows:

DEFINITION 45 *Define* $\mathbf{r}_i^{-1} = [\mathbf{r}_1 \dots \mathbf{r}_{i-1}\mathbf{r}_{i+1} \dots \mathbf{r}_L]$. *NEP* $\mathbf{r}_i$ *is defined as*

$$u_i(\mathbf{r}_i, \mathbf{r}_i^{-1}) \leq u_i(\tilde{\mathbf{r}}_i, \mathbf{r}_i^{-1}), \ \forall i, \ \forall \tilde{\mathbf{r}}_i \in \Omega, \ \mathbf{r}_i^{-1} \in \Omega^{L-1}, \qquad (13.24)$$

i.e., given the other users' rate allocation, no user can reduce its transmitted power alone by changing its rate allocation to different subchannels.

To explain the Nash equilibrium and show the idea of our noncooperative game approach, we next analyze a two-user two-subchannel case. Suppose the ith user puts α_i proportion of its rate R_i to the first subchannel. The second subchannel has a rate of $(1 - \alpha_i)R_i$. Obviously $0 \leq \alpha_i \leq 1$. Let $P_i^j(\alpha_i)$ denote the transmission power of the ith user on the jth subchannel, which is a function of α_i, $\forall i$. For simplicity, we write $P_i^j(\alpha_i)$ as P_i^j. The Lagrangian function of the optimal allocation problem can be represented as

$$J(\alpha_1, \alpha_2) = P_1^1 + P_1^2 + P_2^1 + P_2^2 + \sum_{i=1}^{2} \mu_i \alpha_i + \sum_{i=1}^{2} \lambda_i(1 - \alpha_i), \qquad (13.25)$$

where μ_i and λ_i are the Lagrangian multipliers. By taking derivatives of α_1 and α_2, $\frac{\partial J(\alpha_1)}{\partial \alpha_1} = 0$; $\frac{\partial J(\alpha_2)}{\partial \alpha_2} = 0$, we can obtain the global optimal solution. However, global channel information such as $G_{ki}, k \neq i$, is necessary to solve (13.25), which causes great implementation difficulty in multicell OFDM networks.

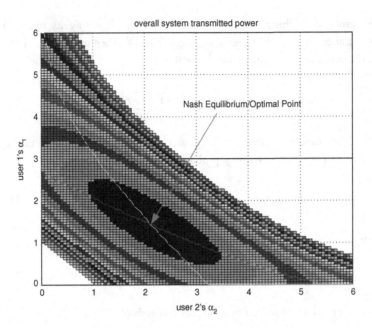

Figure 13.6 Two-user example: unique optimal NEP.

On the other hand, for Nash equilibria, the users perform the water-filling algorithms in a distributed way by using local information only. The Lagrangian functions of optimization problems are illustrated as

$$J_1 = P_1^1 + P_1^2 + \mu_1'\alpha_1 + \lambda_1'(1 - \alpha_1), \tag{13.26}$$

$$J_2 = P_2^1 + P_2^2 + \mu_2'\alpha_2 + \lambda_2'(1 - \alpha_2), \tag{13.27}$$

where μ_i' and λ_i' are the Lagrangian multipliers. The Nash equilibria of the preceding channel-allocation problem can be obtained by solving $\frac{\partial J_1(\alpha_1)}{\partial \alpha_1} = 0$; $\frac{\partial J_2(\alpha_2)}{\partial \alpha_2} = 0$. Obviously, the solutions of global optimum and Nash equilibrium are different. However, from the observation, the two solutions are close when the minimal rates are low.

To compare the Nash equilibria and the optimal solution, a simple two-user two-subchannel example is illustrated as follows. The simulation setup is BER $= 10^{-3}$, $N_0 = 10^{-3}$, $P_{\max} = 10^4$, and

$$G^1 = \begin{bmatrix} 0.0631 & 0.0100 \\ 0.0026 & 0.2120 \end{bmatrix}, G^2 = \begin{bmatrix} 0.4984 & 0.0067 \\ 0.0029 & 0.9580 \end{bmatrix}.$$

Figure 13.6 shows the overall power contour as a function of two users' rate allocations, where $R_1 = R_2 = 6$. The axes are α_1 and α_2, respectively. The two curves show the minimal locations for the two users' own power when the interference from the other user is fixed, respectively. Each user tries to minimize its power by adjusting its rate allocation so that the operating point is closer to the curve. Consequently, the cross is a Nash equilibrium, in which no user can reduce its power alone. We can see that the Nash equilibrium under this setup is unique and optimal for the overall power. (It is worth mentioning that the feasible domain is not convex at all.) Figure 13.7 shows

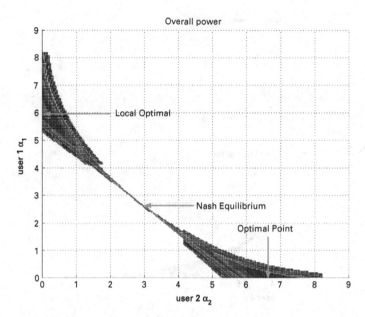

Figure 13.7 Two-user example: multiple local optima.

the situation when $R_1 = R_2 = 8$. Because the rate is increased, the CCIs are increased and the NEP is no longer the optimum. There exist more than one local optimum, and the global optimum occurs when user 1 does not occupy subchannel 1. Figure 13.8 shows the situation in which $R_1 = R_2 = 8.5$. The contour graph is not connected. There are two NEPs and two local optima. Under the preceding two conditions, we need to remove users from using the subchannels. If we further increase $R_1 = R_2 = 10$, there exists no feasible area, i.e., neither user can have a resource allocation that satisfies both power and rate constraints. In this case, the minimal rate requirement should be reduced.

From the preceding observations, we can see that the behaviors of the optimal solution and NEP depend on how severe interferences are. To let the NEP converge to the desired solution, we need to find a criterion as the game rule to decide whether the users can make good use of the subchannels like the situation in Figure 13.6. If not, we should decide which user should be kicked out of using the subchannels. The following two theorems are proved for the properties of NEP.

Next, we study some properties of the Nash equilibrium by the following theorems.

THEOREM 25 *There exists a NEP in the proposed game defined in (13.20) if Ω is not empty.*

Proof 7 In [80], it has been shown a NEP exists if, $\forall\, i$,

1. Ω, the support domain of $u_i(\mathbf{r}_i)$, is a nonempty, convex, and compact subset of some Euclidean space $\mathfrak{R}^L$;
2. $u_i(\mathbf{r}_i)$ is continuous in $\mathbf{r}_i$ and quasi-convex in $\mathbf{r}_i^l$.

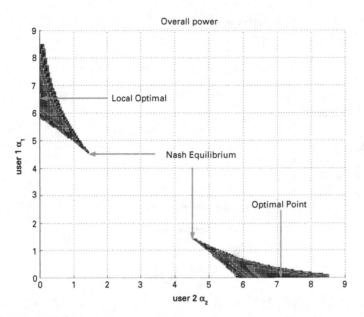

Figure 13.8 Two-user example: Multiple NEPs.

We consider that each user allocates its transmitted power to different subchannnels first. Because each subchannel can be allocated by P_{max} and overall transmitted power for all subchannels is linearly constraint by P_{max}, the supporting domain for power allocation is compact and convex. Because rate is a linear function of transmitted power if the interferences are fixed, the supporting domain Ω for r_i^l, $\forall l$, is a convex and compact subset of some Euclidean space $(\Re^+)^L$. It is worth mentioning that Ω^K is not convex, and one example is shown in Figure 13.6. But our proof needs only that Ω be convex and nonempty.

From (13.14) and (13.16), when the water filling is done for (13.20),

$$u_i = \sum_{l=1}^{L} \left(\frac{(2^{r_i^l} - 1)(\sum_{k \neq i} P_k^l G_{ki}^l + N_0)}{c_3^i G_{ii}^l} - \mu_i r_i^l \right). \tag{13.28}$$

Obviously it is continuous and convex for $\mathbf{r}_i$.   **QED**

THEOREM 26 *If the global minimum of (13.17) occurs when $r_i^l > 0$, $\forall A_{il} \neq 0$ and $\sum_{l=1}^{L} P_i^l < P_{max}$ and $\sum_{l=1}^{L} r_i^l = R_i$, $\forall i$, the NEP satisfies the necessary Karush–Kuhn–Tucker (KKT) condition [19].*

Proof 8 First, if $\sum_{l=1}^{L} P_i^l < P_{max}$ and $\sum_{l=1}^{L} r_i^l = R_i$, $\forall i$ at the NEP, the iterative water filling converges. For each user, the resource allocation is optimal if the interferences are considered noise. By the Lagrangian method, define $\nabla = \frac{\partial}{\partial \mathbf{r}_i}$; the following equation holds at the NEP when power is less than P_{max}:

$$\nabla \left(\sum_{l=1}^{L} P_i^l \right) - \mu_i \nabla \left(\sum_{l=1}^{L} r_i^l - R_i \right) = 0. \tag{13.29}$$

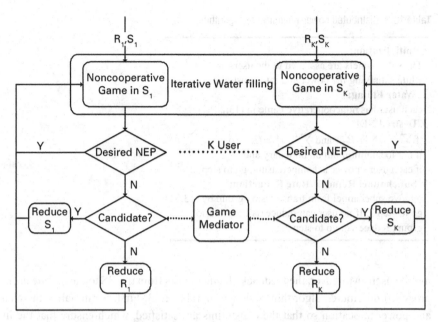

Figure 13.9 Multiuser noncooperative game.

For the problem in (13.17), if $r_i^l > 0$, $\forall A_{il} \neq 0$ and $\sum_{l=1}^{L} P_i^l < P_{\max}$, $\forall i$, the global optima will satisfy the KKT condition without considering the inequality constraints:

$$\sum_{i=1}^{K} \nabla \left(\sum_{l=1}^{L} P_i^l \right) - \sum_{i=1}^{K} \mu_i \nabla \left(\sum_{l=1}^{L} r_i^l - R_i \right) = 0. \qquad (13.30)$$

Obviously, when the iterative water-filling converges, (13.30) will be satisfied from (13.29). So the KKT necessary condition is satisfied for the NEP.   **QED**

Before developing the proposed algorithm, we analyze two extreme cases. In the first case, the groups of subchannels are assigned to different cells without overlapping such that there are no CCIs among cells. We call it the fixed channel-assignment scheme. However, this extreme method has the disadvantage of low spectrum efficiency because of the low-frequency reusage. In the second extreme case, all the users share all the subchannels. We call it the pure water-filling scheme. From Figures 13.7 and 13.8, we can see that the system can be balanced at the undesired point, because of the severe intercell CCIs. So the facts motivate us to believe that the optimal resource allocation is between these two extreme cases, i.e., each subchannel can be shared by only a group of selected users for transmission.

The block diagram of the proposed algorithm is shown in Figure 13.9. We define the subchannel set so that the ith users can allocate their rates as transmission group S_i. Each user plays the noncooperative game to minimize its power. If the game cannot converge to a good solution, a mediator is introduced to kick out some users from using the subchannel. If no candidate can be removed, the required rate has to be reduced. Here, we assume that BSs can accurately measure the channel gains of their associated

Table 13.3 Distributed power-minimization algorithm

---

1. **Initialization:**
   The subchannels are assigned to the users
   while satisfying the power and rate constraints.
2. **Water Filling:**
   each user has noncooperative game in (13.20).
3. **Desired NEP:**
   if $\sum_{l=1}^{L} P_i^l < P_{\max}$ and $\sum_{l=1}^{L} r_i^l = R_i$ and
   if not local minimum on boundary and
   if each user's power is nonincreasing, go to step 2;
4. **Subchannel Removal/Rate Reduction:**
   remove subchannel from transmission group by (13.31)
   go to step 2. If no user can reduce its transmission
   group, reduce R_i, go to step 2.

---

mobile users. Moreover, the feedback channel exists from the BS to the mobile user. The proposed distributed algorithm is shown in Table 13.3. First, we initialize the channel and power allocation so that the constraints are satisfied, which ensure that the initial point is located in the feasible set. This can be obtained by applying any heuristics or greedy approaches. Then, each user minimizes its own utility function, i.e., transmitted power, in a distributed game by applying water filling. The system will be balanced at some NEP.

If the CCIs are not large, the NEP should be the desired solution. Three criteria are used in the proposed algorithm to judge if the NEP is desired: (1) If the constraints of minimal rate and maximal transmitted power are satisfied; (2) if the NEP is not a local optimum; (3) if transmit power is decreasing during NEP convergence. From theorems and the previous observations, if any of these criteria are not satisfied, the system is probably balanced at an undesired solution. So a game mediator needs to redefine the game by reducing the number of users that share the subchannels, i.e., changing S_i.

The criterion to decide which user to be removed from using which subchannel is determined by the SINR level within transmission groups as

$$(l, j) = \arg\min_{l,j} \frac{P_j^l G_{jj}^l}{\sum_{k \neq j} P_k^l G_{kj}^l + N_0}. \tag{13.31}$$

Because of the minimal rate requirement, the selected user might not be able to be removed. Then the algorithm tries to select the next candidate in (13.31). The criterion for whether the user can be removed from the transmission group is determined by three factors: (1) Each user must have at least one subchannel to transmit; (2) no subchannel is wasted, i.e., at least one user is assigned for each subchannel; (3) the user cannot be kicked out from the subchannel, if the user cannot transmit its rate R_i using the rest of subchannels, even though he occupies them alone. If no user can be removed from the transmission group, the minimal rate R_i must be reduced. This situation happens when the system is very crowded and the CCI is large.

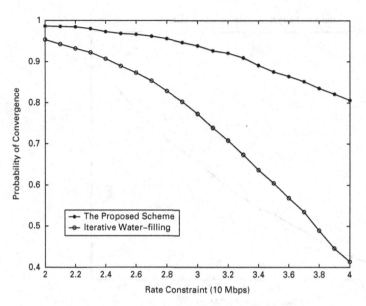

Figure 13.10 Convergence probability vs. rate constraint.

To show the improvements of the proposed algorithm, we set up two simulations consisting of a two-cell case and a seven-cell case. The BS is located at the center of each cell and one cochannel mobile per cell is generated as a uniform distribution within the corresponding cell for each simulation instance. The propagation model takes into consideration path loss and shadowing. The received signal (in dB) at distance d from the BS is $L(d) = L(d_0) + 10\alpha \log_{10} \frac{d}{d_0}$, where $d_0 = 10$ m is used as a reference point in measurements $[L(d_0) = 0$ dB$]$ and α is set to 3.5. Shadow fading for each user is modeled as an independent log-normal random variable with standard deviation $\sigma = 10$ dB. The four-path Rayleigh model is taken into consideration to simulate the frequency-selective fading channels and has an exponential power profile with 100-ns root-mean-square (RMS) delay spread. We consider a multicell OFDM system with 32 subchannels in total. The overall bandwidth is 6.4 MHz. The receiver thermal noise is −70 dBm. The required BER of the transmitted symbols is 10^{-3} for every subchannel and user, which corresponds to $c_3 = 0.2831$. We define the reuse factor R_u as the distance between two BSs D over the cell radius r that is set as 100 m, which is one of the main factors affecting the severity of CCI. To evaluate the performances, we simulate 10^3 sets of frequency-selective fading channels.

First, we show the simulation results for the two-cell case. Let $R_u = 2$ and assume $R_i = R_j, \forall i, j$. In Figure 13.10, we compare the probability of convergence to the desired NEP of the proposed algorithm with that of the pure water-filling algorithm. It shows that the proposed approach achieves a much higher convergence probability, especially with a high rate constraint. The reason is because of the proposed user-removal mechanism, which ensures that each subchannel is well utilized. Note that, if the convergence cannot be achieved, we decrease the rate constraint till convergence. In Figure 13.11, the number

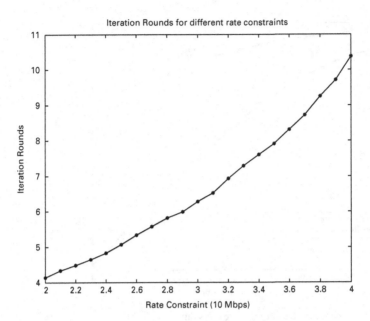

Figure 13.11 Convergence round vs. rate constraint.

of iteration rounds is plotted versus the rate constraint. We can see that the convergence speed is fast for all rates in the range. More rounds are necessary with larger interferences (higher rate constraint).

In Figure 13.12, we show the total transmitted power as the rate constraint R_i increases. Compared with the fixed assignment algorithm, the proposed algorithm reduces about 80% transmission power. The reduction is because the fixed assignment algorithm wastes too many resources by letting only one user occupy one subchannel. Compared with the pure water-filling algorithm, the proposed algorithm reduces about 25% transmission power. The reduction is because the pure water-filling algorithm makes some subchannels overcrowded.

In Figure 13.13, we show the number of users occupying each subchannel when the rate constraint is increased. The fixed channel-assignment algorithm always has only one user per channel (UPC). The proposed algorithm has a lower UPC compared with that of the pure water-filling algorithm. For a pure water-filling algorithm, some subchannels may not be allocated any power when the rate constraint is small, because of the low water-filling level. For the proposed algorithm, more users are kicked off certain subchannels when the rate constraint becomes large.

The simulation results for seven-cell networks are shown as follows. The rate constraint is set as 10 Mbits for each user. In Figures 13.14 and 13.15, the total transmission power and UPC are compared between the proposed scheme and the pure water-filling algorithm as the reuse factor increases, respectively. We can see that the proposed algorithm can reduce the overall power about 90% when the CCIs are severe ($R_u = 2$), which will greatly improve the system performance. Also the proposed

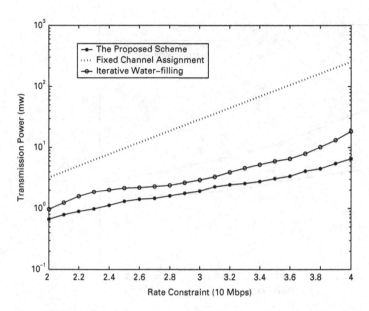

Figure 13.12 Total power vs. rate constraint.

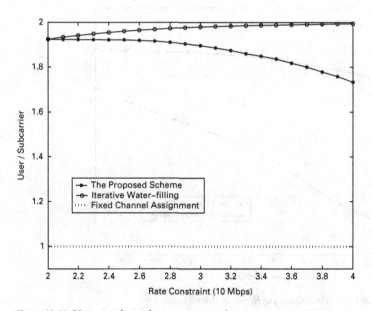

Figure 13.13 User per channel vs. rate constraint.

scheme kicks more users out and reduces the number of users per subchannel for a smaller reuse factor. When R_u increases, the CCIs are reduced. Consequently, two schemes show similar performances.

The goal of this referee-based approach is power minimization under the constraints of minimal rate and maximal transmitted power in multicell OFDM systems. We develop a

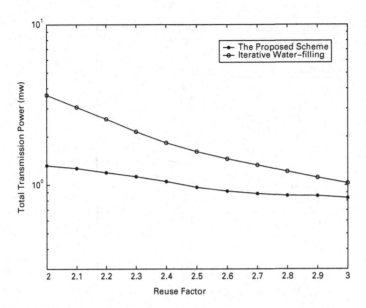

Figure 13.14 Overall power vs. R_u for multicell case.

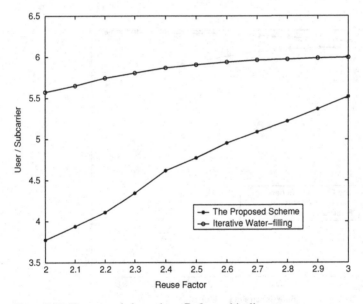

Figure 13.15 User per subchannel vs. R_u for multicell case.

distributed game-theory approach to adaptively assign the subchannels, rate, and power. From the simulation results, the proposed distributed algorithm reduces the overall transmitted power up to 80% compared with the fixed assignment scheme for the two-cell case, and up to 90% compared with the pure water-filling scheme for the seven-cell case in which the CCIs are severe. As a result, the system performance can be greatly improved.

13.2.3   Buyer/Seller Approach

In this subsection, we propose a Stackelberg game-theoretic framework [323, 324] for distributive resource allocation over multiuser cooperative communication networks to improve the system performance and stimulate cooperation. We try to answer two questions: Who should relay and how much power for relaying? We employ a two-level game to jointly consider the benefits of source nodes as buyers and relay nodes as sellers in cooperative communication. From the derived results, the proposed game not only helps the source smartly find relays at relatively better locations but also helps the competing relays ask reasonable prices to maximize their own utilities. From the simulation results, the relays in good locations or good channel conditions can play more important roles in increasing the source node's utility, so the source would like to buy power from these preferred relays. On the other hand, because of competition from other relays and selections from the source, the relays have to set the proper price to attract the source's buying so as to optimize their utility values.

Recently, cooperative communications have gained a lot of attention as an emerging transmit strategy for future wireless networks [175, 272, 273]. The basic idea is that the relay nodes can help the source node's transmission by relaying a copy of the information. The cooperative communications efficiently take advantage of the broadcasting nature of wireless networks, as well as exploit the inherent spatial and multiuser diversities. Many recent works proposed some protocols for different layers of networks. The work in [138] analyzed with more complicated transmitter cooperative schemes involving dirty paper coding. The energy-efficient transmission was considered for broadcast networks in [199]. [195] considered the design of a cooperative relay strategy by exploiting the finite-alphabet property of the source. In [261], the relay-assignment problem is solved for multiuser cooperative communications. In [104], cooperative resource allocation for OFDM is studied.

However, most existing work focuses on a resource allocation by a centralized fashion. To achieve the distributed implementation, game theory is a natural, flexible, and rich tool to study how the autonomous nodes interact and cooperate with each other. Some game-theory literature of the wireless networking is listed as follows. In [196], the behaviors of selfish nodes in the case of random access and power control were examined. In [55] static pricing policies for multiservice networks were proposed to offer the needed incentives for each node to choose the service that best matches her needs, thereby discouraging overallocation of resources and improving social welfare.

Here, we consider how to employ game theory for the distributed nodes to optimize performance within the cooperative communication paradigm. Two main resource-allocation questions about cooperative multiuser wireless networks still remain unanswered: First, among all the distributed nodes, who can help relay and better improve the source's link quality? Second, for the selected relay nodes, how much power do they need to transmit? Both questions need to be answered in a distributed way.

To answer these questions, we employ the Stackelberg game [380] to jointly consider the benefits of source nodes and relay nodes in cooperative communication. The game is divided into two hierarchical levels: The source node plays the *buyer-level* game and the relay nodes play the *seller-level* game. Each player is selfish and wants to maximize its own benefit. Specifically, the source can be viewed as a buyer and it aims to get most benefits at the least possible payment. Each relay can be seen as a seller who aims to earn the payment. The payment not only covers their forwarding cost but also gains as much extra profit as possible. Then we derive the expressions for the proposed game outcomes. We analyze how many relay nodes would be selected by the source to participate in the sale process after the relay nodes announce their optimal prices. In addition, we optimize how much service the source should buy from each relay node. From the seller point of view, the relay nodes set the corresponding optimal price per unit of the service such as relay power so as to maximize its own benefit. From the simulations, because of competition from other relays and selections from the source, the relays have to set the proper price to attract the source's buying so as to optimize their utility values. The source optimally selects the relays and their relaying power, while the relays set the prices that can maximize their utilities.

We use the AF cooperative protocol as an example. Other cooperative protocols such as DF can be applied in a similar way. The relay nodes help the source node by relaying the received information to the destination. The receiver at the destination combines together the directly received signal from the source node and the relayed signals from the relay nodes, using techniques such as MRC. The preceding procedure can be described in two phases, as subsequently discussed.

At Phase 1, without the relay node's help, the SNR that results from direct transmission from the source s to the destination d can be expressed by

$$\Gamma_{s,d} = \frac{P_s G_{s,d}}{\sigma^2},\tag{13.32}$$

where P_s represents the transmit power, $G_{s,d}$ is the channel gain, and σ^2 is the noise variance. The rate at the output of noncooperative transmission is

$$R_{s,d}^{\mathrm{nc}} = W \log_2 \left(1 + \frac{\frac{P_s G_{s,d}}{\sigma^2}}{\Gamma}\right),\tag{13.33}$$

where Γ is a constant for the capacity gap. Without loss of generality, we assume that the noise variance is the same for all links. We also assume the channels are constant over each power-control interval.

At Phase 2, we consider the SNR at the destination that results from relay r_i relaying source s's data to the destination. By assuming that X_{s,r_i} is the transmitted signal from source s to relay r_i, the received signal at relay r_i is

$$R_{s,r_i} = \sqrt{P_s G_{s,r_i}} X_{s,r_i} + \eta,\tag{13.34}$$

where $\eta \sim N(0, \sigma^2)$ and σ^2 is the noise variance. Relay r_i amplifies R_{s,r_i} and relays it to the destination in which the received signal is

$$R_{r_i,d} = \sqrt{P_{r_i} G_{r_i,d}} X_{r_i,d} + \eta,\tag{13.35}$$

where

$$X_{r_i,d} = \frac{R_{s,r_i}}{|R_{s,r_i}|} \qquad (13.36)$$

is the transmitted signal from relay r_i to the destination, and the signal is normalized to have unit energy. Substituting (13.34) into (13.36), we can rewrite (13.35) as

$$R_{r_i,d} = \frac{\sqrt{P_{r_i} G_{r_i,d}}(\sqrt{P_s G_{s,r_i}} X_{s,r_i} + \eta)}{\sqrt{P_s G_{s,r_i} + \eta^2}} + \eta. \qquad (13.37)$$

Using (13.37), the relayed SNR for the source s, which is helped by relay r_i, is given by

$$\Gamma_{s,r_i,d} = \frac{P_{r_i} P_s G_{r_i,d} G_{s,r_i}}{\sigma^2(P_{r_i} G_{r_i,d} + P_s G_{s,r_i} + \sigma^2)}. \qquad (13.38)$$

Therefore, by (13.32) and (13.38), we have the rate at the output of MRC via relay r_i in AF as

$$R_{s,r_i,d}^{AF} = W \log_2 \left(1 + \frac{\frac{P_s G_{s,d}}{\sigma^2} + \frac{P_{r_i} P_s G_{r_i,d} G_{s,r_i}}{\sigma^2(P_{r_i} G_{r_i,d} + P_s G_{s,r_i} + \sigma^2)}}{\Gamma} \right). \qquad (13.39)$$

If there are N relays helping the source, then

$$R_{s,r,d}^{AF} = W \log_2 \left(1 + \frac{\frac{P_s G_{s,d}}{\sigma^2} + \sum_{i=1}^{N} \frac{P_{r_i} P_s G_{r_i,d} G_{s,r_i}}{\sigma^2(P_{r_i} G_{r_i,d} + P_s G_{s,r_i} + \sigma^2)}}{\Gamma} \right). \qquad (13.40)$$

To explore the cooperative diversity for a multiuser system, from (13.40), two fundamental questions should be answered: First, which relay nodes should be included, and second, what is the optimal power P_{r_i}? To answer the questions, we employ the Stackelberg game for buyers and sellers as the following formulated problem.

1. Source/Buyer: The source can be modeled as a buyer, and it aims to get the most benefits for the least possible payment. So the utility function of the source can be defined as

$$U_s = a \Delta R_{tot} - M, \qquad (13.41)$$

where

$$\Delta R_{tot} = R_{s,r,d} - R_{s,d} \qquad (13.42)$$

denotes the total rate increment with the relay nodes helping transmission, a denotes the gain per unit of rate increment at the MRC output, and

$$M = p_1 P_{r_1} + p_2 P_{r_2} + \cdots + p_N P_{r_N} \qquad (13.43)$$

represents the total payment paid by the source to the relay nodes. In (13.43), p_i represents the price per unit of power selling from relay node i to the source s, and P_{r_i} denotes how much power the source would like to buy from relay r_i when the prices are announced from that relay.

We assume the number of relay nodes is N. Without loss of generality the parameter a in (13.41) can be set to 1, and the optimization problem or the buyer's game for the buyer can be formulated as

$$\max_{\{P_{r_i}\}} U_s = \Delta R_{\text{tot}} - M, \tag{13.44}$$

$$\text{subject to } \{P_{r_i}\} \geq 0. \tag{13.45}$$

2. Relays/Seller: Each relay r_i can be seen as a seller who aims to earn the payment that not only covers its forwarding cost but also wants as much extra profit as possible. We introduce one parameter c_i, "the cost of power for relaying data," in our formulation to correctly reflect relays' judgment about whether they can actually profit by the sale. Then relay r_i's utility function can be defined as

$$U_{r_i} = (p_i - c_i)P_{r_i}, \tag{13.46}$$

where c_i is the cost per unit of power in relaying data, p_i has the same meaning as in (13.43), and P_{r_i} is the source's decision by optimizing U_s described in (13.44). It is obvious that determining the optimal p_i depends not only on each relay's own channel condition to the destination but also on its counterpart relays' prices. So, in the sellers' competition, if one relay asks a higher price than what the source expects about it after jointly considering all relays' prices, the source will buy less from that relay or even disregard that relay. On the other hand, if the price is too low, the profit obtained by (13.46) will be unnecessarily low. So there is a trade-off for setting the price. Without loss of generality, set $c_i = c$ in (13.46) and the optimization for relay r_i or the buyer's game is

$$\max_{\{p_i\}>0} U_{r_i} = (p_i - c)P_{r_i}, \quad \forall i. \tag{13.47}$$

Therefore the ultimate goals of the preceding two games are to decide the optimal pricing p_i to maximize relays' profits U_{r_i}, the actual number of relays who will finally get selected by the source, and the corresponding optimal power consumption P_{r_i} to maximize U_s. Notice that the only signalings required to exchange between the source and relays are the price p_i and the information about how much power P_{r_i} to buy. Consequently, the proposed two-level game approach can have distributed resource allocation for the cooperative communication networks. The outcome of the games will be shown in detail.

We give some observations of the U_s function with respect to $\{P_{r_i}\}$. When P_{r_i} is close to 0, less help is received from the relay, so U_s should be close to 0. As P_{r_i} increases, relays sell more power to the source, hence more rate increment is obtained, and U_s increases. If P_{r_i} further increases, the speed of the rate increment should gradually saturate but the cost still keeps growing; hence the utility of the source U_s begins to decrease. Assume the selling price of the relays' power p_i, $i = 1, 2, \ldots, N$, has been announced; then from the first-order optimization condition, the following optimality condition must hold at the optimal point:

$$\frac{\partial U_s}{\partial P_{r_i}} = 0, \quad i = 1, 2, \ldots, N. \tag{13.48}$$

For simplicity, define $C = 1 + \frac{P_s G_{s,d}}{\sigma^2 \Gamma}$, $W' = \frac{W}{\ln 2}$; then, by (13.40) and (13.33),

$$
\begin{aligned}
\Delta R_{\text{tot}} &= R_{s,r,d} - R_{s,d} \\
&= W \log_2 \left(C + \frac{\sum_{i=1}^{N} \frac{P_{r_i} P_s G_{r_i,d} G_{s,r_i}}{\sigma^2 \left(P_{r_i} G_{r_i,d} + P_s G_{s,r_i} + \sigma^2 \right)}}{\Gamma} \right) - W \log_2 C \\
&= W \log_2 \left(1 + \frac{\sum_{i=1}^{N} \frac{P_{r_i} P_s G_{r_i,d} G_{s,r_i}}{\sigma^2 \left(P_{r_i} G_{r_i,d} + P_s G_{s,r_i} + \sigma^2 \right)}}{\Gamma C} \right) \\
&= W \log_2 \left(1 + \frac{\Delta \text{SNR}_{\text{tot}}}{\Gamma C} \right) = W' \ln \left[1 + \Delta \text{SNR}'_{\text{tot}} \right] \\
&= W' \ln \left[1 + \sum_{i=1}^{N} \Gamma'_{r_i,d} \right],
\end{aligned}
\tag{13.49}
$$

where

$$
\Delta \text{SNR}'_{\text{tot}} = \frac{\sum_{i=1}^{N} \frac{P_{r_i} P_s G_{r_i,d} G_{s,r_i}}{\sigma^2 \left(P_{r_i} G_{r_i,d} + P_s G_{s,r_i} + \sigma^2 \right)}}{\Gamma C}
\tag{13.50}
$$

and

$$
\Gamma'_{r_i,d} = \frac{\Gamma_{r_i,d}}{\Gamma C} = \frac{\frac{P_s G_{s,r_i}}{(\Gamma \sigma^2 + P_s G_{s,d})}}{1 + \frac{P_s G_{s,r_i} + \sigma^2}{G_{r_i,d}}} = \frac{A_i}{1 + \frac{B_i}{P_{r_i}}} = \frac{A_i P_{r_i}}{P_{r_i} + B_i},
\tag{13.51}
$$

with $A_i = \frac{P_s G_{s,r_i}}{(\Gamma \sigma^2 + P_s G_{s,d})}$ and $B_i = \frac{P_s G_{s,r_i} + \sigma^2}{G_{r_i,d}}$.

Substituting (13.43) and (13.49) into (13.48), we have

$$
\frac{\partial U_s}{\partial P_{r_i}} = \frac{W'}{\left(1 + \sum_{k=1}^{N} \frac{A_k P_{r_k}}{P_{r_k} + B_k} \right)} \frac{A_i B_i}{\left(P_{r_i} + B_i \right)^2} - p_i = 0,
\tag{13.52}
$$

i.e.,

$$
\frac{W'}{\left(1 + \sum_{k=1}^{N} \frac{A_k P_{r_k}}{P_{r_k} + B_k} \right)} = \frac{p_i}{A_i B_i} \left(P_{r_i} + B_i \right)^2.
\tag{13.53}
$$

So

$$
\frac{p_i}{A_i B_i} \left(P_{r_i} + B_i \right)^2 = \frac{p_j}{A_j B_j} \left(P_{r_j} + B_j \right)^2,
\tag{13.54}
$$

$$
P_{r_j} = \sqrt{\frac{p_i A_j B_j}{p_j A_i B_i}} \left(P_{r_i} + B_i \right) - B_j.
\tag{13.55}
$$

Then, after some manipulation, we have

$$\Gamma'_{r_j,d} = \frac{A_j}{1 + \frac{B_j}{P_{r_j}}} = A_j - \sqrt{\frac{p_j A_i B_i}{p_i A_j B_j}} \frac{A_j B_j}{(P_{r_i} + B_i)}, \tag{13.56}$$

so

$$
\begin{aligned}
\Delta\text{SNR}'_{\text{tot}} &= \sum_{i=1}^{N} \Gamma'_{r_i,d} \\
&= \left[A_1 - \sqrt{\frac{p_1 A_i B_i}{p_i A_1 B_1}} \frac{A_1 B_1}{(P_{r_i} + B_i)} \right] + \left[A_2 - \sqrt{\frac{p_2 A_i B_i}{p_i A_2 B_2}} \frac{A_2 B_2}{(P_{r_i} + B_i)} \right] \\
&\quad + \cdots + \left[A_i - \frac{A_i B_i}{P_{r_i} + B_i} \right] + \cdots + \left[A_N - \sqrt{\frac{p_N A_i B_i}{p_i A_N B_N}} \frac{A_N B_N}{(P_{r_i} + B_i)} \right] \\
&= \sum_{j=1}^{N} A_j - \sqrt{\frac{A_i B_i}{p_i}} \frac{1}{P_{r_i} + B_i} \sum_{j=1}^{N} \sqrt{p_j A_j B_j}.
\end{aligned}
\tag{13.57}
$$

Substituting (13.57) into (13.53), we have a quadratic equation of P_{r_i} as

$$
\left(1 + \sum_{j=1}^{N} A_j \right) \left[\sqrt{\frac{p_i}{A_i B_i}} (P_{r_i} + B_i) \right]^2
$$
$$
- \sum_{j=1}^{N} \sqrt{p_j A_j B_j} \left[\sqrt{\frac{p_i}{A_i B_i}} (P_{r_i} + B_i) \right] - W' = 0.
\tag{13.58}
$$

Therefore we can solve the generalized form solution for the optimal power consumption from each relay node as

$$
P_{r_i} = -B_i + \frac{\sqrt{\frac{A_i B_i}{p_i}}}{2 \left(1 + \sum_{j=1}^{N} A_j \right)} \left[\sum_{j=1}^{N} \sqrt{p_j A_j B_j} \right.
$$
$$
\left. + \sqrt{\left(\sum_{j=1}^{N} \sqrt{p_j A_j B_j} \right)^2 + 4 \left(1 + \sum_{j=1}^{N} A_j \right) W'} \right].
\tag{13.59}
$$

However, the preceding solution may be negative for some relay's high price or bad location. Therefore the optimal price should be modified as follows:

$$P^*_{r_i} = \max(P_{r_i}, 0) = (P_{r_i})^+, \tag{13.60}$$

where P_{r_i} is solved in (13.59).

Substituting (13.59) into (13.47), we have

$$\max_{\{p_i\}>0} U_{r_i} = (p_i - c) P_{r_i}(p_1, \ldots, p_i, \ldots, p_N). \tag{13.61}$$

Note that this is a noncooperative game and there exists a trade-off between the price p_i and relay's utility U_{r_i}. If the relay asks for a relatively lower price p_i at first, the source would be glad to buy more power from the cheaper seller and U_{r_i} will increase as p_i grows. When p_i keeps growing, the source would think it is no longer profitable to buy power from the relay and P_{r_i} will shrink, resulting in a decrease of U_{r_i}. Therefore there is an optimal price for each relay to ask for, and the optimal price is also affected by other relays' prices because the source chooses only the most beneficial relays among all the relays.

From the preceding analysis, by the first-order optimality condition, it follows that

$$
\begin{aligned}
&P_{r_i}(p_1, \ldots, p_i, \ldots, p_N) \\
&+ (p_i - c)\frac{\partial P_{r_i}(p_1, \ldots, p_i, \ldots, p_N)}{\partial p_i} = 0, \quad i = 1, 2, \ldots, N.
\end{aligned}
\tag{13.62}
$$

Solving (13.62) for N unknowns p_i, we have

$$
p_i^* = p_i^*(\sigma^2, \{G_{s,r_i}\}, \{G_{r_i,d}\}), \quad i = 1, 2, \ldots, N.
\tag{13.63}
$$

The problem in (13.62) can be solved by some numerical methods such as sequential quadratic programming (SQP) [218].

As we have mentioned before, substitute (13.63) into (13.59) to see whether P_{r_i} is positive. If it is negative, then the source will disregard the relay with negative P_{r_i} and only the remaining relays constitute the actually relaying subset. Re-solve (13.52) by changing the set of relay nodes to the subset previously solved, and re-solve the new corresponding p_i^*, then check the P_{r_i} until all P_{r_i} are positive. Then we can get the final optimal pricing p_i^* to maximize the relays' utilities U_{r_i}, the actual number of relays that will get selected by the source and the corresponding optimal power consumption $P_{r_i}^*$ to maximize U_s.

To evaluate the performances of the proposed scheme and decide what price each relay should ask for and how much power the source should buy from each relay, we performed simulations for multiple relay systems. In what follows, the simulation results for a one-relay case, a two-relay case, and a multiple-relay case are shown.

We set simulations of the first part as follows. There are one source–destination pair and one relay in the network. The destination is located at coordinate $(0, 0)$, the source is fixed at coordinate $(1, 0)$, and the relay is randomly located within the range of $[-2, 3]$ on the x axis and $[-1, 1]$ on the y axis. The propagation-loss factor is set to 2. The noise level is $\sigma^2 = 10^{-4}$ and we select the capacity gap $\Gamma = 1$, cost per unit of power $c = 0.05$.

In Figure 13.16, we show the optimal price the relay should ask for and the optimal power bought by the source. In this simulation, the relay moves in the rectangular area. We observe that, when the relay is close to the source at $(1, 0)$, it can be more efficient to help the source transmit, so the relay would reduce the price to attract the source to buy more service. When the relay moves close to the destination at $(0, 0)$, it can use a very small amount of power to relay the source's data, so it will set a very high price in to get more profit by selling this small power. When the relay keeps moving away from the destination, the source would stop buying service because the relay is in such a bad location that asking it to help will be no longer beneficial to the source. Similarly, when the relay moves in the opposite direction and locates very far away from the source, the source would not buy service either.

In Figure 13.17, we show the optimal utility the source and the relay can get using the proposed scheme. When the relay is located close to the source, both the relay and the source can get the maximal utility. The reason is that, around this location, the relay

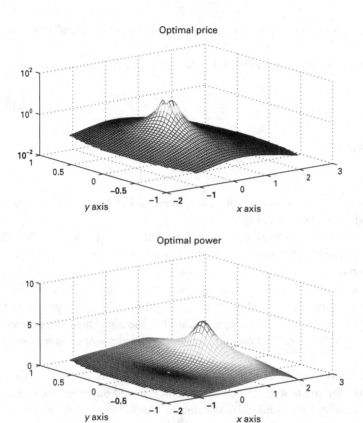

Figure 13.16 Optimal price and power of the relay in different locations.

can most efficiently help the source increase its utility, and the optimal price of the relay is very low compared with that when the relay is at other locations. So the source would like to buy more power. It is worth mentioning that both the x axis and the y axis represent nodes' channel conditions by using locations. As a result, the question of what price the relay should ask for and the question of how much utility both game players can achieve can be clearly illustrated. If channel effects such as shadowing and fading are considered, both the x axis and the y axis can be channel conditions and are not related to the distances.

We set up two-relay simulations to test the proposed scheme. In our simulations, the coordinates of the source and the destination are $(1, 0)$ and $(0, 0)$ respectively. Relay 1 is fixed at the coordinate $(0.5, 0.25)$ and relay 2 moves along the line from $(-2, 0.25)$ to $(3, 0.25)$. Other settings are the same as those for the one-Relay case.

In Figure 13.18 we show the optimal price that each relay should ask to maximize its profit. We can observe that even though only relay 2 moves, the prices of both relays change accordingly. This is because two relays compete and influence each other in the proposed Stackelberg games. When relay 2 is close to the destination [at $(0, 0)$], it can use very little power to relay the source's information. So relay 2 can set a very high price, hoping to get more profit by selling small power. When relay 2 is close to the

Optimal U_r

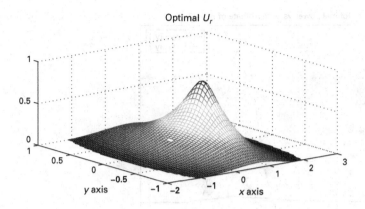

Optimal U_s

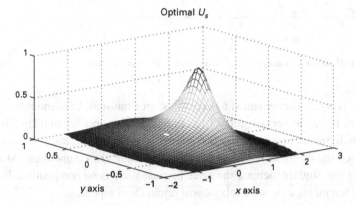

Figure 13.17 Optimal utilities of the relay and the source in different locations.

Optimal prices vs. x coordinate of relay 2

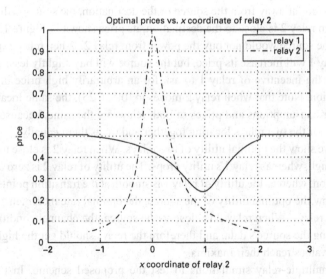

Figure 13.18 Optimal relays' prices when relay 2 moves.

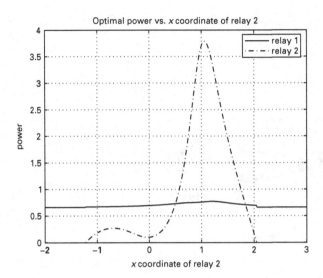

Figure 13.19 Optimal power consumptions of two relays when relay 2 moves.

source at $(1, 0)$, relay 2 is better suited to help the source transmit. Consequently, in to attract the source to buy its service, relay 1 has to reduce the price. When relay 2 is far away, its price will drop because it is less competitive compared with relay 1 at location $(0.5, 0.25)$. When the utility is less than 0, relay 2 will quit the competition. At that moment, relay 1 can slightly increase the price because there is no competition. But it cannot increase too much, otherwise relay 2 will rejoin the competition.

As shown in Figure 13.19, the source will smartly buy different amounts of power from the two relays. When relay 2 moves away from the source, $P_{r_2}^*$ gradually decreases. When relay 2 moves too far away from the source or the destination, the source will not choose relay 2. When relay 2 is close to the destination, its price shown in Figure 13.18 is too high, so that the source would not buy the power from relay 2. When relay 2 quits the competition, relay 1 will increase its price, but the source will buy slightly less. This fact also suppresses the incentive of relay 1 to ask for an arbitrarily high price in the absence of competition. Note that when relay 2 moves to $(0.5, 0.25)$, the same location as relay 1, the power consumptions and prices of both relays are the same, because the source is indifferent for the two relays locating together and treats them equally.

In Figure 13.20, we show the optimal utility of two relays. When relay 2 is close to the source, its utility is high, whereas relay 1's utility drops. The utility of relay 2 is zero after it quits the competition, whereas the utility of relay 1 is smooth at the transition points. In Figure 13.21, we show the optimal utility of the source, the optimal rate increment, and total payment to the relays. When relay 2 is close to the source, the channel conditions are the best in relaying the source's data, and therefore the relays should get the highest profits and all three values reach their maxima.

We then set up multiple-relay simulations to test the proposed scheme. In these simulations, the coordinates of the source and the destination are $(1, 0)$ and $(0, 0)$ respectively, and the relays are randomly located within the range of $[-2, 3]$ in the x

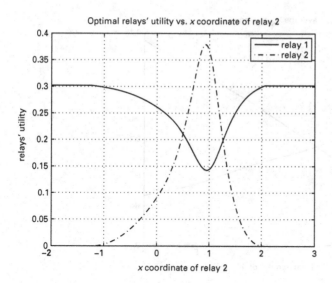

Figure 13.20 Optimal relays' utilities when relay 2 moves.

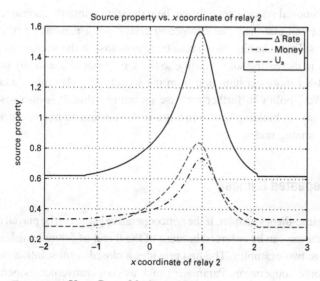

Figure 13.21 U_s, $\Delta Rate$, M when relay 2 moves.

axis and $[-2, 2]$ in the y axis. From Figure 13.22, we can observe that, as the total number of available relays increases, the source will get a higher utility. However, in this way, the competition among relays becomes more severe, which leads to less average payment from the source.

In summary, we propose the game-theory approach for distributive resource allocation over multiuser cooperative communication networks. We target answering two questions: Who will be the relays and how much power is needed for relaying in the AF cooperative scenario? We employ the Stackelberg game to jointly consider the benefits of different

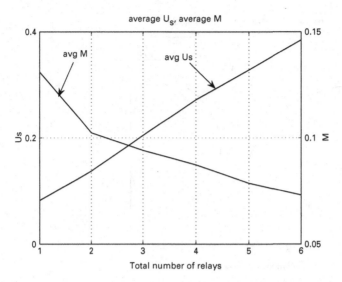

Figure 13.22 Optimal source utility and average money transfer vs. number of relay nodes.

types of nodes. The proposed scheme not only can help the source smartly choose relays at better locations but can also help the competing relays ask a reasonable price to maximize their utilities. From the simulation results, relays close to the source can play a more important role in increasing source utility, so the source would like to buy power from the relay. To attract more consumption from the source, the relay might adapt a "low-price, high-market" policy to further increase its utility value. It is also easy to use the current structures as building blocks in large-scale wireless ad hoc networks to stimulate cooperation among nodes.

13.3   Noncooperative Repeated Games

As mentioned in the game-theory chapter, if the noncooperative game can be played multiple times, the cooperation can be enforced because of the threat of future punishment. In this section, we give two examples. The first provides a closed-form solution on the exact strategy to enforce cooperation. Parameters such as cooperative/noncooperative strategies, punishment method, and punishment time are derived. In the second example, we notice that the distributed nodes might not know how to cooperate even though they are willing to cooperate. We propose some self-learning algorithms to study how to cooperate.

13.3.1   Punishment-Based Approach

In this subsection, we consider wireless networks with multiple distributed users. The users transmit their data packets to a communication node via some multiaccess protocol. Then the communication node transmits the packets to a remote destination via a lossy wireless link. The cost for each user's transmission is modeled as a function of its transmission rate. If a user's packet is successfully transmitted to the destination, a profit will be obtained. However, this profit is reduced by the transmission contention among users and the lossy wireless channel. Each user controls its transmission rate to maximize its profit in a distributed manner. If no rule is applied for rate control, each user will select a large transmission rate, resulting in having a low performance. To improve the system performance, we develop a distributed rate-control algorithm using the optimal Cartel Maintenance strategy. In this strategy, each user prefers to cooperate for rate control, because deviation from cooperation will be punished by other users for a period of time. Because the users want to optimize their performances over time, there is no incentive for them to deviate to gain a benefit while they have to endure more punishment in the future. From the simulation results, we can see that the proposed distributed scheme can enforce cooperation among users and achieve a much better performance than that of the noncooperative transmission.

In wireless networks, there exists competition among users to share the system resources. Efficient resource allocation such as rate control is an important technology used to increase the system performance by controlling users' appetite for resources. In some wireless networks with scattered topology, resource-allocation schemes must be implemented in a distributed way. Moreover, dynamic optimization over time is necessary for the wireless system with fluctuating channels and long-term goals. The preceding aspects challenge the design of a dynamic distributed resource-allocation scheme, which has become an important research topic recently.

Dynamic resource allocation in different wireless networks was discussed in detail in [355]. Because individual mobile users do not have the knowledge of other users' conditions and cannot cooperate with each other, they act selfishly to maximize their own performances in a distributed fashion. Such a fact motivates us to adopt game theory [80]. Repeated game theory analyzes the behaviors of users in multiple stages, so it can be applied to analyze the dynamic optimization of wireless resource allocation. In [173], repeated game theory was applied to routing problems. In [197], multiple-access resource allocation was studied using the game-theory approach. Repeated games were then further applied to physical-layer problems.

Here we consider a special case of wireless networks. Suppose there are many distributed users in a local area. They share a communication node to communicate with a remote destination. There are costs for the users to transmit their packets and also benefits if their packets are successfully transmitted. Successful probability is affected by two factors: the channel condition from the communication node to the destination and competition among users to share the communication node. So the problem is how to control each user's transmission rate in a distributed manner such that the overall profit (i.e., benefit minus cost) will be maximized. The preceding wireless networks fit a variety of practical situations such as wireless sensor networks, in which a strong communication node collects sensors' data and transmits them back.

To solve the preceding problem, we are inspired by the microeconomy approach in [238]. We propose a scheme in which users agree to cooperate at a certain transmission rates first. At each time, users will observe the probability of a successful transmission.

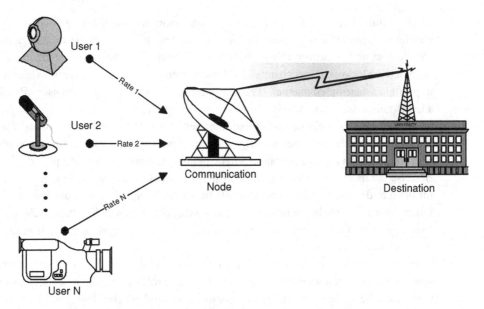

Figure 13.23 Block diagram of wireless networks.

If the probability is lower than some threshold, it probably means some users deviated from the agreed transmission rates and caused more contention in the communication node. Under this condition, the other users will transmit noncooperatively at much higher rates for a period of time. Consequently the probability of successful transmissions drops dramatically. Because users optimize their rates over time, the gain for the deviation will be wiped out by the loss caused by the punishment. For this reason, all users have no incentive to deviate from the agreed transmission rates. So the proposed scheme can force the users to cooperate in a distributed manner.

Figure 13.23 shows the block diagram of multiuser wireless networks. There are many distributed users and one communication node. Each user can transmit its data packets to the communication node by using multiple-access protocols such as Aloha, CSMA, etc. The communication node has the ability to transmit the data packets to the remote destination via a wireless link. We assume there is a reliable feedback channel. So the system can be described as multiple users sharing a communication link. Without loss of generality, we assume users are homogenous. (Heterogeneous cases can be extended in a similar way.) Each user can control its transmission rate. The users need to compete with each other for the communication link that is fluctuating because of the wireless channel. Thus one user's rate affects the performances of other users as well. So it is necessary to find a rate-control algorithm such that the system can operate at the optimal point. Moreover, it is hard to have communication channels among users. A distributed algorithm is required for rate control.

There are N users in wireless networks. The transmission time for the data packets is divided into time slots. The users transmit their packets by Poisson distribution. The average transmission rate vector for all users is denoted by $\vec{\lambda}^t = [\lambda_1^t, \ldots, \lambda_N^t]'$, where

λ_i^t is the rate of user i in period t. The total arriving rate at the communication node is then $\Lambda^t = \sum_{i=1}^{N} \lambda_i^t$. Each user intends to increase its transmission rate. However, arbitrary increases of the transmission rates will result in a higher probability of collision at the communication node and reduce the system throughput. In addition, the probability of successful transmission at the communication node is also affected by the wireless link quality from the communication node to the destination. The overall probability of successful transmission can be observed by all users and can be expressed as

$$\hat{P}_t = P(\Lambda^t)\theta_t, \tag{13.64}$$

where $P : \Re_+ \to \Re_+$ denotes the system throughput that is a function of Λ^t for the multiple-access protocol. θ_t is the channnel-error probability, which is an identically and independently distributed sequence of random variables with mean μ, PDF f, and cumulative distribution function (CDF) F. For simplicity, we approximate P as a linear and decreasing function in total arriving rate Λ^t as

$$P(\Lambda^t) = a - b\Lambda^t, \tag{13.65}$$

where a and b are positive constants. This approximation fits the scenarios of highly loaded wireless networks, for which the attempted transmission at each time slot is high. Other approximations can also be applied in a similar way.

For each user, the cost function for the transmissions is hypothesized to be homogeneous as

$$C_t = c_0 + c_1\lambda_i^t, \tag{13.66}$$

where c_0 is the basic cost to maintain the link and c_1 is the cost per transmission rate.

Then we omit subscript t for simplicity. For each successful transmission, the user has the benefit of c_2. User i has the profit as the benefit minus cost as

$$\pi_i(\overrightarrow{\lambda}) = c_2\hat{P}_t\lambda_i - C = c_2\theta(a - b\Lambda)\lambda_i - c_0 - c_1\lambda_i. \tag{13.67}$$

Given $\Lambda_{-i} = \Lambda - \lambda_i = \sum_{j \neq i} \lambda_j$, the total arrival rate of the other users, the single-time-slot expected profit can be further represented as

$$\pi_i(\overrightarrow{\lambda}) = [A - B(\Lambda_{-i} + \lambda_i)]\lambda_i - c_0, \tag{13.68}$$

where $A = \mu a - c_1$ and $B = \mu b$. It is assumed that

$$0 < c_1 < \mu a, \tag{13.69}$$

$$0 < c_0 < (\mu a - c_1)^2/\mu b(N + 1)^2. \tag{13.70}$$

(13.69) and $b > 0$ imply that A and B are positive constants.

Next, we first present the motivation of our proposed strategy. Then we formulate the problem and construct a distributed algorithm for each user. Finally, the optimal parameters of the algorithm are deduced.

Because users are located distributively, they act noncooperatively and independently to increase their profits by adapting their rates, i.e.,

$$\arg \max_{\lambda_i} \pi_i. \tag{13.71}$$

Let $\vec{s} = [s_1, \ldots, s_N]'$ denote the optimal rate vector for the preceding noncooperative optimization. By taking the derivative, we have

$$s_i = s = A/B(N+1), \ \forall i, \tag{13.72}$$

and

$$\pi_i(\vec{s}) = [A^2/B(N+1)^2] - c_0, \ \forall i. \tag{13.73}$$

(13.70) guarantees that users earn positive profits.

On the other hand, if there exists a centralized control, users can cooperate to maximize the system's overall profit. Then the optimization goal is

$$\arg \max_{\vec{\lambda}} \sum_{i=1}^{N} \pi_i. \tag{13.74}$$

Denote the rate vector that maximizes expected joint overall profit by $\vec{r} = [r_1, \ldots, r_N]'$. Given Λ_{-i}, we have the solutions as

$$r_i = r = A/2BN, \ \forall i \tag{13.75}$$

and

$$\pi_i(\vec{r}) = (A^2/4BN) - c_0, \ \forall i. \tag{13.76}$$

From (13.72), (13.73), (13.75), and (13.76), we can see that, as long as there is more than one user, a single user's expected profits with centralized control will be higher than those of noncooperative results. However, there exist two problems for centralized control. First, the network topology is distributed and it is hard to implement the centralized control. Second, it can be shown that, if any user deviates from (13.75) and transmits at a higher rate, it can get a greater benefit in (13.76). So instead of cooperating, users will deviate because of their greediness. As a result, the competition for the communication node becomes intense and P will decrease. Consequently, each user's profit will drop to the noncooperative value in (13.72). So *our goal* is to construct a distributed scheme such that users have to cooperate and have no incentive to deviate.

One possible solution comes from game-theory literature [238]. In staged repeated games, the player's overall payoff is weighted average payoffs over time. At each stage, the "noncooperative" game can be played, which means that the players' choices are based on only their perceived self-interest. On the other hand, players can also play "cooperative" games, which allow cooperation among users to achieve better payoffs. If some consequences of the player's actions can be observed at the end of each stage, it becomes possible for players to adjust their strategy, which can lead to better equilibrium outcomes that do not arise when the game is played only once. So we want to construct the scheme based on the following underlying rationale: At the beginning, all players

Table 13.4 Cartel Maintenance algorithm

Initialization:
$t = 0$ is a cooperative period;

Strategy:
If t is a cooperative period and $\hat{P}_t \geq P^*$,
    then $t + 1$ is a cooperative period;
If t is a cooperative period and $\hat{P}_t < P^*$,
    then $t + 1, \ldots, t + T - 1$ are noncooperative periods
    with $\lambda_i = s$, $\forall i$, and $t + T$ is a cooperative period.

---

agree to operate in the cooperative way. If any player deviates, in the next stage, other players will observe the deviation and play in the noncooperative way instead. So this deviating player will get less utility because of the punishment from the other players. Because each player tries to optimize the payoff over time, no player will have the incentive to deviate from the cooperative stage. So using the threat of punishment from other users, the system forces all users to act cooperatively in a distributed manner.

For the system shown in Figure 13.23, each user can play either noncooperatively or cooperatively. Each user tries to optimize its rate such that the overall profit over time can be optimized. At each time slot, each user can observe the transmission status such as the successful transmission probability in the communication node. The preceding facts motivate us to apply the repeated game approach.

In the wireless network, each user's goal is to maximize a discounted expected payoff over time slots. We define the discount factor as β. For most applications of wireless networks, β is close to 1. The optimization problem can be represented as

$$\max_{\lambda_i^t} \sum_{t=0}^{\infty} \beta^t \pi_i(\lambda_i^t, \Lambda_{-i}^t), \ \forall i. \tag{13.77}$$

Intuitively, every user wants to maximize its own expected profit by increasing its transmission rate. It will result in too many collisions among users, which limit each user's profit. Thus we need to introduce a certain mechanism, namely, the game rule, to force the users to act cooperatively to achieve better profits and to be robust to the cheating phenomenon.

In [238], Porter developed a Cartel trigger-price strategy for a dynamical industry model, in which a company deters others from deviating from collusive output levels by threatening to produce at noncooperative quantities for a period of duration whenever the market price falls below some trigger price. From a similar idea, we develop the Cartel Maintenance algorithm in Table 13.4.

In the beginning, all users are in a cooperative period with rate $\overrightarrow{\lambda}$. Then they will monitor the overall probability of successful transmission in (13.64). If the probability is higher than some threshold P^*, it probably means that all users transmit at the cooperative operating rate. On the other hand, if the probability drops lower than the threshold, it means that some users may cheat. Then the other users will play punishment by transmitting noncooperatively according to (13.72) for a period of T. Then they will come back to play the cooperative period again. Because users are afraid of future punishment, they are inclined to play cooperatively.

The probability of successful transmission in the communication node is determined by two factors: the users' rates and the wireless channel-link condition. It is possible that all users act cooperatively but the probability is still under the threshold because of the bad channel. Under this situation, the users will play the noncooperative period, because they cannot tell if the low successful transmission probability is caused by the deviations or the bad channels. This is a penalty for the distributed implementation.

The remaining problem is how to find the optimal values of the cooperative rate $\overrightarrow{\lambda}$, threshold P^*, and punishment duration T, which we show as follows.

In cooperative periods, the expected discounted profit of user i is given by

$$V_i(\overrightarrow{\lambda}) = \pi_i(\overrightarrow{\lambda}) + P_r\{\hat{P}_t \geq P^*\}\beta V_i(\overrightarrow{\lambda})$$
$$+ P_r\{\hat{P}_t < P^*\}\left(\sum_{\tau=1}^{T-1}\beta^\tau \pi_i(s) + \beta^T V_i(\overrightarrow{\lambda})\right), \qquad (13.78)$$

where the first term on the right-hand side (RHS) is the current expected value, the second term on the RHS is the expected value of the next period if cooperation is conducted, and the third term on the RHS is the expected value of next period if noncooperation (13.78) can be rewritten as

$$V_i(\overrightarrow{\lambda}) = \frac{\pi_i(s)}{1-\beta} + \frac{\pi(\overrightarrow{\lambda}) - \pi(s)}{1 - \beta + (\beta - \beta^T)F(P^*/P)}, \qquad (13.79)$$

where F is the CDF of random variable θ. To find the optimal transmission rate vector, threshold, and punishment duration, the following derivatives are set to zeros:

$$\frac{\partial V_i}{\partial \overrightarrow{\lambda}} = 0, \quad \frac{\partial V_i}{\partial P^*} = 0, \text{ and } \frac{\partial V_i}{\partial T} = 0. \qquad (13.80)$$

Because the system is homogenous, the optimal transmission rate is the same for all users. A heterogeneous case can be analyzed in a similar way. The detailed deductions are similar in [238]. The optimal λ^*, P^*, and T^* can be determined by the following steps:

$$\lambda^* = \frac{A}{2BN}\left(\frac{N + \eta^* + (N+1)(a/A)}{N + 1 + \eta^*}\right), \text{ if } \eta^* > \eta^0,$$
$$= s, \text{ otherwise}, \qquad (13.81)$$

where

$$\eta^0 = \frac{(N+1)[(N+1)(a/A) - N]}{(N-1)}$$

and

$$\eta^* = \frac{f[P^*/P(\lambda^*)]}{F[P^*/P(\lambda^*)]}\frac{P^*}{P(\lambda^*)}.$$

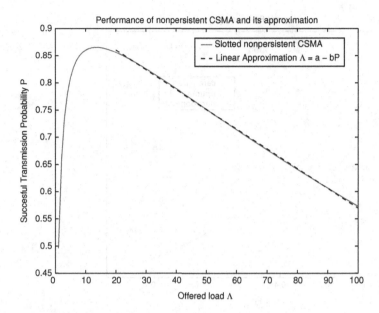

Figure 13.24 Linear approximation.

P^* is determined by

$$\frac{f[P^*/P(\lambda^*)]}{F[P^*/P(\lambda^*)]}\frac{P^*}{P(\lambda^*)} - \frac{f'[P^*/P(\lambda^*)]}{f[P^*/P(\lambda^*)]}\frac{P^*}{P(\lambda^*)} = 1, \qquad (13.82)$$

where f and f' are the PDF and its derivative of θ, respectively. The optimal P^* and λ^* are calculated by (13.81) and (13.82) iteratively. Obviously we have $\lambda^* \in (r, S]$, and $\lambda^* \to r$ when $\eta^* \to \infty$. After the values converge, we can calculate the optimal punishment period as

$$T^* = \frac{1}{\ln\beta}\ln\left\{\beta - \frac{(1-\beta)[A-(N+1)B\lambda^*]}{f(\theta^*)[b\theta^*/P(\lambda^*)]\Delta - F(\theta^*)[A-(N+1)B\lambda^*]}\right\}, \qquad (13.83)$$

where $\theta^* = P^*/P(\lambda^*)$ and $\Delta = \pi_i(\lambda^*) - \pi_i(s)$.

For simulations, we assume there are $N = 10$ users in the networks. The basic cost is $c_0 = 0.001$, profit per success transmission is $c_2 = 1$, and β is a number very close to 1. We assume the slotted nonpersistent CSMA as the multiaccess protocol in the communication node. Without loss of generality, we assume unit service rate. The probability of successful transmission is represented as [28, 165]

$$P = \frac{\alpha\Lambda e^{-\alpha\Lambda}}{1 - e^{-\alpha\Lambda} + \alpha}, \qquad (13.84)$$

where Λ is the offered load and α is the one-way normalized propagation delay that is assumed equal to 0.01. In Figure 13.24, we show the linear approximation of P using function $P \approx a - b\Lambda$. The linear approximation is shown to be accurate when $\Lambda > 20$, i.e., when the system is overloaded. Here $a = 0.9331$ and $b = -0.0036$.

The distribution of θ is approximated by a binomial distribution, which gives the probability of packet loss over Bernoulli trails. The binomial distribution can be

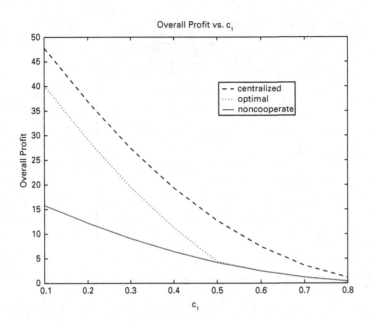

Figure 13.25 Overall profit vs. cost per transmission.

approximated by normal distribution. Here we fix the mean of normal distribution as unit and change the variance to show the performances. Any probability larger than 1 or less than 0 will be truncated. Other probability distributions of packet errors can be applied in a similar way.

In Figure 13.25, we show the overall profit as a function of cost per transmission rate c_1 for variance equal to 0.015. Obviously the overall profit will decrease when increasing c_1, which is because of the increases of the cost. If users play with centrally controlled rate r, the overall profit is much greater than the profit if the users play noncooperatively with rate s. The proposed scheme has the performance in between, which is because the users prefer to send the packets at the rate λ^* rather than s so as to avoid future punishment. When c_1 is a small number, the cost for transmission is low and the users would like to transmit at higher rate. If users transmit noncooperatively, the competitions for the communication nodes will be high. As a result, the overall profit is much less than that if they play cooperatively. For the proposed scheme, because users are afraid of others' punishment, they show a behavior of cooperation. So the overall profit is close to the centralized control result.

We change the variance of the distribution of θ. The channel condition becomes worse when the variance increases. In Figure 13.26, we show the effects of channels on the overall profit. In Figure 13.27, we show the probability of successful transmission versus the channel variance. We also show the threshold P^*. When the variance is small, the proposed scheme has an overall profit similar to that of centralized control. On the other hand, if the channel becomes worse, the performance of the proposed scheme drops to the noncooperative case. Under this condition, the users cannot distinguish whether the probability of successful transmission drops because of the other user's deviation

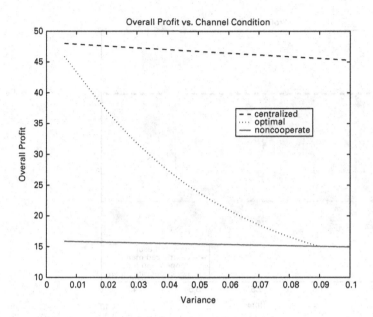

Figure 13.26 Overall profit vs. channel condition.

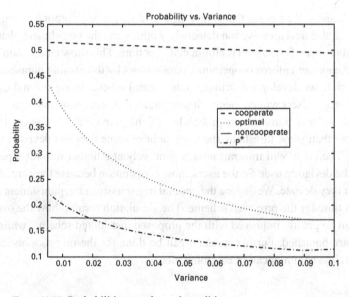

Figure 13.27 Probabilities vs. channel condition.

or the channel deterioration. Consequently, they will prefer to transmit more and play noncooperatively.

In Figure 13.28, we show how the scheme punishes the cheating user. Here channel variance is 0.015 and $c_1 = 0.1$. We assume that one user deviates from the optimal λ^* and transmits at the higher rate s, while others transmit at λ^*. We show the profit of this deviating user over time. For comparison, we also show the average profit when the user

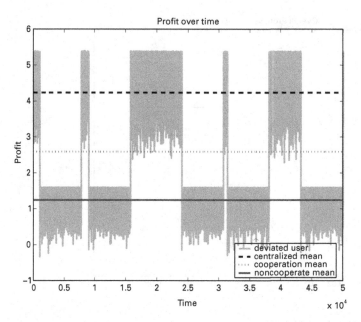

Figure 13.28 Punishment for deviation.

transmits at r, λ^*, and s. We can see that at first the user does get more profit by deviating from λ^*. However, this deviation is soon detected by others and the punishment phase is performed by other users. So the gain is eliminated over time. This shows the reason why the proposed scheme can enforce cooperation among users by threatening punishment.

In this subsection, we develop a distributed rate-control scheme using optimal Cartel Maintenance strategy. Users will cooperate at some agreed transmission rate first. Then they monitor the success transmission probability of the communication node. If the probability is less than some threshold, the users believe some user has deviated from the agreed rate. Then they will transmit noncooperatively at a higher rate for a period time to punish the deviating user. So the users show cooperation because they are afraid of punishment if they deviate. We deduce the optimal transmission rate, punishment rate, and punishment time for the proposed scheme. The simulation results show the overall system profit can be greatly improved with the proposed distributed scheme, while the cheating users are punished. Further research can be done for the more sophisticated and realistic wireless communication environments.

13.3.2   Self-Learning-Based Approach

For the networks with packet forwarding, distributed control to enforce cooperation for nodes' packet-forwarding probabilities is essential to maintain the connectivity of networks. In this subsection, we propose a novel self-learning repeated game framework to optimize packet-forwarding probabilities of distributed users. The framework has two major steps: First, an adaptive repeated game scheme ensures the cooperation among users for the current cooperative packet-forwarding probabilities. Second, a self-learning scheme tries to find better cooperation probabilities. Some special cases are analyzed to evaluate the proposed framework. From the simulation results, the proposed framework demonstrates the near-optimal solutions in both symmetric and asymmetric network.

Recent advances in wireless communication have made possible the large-scale deployment of wireless ad hoc and sensor networks, which consist of small, low-cost nodes with simple processing and networking capabilities. The nodes in the scenarios preceding mentioned may not be willing to fully cooperate. And in fact, it is reasonable to assume that the nodes are selfishly maximizing their own benefits for the following reason. From the node perspective, forwarding the arriving packets consumes its limited battery power. Therefore it may not be in the node's interest to forward all the arriving packets. In contrast, not forwarding other's packets will adversely affect the network connectivity. Hence it is very crucial to design a distributed control mechanism that enforces cooperation among participating nodes.

Because the distributed nodes do not have information of others, they act selfishly to optimize their own performances. This motivates us to apply the game-theory approach [227] to packet-forwarding problems. In [173], repeated game theory is applied to routing problems. In [197], multiple-access resource allocation is studied using the game-theory approach. Repeated games are further applied to physical-layer problems. The distributed control mechanism that enforces collaboration/cooperation has also been studied within the context of game theory in several existing literatures. Srinivasan et al. [287] provide a mathematical framework for cooperation in ad hoc networks that focuses on the energy-efficient aspects of cooperation. In [206], the authors focus on the properties of the cooperation enforcement mechanisms used to detect and prevent selfish behavior of nodes in ad hoc networks. They show that the formation of large coalitions of cooperating nodes is possible when a mechanism similar to CORE [207] is used. In [9], the authors consider a less aggressive punishment policy. In this scheme, the node uses the minimum forwarding probability among its neighborhood as its forwarding probability after detecting the misbehavior. Felegyhazi et al. [71] consider a model to show cooperation among participating nodes and provide sufficient condition on the network topology under which each node employing the punishment strategy results in a Nash equilibrium. In [207] and [37], the authors define protocols that are based on a reputation system. In [109], the authors propose a repeated game framework for multiple access using Cartel Maintenance. Other work of applying cooperative game theory and noncooperative game theory to OFDMA networks can be found in [108].

Here, we propose a self-learning repeated game framework for users to obtain the optimal packet-forwarding probabilities distributively. The framework has two major steps. First, to ensure cooperation among users, the users apply an adaptive repeated game scheme to punish the greedy users for deviation and noncooperation. Second, the users try to learn the better packet-forwarding probabilities that generate better performances. From the simulation results, the proposed scheme can find the optimal solutions or near-optimal solutions in both symmetric and asymmetric networks.

The proposed scheme has an analogy to human society. Before civilization, there were no rules in society to enforce cooperation. People fought each other greedily and noncooperatively. The consequences were low social productivity and low living standards for the people themselves. Then, through revolutions, new relationships among human beings were proposed, such as slavery, feudalism, and capitalism, etc. To maintain the new relationship, rules such as laws are defined to enforce cooperation under the new relationship. Similarly, in a packet-forwarding network, it has been proved from previous works that the network performance will degrade to zero asymptotically if no cooperation is enforced. If we can enforce cooperation among distributed and greedy users and if we can find the better relationship that users forward others' packets, the system efficiency as well as all users' performances can be improved.

The packet-forwarding problem is essential for distributed users to get connected to the destinations. Suppose there are a total of K users. The kth user has a total of N_k routes for its packet transmission. We assume the routes have been determined and known. Let's define I_k^i as the set of the nodes on the ith route for the kth user. Suppose each user has the willingness to forward another user's packet with probability of α_i. For each user, the successful transmission or reception of one packet will have the benefit of G, and forwarding others' packet will cost F per packet. Suppose the kth user transmits its packet with probability of P_k^i to the ith route. Obviously, we have $\sum_{i=1}^{N_k} P_k^i = 1$. So the utility function U_k for the kth user can be expressed as

$$U_k = \sum_{i=1}^{N_k} P_k^i G \Pi(\alpha_j, j \in I_k^i) - F\alpha_k B_k, \tag{13.85}$$

where Π is the successful transmission probability that is a function of packet-forwarding probabilities along the routes. B_k is the forward-request probability from other users. The first term on the RHS of the preceding equation is the average benefit for the kth user, which depends on other users' willingness for forwarding. The second term shows the cost of forwarding other users' packets, which depends on its own willingness for packet forwarding.

Because an individual distributed user has less information about others and may selfishly optimize its own performance, the packet-forwarding problem is, in some sense, analogous to the economy system of human society. Game theory is a successful economy approach, which studies the mathematical models of conflict and cooperation between intelligent and rational decision makers. We can formulate this problem as a noncooperative game problem in which each user adjusts its forward probability to maximize its own utility function:

$$\max_{0 \le \alpha_k \le 1} U_k(\alpha_k, \alpha_{-k}), \tag{13.86}$$

where $\alpha_{-k} = [\alpha_1, \ldots, \alpha_{k-1}, \alpha_{k+1}, \alpha_K]^T$ is the other users' behaviors of packet forwarding. To analyze the outcome of the game, Nash equilibrium is a well-known concept, which states that in equilibrium every user will select a utility-maximizing strategy given the strategy of every other user.

DEFINITION 46 *Define feasible range Ω as $[0, 1]$. Nash equilibrium $[\hat{\alpha}_1, \ldots, \hat{\alpha}_K]^T$ is defined as*

$$U_k(\hat{\alpha}_k, \alpha_{-k}) \le U_k(\tilde{\alpha}_k, \alpha_{-k}), \ \forall k, \ \forall \tilde{\alpha}_k \in \Omega, \ \alpha_{-k} \in \Omega^{K-1}, \tag{13.87}$$

i.e., given the other users' packet-forwarding probability, no user can increase its utility alone by changing its own packet-forwarding probability.

Unfortunately, the Nash equilibrium for the packet-forwarding game in (13.87) is usually $\hat{\alpha}_k = 0,\ \forall k$, because each user's benefit depends on other users' willingness for forwarding and does not depend on its own behavior, while the user's cost solely depends on its willingness for packet forwarding. So each user will greedily drop its packet-forwarding probability to reduce the cost and increase the utility. However, if no user forwards, the successful packet-transmission probabilities might become zero. Consequently, the benefits for users are zeros and the whole system shuts down. So if the users play noncooperatively and have Nash equilibrium, all users' utility might be zero. On the other hand, if the users can cooperate and have some positive packet-forwarding probability, all users can have benefits.

So the problem can be formulated to design a method to enforce cooperation among users in packet forwarding. First, we want to find the best packet-forwarding vector such that the utilities of all users are strictly better than those of the Nash equilibrium. Moreover, we want to design a mechanism to enforce such cooperation among users. Because this problem is very similar to some problems in human society, we use the economical approach called repeated game to solve the proposed problem.

To solve the proposed problem, we apply the repeated game approach to the packet-forwarding problem. Repeated game is a special case of a dynamic game (a game that is played multiple times). When players interact by playing a similar static game that is played only once like (13.86) numerous times, the game is called a repeated game. Unlike a game played once, a repeated game allows a strategy to be contingent on past moves, thus allowing reputation effects and retribution, which give the possibility for cooperation.

DEFINITION 47 *For T-period repeated game, at each period t, the moves during periods $1, \ldots, t - 1$ are known to every player. β is the discount factor. The total discounted payoff for each player is computed by*

$$\sum_{t=1}^{T} \beta^{t-1} U_k(t), \qquad (13.88)$$

where $U_k(t)$ denotes the payoff to player k in period t. If $T = \infty$, the game is referred as the infinitely repeated game. The average payoff to player k is then given by

$$\bar{U}_k = (1 - \beta) \sum_{t=1}^{\infty} \beta^{t-1} U_k(t). \qquad (13.89)$$

From game-theory literature, the repeated game can enforce the greedy user to show cooperation, because the user will get punished by other users in the near future if the user acts greedily. The benefit of greediness will be eliminated by the punishment loss in the future. So the users would rather act cooperatively. Therefore the problem remains of defining a good rule to enforce cooperation. From Folk Theorem, we know that, in

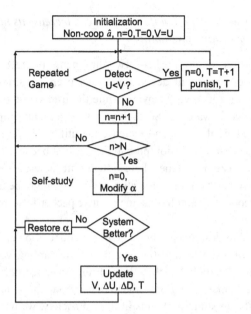

Figure 13.29 Proposed self-learning repeated game framework.

an infinitely repeated game, any feasible outcome that gives each player a better payoff than the Nash equilibrium can be obtained.

THEOREM 27 Folk Theorem *[227]: Let $(\hat{\alpha}_1, \ldots, \hat{\alpha}_n)$ be the payoffs from a Nash equilibrium of G and let $(\alpha_1, \ldots, \alpha_n)$ be any feasible payoffs. There exists an equilibrium of the infinitely repeated game that attains $(\alpha_1, \ldots, \alpha_n)$ for $\alpha_i > \hat{\alpha}_i, \forall i$ as the average payoff, provided that β is sufficiently close to 1.*

Now we know that, by using the repeated game, the greedy users can be forced to cooperate and have better payoffs. So the remaining problem is defining a good mechanism to enforce cooperation, which we will show later.

The basic idea for the proposed self-learning cooperation enforcing framework is to let distributed users learn the optimal packet-transmission probability step by step, while, within each step, the strategy of repeated game is applied to ensure the cooperation among users. For simplicity, we omit the user index. The block diagram of the algorithm is shown in Figure 13.29 and detailed descriptions are as follows:

During initialization, all users play a noncooperative game and all users are balanced in an inefficient Nash equilibrium $\hat{\alpha}$. We set the time counter $n = 0$, the punishment time $T = 0$, and trigger threshold $V = \hat{\alpha}$.

In the next step, we play a repeated game strategy. If all users play cooperatively, every user will have some benefits. However, from (13.85), if any user deviates from cooperation by playing noncooperatively and other users still play cooperatively, this user will have more benefits, while others suffer with lower benefits because of this user's greediness. To prevent users from deviation, the repeated game strategy provides a punishment mechanism. The basic idea is that each user sees if the utility function is lower than the threshold V. If this is the case, that means some user may have deviated; then this user also plays noncooperatively for a period of time T. By doing this, the greedy user's short-term benefit will be eliminated by the long-term punishment. If all users are concerned with the long-term payoff such as (13.89), which is true by the assumption of rational users, then none of the users will have incentive to deviate from cooperation.

In detail, the repeated game scheme with parameter (V, T) for all users is explained as follows: Each user's utility U is compared with the threshold V. If $U < V$, i.e., someone deviates, the time counter n is set to zero, punish time is increased by one, and the user plays noncooperatively for a T period of time. Because we assume all users are rational, with increasing of T, the benefit of one time deviation will be eliminated out sooner or later. So, finally, no user wants to deviate and $U \geq V$. At this time, the counter n starts increasing. If the system is stable in the cooperation for a period of time N, where N is a predefined constant, the algorithm assumes that the cooperation is enforced, and changes to the next step to improve the current cooperation.

In the next step, the algorithm tries to self-learn the optimal forward probabilities by modifying α with the goal of optimizing the performances. The simplest way is to randomly generate $\alpha \in [0, 1]$, where different users may have different α. In the next time slot, all users observe whether their performances become better. If not, the α is changed to the previous value. Otherwise, each user selects its packet-forwarding probability as α, updates its threshold to current benefit $V = U$, calculates the difference of cooperation and noncooperation for the utility ΔU as

$$\Delta U = U(\text{new } \alpha) - U(\hat{\alpha}), \tag{13.90}$$

and calculates the deviation benefit ΔD. If the network is symmetric, the optimal punishment time can be written as

$$T = \frac{\Delta D}{\Delta U}, \tag{13.91}$$

where T is the estimated punishment time that prevents the users from deviation. Then the algorithm goes back to the repeated game case to update the punishment time T such that all users are willing to cooperate.

Notice that, during the first time slot after α is modified, all users will act cooperatively, because of the rationale that deviation eliminates the chance of utility improvement in the future. In the repeated game step, the benefit of instantaneous deviation is eliminated sooner or later as long as the discount factor β is close enough to 1. So T will converge to some value. In the self-learning step, if the new sets of α are not good for all users, the original value of α will be restored. If the new sets of α are good, the cooperation can be enforced by the future repeated game step. So the framework will converge.

In summary, the framework uses the threat of punishment to maintain the cooperation for the current α and tries to learn if there is a better α for cooperation. Next, we give some cases to analyze and evaluate the behaviors of the proposed framework.

In the previous analysis, we assume the networks are synchronous, i.e., each user's utility can be observed instantaneously whenever other users deviate. This might not

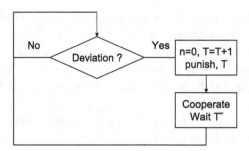

Figure 13.30 Modified repeated game step.

be true in real networks. Next, we discuss the problem introduced by asynchronous networks and some possible solutions.

When the network is asynchronous, the deviation of users will be detected by other users with some time delay. This is not the problem. The problem is that, when the punishment period is over, the users may return to the cooperation phase in a different time. This may trigger some users to continue punishment because they cannot distinguish whether the users are deviating or whether the users are still in the punishment phase. The problem will make the network fluctuate and the punishment time T cannot converge. To solve such a problem, we propose the following modification to the repeated game step of the proposed framework.

The modified repeated game step is shown in Figure 13.30, in which an extra step is added at the end of the punishment period. After switching back to cooperation, the user will wait for time T' and then observe whether others deviate. This time T' is reserved for the other users to return to cooperation. This value is determined by the scale and topology of the networks. If the value of T' is too small, the network will not be stable and the punishment period always prevails because some users' delayed return to cooperation triggers others' new punishment periods. If the value of T' is too large, it gives the opportunities for greedy users to deviate to gain the benefits without detecting from other users. There are trade-offs for selection of the value for T'.

The other concern is during the step when α is modified after the system is stable $n > N$. The message for all users to modify α can be transmitted by a protocol like flooding. This message will take time to reach every node. So to judge whether the system becomes better, a user needs to wait for a period of time that may be similar to T'.

Next, we analyze two cases: symmetric networks and asymmetric networks. Some simple examples are given and analytical optimal results are deduced. Simulation results of the proposed framework are conducted to evaluate the performances.

First we analyze the characteristic of the symmetric network. The topology of such a network is symmetric; consequently, the resulting Nash equilibrium and the optimum of the packet-forwarding probabilities should be the same for all users, i.e., $\hat{\alpha}_k = \hat{\alpha}_j$, $\alpha_k = \alpha_j$, $\forall k, j$. In general, the networks are asymmetric. However, at the edges of networks where some nodes may equally access the networks, symmetric topology may exist and symmetric analysis can be applied.

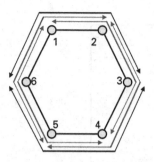

Figure 13.31 Example of symmetric network.

Next, we give an example of the analysis of the synchronous symmetric networks. Suppose the considered network is shown in Figure 13.31. In this network, there are six fixed routes: $1 \leftrightarrow 4$, $2 \leftrightarrow 5$, and $3 \leftrightarrow 6$. All the destinations are three hops away from the source. We consider the node's utility function as the reward obtained from successfully transmitting or receiving a packet. We also assume the forwarding others' packets consume resources such as energy; therefore forwarding contributes a cost (negative reward) to the utility function. The utility functions for each of the node in Figure 13.31 are represented as follow:

$$U_1 = 2G[1 - (1 - \alpha_2\alpha_3)(1 - \alpha_5\alpha_6)] - F[\alpha_1 + \alpha_1\alpha_2],$$
$$U_2 = 2G[1 - (1 - \alpha_1\alpha_6)(1 - \alpha_3\alpha_4)] - F[\alpha_2 + \alpha_2\alpha_3],$$
$$U_3 = 2G[1 - (1 - \alpha_1\alpha_2)(1 - \alpha_4\alpha_5)] - F[\alpha_3 + \alpha_3\alpha_4],$$
$$U_4 = 2G[1 - (1 - \alpha_2\alpha_3)(1 - \alpha_5\alpha_6)] - F[\alpha_4 + \alpha_4\alpha_5],$$
$$U_5 = 2G[1 - (1 - \alpha_1\alpha_6)(1 - \alpha_3\alpha_4)] - F[\alpha_5 + \alpha_5\alpha_6],$$
$$U_6 = 2G[1 - (1 - \alpha_1\alpha_2)(1 - \alpha_4\alpha_5)] - F[\alpha_6 + \alpha_1\alpha_6],$$

where α_i is the probability that node i is willing to forward others' packets, G is the reward for successfully transmitting and receiving a packet, and F is the cost for forwarding others' packet. We also assume that nodes are greedy and rational but not malicious, that is every node decides its forwarding probability to maximize its own utility function. If we consider the Nash equilibrium obtained noncooperatively from (13.86), obviously, to the best of each node's interest, every node selects zero forwarding probability (i.e. $\alpha_k = 0$, $\forall k$) to minimize its forwarding cost in the utility function. However, the overall network becomes disconnected as all the nodes act in noncooperative manner.

Note that, because of the symmetric property of the network in Figure 13.31, the optimal forwarding probability and the corresponding utility of each node will be the same. We omit the subscript for simplicity. Consider the *system-wide* optimal solution to maximize everybody's utility; we can formulate the problem as

$$\max_{\alpha} U = 2G(2\alpha^2 - \alpha^4) - F(\alpha + \alpha^2),$$
$$\text{s.t.} \qquad 0 \le \alpha \le 1. \tag{13.92}$$

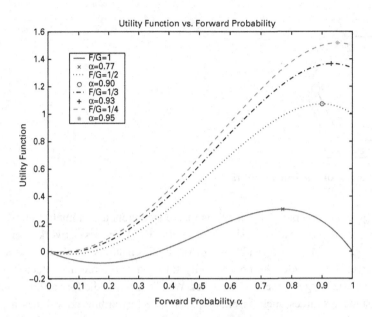

Figure 13.32 Effect of forwarding probability on utility.

By differentiating the preceding equation, we obtain

$$\frac{\partial U}{\partial \alpha} = 8G(\alpha - \alpha^3) - F(1 + 2\alpha) = 0, \tag{13.93}$$

$$\alpha^3 - \left(1 - \frac{F}{4G}\right)\alpha + \frac{F}{8G} = 0. \tag{13.94}$$

The optimal forwarding probability in the symmetric network can be obtained by solving (13.94). Figure 13.32 shows the effects of forwarding probability α on the utility function for different normalized forwarding costs, F/G. We also show the optimal forwarding probabilities for different cases. It is obvious that, as the cost for forwarding F is smaller compared with the transmitting/receiving reward G, the optimal forwarding probability will approach the unity forwarding probability and the corresponding utility will also be high. On the other hand, when F/G is large, then every node has a lower incentive to forward the others' packet and utility is low. This phenomenon is reasonable because, when the cost for forwarding is very large, it is better for the node to save the energy for its own transmission. The goal of the cooperation mechanism design is to design the incentive for the nodes to avoid the noncooperative solution and to result in the *system-wide* optimal forwarding probability. It is also worth mentioning that not every positive packet-forwarding probability will generate the larger utility than the full noncooperation case in which the packet-forwarding probability is zero. For example, when $F/G = 1$, the utility is higher than 1 only when $\alpha \geq 0.37$. So, in the self-learning step, if α is modified less than 0.37, the system will have worse performance than noncooperation. As a result, the new α will be discarded, and the original α is restored.

In Figure 13.33, we show the simulation results of the proposed framework for utility and packet-forwarding probability over time. Here $F/G = 1$ and $N = 200$. Initially,

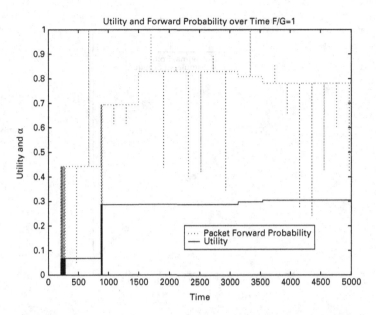

Figure 13.33 Utility function and forwarding probability over time.

$\alpha = 0$, because of the noncooperation transmission. Then the system tries to find a better packet-transmission rate. When it finds a better solution, all users adopt its α to the value. However, because the punishment period T is not adjusted to an optimal value, the deviation can have benefits. So there exists a period that the utility and α switch from cooperation to noncooperation. In this period, T is increased until everybody realizes that there is no benefit for deviation because of the long period of punishment. If the system is stable for time N, a new α is attemped to see if the performance can be improved. If yes, the new value is adopted; otherwise the original value is restored. So the packet-forwarding probability is adjusted until the optimal solution is found, and the learned utility function is a nondecreasing function. Notice that users are less reluctant to deviate when α is close to the optimal solution, because the benefit of deviation becomes smaller and users already have the estimated punishment time according to (13.91).

Figures 13.34 and 13.35 show the packet-forwarding probability and utility versus normalized packet-forwarding cost F/G for the optimal solutions and the solutions studied by the proposed framework, respectively. Here the system tries to find the new α only 250 times. From the simulation results, we can see that the proposed framework can find the optimal packet-forwarding probability and the optimal utility with maximum of 0.7% and 0.04% difference, respectively. This proves that the proposed framework can find the optimal packet-forwarding probability very efficiently.

Practical networks are generally asymmetric in nature. Next we turn our attention to analyzing the performance of the proposed framework over the asymmetric networks. An example of synchronous asymmetric networks is shown in Figure 13.36. In this case, node 1 and node 6 act as the sinks of the information. The red arrows indicate the flow

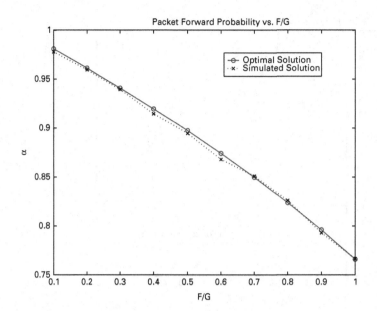

Figure 13.34 Effect of F/G on optimal forwarding probability.

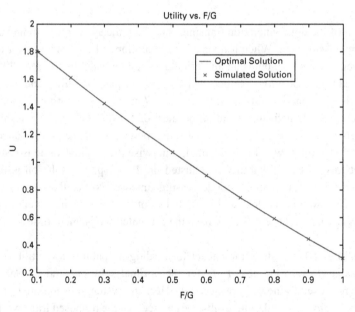

Figure 13.35 Effect of forwarding cost on the optimal utility.

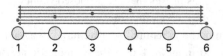

Figure 13.36 Example of asymmetric network.

directions of routes, in which node 6 is the sink. In this case, nodes 1–5 want to transmit to node 6. Similarly, the blue arrows indicate the flow direction where node 1 is the sink. In this case, nodes 2–6 want to transmit to node 1. Notice that node 2 and node 3 are asymmetric. We formulate the utility function for nodes 1–6 as follows:

$$U_1 = 2G\alpha_2\alpha_3\alpha_4\alpha_5,$$
$$U_2 = G[1 + \alpha_3\alpha_4\alpha_5] - F[2\alpha_2 + \alpha_2\alpha_3 + \alpha_2\alpha_3\alpha_4 + \alpha_2\alpha_3\alpha_4\alpha_5],$$
$$U_3 = G[\alpha_2 + \alpha_4\alpha_5] - F[2\alpha_3 + \alpha_2\alpha_3 + \alpha_3\alpha_4 + \alpha_3\alpha_4\alpha_5],$$
$$U_4 = G[\alpha_5 + \alpha_2\alpha_3] - F[2\alpha_4 + \alpha_3\alpha_4 + \alpha_4\alpha_5 + \alpha_2\alpha_3\alpha_4],$$
$$U_5 = G[1 + \alpha_2\alpha_3\alpha_4] - F[2\alpha_5 + \alpha_4\alpha_5 + \alpha_3\alpha_4\alpha_5 + \alpha_2\alpha_3\alpha_4\alpha_5],$$
$$U_6 = 2G\alpha_2\alpha_3\alpha_4\alpha_5.$$

We can see that the noncooperative solution for each node is to use zero forwarding probability. Notice that, because of the symmetry in the network flow, node 2 and node 5 have the same forwarding probability, and node 3 and node 4 have the same forwarding probability. Moreover, node 1 utility and node 6 utility are totally dependent on the packet-forwarding probabilities of other nodes. So the optimization parameters are α_2 ($\alpha_2 = \alpha_5$) and α_3 ($\alpha_3 = \alpha_4$) only. Because node 2 and node 4 have their own optimization goals, from the point of view of system optimization, this is a multiple-objective optimization. To quantify the optimality, we need to define the following concept:

DEFINITION 48 *Pareto optimality is a measure of optimality. An outcome of a game is Pareto optimal if there is no other outcome that makes every node at least as well off and at least one node strictly better off. That is, a Pareto optimal outcome cannot be improved on without hurting at least one node. Often, a Nash equilibrium is not Pareto optimal, implying that the players' payoffs can all be increased.*

In Figure 13.37, we show the Pareto optimal region and the simulated results obtained by the proposed framework. The x axis and the y axis are α_2 and α_3, respectively. Here the system tries to find the new packet-forwarding probability 250 times. Any point within the shaded area is Pareto optimal. Most of the simulated points are within this region. Very few points are located outside, because of the failure of searching the optimal packet-forwarding probability within 250 times. We can see that the proposed framework is effective for finding the Pareto optimum for asymmetric networks.

Overall, we proposed a self-learning repeated game framework for packet-forwarding networks. The cooperation within users for packet forwarding is obtained by threat of punishment in the future, whereas the optimal packet-forwarding probability of each user can be studied distributively. From the simulation results for symmetric and asymmetric networks, we can see that the proposed framework can effectively find the solutions very close to the optimal solutions in a distributed way. The proposed framework can have impacts on the designs of future communication networks such as wireless networks, wired networks, ad hoc networks, and sensor networks.

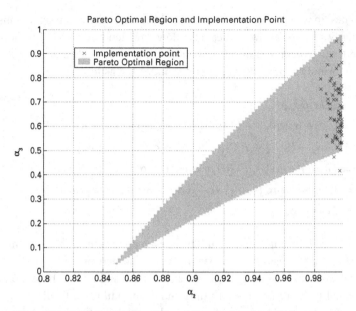

Figure 13.37 Pareto optimal region and the simulated results.

13.4   Games with Cooperation

The philosophy of a cooperative game is significantly different compared with the non-cooperative game. Instead of fiercely competing for the limited resources, some contracts can be negotiated between users to more efficiently share the limited radio resources. The basic idea is the mutual benefit, i.e., each user can obtain better performances than those by noncooperation. In this section, we study an OFDMA resource-allocation example first.

Another interesting problem is the information. If the players have enough information about the environment and the components' moves, a certain equilibrium with mutual benefits can be achieved. Unfortunately, the players usually have less information and consequently select the strategies that result in low system efficiency. In the second example, we investigate the correlated equilibrium that can bring mutual benefits and construct the learning algorithms that can study the correlated equilibrium.

13.4.1   Bargaining-Based Approach

In this example, a fair scheme to allocate subcarrier, rate, and power for multiuser OFDMA systems is proposed. The problem is to maximize the overall system rate, under each user's maximal power and minimal rate constraints, while considering fairness among users. The approach considers a new fairness criterion, which is a generalized proportional fairness based on NBSs and coalitions. First, a two-user algorithm is developed to bargain subcarrier usage between two users. Then a multiuser bargaining algorithm is developed based on optimal coalition pairs among users. The simulation results show that the proposed algorithms not only provide fair resource allocation among users, but also have an overall system rate comparable with the scheme without considering fairness. They also have much higher rates than that of the scheme with max–min fairness. Moreover, the proposed iterative fast implementation has a complexity for each iteration of only $O(K^2 N \log N + K^4)$, where N is the number of subcarriers and K is the number of users.

OFDMA is a promising multiple-access technique for high-data-rate transmissions over wireless radio channels. Efficient resource allocation, which involves bit loading, transmission power allocation, and subcarrier assignment, can greatly improve system performance and so has drawn a great deal of attention in recent research.

The resource allocation for a single user across parallel orthogonal channels maximizes the total achievable rate subject to a total power constraint, which can be optimally solved by means of the water-filling method [52]. The rate allocation in each subcarrier is then determined by the corresponding power allocation. In single-cell multiuser systems with a given set of allocated subcarriers to each user, the water-filling solution can also be applied because resource allocation for each user can be considered independently.

However, if we consider the different users' link qualities and the discrete nature of the subcarrier assignment problem, it is more difficult to optimally assign the subcarriers to different users in a multiuser environment. By adaptively assigning subcarriers of various frequencies, we can take advantage of channel diversity among users in different locations, which is called multiuser diversity. Such multiuser diversity stems from channel diversity, including independent path loss and fading of users. Most of the existing works focus on improving the system efficiency by exploring multiuser diversity [312, 335]. In [335], the authors studied the dual problem, namely, to find the optimal subcarrier allocation so as to minimize the total transmitted power and satisfy a minimum rate constraint for each user. The dual problem is further formulated as an integer programming problem, and a suboptimal solution is found by use of continuous relaxation. In [349], a low-complexity suboptimal algorithm is proposed, which divides the problem into two subproblems: (i) find the required power and the number of subcarriers for each user and (ii) find the exact subcarrier and rate allocation. In [350], the discrete subcarrier allocation problem is relaxed into a constrained optimization problem with continuous variables. The problem is shown to belong to the class of convex programming problems, thus allowing the optimal assignment to be found with numerical methods. In [252], the problem is formulated using a max–min criterion for downlink application. The optimal channel-assignment problem is formulated as a convex optimization problem, and a low-complexity suboptimal algorithm is developed. Real-time subcarrier allocation schemes are studied in [236, 336], which use subcarrier allocation only to enhance the performance while fixing modulation levels. The Hungarian method [169] can be used to solve such problems with a high computational complexity of $O(N^4)$, where N is the number of subcarriers. The suboptimal algorithms are developed in [236, 336] to simplify the Hungarian algorithm and achieve similar performances. In [312], adaptive modulation is applied for the uplink OFDMA system.

Most of the previous approaches study how to efficiently maximize the total transmission rate or minimize the total transmitted power under some constraints. The formulated problems and their solutions are focused on the efficiency issue. But these approaches benefit the users closer to the BS or with a higher power capability. The fairness issue has been mostly ignored. On the other hand, as for fairness among users, the max–min criterion has been considered for channel allocation in multiuser OFDM systems [252]. However, by using this criterion, it is not easy to take into account the notion that users might have different requirements. Moreover, because the max–min approach deals with the worst-case scenario, it penalizes users with better channels and reduces the system efficiency. In addition, most of the existing solutions have high complexities, which prohibit them from practical implementation. Therefore it is necessary to develop an approach that considers altogether fairness of resource allocation, system efficiency, and complexity.

In daily life, a market serves as a central gathering point, where people can exchange goods and negotiate transactions, so that people can be satisfied through bargaining. Similarly, in single-cell multiuser OFDMA systems, there is a BS that can serve as a function of the market. The distributed users can negotiate via the BS to cooperate in making the decisions on subcarrier usage, such that each of them can operate at its optimum and joint agreements are made about their operating points. Such a fact motivates us to apply game theory [95, 227, 343] and especially cooperative game theory, which can achieve the crucial notion of fairness and maximize the overall system rate. The concepts of the NBS and coalitions are taken into consideration, because they provide a fair operation point in a distributed implementation.

Motivated by the preceding reasons, we apply cooperative game theory for resource allocation in OFDMA systems. The goal is to maximize the overall system rate under the constraints of each user's minimal rate requirement and maximal transmitted power. First we develop a two-user bargaining algorithm to negotiate the usage of subcarriers. The approach is based on the NBS, which maximizes the system performance while keeping the NBS fairness, which is a generalized proportional fairness. Then we group the users into groups of size two. The group is defined as a coalition. Within each coalition, we use the two-user algorithm to improve the performance. In the next iteration, new coalitions are formed and subcarrier allocation is optimized until no improvement can be obtained. By using the Hungarian method, optimal coalitions are formed and the number of iterations can be greatly reduced. A significant point for the proposed iterative algorithm is that the complexity for each iteration is only $O(K^2 N \log N + K^4)$, where K is the number of users. From the simulation results, the proposed algorithms allocate resources fairly and efficiently compared with the other two schemes: maximal rate and max–min fairness. The NBS fairness is demonstrated by the fact that a user's rate is not influenced by the interfering users.

For the system model, we consider an uplink scenario of a single-cell multiuser OFDMA system. There are K users randomly located within the cell. The users want to share their transmissions among N different subcarriers. Each subcarrier has a bandwidth of W. The ith user's transmission rate is R_i and is allocated to different subcarriers as $R_i = \sum_{j=1}^{N} r_{ij}$, where r_{ij} is the ith user's transmission rate in the jth subcarrier. Define the rate-allocation matrix $\mathbf{r}$ with $[\mathbf{r}]_{ij} = r_{ij}$. Define the subcarrier assignment matrix $[\mathbf{A}]_{ij} = a_{ij}$, where

$$a_{ij} = \begin{cases} 1, & \text{if } r_{ij} > 0 \\ 0, & \text{otherwise} \end{cases}. \tag{13.95}$$

For single-cell multiuser OFDMA, no subcarrier can support the transmissions for more than one user, i.e., $\sum_{i=1}^{K} a_{ij} = 1, \forall j$.

Adaptive modulation provides each user with the ability to match each subcarrier's transmission rate r_{ij} according to its channel condition. MQAM is a modulation method with a high spectrum efficiency, which is adopted in our system without loss of generality. In [49], the BER of MQAM as a function of rate and SNR is approximated by

$$\text{BER}_{ij} \approx c_1 e^{-c_2 \frac{\Gamma_{ij}}{2^{r_{ij}} - 1}}, \tag{13.96}$$

where $c_1 \approx 0.2$, $c_2 \approx 1.5$, and Γ_{ij} is the ith user's SNR at the jth subcarrier, given by

$$\Gamma_{ij} = \frac{P_{ij} G_{ij}}{\sigma^2}, \tag{13.97}$$

where G_{ij} is the subcarrier channel gain and P_{ij} is the transmitted power for the ith user in the jth subcarrier. The thermal-noise power for each subcarrier is assumed to be the same and equal to σ^2. Define the power-allocation matrix $[\mathbf{P}]_{ij} = P_{ij}$. From (13.96), without loss of generality, we assume the same BER for all users in all subcarriers. Then we have

$$r_{ij} = W \log_2 \left(1 + \frac{P_{ij} G_{ij} c_3}{\sigma^2} \right), \tag{13.98}$$

where $c_3 = c_2 / \ln(c_1 / \text{BER})$ with $\text{BER} = \text{BER}_{ij}, \forall i, j$.

We assume slow fading in which the channel is stable within each OFDM frame. The channel conditions of different subcarriers for each user are assumed perfectly estimated. There exist reliable feedback channels from BS to mobile users without any delay. Moreover, for a practical system, the OFDM frequency offset between the mobile user and the BS is around several tenths of a hertz. The intercarrier interference caused by the frequency offset may cause some error floor increase. However, this is not the bottleneck limiting the system performance, and this offset can be fed back to the mobile for adjustment. In [23], a guard subcarrier is put at the edge of each subcarrier such that multiple-access interference can be minimized and a synchronized algorithm can be applicable to each subcarrier. So we assume that mobiles and BSs are synchronized.

In Figure 13.38, an illustrative three-user example is given for the system setup. The number of subcarriers for communication is eight. Each subcarrier is occupied by one user. According to the channel conditions, a user selects an adaptive modulation level and adjusts its rate for this subcarrier. The conflicts are that a certain subcarrier is good for more than one user, and the problem is to decide the user to which this subcarrier should be assigned. So our goal is to assign the subcarriers by negotiating with other users via BS so that each user can obtain its minimal rate while the system overall performance is optimized. In the following, we discuss in detail how to implement the negotiation process.

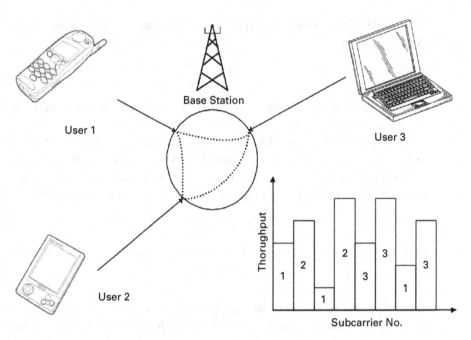

Figure 13.38 System model.

Next, we briefly review the basic concepts and theorems for NBS. Then we give an overview on how to apply these ideas to OFDMA resource allocation.

The bargaining problem of cooperative game theory can be described as follows [95, 227, 343]: Let $\mathbf{K} = \{1, 2, \ldots, K\}$ be the set of players. Let $\mathbf{S}$ be a closed and convex subset of $\Re^K$ to represent the set of feasible payoff allocations that the players can get if they all work together. Let $R_{\min}^i$ be the minimal payoff that the ith player would expect; otherwise, he will not cooperate. Suppose $\{R_i \in \mathbf{S} | R_i \geq R_{\min}^i, \forall i \in \mathbf{K}\}$ is a nonempty bounded set. Define $\mathbf{R}_{\min} = (R_{\min}^1, \ldots, R_{\min}^K)$; then the pair $(\mathbf{S}, \mathbf{R}_{\min})$ is called a K-person bargaining problem.

Within the feasible set $\mathbf{S}$, we define the notion of the Pareto optimal as a selection criterion for the bargaining solutions.

DEFINITION 49 *The point $(R_1, \ldots, R_K)$ is said to be Pareto optimal if and only if there is no other allocation R_i' such that $R_i' \geq R_i, \forall i$, and $R_i' > R_i, \exists i$, i.e., there exists no other allocation that leads to superior performance for some users without inferior performance for some other users.*

There might be an infinite number of Pareto optimal points. We need further criteria to select a bargaining result. A possible criterion is fairness. One commonly used fairness criterion is max–min [252], in which the performance of the user with the worst channel conditions is maximized. This criterion penalizes the users with good channels, and as a result generates inferior overall system performance. Here we use the criterion of fairness NBS. The intuitive idea is that, after the minimal requirements are satisfied for all users, the rest of the resources are allocated proportionally to users according to their conditions. We will discuss the *proportional fairness* concept, which is a special case of NBS fairness and show, the fair results later. There exist many kinds of cooperative game solutions [227]. Among them, NBS provides a unique and fair Pareto optimal operation point under the following conditions. NBS is briefly explained as follows:

DEFINITION 50 $\bar{\mathbf{r}}$ *is said to be a Nash Bargaining Solution in* $\mathbf{S}$ *for* $\mathbf{R}_{\min}$, *i.e.*, $\bar{\mathbf{r}} = \phi(\mathbf{S}, \mathbf{R}_{\min})$, *if the following Axioms are satisfied:*

1. *Individual Rationality:* $\bar{R}_i = \sum_{j=1}^{N} \bar{r}_{ij} \geq R_{\min}^i, \forall i.$
2. *Feasibility:* $\bar{\mathbf{r}} \in \mathbf{S}.$
3. *Pareto Optimality: For every* $\hat{\mathbf{r}} \in \mathbf{S}$, *if* $\sum_{j=1}^{N} \hat{r}_{ij} \geq \sum_{j=1}^{N} \bar{r}_{ij}, \forall i$, *then* $\sum_{j=1}^{N} \hat{r}_{ij} = \sum_{j=1}^{N} \bar{r}_{ij}, \forall i.$
4. *Independence of Irrelevant Alternatives: If* $\bar{\mathbf{r}} \in \mathbf{S}' \subset \mathbf{S}$, $\bar{\mathbf{r}} = \phi(\mathbf{S}, \mathbf{R}_{\min})$, *then* $\bar{\mathbf{r}} = \phi(\mathbf{S}', \mathbf{R}_{\min}).$
5. *Independence of Linear Transformations: For any linear scale transformation* ψ, $\psi[\phi(\mathbf{S}, \mathbf{R}_{\min})] = \phi[\psi(\mathbf{S}), \psi(\mathbf{R}_{\min})].$
6. *Symmetry: If* $\mathbf{S}$ *is invariant under all exchanges of agents,* $\phi_j(\mathbf{S}, \mathbf{R}_{\min}) = \phi_{j'}(\mathbf{S}, \mathbf{R}_{\min}), \forall j, j'.$

Axioms 4–6 are called axioms of fairness. The irrelevant alternative axiom asserts that eliminating the feasible solutions that would not have been chosen should not affect the NBS solution. Axiom 5 asserts that the bargaining solution is scale invariant. The symmetric axiom asserts that, if the feasible ranges for all users are completely symmetric, then all users have the same solution.

The following theorem shows that there is exactly one NBS that satisfies the preceding axioms [227].

THEOREM 28 *Existence and uniqueness of NBS: There is a unique solution function* $\phi(\mathbf{S}, \mathbf{R}_{\min})$ *that satisfies all six axioms in Definition* 50. *And this solution satisfies* [227]:

$$\phi(\mathbf{S}, \mathbf{R}_{\min}) \in \arg \max_{\bar{\mathbf{r}} \in \mathbf{S}, \bar{R}_i \geq R_{\min}^i, \forall i} \prod_{i=1}^{K} (\bar{R}_i - R_{\min}^i). \qquad (13.99)$$

As previously discussed, the cooperative game in the multiuser OFDMA system can be defined as follows: Each user has R_i as its objective function, where R_i is bounded above and has a nonempty, closed, and convex support. The goal is to maximize all R_i simultaneously. $\mathbf{R}_{\min}$ represents the minimal performance and is called the initial agreement point. Define $\mathbf{S}$ as the feasible set of rate-allocation matrix $\mathbf{r}$ that satisfies $R_i \geq R_{\min}^i, \forall i$. The problem, then, is to find a simple way to choose the operating point in $\mathbf{S}$ for all users, such that this point is optimal and fair.

Because the channel conditions for a specific subcarrier may be good for more than one user, there is a competition among users for their transmissions over the subcarriers with large G_{ij}. Moreover, the maximal transmitted power for each user is bounded by the maximal transmitted power $P_{\max}$ and each user has a minimal rate requirement $R_{\min}^i$ if it is admitted to the system. The optimization goal is to determine different users' transmission function **A** and **P** for the different subcarriers such that the cost function can be maximized, i.e.,

$$\max_{\mathbf{A},\mathbf{P}} U, \tag{13.100}$$

$$\text{subject to} \begin{cases} \sum_{i=1}^{K} a_{ij} = 1, \forall j; \\ R_i \geq R_{\min}^i, \forall i; \\ \sum_{j=1}^{N} P_{ij} \leq P_{\max}, \forall i, \end{cases}$$

where U can have three definitions in terms of the objectives:

$$\text{Maximal rate: } U = \sum_{i=1}^{K} R_i, \tag{13.101}$$

$$\text{Max–min fairness: } U = \min R_i, \tag{13.102}$$

$$\text{Nash bargaining solutions: } U = \prod_{i=1}^{K} \left(R_i - R_{\min}^i \right). \tag{13.103}$$

For maximal rate optimization, the overall system rate is maximized. For max–min fairness optimization, the worst-case situation is optimized with strict fairness. We proposed the NBSs for the following two reasons. First, it will be shown later that this form can ensure fairness of allocation in the sense that NBS fairness is a generalized proportional fairness. Second, cooperative game theories prove that there exists a unique and efficient solution under the six axioms. The difficulty in solving (13.100) by traditional methods lies in the fact that the problem itself is a constrained combinatorial problem and the constraints are nonlinear. Thus the complexities of the traditional schemes are high especially with large number of users. Moreover, distributed algorithms are desired for uplink OFDMA systems, while centralized schemes are dominant in literature. In addition, most of the existing work does not discuss the issue of fairness. We use the bargaining concept to develop simple and distributed algorithms with limited signaling that can achieve an efficient and fair resource allocation.

Figure 13.39 illustrates a two-user example in which $\mathbf{R}_{\min}$ is assumed to be zero. The shaded area **S** is the feasible range for $\mathbf{R}_1$ and $\mathbf{R}_2$. For the NBS cost function, the optimal point is B at $(\tilde{R}_1, \tilde{R}_2)$ with $R_1 R_2 = \tilde{C}$, where $\tilde{C}$ is the largest constant for feasible set **S**. The physical meaning of this is that "after the users are assigned with the minimal rate, the remaining resources are divided between users in a ratio equal to the rate at which the utility can be transferred" [227]. The geometrical interpretation is that an isosceles triangle ABC can be drawn with $(\tilde{R}_1, \tilde{R}_2)$ as the apex, such that its one side is tangent to the set **S** and the other side passes $(R_{\min}^1, R_{\min}^2)$, i.e., the origin. Because line BC is also tangent to curve $R_1 R_2 = \tilde{C}$, the ratio that two rates can be exchanged within the set **S** is equal to the ratio of the two rates. The maximal rate approach has the optimal point at $R_1^* + R_2^* = C^*$, which is the point within the feasible set **S** where the sum C^* of R_1 and R_2 is maximized. Compared with the maximal rate approach, the overall rate of the NBS is C_{NBS}, which is slightly smaller than C^*. So the NBS has a small overall rate loss, but keeps fairness. The max–min approach considers the worst-case scenario and

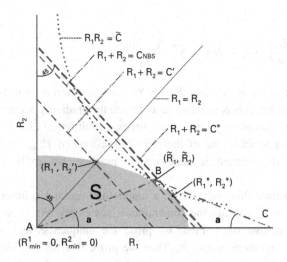

Figure 13.39 Two-user illustrative example.

has the optimal point with $R_1' = R_2' = C'$, where C' is the largest constant for feasible set **S**. The overall rate for the max–min approach is $2C'$. Compared with the max–min algorithm, the NBS has a much higher overall rate, i.e., $2C' \ll C_{\text{NBS}}$.

In addition, we will show in the following definition and theorem that proportional fairness [149], which is widely used in wired networks, is a special case of fairness provided by the NBS.

DEFINITION 51 *We call the rate distribution proportionally fair when any change in the distribution of rates results in a sum of the proportional changes of the utilities that is nonpositive, i.e.,*

$$\sum_i \frac{R_i - \tilde{R}_i}{\tilde{R}_i} \leq 0, \ \forall R_i \in \mathbf{S}, \tag{13.104}$$

where $\tilde{R}_i$ and R_i are the proportionally fair rate distribution and any other feasible rate distribution for the ith user, respectively.

THEOREM 29 *When $R_{\min}^i = 0, \forall i$, the NBS fairness is the same as the proportional fairness.*

Proof: Because the function of ln is concave and monotonic, when $R_{\min}^i = 0, \forall i$, the NBS in (13.99) is equivalent to

$$\max_{\mathbf{r} \in \mathbf{S}} \sum_{i=1}^{K} \ln(R_i). \tag{13.105}$$

Define $\hat{U}_i = \ln(R_i)$. The gradient of $\hat{U}_i$ at the NBS point $\tilde{R}_i$ is $\frac{\partial \hat{U}_i}{\partial R_i}|_{\tilde{R}_i}$. Because the NBS point optimizes (13.105), for any point deviating from the NBS point, the following optimality condition holds:

$$\sum_i \frac{\partial \hat{U}_i}{\partial R_i}|_{\tilde{R}_i}(R_i - \tilde{R}_i) = \sum_i \frac{R_i - \tilde{R}_i}{\tilde{R}_i} \leq 0. \tag{13.106}$$

The preceding equation means for all feasible $R_i \in S, \forall i$ that are different from NBS point $\tilde{R}_i$, the overall change of benefits is negative according to the gradients. Moreover, the preceding equation is the same as the proportional fairness definition in (13.104). So the proportional fairness is a special case of the NBS fairness when $R^i_{\min} = 0, \forall i$. Because the minimal rate requirement is necessary in practice, we apply NBS fairness.

Next, we want to demonstrate that there exists a unique and optimal solution in (13.100), when the feasible set satisfying the constraints is not empty. We show the uniqueness and optimality in two steps. First, we prove the uniqueness and optimality with fixed channel-assignment matrix $\mathbf{A}$. Then we prove that the probability that there exists more than one optimal point is zero for different channel-assignment matrix $\mathbf{A}$.

First, under fixed channel-assignment matrix $\mathbf{A}$, each user tries to maximize its own rate under the power constraint independently, because any subcarrier is not shared by more than one user. This is similar to the single-user case.   **QED**

Because we assume the feasible set that satisfies the constraints in (13.100) is not empty, within the feasible set, each user can get its minimal rate requirement $R^i_{\min}$ by allocating its power to the assigned channel set. For all three cost functions in (13.101), (13.102), and (13.103), the problem in (13.100) is reduced to the following problem:

$$\max_{\mathbf{P}} U_i = R_i, \tag{13.107}$$

$$\text{subject to } \sum_{j=1}^{N} P_{ij}a_{ij} \leq P_{\max}, \forall i.$$

Obviously, the preceding problem is a water-filling problem [52] and has a unique optimal solution. Define

$$I_{ij} = \frac{\sigma^2}{c_3 G_{ij}}. \tag{13.108}$$

The unique optimal solution is

$$P_{ij} = (\mu_i - I_{ij})^+ \text{ and } r_{ij} = W \log_2\left(1 + \frac{P_{ij}a_{ij}}{I_{ij}}\right), \tag{13.109}$$

where $y^+ = \max(y, 0)$. Here μ_i is the water level and can be solved by bisection search of

$$\sum_{j=1}^{N} P_{ij}a_{ij} = P_{\max}. \tag{13.110}$$

We have proved the optimality and uniqueness with fixed channel assignment. The channel assignment is a combinatorial problem with a finite number of combinations. For example, the total number of combinations for the system with K users and N subcarriers is K^N. So we can obtain the optimal solution by solving the following problem:

$$\arg\max_{\mathbf{A}} U, \qquad\qquad (13.111)$$

where U is obtained by solving (13.107) with respect to each $\mathbf{A}$.

The preceding problem can be solved by means of a full search to get the optimal channel assignment and power allocation. Among all the implementations of $\mathbf{A}$, we select the one that generates the largest U. Because the optimization goal U, the channel gains, and the rates are continuous random values, it has zero probability of having two channel assignment matrices that generate the same value of optimization goal. So, with probability one, there exists a unique channel-assignment matrix that generates the optimal solution in (13.100).

Next, we consider the case in which $K = 2$ and we will develop a fast two-user bargaining algorithm. Similar to bargaining in a real market, the intuitive idea to solve the two-user problem is to allow two users to negotiate and exchange their subcarriers such that mutual benefits can be obtained. The difficulty is to determine how to optimally exchange subcarriers, which is a complex integer programming problem. An interesting low-complexity algorithm was given in [350]. The idea is to sort the order of subcarriers first and then to use a simple two-band partition for the subcarrier assignment. When the SNR is high, the two-band partition for two-user subcarrier assignment is near optimal for the optimization goal of maximizing the weighted sum of both users' rates.

We propose a fast algorithm between two users for the optimization goals by exchanging their subcarriers, as shown in Table 13.5. First all subcarriers are initially assigned. Then, two users' subcarriers are sorted and a two-band partition algorithm is applied for them to negotiate the subcarriers. For a maximal rate optimization goal, only one iteration is necessary. For a NBS optimization goal, an intermediate parameter needs to be updated for every iteration. From the simulations, the iterations between step 2 and step 5 converge within 2 to 3 rounds. The algorithm has the complexity of $O(N^2)$ for each iteration and can be further improved by using a binary search algorithm with a complexity of only $O(N \log N)$ for each iteration. It is worth mentioning that all the iterations in Table 13.5 happen within the BS. So there is no need for signaling between users and base stations.

PROPOSITION 3 *The algorithm in Table 13.5 is near optimal for both the problem of maximal rate and NBS goals in (13.100) with the number of users equal to two, when the SNR of each subcarrier for all users in (13.97) is much greater than 1 and there exists a feasible solution.*

Proof: In [350], the authors proved that if, at the optimal subcarrier partition, the SNR is large in every subcarrier for all users, and if the subcarriers are sorted according

Table 13.5 Two-user algorithm

---

1. Initialization:
Initialize the subcarrier assignment with the minimal rate requirements.
For *Maximal rate*, $\varrho_1 = \varrho_2 = 1$;
For *NBS*, calculate ϱ_1 and ϱ_2.

2. Sort the subcarriers:
  Arrange the index from the largest to smallest $g_{1j}^{\varrho_1}/g_{2j}^{\varrho_2}$.

3. For $j = 1, \ldots, N - 1$
  User 1 occupies and water fills subcarrier 1 to j;
  User 2 occupies and water fills subcarrier j + 1 to N.
  Calculate U.
  End

4. Choose the two-band partition (the corresponding j)
  that generates the largest U satisfing the constraints.
  Calculate **A**, **P**, R_1, and R_2.

5. Update channel assignment
  $-$*Maximal rate*: Return
  $-$*NBS*:
  If U can not be increased by updating ϱ_1 and ϱ_2, the iteration ends;
  else, update $\varrho_1 = 1/(R_1 - R_{\min}^1)$, $\varrho_2 = 1/(R_2 - R_{\min}^2)$; go to Step 2.

---

to users' subcarrier channel gain, then the optimal subcarrier partition that maximizes $\alpha_1 R_1 + \alpha_2 R_2$ consists of two contiguous frequency bands with each user occupying one band. Here R_1 and R_2 are the two users' rates, and α_1 and α_2 are the relative priorities for both users. For maximal rate optimization goal, the theorem is proved by letting $\alpha_1 = \alpha_2 = 1$ [350].

For NBS, the optimization goal is $U = (R_1 - R_{\min}^1)(R_2 - R_{\min}^2)$, which contains a term of $R_1 R_2$. Similar to the approach in [350], here we relax the channel-assignment matrix **A** to continuous values with $0 \le a_{ij} \le 1$, $\forall i, j$. We write the Lagrangian function of (13.100) as a function of a_{ij} and P_{ij}:

$$
\begin{aligned}
L = & \left[\sum_{j=1}^{N} a_{1j} W \log_2 \left(1 + \frac{P_{1j} G_{1j} c_3}{a_{1j} \sigma^2} \right) - R_{\min}^1 \right] \\
& \left[\sum_{j=1}^{N} a_{2j} W \log_2 \left(1 + \frac{P_{2j} G_{2j} c_3}{a_{2j} \sigma^2} \right) - R_{\min}^2 \right] \\
& + \sum_{j=1}^{N} \lambda_j \left(\sum_{i=1}^{2} a_{ij} - 1 \right) + \sum_{i=1}^{2} \kappa_i \left(\sum_{j=1}^{N} P_{ij} - P_{\max} \right) \\
& - \sum_{i=1}^{2} \sum_{j=1}^{N} v_{ij}^1 P_{ij} - \sum_{i=1}^{2} \sum_{j=1}^{N} v_{ij}^2 a_{ij},
\end{aligned}
\tag{13.112}
$$

where λ_j, κ_i, v_{ij}^1, and v_{ij}^2 are Lagrangian multipliers. By using the KKT condition [27], we take the derivative of (13.112) with respect to a_{ij} and have

$$\frac{\log_2\left(1 + \frac{P_{1j}G_{1j}c_3}{a_{1j}\sigma^2}\right) - \frac{\frac{P_{1j}G_{1j}c_3}{a_{1j}\sigma^2}}{1 + \frac{P_{1j}G_{1j}c_3}{a_{1j}\sigma^2}}}{\left[\sum_{j=1}^N a_{1j} W \log_2\left(1 + \frac{P_{1j}G_{1j}c_3}{a_{1j}\sigma^2}\right) - R_{\min}^1\right]}$$

$$= \frac{\log_2\left(1 + \frac{P_{2j}G_{2j}c_3}{a_{2j}\sigma^2}\right) - \frac{\frac{P_{2j}G_{2j}c_3}{a_{2j}\sigma^2}}{1 + \frac{P_{2j}G_{2j}c_3}{a_{2j}\sigma^2}}}{\left[\sum_{j=1}^N a_{2j} W \log_2\left(1 + \frac{P_{2j}G_{2j}c_3}{a_{2j}\sigma^2}\right) - R_{\min}^2\right]}. \tag{13.113}$$

From (13.109), we have water-filling results for discrete a_{ij}. Define the positive weight factor

$$\varrho_i = \begin{cases} 1/\left(\sum_{j=1}^N a_{ij} W \log_2\left(1 + \frac{P_{ij}G_{ij}c_3}{a_{ij}\sigma^2}\right) - R_{\min}^i\right), \\ \quad \text{if } \sum_{j=1}^N a_{ij} W \log_2\left(1 + \frac{P_{ij}G_{ij}c_3}{a_{ij}\sigma^2}\right) \geq R_{\min}^i + \epsilon; \\ 1/\varepsilon, \text{ otherwise,} \end{cases} \tag{13.114}$$

where ϵ is a small positive number and ε is a small positive value to ensure the large weight for the user whose rate is less than $R_{\min}^i + \epsilon$. We put (13.108), (13.109), and (13.114) into (13.113); at a high SNR, we have

$$\varrho_1\left[\log_2\left(\frac{\mu_1}{I_{1j}}\right) + \frac{I_{1j}}{\mu_1} - 1\right] = \varrho_2\left[\log_2\left(\frac{\mu_2}{I_{2j}}\right) + \frac{I_{2j}}{\mu_2} - 1\right]. \tag{13.115}$$

If a subcarrier is used by user 1, i.e., $a_{1j} = 1$ and $a_{2j} = 0$, in which case the left-hand side should be strictly greater than the right-hand side. At high SNR, the fraction I_{ij}/μ_i on either side of (13.115) can be approximated by zero. Let $g_{ij} = 1/I_{ij}$. Take the difference between the left-hand side and the right-hand side of (13.113) and define function f as

$$f\left(\frac{g_{1j}^{\varrho_1}}{g_{2j}^{\varrho_2}}\right) \approx \log_2\left(\frac{g_{1j}^{\varrho_1}}{g_{2j}^{\varrho_2}}\right) + \log_2\left(\frac{\mu_1^{\varrho_1}}{\mu_2^{\varrho_2}}\right) + \varrho_2 - \varrho_1. \tag{13.116}$$

We are able to decide whether a subcarrier is used by user 1 or user 2 by checking whether the function is greater than zero or less than zero. We arrange the index of subcarriers to make $g_{1j}^{\varrho_1}/g_{2j}^{\varrho_2}$ decreased in j. With fixed ϱ_1 and ϱ_2, $f(g_{1j}^{\varrho_1}/g_{2j}^{\varrho_2})$ is a monotonous function of j. Then (13.116) is similar to the weighted maximization in [350] and the optimum partition is a two-band solution.

The left-hand side and the right-hand side of (13.115) illustrate the marginal benefits of extra bandwidth for user 1 and user 2 on the subcarrier j, respectively. Within each iteration, ϱ_i is fixed. Then the algorithm achieves the boundary point of the feasible region [350]. Then in the next iteration, the new ϱ_i is updated. Remember that $\tilde{R}_i$ is the NBS. If $R_1 > \tilde{R}_1$ and $R_2 < \tilde{R}_2$, from (13.114), ϱ_1 is small and ϱ_2 is large. Consequently, the marginal benefit of user 1 will be reduced and it will have disadvantage for channel allocation in the next iteration, and vice versa. This is one explanation of why the proposed scheme converges to the NBS. The iterative algorithm converges when (13.113)

Table 13.6 Multiuser algorithm

---

1. Initialize the channel assignment:
Assign all subcarriers to users.
2. Coalition Grouping:
If the number of users is even, the users are grouped into coalitions;
else, a dummy user is created to make the total number of users even.
No user can exchange its resource with this dummy user.
— *Random method*: Randomly form two-user coalition.
— *Hungarian method*: Form user coalitions by the Hungarian algorithm.
3. Bargain within Each Coalition:
Negotiate between two users in all coalitions to exchange the subcarriers
using the two-user algorithm in Table 13.5.
4. Repeat:
Repeat step 2 and step 3, until no further improvement can be achieved.

---

is held. It is worth mentioning that the proposed two-user algorithm might not converge toward the NBS, because of the nonlinear and combinatorial nature of the formulated problem.   **QED**

For the case in which the number of users is larger than two, most work in the literature concentrates on solving the OFDMA resource allocation problem for all users together in a centralized way [312, 335]. Because the problem itself is combinatorial and nonlinear, the computational complexity is very high with respect to the number of subcarriers by the existing methods [312, 335]. Here, we propose a two-step iterative scheme: First, users are grouped into pairs, which are called coalitions. Then for each coalition, the algorithm in Table 13.5 is applied for two users to negotiate and improve their performances by exchanging subcarriers. Further, the users are regrouped and renegotiate again and again until convergence. By use of this scheme, the computational cost can be greatly reduced. First, we give the strict definition of coalition as follows.

DEFINITION 52 *For a K-person game, any nonempty subset of the set of players is called a coalition.*

The question now is how to group users into coalitions with size 2. A straightforward algorithm forms the coalition randomly and lets the users bargain arbitrarily. We call this algorithm the *random method*, which can be described by the steps in Table 13.6. During the initialization, the goal is to assign all subcarriers to users and try to satisfy the minimal rate and maximal power constraint. We develop a fast algorithm. Starting from the user with the best channel conditions, if the user has a rate higher than or equal to $R_{\min}^i$, it is removed from the assignment list. After every user has a high enough rate, the remaining subcarriers are greedily assigned to the users according to their channel gains. Note that there is no need for the initial assignment to satisfy all constraints. The constraints can be satisfied during the iterations of negotiations.

We quantify the convergence speed by the round of negotiations. The convergence speed of the random method becomes slow with the number of users increasing, because the negotiations within arbitrarily grouped coalitions are less effective and most negotiations turn out to be the same or have little improvement compared with the performance of the channel allocation before the negotiations. So the optimal cooperation grouping among subsets of the users should be taken into consideration. To speed up convergence, each user needs to carefully select whom it should negotiate with.

Each user's channel gains vary over different subcarriers. A user may be preferred by many users to form coalitions with, while only a two-user coalition is allowed. Thus the problem of deciding the coalition pairs can be stated as an assignment problem [169]: "a special structured linear programming which is concerned with optimally assigning individuals to activities, assuming that each individual has an associated value describing its suitability to execute that specific activity."

Now we formulate the assignment problem in detail. Define the benefit for the ith user to negotiate with the jth user as b_{ij}. Obviously $b_{ii} = 0, \forall i$. For the other cases, from (13.100), each element of the cost table $\mathbf{b}$ can be expressed as

$$b_{ij} = \max[U(\tilde{R}_i, \tilde{R}_j) - U(\hat{R}_i, \hat{R}_j), 0], \tag{13.117}$$

where $\tilde{R}_i$ and $\tilde{R}_j$ are the rates if the negotiation happens, and $\hat{R}_i$ and $\hat{R}_j$ are the original rates, respectively. Obviously $\mathbf{b}$ is also symmetric. The proposed two-user algorithm discussed previously can calculate each $b_{ij}, \forall i, j$. The total complexity is $O(K^2 N \log_2 N)$.

Define a $K \times K$ assignment table $\mathbf{X}$. Each component represents whether or not there is a coalition between two users:

$$X_{ij} = \begin{cases} 1 & \text{if user i negotiates with user j} \\ 0 & \text{otherwise} \end{cases}. \tag{13.118}$$

Obviously matrix $\mathbf{X}$ is symmetric, $\sum_{i=1}^{K} X_{ij} = 1, \forall j$, and $\sum_{j=1}^{K} X_{ij} = 1, \forall i$.

So the assignment problem is how to select the pairs of negotiations such that the overall benefit can be maximized, which is stated as

$$\max_{\mathbf{X}} \sum_{i=1}^{K} \sum_{j=1}^{K} X_{ij} b_{ij}, \tag{13.119}$$

$$\text{s.t.} \begin{cases} \sum_{i=1}^{K} X_{ij} = 1 & j = 1 \ldots K, \forall i; \\ \sum_{j=1}^{K} X_{ij} = 1 & i = 1 \ldots K, \forall j; \\ X_{ij} \in \{0, 1\} & \forall i, j. \end{cases}$$

One of the solutions for (13.119) is the Hungarian method [169], which can always find the optimal coalition pairs. The Hungarian method has the minimization optimization goal. So we change the maximization problem in (13.119) into a minimization problem by defining $B_{ij} = -b_{ij} + \max(b_{ij})$. The Hungarian algorithm is briefly explained in Table 13.7.

In each round, the optimal coalition pairs $\mathbf{A}$ are determined by the Hungarian method, and then the users are set to bargain together using the two-user algorithm in Table 13.5. The whole algorithm stops when no bargaining can further improve the performance, i.e., $\mathbf{b}$ is equal to a zero matrix. From the preceding explanations, we develop the multiuser resource allocation in the multiuser OFDMA systems in Table 13.6.

Table 13.7 Hungarian method

---

1. Subtract the entries of each row of **B** by the row minimum,
   so that each row has at least one zero and all entries are positive or zero.
2. Subtract the entries of each column by the column minimum,
   so that each row and each column has at least one zero.
3. Select rows and columns across which lines are drawn, in such a way
   that all the zeros are covered and that no more lines have been drawn
   than necessary.
4. A test for optimality.
   (i) If the number of the lines is K, choose a combination **A**
       from the modified cost matrix in such a way that the sum is zero.
   (ii) If the number of the lines is less than K, go to step 5.
5. Find the smallest element that is not covered by any of the lines.
   Then subtract it from each entry which is not covered by the lines and
   add it to each entry that is covered by a vertical and a horizontal line.
   Go back to 3.

In each iteration, the optimization function U is nondecreasing in step 2 and step 3 and the optimal solution is upper bounded. Consequently, the proposed multiuser algorithm is convergent. However, because the proposed problem in (13.100) is nonlinear and nonconvex, and also because of the combinatorial nature of the formulated problem, there might be some local optima that the proposed scheme may fall into, even though the Hungarian method can find optimal **X**. From the simulation results, we will show that the problem of local optima is not severe.

The complexity of the Hungarian method is $O(K^4)$. So the overall complexity for each iteration of the proposed scheme is $O(K^2 N \log_2 N + K^4)$. Because the number of users is much less than the number of subcarriers, the complexity of the proposed algorithm is much lower than the schemes that apply the Hungarian method directly to the subcarrier domain. For example, for IEEE 802.11a, there are 48 subcarriers. For the schemes previously mentioned, the complexity is $N^4 = 5308416$. When $K = 4$, the proposed scheme has the complexity of $K^2 N \log_2 N + K^4 = 4545$. Suppose the number of iterations is 10 and the complexity is only 0.86% of the complexity of $O(N^4)$. Moreover, as shown in the simulation, the convergence is mostly obtained within four to six rounds.

When we apply the algorithm in Table 13.7 to the system shown in Figure 13.38, each mobile unit tries to negotiate with other mobile units to exchange resources via the BS, which serves as a mediator. The whole system is similar to the market in the real world: People (mobile units) gather in the market place (BS) to exchange their goods (resources such as subcarriers). Because the channel responses for each user over different subcarriers are known in the BS, the bargaining process is performed within the BS without costing bandwidth for signaling between the users and the BS. The random method can be implemented in a distributed manner with limited signaling to form the coalition pairs, whereas the Hungarian method needs some limited centralized control within the BS to determine the optimal coalition pairs.

To evaluate the performance of the proposed schemes, we consider two-user and multiple-user simulation setups. Three different optimization goals (maximal rate, max–min, and NBS) are compared.

First, a two-user OFDMA system is taken into consideration. We simulate the OFDMA system with 128 subcarriers over the 3.2-MHz band. To make the tones orthogonal to each other, the symbol duration is 40 μs. An additional 10-μs guard interval is used to avoid intersymbol interference because of channel delay spread. This results in a total block length of 50 μs and a block rate of 20 k. The maximal power is $P_{max} = 50$ mW, and the desired BER is 10^{-2} (without channel coding). The thermal-noise level is $\sigma^2 = 10^{-11}$ W. The propagation-loss factor is 3. The distance between user 1 and BS is fixed at $D_1 = 100$ m, while D_2 is varying from 10 to 200 m. $R_{min}^i = 100$ Kbps, $\forall i$. The Doppler frequency is 100 Hz.

To evaluate the performances, we tested 10^5 sets of frequency-selective fading channels, which are simulated by the four-ray Rayleigh model [247] with the exponential power profile and 100-ns RMS delay spread. Thus the impulse response of the model can be represented as

$$g(t) = \sum_{l=1}^{L_p} A_l \alpha_l(t) \delta(t - \tau_l), \tag{13.120}$$

where $L_p = 4$, A_l and τ_l are the amplitude and time delay for the lth ray, respectively, and $\alpha_l(t)$ is the channel gain of a flat Rayleigh fading channel, which can be simulated by Jake's model [148]. Note that the simulated power of each ray is decreasing exponentially according to its delay and the total power of all rays is normalized as 1. The RMS delay spread is the square root of the second central moment of the power-delay profile, which is defined as

$$\sigma_\tau = \sqrt{\overline{\tau^2} - (\overline{\tau})^2}, \tag{13.121}$$

where

$$\overline{\tau^2} = \frac{\sum_{l=1}^{L} A_l^2 \tau_l^2}{\sum_{l=1}^{L} A_l^2}, \quad \overline{\tau} = \frac{\sum_{l=1}^{L} A_l^2 \tau_l}{\sum_{l=1}^{L} A_l^2}. \tag{13.122}$$

In Figure 13.40, the rates of both users for the NBS, maximal rate, and max–min schemes are shown versus D_2. For the maximal rate scheme, the user closer to the BS has a higher rate and the rate difference is very large when D_1 and D_2 are different. For the max–min scheme, both users have the same rate, which is reduced when D_2 is increasing, because the system has to accommodate the user with the worst channel condition. For the NBS scheme, user 1's rate is almost the same regardless of D_2, and user 2's rate is reduced when D_2 is increasing. This shows that the NBS algorithm is fair in the sense that the user's rate is determined only by its channel condition and not by other interfering users' conditions.

In addition, the ratio of two users' rates is shown in Figure 13.41. For the max–min scheme, the ratio is always equal to 1, which is strictly fair but inefficient. For the maximal rate scheme, the ratio changes greatly for different D_2, which is very unfair.

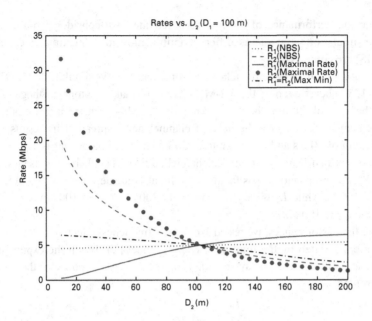

Figure 13.40 Each user's rate (Mbps) vs. D_2.

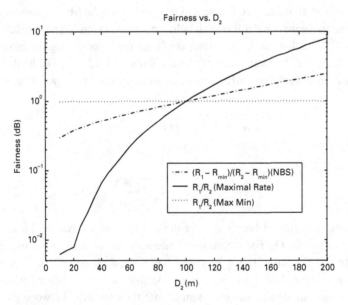

Figure 13.41 Fairness for three schemes.

The user with a better channel condition dominates the resource allocation, while the other user has to starve. The channel gain is mainly determined by the distance and the propagation-loss factor. For the proposed NBS scheme, the ratio of $(R_1 - R_{min}^1)$ over $(R_2 - R_{min}^2)$ changes almost linearly with D_2 in log scale, which shows the NBS fairness.

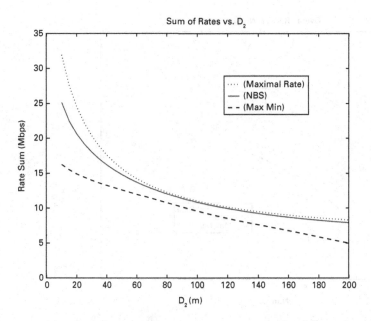

Figure 13.42 Overall rate (Mbps) $R_1 + R_2$.

After each user has the minimal rate requirement, the rest of the resources are allocated to users proportionally according to their channel conditions.[3]

In Figure 13.42, we show the overall rate of two users for three schemes versus D_2. Because the max–min algorithm is for the worst-case scenario, it has the worst performance, especially when the two users have very different channel conditions, because the user with the worse channel conditions limits the usage of the system resources. The NBS scheme has the performance between the maximal rate scheme and the max–min scheme, while the maximal rate scheme is extremely unfair. Moreover, the performance loss of the NBS scheme to that of the maximal rate scheme is small. As we mentioned before, the NBS scheme maintains fairness in a way that one user's performance is unchanged if the other user's channel conditions change. So the proposed algorithm is a good trade-off between fairness and overall system performance.

We set up the simulations with more users to test the proposed algorithms. All the users are randomly located within the cell of radius 200 m. One BS is located in the middle of the cell. Each user has the minimal rate $R_{\min}^i = 25$ kbps. The other settings are the same as those of the two-user case simulations.

In Figure 13.43, we show the sum of all users' rates versus the number of users in the system for the three schemes. We can see that all three schemes have better performances when the number of users increases. The gain is because of multiuser diversity, provided by the independent varying channels across the different users. Performance improvement is satiated gradually. The NBS scheme has a similar performance to that of the maximal rate scheme and has a much better performance than that of the

---

[3] The bumpy part of the max rate scheme curve when D_2 is small is because of the minimal rate constraint.

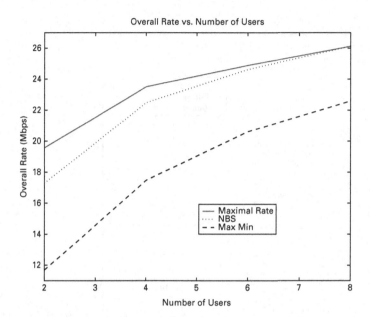

Overall Rate vs. Number of Users

Figure 13.43 Overall rate (Mbps) vs. number of users.

max–min scheme. The performance gap between the maximal rate scheme and the NBS scheme reduces when the number of users is large, because more bargain pair choices are available to increase the system performance.

In Figure 13.44(a), we show the histogram of the number of rounds that is necessary for convergence of the random method and the Hungarian method with eight users. The Hungarian method converges in about one to six rounds, whereas the random method may converge very slowly. The average convergence rounds for the random method is 4.25 times that of the Hungarian method. By use of the Hungarian method, the best negotiation pairs can be found. Consequently, the convergence rate is much quicker and the computation cost is reduced.

In Figure 13.44(b), we show the PDF of the ratio of $\Pi_{i=1}^{K}(R_i - R_{\min}^i)$ of the Hungarian method over that of the random method with eight users. If the ratio is larger than one, the Hungarian method converges to a better solution than the random method. From the curve, the Hungarian method converges to a better solution in most of cases because the random algorithm finds an arbitrary path for convergence and may fall into different local optima. Notice that, for most of cases, the ratio is a small number, so the problem of local optima is not severe. On the other hand, there is a small probability (shown as the shaded area) that the random algorithm has a better performance than that of the Hungarian method, because not all six Nash axioms would be satisfied and the two-user algorithm is suboptimal under low SNR conditions. Therefore, by using the Hungarian method to find the optimal coalition, we can achieve a better and fast NBS for the multiuser situation. Note that the disadvantage of the Hungarian method needs a limited central control in the BS.

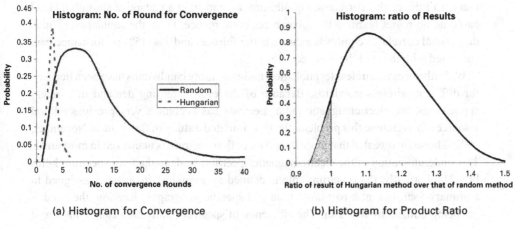

Figure 13.44 Histogram.

We use cooperative game theory, including NBS and coalitions, to develop a fair algorithm for adaptive subcarrier, rate, and power allocation in the multiuser OFDMA systems. The optimization problem takes into consideration fairness and the practical implementation constraints. The proposed algorithm consists of two steps. First a Hungarian method is constructed to determine optimal bargaining pairs among users. Then a fast two-user bargaining algorithm is developed for two users to exchange their subcarriers. The preceding two steps are taken iteratively for users to negotiate the optimal resource allocation. The proposed fast implementation has a low complexity of $O(K^2 N \log N + K^4)$ for each iteration, which is much lower than that of the existing schemes.

From the simulation results, the proposed algorithm shows a similar overall rate to that of the maximal rate scheme and much better performance than that of the max–min scheme. The NBS fairness is demonstrated by the fact that a user's rate is not determined by the interfering users. The proposed algorithm provides a near-optimal fast solution and finds a good trade-off between the overall rate and fairness. The significance of the proposed algorithm is the bargaining and NBS fairness that result in the fair individual performance and good overall system performance.

13.4.2   Opportunistic Spectrum Access for Cognitive Radio

Cognitive radio has been proposed as a novel approach for improving the utilization of the precious limited radio resources by dynamically accessing the spectrum. One of the major design challenges is to coordinate and cooperate in accessing the spectrum opportunistically among multiple distributive users with only local information. In this subsection, we propose a game-theoretical approach with a new solution concept, the correlated equilibrium, which is better compared with the noncooperative Nash equilibrium in terms of spectrum utilization efficiency and fairness among the distributive users. To achieve this correlated equilibrium, we construct an adaptive algorithm [121] based on no-regret learning that guarantees convergence. From the simulation results, the optimal correlated equilibria achieve better fairness and 5%–15% performance gain compared with these of the Nash equilibria.

With advances in wireless technologies, more and more bandwidths have been licensed for different wireless standards. Because of the ever-increasing demand in wireless applications, the electromagnetic radio spectrum has become a very precious natural resource. To overcome this problem, the FCC initiated a study of this issue in November, 2002. The study revealed that more than 70% of the spectrum is unutilized in most areas. This underutilization of the electromagnetic spectrum leads to the term spectrum holes [4, 131]. Formally, the spectrum hole is defined as a band of frequencies assigned to a primary user, but, at a particular time and specific geographic location, the band is not being utilized by that user. The efficiency of spectrum utilization can be improved significantly by having a secondary user to access a spectrum hole that is unoccupied by the primary user at the right place and at the right time. Cognitive radio, which can sense the spectrum utilization of the primary user and opportunistically access the spectral holes, has been proposed to promote the efficient spectrum usage [4, 131]. Some major problems in cognitive radio are the opportunistic spectrum access, coordination among secondary users, and cognitive MAC.

Recent literature targeting the preceding problems is as follows. In [189], the authors study metrics to quantify the characteristics of opportunistic spectrum access, namely equivalent nonopportunistic bandwidth and space–bandwidth product. Sensing-based opportunistic channel access is proposed in [188]. In this work, they address whether an accessible channel is a good opportunity for a secondary user. A decentralized cognitive medium access based on a partial observable Markov decision process (POMDP) is presented in [364]. Spectrum sharing with distributed interference compensation by means of pricing is presented in [142]. Finally, the rule-based device-centric spectrum management scheme is proposed in [366].

Most of the existing works assume the available signaling among cognitive users to coordinate the spectrum usage. To achieve the distributed implementation, game theory is a natural, flexible, and rich tool to study how the autonomous nodes interact and cooperate with each other. There are many applications of game theory in wireless networking, such as noncooperative power-control game [268], cooperative game for OFDMA [107], self-learning repeated game in ad hoc networks [120], and a general survey [288]. In the cognitive radio literature, spectrum sharing for unlicensed band using the one-shot game and repeated game is proposed in [67]. In [223], the resource allocation for secondary users is formulated as a potential game. A survey is studied in [219].

In this subsection, we study the behavior of an individual distributed secondary user to control its rate when the prime user is absent. Each secondary user seeks to maximize its rates over different channels. However, excessive transmissions can cause collisions with other secondary users. The collisions reduce not only the system throughput but also individual performances. From the game-theory perspective, we propose a distributed protocol based on an adaptive learning algorithm for multiple secondary users using only local information. We study a new concept, correlated equilibrium, which is a better solution compared with the noncooperative Nash equilibrium in terms of spectrum utilization efficiency and fairness among the users. Using the correlated equilibrium concept, the distributive users adjust their transmission probabilities over the available channels, so that the collisions are avoided and the users' benefits are optimized. We exhibit the adaptive regret-matching (no-regret) algorithm to learn the correlated equilibrium in a distributed manner. We show that the proposed learning algorithm converges to a set of correlated equilibria with a probability of one. From the simulations, the optimal correlated equilibria achieve better fairness and 5%–15% better performances compared with those of the Nash equilibrium.

For the system model, we consider the general models for dynamic opportunistic spectrum access for cognitive radio, in which there exist several primary users with a set of available channels and a large number of secondary users. The primary users are the rightful owners and have strict priority on the spectrum access. The secondary users are equipped with the spectrum-agile devices, which are used by the secondary users to sense the environment (spectrum usage of the primary users) and adapt the suitable frequency, power, and transmission scheme. The secondary users can opportunistically access the spectrum while the primary users are idle. The primary users are the legacy users that communicate in the traditional way and do not retrofit the secondary users. Hence, the channel availability of secondary users inherently depends on the activities of the primary users. Moreover, the secondary users have to compete for the idle channels among the interfering secondary users. If collisions occur, there are some penalties in the forms of packet loss and power waste. This is the major focus here.

We consider there are N channels in the wireless network. Without loss of generality, each channel has a unit bandwidth. These channels are shared among M primary users and K secondary users seeking for channel access opportunistically.

For adjacent secondary users, they can interfere with each other. We use *interference matrix* $\mathbf{L}$ to depict the interference graph. The interference matrix has the dimension of K by K, and its elements are defined as

$$\mathbf{L}_{ij} = \begin{cases} 1, & \text{if } i \text{ and } j \text{ interfere with each other} \\ 0, & \text{otherwise} \end{cases} \tag{13.123}$$

The interference matrix depends on the relative location of the secondary users.

Next, we define the *channel-availability matrix* as a $K \times N$ matrix, $\mathbf{A}(t)$. Each user can transmit over a specific channel with a set of different rates. The elements of the matrix are defined as

$$\mathbf{A}_{in}(t) = \begin{cases} 1, & \text{if channel } n \text{ is available for secondary user } i \text{ at time } t \\ 0, & \text{otherwise} \end{cases} . \tag{13.124}$$

We note that the channel-availability matrix $\mathbf{A}(t)$ varies over time. This matrix is the result of a sensing task done by secondary users and depends on the primary users' traffic, relative location between the secondary users and the primary users. Notice that each individual user knows only its corresponding row of matrix $\mathbf{L}$ and $\mathbf{A}(t)$.

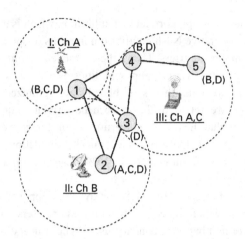

Figure 13.45 Illustration of channel availability for secondary users.

We illustrate an instantaneous example of the channel availability to the secondary users using the footprint abstraction [189] as the space occupancy of the primary users in Figure 13.45. There are four channels (namely channels A, B, C, and D), three sets of primary users, and five secondary users. The channel availability of secondary users is also determined by its location with respect to the primary users.[4] From this figure, primary uses I, II, and III occupy channels A, B, and (A,C), respectively; hence secondary users 1 to 5 can only opportunistically access channels (B,C,D), (A,C,D), (D), (B,D), and (B,D), respectively. Furthermore, two secondary users that interfere with each other are connected with an edge. Hence they conflict with each other on using the same spectrum.

Define the set of secondary user i as I, which is the finite set $\{1, 2, \ldots, K\}$. For each available channel, a secondary user can select $L+1$ discrete rates $\Upsilon = \{0, \upsilon_1, \ldots, \upsilon_L\}$. The strategy space Ω_i for secondary user i is on the available channels and can be denoted as

$$\Omega_i = \prod_{n=1}^{N} \Upsilon^{\mathbf{A}_{in}}. \tag{13.125}$$

The action of user is $r_i^n = \upsilon_l$ representing user i occupies channel n by rate υ_l. We define the strategy profile $\mathbf{r}^n = (r_1^n, r_2^n, \ldots, r_K^n)'$, and we define $\mathbf{r}_{-i}^n$ as the strategies of user i's opponents (interference neighbors defined in $\mathbf{L}$) for channel n. We also define $\mathbf{r}_i = (r_i^1, \ldots, r_i^N)'$ as the action of users over all channels, and $\mathbf{r}_{-i}$ as the user i's opponents' actions.

The utility function U_i measures the outcome of secondary user i for each strategy profile $\mathbf{r}^1, \ldots, \mathbf{r}^N$ over different channels. We define the utility function as the maximum achievable rate for the secondary users over all the available channels as

$$U_i = \sum_{n=1}^{N} \mathbf{A}_{in} R_i(r_i^n, \mathbf{r}_{-i}^n), \tag{13.126}$$

---

[4] In this example, the interference matrix is assumed symmetric.

where $R_i(r_i^n, r_{-i}^n)$ is the outcome of resource competition for user i and the other users. Notice that the utility function represents the maximum achievable rate. In practice, the secondary users need not occupy all the available channels.

We consider unslotted 1-persistent CSMAs as the random multiple-access protocols for the secondary users. Because the channel can be occupied by the prime user again in the near future, each secondary user transmits whenever the channel is idle. From [28], we have

$$R_i(r_i^n, r_{-i}^n) = \begin{cases} \frac{r_i^n S^n}{\sum_i r_i^n}, & \text{if } G \leq G_0 \\ 0 & \text{otherwise} \end{cases}, \tag{13.127}$$

where

$$S^n = \frac{G^n[1 + G^n + \tau G^n(1 + G^n + \tau G^n/2)]e^{-G^n(1+2\tau)}}{G^n(1+2\tau) - (1 - e^{-\tau G^n}) + (1 + \tau G^n)e^{-G^n(1+\tau)}}, \tag{13.128}$$

$G^n = \sum_i r_i^n$, and τ is the propagation delay over the packet-transmission time. When the network payload increases, more collisions happen and consequently the average delay for each packet increases. For some types of payloads like multimedia services, the delayed packets can cause significant QoS loss. In [345], it has been shown that the average delay can be unbounded for a sufficiently large load. Moreover, for cognitive radio, because the prime users can reoccupy the channel in the near future, a certain delay can cause the second user to lose the opportunity for transmission entirely. So we define G_0 as the maximum network payload. Any network payload larger than G_0 will cause an unacceptable average delay. As a result, the utility function is zero.

Next, we first propose a new solution concept, correlated equilibrium. Then we investigate a linear programming method to calculate the optimal correlated equilibrium. Finally, we utilize a no-regret algorithm to learn the correlated equilibria in a distributed way.

To analyze the outcome of the game, Nash equilibrium is a well-known concept, which states that in the equilibrium every user will select a utility-maximizing strategy given the strategies of every other user.

DEFINITION 53 *Nash equilibrium* $\mathbf{r}_i^*$ *is defined as*

$$U_i(\mathbf{r}_i^*, \mathbf{r}_{-i}) \geq U_i(\mathbf{r}_i', \mathbf{r}_{-i}), \ \forall i, \ \forall \mathbf{r}_i' \in \Omega_i, \tag{13.129}$$

i.e., given the other users' actions, no user can increase its utility alone by changing its own action.

If a user follows an action in every possible attainable situation in a game, the action is called pure strategy, in which the probability of using action v_l, $p(r_i^n = v_l)$, has only one nonzero value 1 for all l. In the case of mixed strategies, the user will follow a probability distribution over different possible action, i.e., different rate l. In Table 13.8, we illustrate an example of two secondary users with different actions. In Table 13.8(a), we list the utility function for two users taking action 0 and 1. We can see that, when two users take action of 0, they have the best overall benefit. We can see this action as a cooperative action (in our case the users transmit less aggressively). But if any user

Table 13.8 Two secondary users' game: (a) reward table, (b) Nash equilibrium, (c) mixed Nash equilibrium, (d) correlated equilibrium

	0	1
0	(5,5)	(6,3)
1	(3,6)	(0,0)

(a)

	0	1
0	0	(0 or 1)
1	(1 or 0)	0

(b)

	0	1
0	9/16	3/16
1	3/16	1/16

(c)

	0	1
0	0.6	0.2
1	0.2	0

(d)

plays more aggressively using action 1 while the other still plays action 0, the aggressive user has a better utility, but the other user has a lower utility and the overall benefit is reduced. In our case, the aggressive user can achieve a higher rate. However, if both users play aggressively using action 1, both users obtain the very low utilities. This situation represents the congested network with low throughput of CSMA. In Table 13.8(b), we show two Nash equilibria, where one of the user dominates the other. The dominating user has the utility of 6 and the dominated user has the utility of 3, which is unfair. In Table 13.8(c), we show the mixed Nash equilibrium in which two users have the probability 0.75 for action 0 and 0.25 for action 1, respectively. The utility for each user is 4.5.

Next, we study a new concept of correlated equilibrium that is more general than Nash equilibrium and was first proposed by Nobel Prize winner Robert J. Aumann [14] in 1974. The idea is that a strategy profile is chosen randomly according to a certain distribution. Given the recommended strategy, it is in the players' best interests to conform with this strategy. The distribution is called the correlated equilibrium.

We assume that $N = 1$ and we omit the notation n. Let $\mathcal{G} = \{K, (\Omega_i)_{i \in K}, (U_i)_{i \in K}\}$ be a finite K-user game in strategic form, where Ω_i is the strategy space for user i and U_i is the utility function for user i. Define Ω_{-i} as the strategy space for user i's opponents. Denote the action for user i and its opponents as $\mathbf{r}_i$ and $\mathbf{r}_{-i}$, respectively. Then the correlated equilibrium is defined as follows:

DEFINITION 54 *A probability distribution p is a correlated strategy of game $\mathcal{G}$, if and only if, for all $i \in K$, $\mathbf{r}_i \in \Omega_i$, and $\mathbf{r}_{-i} \in \Omega_{-i}$,*

$$\sum_{\mathbf{r}_{-i} \in \Omega_{-i}} p(\mathbf{r}_i, \mathbf{r}_{-i})[U_i(\mathbf{r}'_i, \mathbf{r}_{-i}) - U_i(\mathbf{r}_i, \mathbf{r}_{-i})] \leq 0, \forall \mathbf{r}'_i \in \Omega_i. \tag{13.130}$$

By dividing inequality in (13.130) by $p(\mathbf{r}_i) = \sum_{\mathbf{r}_{-i} \in \Omega_{-i}} p(\mathbf{r}_i, \mathbf{r}_{-i})$, we have

$$\sum_{\mathbf{r}_{-i} \in \Omega_{-i}} p(\mathbf{r}_{-i}|\mathbf{r}_i)[U_i(\mathbf{r}'_i, \mathbf{r}_{-i}) - U_i(\mathbf{r}_i, \mathbf{r}_{-i})] \leq 0, \forall \mathbf{r}'_i \in \Omega_i. \tag{13.131}$$

Inequality (13.131) means that when the recommendation to user i is to choose action $\mathbf{r}_i$, then choosing action $\mathbf{r}'_i$ instead of $\mathbf{r}_i$ cannot obtain a higher expected payoff to i.

We note that the set of correlated equilibria is nonempty, closed, and convex in every finite game. Moreover, it may include the distribution that is not in the convex hull of the Nash equilibrium distributions. In fact, every Nash equilibrium is a correlated equilibrium, and Nash equilibria correspond to the special case in which $p(\mathbf{r}_i, \mathbf{r}_{-i})$ is a product of each user's probability for different actions, i.e., the play of the different players is independent [14, 15, 129]. In Table 13.8(b) and 13.8(c), the Nash equilibria and mixed Nash equilibria are all within the set of correlated equilibria. In Table 13.8(d), we show an example in which the correlated equilibrium is outside the convex hull of the Nash equilibrium. Notice that the joint distribution is not the product of two users' probability distributions, i.e., two users' actions are not independent. Moreover, the utility for each user is 4.8, which is higher than that of the mixed strategy.

The characterization of the correlated equilibria set illustrates that there are solutions of correlated equilibria that achieve strictly better performances compared with those of the Nash equilibria in terms of the spectrum utilization efficiency and fairness. However, the correlated equilibrium defines a set of solutions that is better than Nash equilibrium, but it does not give any more information regarding which correlated equilibrium is most suitable in practice. We propose two refinements. The first one is the maximum sum correlated equilibrium that maximizes the sum of utilities of the secondary users. The second is the maximin fair correlated equilibrium that seeks to improve the worst-case situation. The problem can be formulated as a linear programming problem as

$$\max_{p} \sum_{i \in K} E_p(U_i) \text{ or } \max_{p} \min_{i} E_p(U_i), \tag{13.132}$$

$$\text{s.t.} \begin{cases} p(\mathbf{r}_i, \mathbf{r}_{-i})[U_i(\mathbf{r}'_i, \mathbf{r}_{-i}) - U_i(\mathbf{r}_i, \mathbf{r}_{-i})] \leq 0, \\ \forall \mathbf{r}_i, \mathbf{r}'_i \in \Omega_i, \forall i \in K, \end{cases}$$

where $E_p(\cdot)$ is the expectation over p. The constraints guarantee that the solution is within the correlated equilibrium set.

Next, we exhibit a class of algorithm called the regret-matching algorithm [129]. The algorithm was named the regret-matching (no-regret) algorithm because the stationary solution of the learning algorithm exhibits no regret and the play probabilities are proportional to the "regrets" for not having played other actions. In particular, for any two distinct actions $\mathbf{r}_i \neq \mathbf{r}'_i$ in Ω_i and at every time T, the regret of user i at time T for not playing $\mathbf{r}'_i$ is

$$\mathcal{R}_i^T(\mathbf{r}_i, \mathbf{r}'_i) := \max\{D_i^T(\mathbf{r}_i, \mathbf{r}'_i), 0\}, \tag{13.133}$$

where

$$D_i^T(\mathbf{r}_i, \mathbf{r}'_i) = \frac{1}{T} \sum_{t \leq T} [U_i^t(\mathbf{r}'_i, \mathbf{r}_{-i}) - U_i^t(\mathbf{r}_i, \mathbf{r}_{-i})]. \tag{13.134}$$

$D_i^T(\mathbf{r}_i, \mathbf{r}'_i)$ has the interpretation of average payoff that user i would have obtained if it had played action $\mathbf{r}'_i$ every time in the past instead of choosing $\mathbf{r}_i$. The expression $\mathcal{R}_i^T(\mathbf{r}_i, \mathbf{r}'_i)$ can be viewed as a measure of the average regret. The probability $p_i(\mathbf{r}_i)$ for user i to take action $\mathbf{r}_i$ is a linear function of the regret. The details of the regret-matching algorithm are shown in Table 13.9. The complexity of the algorithm is $O(L)$.

Table 13.9 The regret-matching learning algorithm

Initialize arbitrarily probability for taking action of user i,
$p_i^1(\mathbf{r}_i), \forall i \in K$
for $t = 1, 2, 3, \ldots$
  1. Find $D_i^t(\mathbf{r}_i, \mathbf{r}_i')$ as in (13.134)
  2. Find average regret $\mathcal{R}_i^t(\mathbf{r}_i, \mathbf{r}_i')$ as in (13.133)
  3. Let $\mathbf{r}_i \in \Omega_i$ be the strategy last chosen by user i,
    i.e., $\mathbf{r}_i^t = \mathbf{r}_i$. Then probability distribution action for
    next period, p_i^{t+1}, is defined as
    $p_i^{t+1}(\mathbf{r}_i') = \frac{1}{\mu} \mathcal{R}_i^t(\mathbf{r}_i, \mathbf{r}_i') \quad \forall \mathbf{r}_i' \neq \mathbf{r}_i$
    $p_i^{t+1}(\mathbf{r}_i) = 1 - \sum_{\mathbf{r}_i' \neq \mathbf{r}_i} p_i^{t+1}(\mathbf{r}_i')$,
  where μ is a certain constant that is sufficiently large.

Forvery period T, let us define the relative frequency of users' action $\mathbf{r}$ played till T periods of time as follows:

$$z_T(\mathbf{r}) = \frac{1}{T} \#\{t \leq T : \mathbf{r}_t = \mathbf{r}\}, \tag{13.135}$$

where $\#(\cdot)$ denotes the number of times the event inside the braces happens and $\mathbf{r}_t$ is all users' action at time t. The following theorem guarantees that the adaptive learning algorithm shown in Table 13.9 has the property that z_T converges almost surely to a set of the correlated equilibria.

THEOREM 30 *[129] If every player plays according to the adaptive learning algorithm in Table 13.9, then the empirical distributions of play z_T converge almost surely to the set of correlated equilibrium distributions of the game G, as $T \to \infty$.*

In the simulations, we employ the maximal sum utility function as the objective. In the first simulation, we study the convergence of the no-regret algorithm. In Figure 13.46, we show a two-user case with actions of $[0.5, 1]$. Here p is the joint probability of a user's taking action of 1 and 2. We can see that the learning algorithm converges to the set of the correlated equilibria with about 100 iterations. The fast convergence of the learning algorithm can ensure that the second users obtain the performance gain before the prime users retake the channels.

In Figure 13.47, we show the different equilibria as functions of G_0 for the three-user game. We show the results of the gain obtained by the greedy user in the NEP, the gain obtained by the victim of the greedy user in the NEP, the learning result, and the optimal correlated equilibrium calculated by linear programming. Here the action space is $[0.1, 0.2, \ldots, 1.5]$. When G_0 is large, there is less penalty for greedy behaviors. So all users tend to transmit as aggressively as possible. This results in the Prisoner's Dilemma [213], in which all users suffer. When G_0 is less than 2.8, the greedy user can have a better performance (NEP best) than that (NEP worst) of the cooperative user. Because of the significant penalty if all users transmit aggressively, the game will not degrade to the Prisoner's Dilemma. However, the performances are quite unfair for the greedy users and the cooperative users. All users have the same utility in the correlated equilibrium

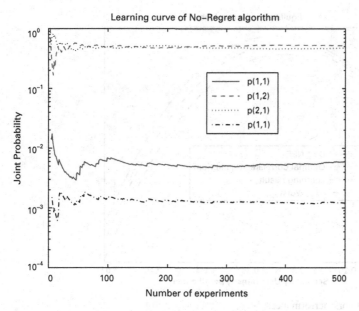

Figure 13.46 Learning curves for no-regret algorithm.

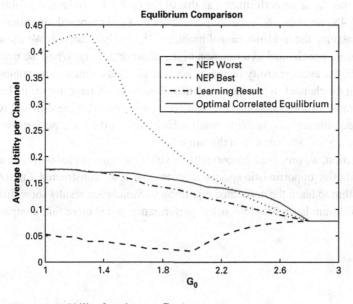

Figure 13.47 Utility function vs. G_0, three users.

and learning result. So fairness is better than that of the NEP. When G_0 is from 2.2 to 2.8, the correlated equilibrium has a better performance even than that of the greedy user (NEP best). When G_0 is from 1.4 to 2.8, the optimal correlated equilibrium has a better performance than that of the learning result. When G_0 is sufficiently small, most

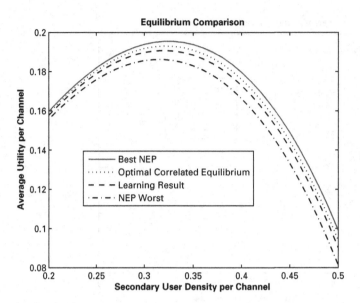

Figure 13.48 Network performances.

of the uncooperative strategies are eliminated by a significant penalty. Consequently the learning result has the same performance as that of the optimal correlated equilibrium.

In Figure 13.48, we show the network performance of the proposed algorithm. For simplicity, we assume the hidden-terminal problem [28] has been solved. We show the average user utility per channel as a function of the network density. When the network density is small, the average utility increases because there is an increasing number of users occupying the channel. When the user density is sufficiently large, the utility begins to decrease because of the collisions. The best NEP and worst NEP are different while the correlated equilibrium and learning result achieve almost the same performance as the best NEP and 5%–15% better than the worst NEP.

In this subsection, we proposed the correlated equilibrium concept for cognitive users to have a distributive opportunistic spectrum access. Then we constructed a no-regret learning algorithm to learn the correlated equilibrium. Simulation results show that the correlated equilibrium has a 5%–15% better performance and is more fair, comparable with the NEP.

14 Ad Hoc/Sensor/Personal-Area Networks

14.1 Introduction

Over the past few decades, increasing demands from military, national security, and commercial customers have been driving the large-scale deployment of ad hoc networks, sensor networks, and personal area networks. Unlike the cellular network or the WiMAX, there are no sophisticated infrastructures such as base stations for these wireless networks. In these scenarios, the mobile users have to set up the network functionality on their own. For example, an individual sensor can sense its immediate environment, process what it senses, communicate its results to others over a wireless link, and possibly take an action in response. Although a single sensor has very limited use, a network of sensors can be used to manage large environments and systems.

These types of wireless networks contain new types of computing machines, run different kinds of network applications, execute in different physical environments, and possess large numbers of nodes. Moreover, there are other challenges such as battery life, maximal power, interferences, limited bandwidth, and connectivity for the different applications of such networks. Significant scientific and technical progress is required to realize the potential of these networks. In short, building such networks requires overcoming many challenges, especially from an optimization design point of view. In this chapter, we study three examples for ad hoc networks, sensor networks, and ultrawide-band (UWB) networks, respectively.

1. First, we investigate unmanned air vehicles (UAVs) that can provide important communication advantages to ground-based wireless mobile ad hoc networks (MANETs). The location and movement of UAVs are optimized such that the network connectivity can be improved. Two types of network connectivity are quantified: global message connectivity and worst-case connectivity. The problems of UAV deployment and movement are formulated for these applications. The optimization problems are NP-hard, and some heuristic adaptive schemes are proposed to yield simple solutions. From the simulation results, by deploying only a single UAV, global message network connectivity and worst-case network connectivity can be improved by up to 109% and 60%, respectively.

2. For distributed nodes in wireless networks such as sensor networks, extending the lifetime of battery-operated devices is a key design issue that allows uninterrupted information exchange. Cooperative communication has recently emerged as a new

communication paradigm that enables and leverages effective resource sharing among cooperative nodes. In this example, a general framework for lifetime extension of battery-operated devices by exploiting cooperative diversity is proposed. The framework efficiently takes advantage of different locations and energy levels among distributed nodes. First, a lifetime maximization problem via cooperative nodes is considered and performance analysis for M-ary PSK (MPSK) modulation is provided. With an objective of maximizing the minimum device lifetime under a constraint on BER performance, the optimization problem determines which nodes should cooperate and how much power should be allocated for cooperation. Because the formulated problem is NP-hard, a closed-form solution for a two-node network is derived to obtain some insights. Based on the two-node solution, a fast suboptimal algorithm is developed for multinode scenarios. Moreover, the device lifetime is further improved by a deployment of cooperative relays in order to help forward information of the distributed nodes in the network. Optimum location and power allocation for each cooperative relay are determined with an aim to maximize the minimum device lifetime. A suboptimal algorithm is developed to solve the problem with multiple cooperative relays and cooperative nodes. Simulation results show that the minimum device lifetime of the network with cooperative nodes doubles the lifetime of the noncooperative network. In addition, deploying a cooperative relay in a proper location leads to up to a lifetime that is 12 times longer than that of the noncooperative network.

3. Finally, emerging personal-area network technology such as UWB offers great potential for the design of high-speed short-range communications. However, for a UWB device to coexist with other devices, the transmitted power level of the UWB is strictly limited by the FCC spectral mask. Such a limitation poses a significant design challenge to any UWB system. Efficient management of the limited power is thus a key feature used to fully exploit the advantages of UWB. A cross-layer multiuser multiband UWB scheme is proposed to obtain the optimal subband and power-allocation strategy. Optimization criteria involve minimization of power consumption under the constraints on the packet error rate (PER), the data rate, and the FCC limit. To ensure system feasibility in variable channel conditions, an algorithm to jointly manage the rate assignment of UWB devices, subband allocation, and power control is proposed. A computationally inexpensive suboptimal approach is also developed to reduce the complexity of the problem, which is found to be NP-hard. Simulation results under UWB channel model specified in the IEEE 802.15.3a standard show that the proposed algorithm achieves performances comparable with those of the complex optimal full-search approach, and it can save up to 61% of transmit power compared with the current multiband scheme in the standard proposal. Moreover, the proposed algorithm can obtain the feasible solutions adaptively when the initial system is not feasible for the rate requirements of the users.

14.2 Connectivity Improvement for MANETs

UAVs are playing increasingly prominent roles in the nation's defense programs and strategy. Although drones have been employed in military applications for many years,

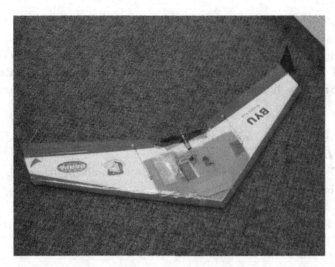

Figure 14.1 A miniature UAV built and flown at Brigham Young University.

technological advances in microcontrollers, sensors, and batteries have dramatically increased their utility and versatility. Traditionally, emphasis has been placed on relatively large platforms such as *Global Hawk* and *Predator*, but increasing attention has recently been focused on "mini-UAVs" (MUAVs) that offer advantages in flexibility and cost [21, 22, 42]. An example of an experimental MUAV built and tested at Brigham Young University is depicted in Figure 14.1. Because of their small size, they are difficult for others to detect and track, and they are able to more easily avoid threats in the environment they fly through. As a result, they can fly at much lower altitudes, of the order of tens or hundreds of feet, and collect much more precise, "localized" data. They are significantly cheaper and easier to fly and can often be launched by an individual in any kind of terrain without a runway or special launching device.

Because of their mobility and elevation, UAVs equipped with communication capabilities can provide important advantages for ground-based ad hoc networks. Their use in routing, MAC, and scheduling applications has been detailed in [97, 98, 257, 342]. These studies have been primarily heuristic and have focused on simulations to qualitatively assess the benefits of UAV-assisted networks. We take a mathematical approach to positioning and flying a UAV over a MANET in order to optimize the network's connectivity for better QoS and coverage. We assume a single UAV flying over a connected network with knowledge of the positions and velocities of the network nodes. The UAV itself acts as a node in the network, and can generate, receive, or forward data packets to other users.

We quantify two types of network connectivity [123]. First, *global message connectivity* is defined as the highest possible probability of successfully propagating one message to all users in the network. Second, *worst-case connectivity* is defined to measure the severity that a network will be divided into two. From these two definitions, we formulate the UAV deployment problem and the UAV movement problem, which are NP-hard. Then we develop algorithms for optimally governing the UAV's position and velocity in the network. From the simulation results, one UAV can improve global message connectivity and worst-case connectivity by up to 109% and 60%, respectively.

We assume a single UAV flying over a wireless MANET that is able to obtain (e.g., through sharing of GPS data) the locations and velocities of the randomly distributed mobile users in the network. In particular, we assume the UAV to possess the following information:

- Locations of all users (x_i, y_i), from which the distance between any two nodes is calculated to be $D_{ij} = \sqrt{|x_i - x_j|^2 + |y_i - y_j|^2}$.
- Using the users' locations at different times, the UAV can obtain the speeds and directions of mobile users:

$$S_i = \frac{dx_i}{dt} + z\frac{dy_i}{dt}, \tag{14.1}$$

where $z = \sqrt{-1}$.

Suppose there are K mobile users denoted as $N_1, \ldots, N_K$, plus one UAV denoted as N_0 in the wireless MANET. The wireless channel response between any two nodes is $G_{ij}, i \neq j$. Suppose the transmitted power for each node is P_i and the noise variance σ^2 is the same for all users. The received SNR Γ_{ij} for the signal transmitted by the ith node and received by the jth node is

$$\Gamma_{ij} = \frac{P_i G_{ij}}{\sigma^2}. \tag{14.2}$$

Using a Raleigh statistical model, the channel gain can be expressed as

$$G_{ij} = \frac{C_{ij}|h_{ij}|^2}{(D_{ij})^\alpha}, \tag{14.3}$$

where C_{ij} is a constant that takes into account the antenna gains and any propagation obstructions (shadowing), $|h_{ij}|^2$ is the squared magnitude of the channel fade and follows an exponential distribution with unit mean, D_{ij} is the distance between user i and user j, and α is the propagation-loss factor. Here we assume the channels among different users are orthogonal. This assumption is valid for most military applications.

A sufficiently high SNR will guarantee that the receiver will have an acceptably small packet loss and successfully receive a transmitted packet, so that minimal link quality can be maintained. Suppose that the SNR threshold for successful packet reception is γ, so that, by use of (14.2), (14.3), and the Raleigh statistical model, the probability of a successful transmission is given by

$$P_r^{ij}(\Gamma_{ij} \geq \gamma) = \exp^{-\frac{\sigma^2 \gamma (D_{ij})^\alpha}{C_{ij} P_i}}. \tag{14.4}$$

Because transmission power is bounded, each user can communicate with other users within only a certain radius. We will say that two nodes are connected if the probability defined in (14.4) is greater than or equal to some threshold δ. From this, we define a graph $G(K, \mathbf{A})$ to describe the connectivity of the network, where the matrix $\mathbf{A}$ has the following definition:

$$[\mathbf{A}]_{ij} = \begin{cases} 1, & \text{if } P_r^{ij} \geq \delta \\ 0, & \text{otherwise} \end{cases}. \tag{14.5}$$

We assume that the network is connected, i.e., $\sum_{j=1}^{K}[\mathbf{A}]_{ij} \geq 1, \forall i$, and we concentrate on how to improve the connectivity. If the network is not connected, then other methods, such as those based on Steiner trees [381], must be employed. To quantify each link's connectivity, we define the weight for each link as a function of the probability of successful transmission:

$$W_{ij} = -\log P_r^{ij}, \tag{14.6}$$

where the minus sign is added to make the weight positive. Suppose user i tries to communicate with user j via a relay with user k. Because of the log form, the sum of weights W_{ik} and W_{kj} will represent the probability of successful transmission between i and j as $P_{ik}P_{kj}$. The smaller the weight, the higher the connectivity.

In some applications, such as military ad hoc networks, it is important to keep all the users connected. For example, in battlefield scenarios, it is essential to propagate commands to the distributed soldiers and vehicles. Given that the quality of each link can be represented by different weights as in (14.6), a natural question is that of how to select the links such that all the nodes are connected and the overall weights are minimized. The concept of a minimal spanning tree (MST) from graph theory provides the solution to this question:

DEFINITION 55 *Given a graph, a spanning tree of that graph is a subgraph that is a tree and connects all the vertices together. A single graph can have many different spanning trees. We can also assign a weight to each edge, which is a number representing how unfavorable it is, and use this to assign a weight to a spanning tree by computing the sum of the weights of the edges in that spanning tree. A MST or minimum-weight spanning tree is then a spanning tree with a weight less than or equal to the weight of every other spanning tree. [381]*

One example of such a MANET is shown in Figure 14.2. First, without considering the Steiner point, which we will discuss later, there are a total of 10 nodes. The possible connections between each node are marked with different costs. The bold link shows the MST that connects all the nodes. To find the MST solution, Prim's algorithm, Kruskal's algorithm, and the Chazelle algorithm [381] can be utilized with polynomial time.

MSTs are widely used in wired networks to minimize the cost of transmission. Because of the broadcast nature of wireless communications, the transmissions of one user can be heard by many others. In [333], a pruning MST is proposed to yield energy-efficient broadcast and multicast trees. In our work, however, we concentrate on how to improve the connectivity and not on how to construct the spanning tree. The approaches and discussions can be employed for any tree like those in [333]. Suppose the matrix $\mathbf{A}'$ represents the MST, where $[\mathbf{A}']_{ij} = 1$ if the link from user i to user j is in the MST, and $[\mathbf{A}']_{ij} = 0$, otherwise. We have two different definitions of connectivity:

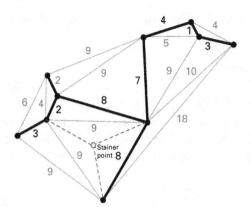

Figure 14.2 Example of an MST, Steiner point, and Steiner tree [381].

1. Global Message Connectivity

    If we define *global message connectivity* as the probability that a message can be successfully transmitted to all users in the network under the unlimited energy but limited power condition, then the MST is the optimal solution for connectivity. Because the received SNR in (14.2) is a monotonically increasing function of power, to maximize the probability in (14.4), each user will transmit with maximal power. Because there is abundant energy, we do not have an energy-efficiency problem. Suppose the MST has already been constructed with the weight defined in (14.6). The sum of the weights in the MST represents the overall probability that a message is successfully transmitted via this MST. Suppose there exists a tree that yields a higher probability to connect all users. We can convert the probability of successful transmission for each link in the tree to a corresponding weight for the link. Because the weights in (14.6) are monotonously decreasing with P_r^{ij}, the sum of the weights for the new tree will be smaller than that for the MST, which is a contradiction.

    The probability that a command is transmitted to all the users can be computed as the product of successful transmission probabilities of the links on the MST. Maximizing this probability is equivalent to minimizing the MST cost:

$$U = \sum_{i=0}^{K} \sum_{j=0}^{K} [\mathbf{A}']_{ij} W_{ij}. \qquad (14.7)$$

2. Worst-Case Connectivity

    If we define the *worst-case connectivity* as the lowest probability that part of the network can communicate to the rest of the network, this connectivity measures how severely the network will be divided into two parts. Then, under the unlimited energy but limited power condition, the largest weight in the MST can be used to measure this connectivity, i.e, maximizing this connectivity is equivalent to minimizing the weight of the worst-case MST edge:

$$U = \min_{[\mathbf{A}']_{ij}=1, \forall i, j} \max W_{ij}. \qquad (14.8)$$

Because the UAV will be an additional node to the existing MST, another concept is defined to improve the MST, as follows:

DEFINITION 56 *A node added to the network that minimizes the length of the spanning tree is called a Steiner point. The resulting tree is called a Steiner tree.*

A Steiner point example is shown in Figure 14.2, in which a Steiner point is added and the overall weight of the MST is reduced.

Next, we formulate the optimization problems for deployment and movement of the UAV. First, the UAV optimizes its location so that better network connectivity U can be obtained. Second, the UAV needs to decide in which direction and at what speed it should move so that the connectivity can be better maintained. Recall that the UAV is denoted as node 0. The two problem formulations are given by the following:

1. Formulation 1: UAV deployment:

$$\min_{(x_0, y_0)} U. \tag{14.9}$$

This is a minimal Steiner tree problem and has been shown to be NP-hard.
2. Formulation 2: UAV movement:

$$\max_{S_0} \lim_{\Delta t \to 0} -\frac{\Delta U}{\Delta t}, \tag{14.10}$$

$$\text{s.t. } v_{min} \leq \|S_0\|^2 \leq v_{max}^2,$$

where v_{min} and v_{max} are the minimum and maximum speeds of the UAV, respectively, and

$$\Delta U = U[x_i(t + \Delta t), y_i(t + \Delta t)] - U[x_i(t), y_i(t)]. \tag{14.11}$$

Next, we analyze the two-user case first. Then we propose multiuser algorithms for the formulations in (14.9) and (14.10). The main approach is based on gradient methods and some engineering heuristics to reduce the complexity.

For the performances for a two-user case, we suppose two users are uniformly randomly located within a radius of R, as shown in Figure 14.3. Suppose the distance between two users is d. Obviously $0 \leq d \leq 2R$. We try to find the PDF of d. Suppose user 1 is located at the distance of r to the center. There are three cases to discuss:

1. $I(r, d) = 1$, if $r + d \leq R$.
   Because of the uniform distribution assumption, we have the probability that the distance between two users is greater than d:

   $$P_r(d|I = 1, r) = \frac{\pi R^2 - \pi d^2}{\pi R^2} = 1 - \frac{d^2}{R^2}. \tag{14.12}$$
2. $I(r, d) = 2$, if $r + d > R$, $R - d > 0$.
   Within the distance of d to user 1, there are some areas where user 2 cannot be located, because it is outside the radius R. An example is shown in Figure 14.3. As a result, the probability is proportional to the size of the area of the round dish with radius R minus the area of the part, inside the dish with radius R, of the round dish

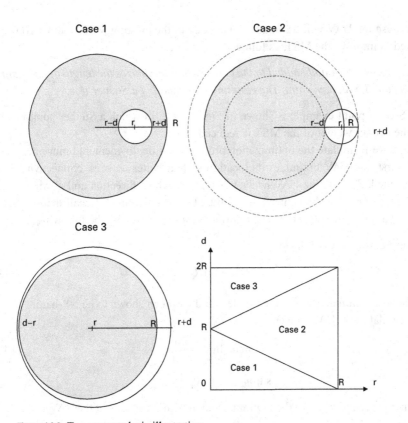

Figure 14.3 Two-user analysis illustration.

with radius d. The shape looks like a waning crescent moon. We have the probability that user 2 is located in that area as

$$P_r(d|I = 2, r) = 1 - \frac{1}{\pi R^2} \left\{ d^2 \arccos\left(\frac{r^2 + d^2 - R^2}{2rd}\right) \right.$$

$$+ R^2 \arccos\left(\frac{r^2 + R^2 - d^2}{2rR}\right)$$

$$\left. - Rr \sin\left[\arccos\left(\frac{r^2 + R^2 - d^2}{2rR}\right)\right]\right\}.$$

3. $I(r, d) = 3$, if $d - r \geq R$.

Under this condition, there is no possibility that the second users can be located within radius R:

$$P_r(d|I = 3, r) = 0. \tag{14.13}$$

As a result, the CDF can be calculated by

$$\text{CDF}(d) = 1 - \int_0^R P_r[d|I(d, r), r]\frac{2\pi r}{\pi R^2} dr. \tag{14.14}$$

The PDF can be obtained by differentiating the CDF function. The probability for two users to communicate without UAV can be calculated by

$$P_r^{12} = \int_0^{2R} \exp^{-\frac{\sigma^2 \gamma(d)^{\alpha}}{C_{ij} P_i}} \text{PDF}(d) d(d).$$
(14.15)

Assume that the channels between two users are symmetric, i.e., $C_{ij} = C_{ji}$. By adding one UAV optimally in the middle of the two users, the estimated average global message connectivity is given by

$$P_r^0 = \int_0^{2R} \exp^{-\frac{2\sigma^2 \gamma(\frac{d}{2})^{\alpha}}{C_{i0} P_i}} \text{PDF}(d) d(d).$$
(14.16)

The two-user case can represent a scenario in which a link between two mobile users is critical to be maintained and improved. In this situation, the analytical result shows how much a UAV can improve the connectivity.

Next, we determine the UAV deployment, i.e., what the optimal (x_0, y_0) is. The problem is NP-hard, and we propose an adaptive algorithm to find a local optimum. Starting from any initialization point, we want to find out how to change the UAV's location in some neighborhood so that a better MST can be obtained. The gradient for such a search can be written as

$$g_0 = \frac{dU(x_0, y_0)}{dx_0} + z \frac{dU(x_0, y_0)}{dy_0}.$$
(14.17)

A linear search algorithm [34] can be utilized to reduce the complexity of the gradient method. The stopping criteria can be $\|g_0\|^2 \leq \varepsilon$, where ε is a small positive number or where the KKT condition holds [34], i.e., where the local optimum is achieved.

The problem in (14.9) has many local optima. This can be shown by the following extreme example. Suppose there are only three users and one UAV in the network. Three users are located in a line with locations $(0, 0)$, $(1, 0)$, and $(2, 0)$. With some simple calculations, we can see that there are two locally optimal locations for the UAV at $(0.5, 0)$ and $(1.5, 0)$, respectively. To overcome the local optimum problem, we propose the following two initialization methods.

1. Random initialization
   This approach generates a number of seeds within the area of the MANET and lets the gradient method find the local optima. From the local optima, the global optimum is selected with the minimal overall weight. The advantage of this initialization method is that the global optimum can be obtained with high probability when the density of the initialization seeds is high enough. The disadvantage is that the computational complexity is high, especially when the number of users is large.
2. Heuristic initialization
   Suppose the MST without the UAV is constructed and the maximal link weight occurs between node i and node j. The links between node i an node j are symmetric. The heuristic initialization for the UAV is at the middle of these two nodes, i.e.,

$$x_0^0 = \frac{x_i + x_j}{2} \text{ and } y_0^0 = \frac{y_i + y_j}{2}.$$
(14.18)

Table 14.1 Algorithm to find best deployment

Initialization	Initialize (x_0^0, y_0^0), $t = 0$.
Iteration	1. Calculate gradient in (14.17).
	2. $t = t + 1$.
	3. line search $\zeta > 0$ so that U is optimized with $x_0^t + zy_0^t = x_0^{t-1} + zy_0^{t-1} - \zeta g_0^t$.
Stopping criteria	$\|g_0\|^2 \leq \varepsilon$ or KKT condition holds.
Return (x_0, y_0)	If random initialization, select the best.

The rationale is to improve the worst-case link, so that the initial performance improvement can be good before the gradient method is applied. This initialization cannot be guaranteed to be globally optimal.

Overall, the algorithm to find the best location to deploy the UAV is shown in Table 14.1. The complexity of the algorithm for each iteration is $O(K^3 \log K)$. Because the MST cost at each iteration of the algorithm is nonincreasing and the solution has a lower bound, the algorithm always converges.

Next, we assume that the initial UAV deployment has been done as described previously; i.e., (x_0, y_0) is known. We try to determine the movement of the UAV so that the network connectivity can be improved in the future.

First, we assume that within a short period of time dt, the network and MST topologies do not change. The movement of the UAV will affect only the weights where the links are connected to the UAV. Define the set of nodes that are connected to the UAV as **V**. The UAV needs to monitor only the nearby nodes in **V**. The estimation, signaling, and overheard burdens can thus be greatly reduced.

From (14.3), (14.4), and (14.6), the gradient for the utility change can be written as

$$\frac{dU}{dt} = \frac{d}{dt}[U(D_{ij}^*) - U(D_{ij})], \tag{14.19}$$

where $D_{ij}^* = \|x_i + zy_i + S_i dt - x_j - zy_j - S_j dt\|^2$.

Because the UAV is airborne, we need to consider the issue of speed constraints. If the gradient in (14.19) is small, the connected nodes are hardly moving, and hence the UAV should not change position. Under this condition, the UAV must fly in a small circle. When the gradient is large enough, the UAV flies against the gradient direction with the speed proportional to the magnitude of the gradient. When the gradient is too large, the UAV can only fly in the direction of the gradient with its maximum speed v_{max}. The speed of the UAV S_0 can be calculated from Table 14.2, where μ is a constant that can be determined experimentally.

The two algorithms in Tables 14.1 and 14.2 can be utilized in turn. First, the deployment algorithm is used to find the best location the UAV should initially fly to. Then the movement algorithm keeps track of the mobility of the distributed users. Occasionally, the network topology has been changed too much and the UAV falls to some local optimum. Under this condition, the deployment algorithm is reapplied to relocate the

Table 14.2 Algorithm to find best movement

Monitor	Measure (x_i, y_i) and S_i for $i \in \mathbf{V}$.
Movement	Circle:
	$\quad$ If $\lvert \mu \frac{dU}{dt} \rvert < v_{min}$.
	Fly:
	$\quad$ If $\lvert \mu \frac{dU}{dt} \rvert \geq v_{min}$
	$$S_i = \begin{cases} -\mu \frac{dU}{dt}, & \text{if } \lvert \mu \frac{dU}{dt} \rvert \leq v_{max} \\ -v_{max} \frac{\frac{dU}{dt}}{\lvert \frac{dU}{dt} \rvert}, & \text{otherwise.} \end{cases}$$
Update	Detect local optimum. If this is the case, performance Table I.

UAV to some better position. The frequency for employing the deployment algorithm depends on the mobility of the users.

To demonstrate the effectiveness of the proposed algorithms, we use the following simulation study: a total of K users are randomly located within a square region of $1000\,\text{m} \times 1000\,\text{m}$. The transmission power is 300 dB mW, the noise value $\sigma^2 = 10^{-7}$ dB mW, the SNR requirement $\gamma = 10$ dB, and the propagation-loss factor is $\alpha = 3$. Without loss of generality, we assume $C_{ij} = C_1, \forall i, j \in \{1, 2, \ldots, K\}$; for the communication link between the UAV and the mobile users, $C_{ij} = C_0$, i or $j = 0$, for the communication link between different mobile users. Here, because the UAV is in the sky, $C_0 > C_1$. For the simulations conducted here, we assumed that $C_0 = 2$ and $C_1 = 1$. 2500 initializations are realized for the random initialization methods.

In Figure 14.4, we show the global message connectivity as a function of the UAV location (x_0, y_0). Here the number of users is $K = 10$. On the z axis, we show the connectivity of the network without the UAV as a star with a value of 0.2966. By deploying the UAV at the best location $(x_0, y_0) = (305, 951)$, the connectivity probability is improved to 0.4964, as shown by a diamond on the z axis. On the x, y plane, we show the MST with the users denoted by crosses and the UAV denoted by a circle. We can see that the UAV tries to improve the link from a faraway user to the rest of the network in this case. Moreover, from the curve, we can see that there are many local optima for (x_0, y_0).

In Figure 14.5, we show one example of the UAV flying tracks with different types of initialization. Here the number of users is $K = 10$, and there are five different initial seeds for the random initialization. We can see that different initializations lead to different local optima. From the simulations, the heuristic initialization leads to global optimization most of time. But there are cases in which the random initialization leads to the global optimum, whereas the heuristic provides only a local optimum. On the other hand, there are also cases in which the heuristic approach has a better solution, because the number of random initializations is not large enough. Moreover, the flying tracks are not smooth and the UAV may change directions. This is because the derivative of U is not continuous, which can be easily observed from Figure 14.4.

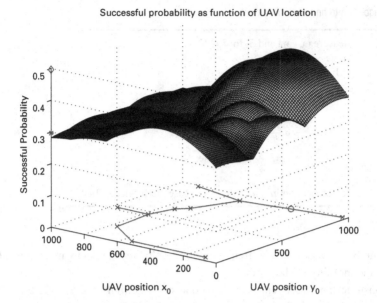

Figure 14.4 Global message connectivity as a function of UAV location.

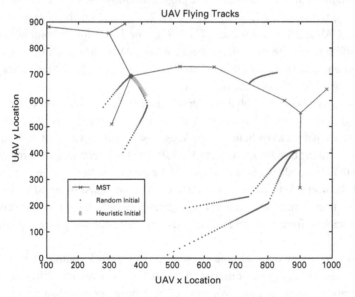

Figure 14.5 UAV flying tracks with different initials.

In Figure 14.6, we show network connectivity for different numbers of users. For both global message connectivity and worst-case connectivity, we show the performance of no UAV, random initialization, and heuristic initialization, respectively. We can see that the performance drops first when the number of users increased from a small number. This is because the users have to transmit over long distances for the message to

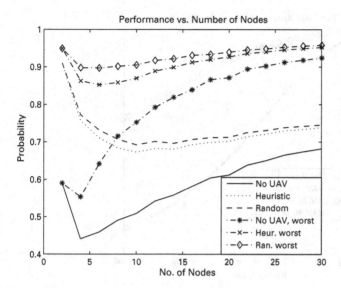

Figure 14.6 Connectivity vs. number of users.

propagate. When the number of users becomes large, the higher density of users makes the connectivity better. The addition of a single UAV can improve the connectivity of the network by 109% when the number of users is 4. This is because the size of **V** (the users connected to the UAV) is limited, while the rest of the links are kept the same. The improvement shrinks with larger number of users, because a higher node density means that most of the links already have good connection probabilities and the addition of the UAV offers only a slight improvement. For worst-case connectivity, the UAV can improve performance by up to 60%. Heuristic initialization has a slightly worse performance compared with random initialization because of the local optima, but the complexity of the heuristic approach is much lower.

We set up another simulation to test the analysis results. The users are uniformly randomly located within a cell with radius R. The remaining simulation settings are the same. In Figure 14.7, we show the global message connectivity for the different radius R. We can see that the analysis results match the numerical results well. The larger the cell size, the more improvement a UAV can provide. At a cell radius of 1000 m, the improvement is up to 240%.

In Figure 14.8, we show the average UAV speed and the probability that the UAV falls into a local optimum. Here $K = 5$. The mobile users move in arbitrary directions with the speeds uniformly distributed from zero to the value on the x axis. The total time for each network situation is 300 s, and the UAV updates its direction in every 10 s. $v_{max} = 30$ m/s. We average 500 different simulations. We can see that the average speed of the UAV increases according to the users' mobility. The probability that the UAV falls into a local optimum during 300 s increases faster when the users' speed is higher. According to the different users' speeds, the frequency for applying the deployment algorithm in Table 14.1 should vary.

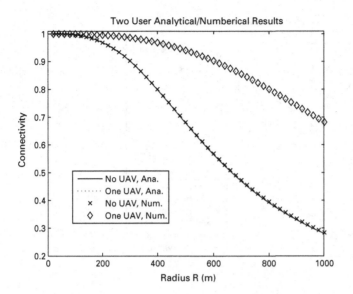

Figure 14.7 Two-user, one-UAV analytical/numerical results.

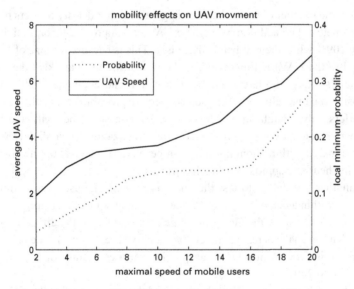

Figure 14.8 UAV speed and local optima probability vs. mobility.

Overall, we studied how to utilize UAVs to improve the network connectivity of a MANET. We defined two types of connectivity, global message connectivity and worst-case connectivity. Then we formulated the deployment and movement problems for the UAV. Adaptive heuristic algorithms are proposed to provide a simple solution as well as good performance. From the simulation results, a UAV can improve the two types of connectivity by 109% and 60%, respectively.

14.3   Lifetime Extension for Sensor Networks

In many applications of wireless networks, extending the lifetime of battery-operated devices is a key design issue that ensures uninterrupted information exchange and alleviates the burden of replenishing batteries. Several approaches for lifetime extension of battery-limited devices have been proposed in the literature. In [307], a data routing algorithm is proposed with an aim to maximize the minimum lifetime among nodes in wireless sensor networks. Considerable research efforts have been devoted to maximizing such minimum lifetimes, which are also referred to as network lifetimes. For example, upper bounds on the lifetime of various wireless networks with energy-constrained nodes are derived in [30, 140] and references therein. The problem of finding an energy-efficient tree for network lifetime maximization is considered in [30, 200] for a broadcasting scenario and in [75] for a multicasting scenario. The works in [3, 183] consider a problem of minimum-energy broadcasting, which is proved to be NP-complete. In [347], a technique for network lifetime maximization by employing accumulative broadcast strategy is considered. The proposed work relies on the assumption that nodes cooperatively accumulate energy of unreliable receptions over the relay channels. The work in [139] consider provisioning additional energy on existing nodes and deploying relays to extend the network lifetime.

Recently, the cooperative diversity concept was introduced as a promising alternative to combat fading in wireless relay channels. The basic idea of cooperative diversity is to allow distributed users in the network help relay information of each other so as to explore inherent spatial diversity that is available in the relay channels. Several cooperation protocols have been proposed, e.g., AF and DF protocols [175], user cooperation protocol [272, 273], and coded cooperation protocol [145]. In [175, 263], physical-layer issues such as outage probability analysis and symbol error rate (SER) analysis for different cooperation systems are considered. The channel capacity of cooperative networks is investigated in [137, 317]. The outage probability of the coded cooperation protocol is analyzed in [145], and the exact SER analysis, as well as optimum power allocation for the DF protocol, is provided in [263, 298]. Later, higher-layer issues are considered in [104, 195] to determine which nodes are appropriate for cooperation and how much power should be facilitated. The work in [32] considers a distributed relay-selection scheme by which one relay out of multiple relays is selected for cooperation. The scheme requires limited network knowledge, and the relay-selection strategy relies on instantaneous SNR.

The research works in [32, 175] have proved the significant potential of using cooperative diversity in wireless networks. However, most of the existing works focus on improving physical-layer performance or minimizing energy consumption. On the other hand, most of previous works on extending lifetime [45, 139] concentrate on noncooperative transmissions in which received signals from a source and each relay are not combined to explore the cooperative diversity. Consider contemporary wireless networks that comprise heterogeneous devices such as mobile phones, laptop computers, and personal digital assistants (PDAs). These devices have limited lifetimes; nevertheless, some of them may have longer lifetimes because of their location or energy advantages. For

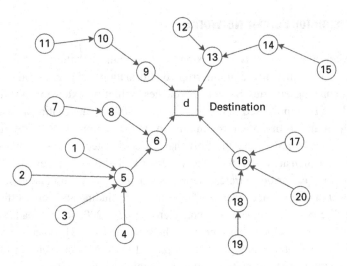

Figure 14.9 An example of a wireless network with a destination (d) and N distributed nodes ($N = 20$).

instance, the devices in some ideal locations may have a location advantage, whereas the other devices may have an energy advantage if they are equipped with high initial energy. By introducing a cooperation protocol among distributed nodes, a portion of energy from these devices can be allocated to help forward information of other energy-depleting devices in the network. In this way, the lifetime of the energy-depleting devices can be greatly improved, and hence the minimum device lifetime of the network is increased.

We propose to increase the device lifetime by exploiting the cooperative diversity and taking both location and energy advantages in wireless networks. The framework [132, 133] is based on the DF cooperation protocol, which is well suited for WLAN or cellular settings; nevertheless, other cooperation protocols such as the AF protocol can be used as well. We first describe a signal model for a noncooperative network. Then we present a signal model for cooperative networks employing the DF protocol. After that, we formulate an optimization problem with an objective to maximize the minimum device lifetime under the BER constraint. The analysis based on MPSK modulation is provided. We derive an analytical solution for a two-node cooperative network to obtain some insights on the formulated problem, which is NP-hard. From the two-node solution, we develop a fast suboptimal algorithm to reduce the complexity of the formulated problem. Furthermore, we propose to improve the device lifetime by deploying additional relays over a network with energy depleting nodes. We determine in which locations to place the relays and how much power these relays use for cooperation in order to maximize the device lifetime. To reduce complexity of the formulated problem, we also develop an efficient suboptimal algorithm that solves the problem of using at most one relay per node. Simulation results are given to show the merit of the proposed work and support our theoretical analysis.

Consider a wireless network with N randomly deployed nodes, as shown in Figure 14.9. Each node knows its next node in a predetermined route by which its information can be delivered to the destination. The destination node can be a base station or an access point in WLANs, a piconet coordinator in WPANs, or a data-gathering unit in wireless sensor networks. Next, a system model of a noncooperative network is described. Then a system model of a cooperative network is considered.

In a noncooperative wireless network, each source node transmits its own information to the destination node through only a predetermined route. Figure 14.9 shows an example of a wireless network with several randomly deployed nodes. Suppose there are N nodes in the network, and let x_j denote a symbol to be transmitted from node j to its next node, defined as n_j, in its predetermined route. The symbol x_j can be the information of node j itself or it can be the information of other nodes that node j routes through the destination. The signal received at n_j because of the transmitted information from node j can be expressed as

$$y_{jn_j} = \sqrt{P_j} h_{jn_j} x_j + w_{jn_j}, \tag{14.20}$$

where P_j is the transmit power of node j, h_{jn_j} is the fading coefficient from node j to n_j, and w_{jn_j} is an additive noise. The channel coefficient h_{jn_j} is modeled as a complex Gaussian random variable with zero mean and variance $\sigma_{jn_j}^2$, i.e., $CN(0, \sigma_{jn_j}^2)$, and w_{jn_j} is $CN(0, N_0)$ distributed. The channel variance $\sigma_{jn_j}^2$ is modeled as

$$\sigma_{jn_j}^2 = \eta D_{jn_j}^{-\alpha}, \tag{14.21}$$

where D_{jn_j} denotes distance between node j and n_j, α is the propagation-loss factor, and η is a constant whose value depends on the propagation environment. Considering an uncoded system and using a BER formulation in [298], the average BER performance for a noncooperative node with MPSK modulation is upper bounded by

$$\text{BER}_j \leq \frac{N_0}{4b P_j \sigma_{jn_j}^2 \log_2 M}, \tag{14.22}$$

where $b = \sin^2(\pi/M)$.

Let the performance requirement of node j be $\text{BER}_j \leq \varepsilon$ in which ε represents the maximum allowable BER. We assume that ε is the same for every node. Accordingly, the optimum transmit power of a noncooperative node is given by

$$P_j = \frac{N_0}{4b\varepsilon\sigma_{jn_j}^2 \log_2 M}. \tag{14.23}$$

We denote E_j as an initial battery of node j, and denote P_s as an amount of processing power (i.e. power used for encoding information, collecting data, and etc.) at the source node. Let λ_{lj} ($l = 1, 2, \ldots, N$ and $l \neq j$) be the data rate at which node l sends information to node j, and λ_j be a data rate at which node j sends information to its next node n_j. Then, the total power that node j uses to send information to n_j is $\lambda_j P_s + \sum_{l=1}^{N} \lambda_{lj} P_j$, where $\lambda_j P_s$ is the total processing power at node j, $\lambda_j P_j$ represents the power that node j sends its own formation, and $\sum_{l=1,l\neq j}^{N} \lambda_{lj} P_j$ corresponds to the power that node j routes information of other nodes.

We consider a cooperative wireless network in which all nodes can transmit information cooperatively. Each node can be a source node that transmits its information or it can be a relay node that helps forward information of other nodes. The cooperation

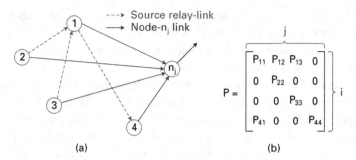

Figure 14.10 Illustration of a cooperative wireless network: (a) a network with four nodes, (b) the corresponding power-allocation matrix.

strategy is based on the DF protocol that comprises two transmission phases. In Phase 1, the source node sends the information to its next node on the route. In Phase 2, the relay node decodes the information it receives from the source and helps forward the correctly decoded information. We assume that each signal transmission is constrained to half-duplex mode, the system is uncoded, and the source and relay transmit the signals through orthogonal channels by using existing TDMA, FDMA, or CDMA schemes.

For subsequent derivations, we define a power allocation matrix $\mathbf{P}$ as an $N \times N$ matrix with the following properties:

1. Each element $P_{ij} \geq 0$, for $i, j = 1, 2, \ldots, N$.
2. P_j represents a power that node j uses to transmit its own information to its next node n_j and the relays.
3. P_{ij} represents a power that node i helps forward information of node j (information of other nodes) to the next node n_j.

Assuming that all nodes have their information to be transmitted, then $P_j > 0$ for all j. Figure 14.10(a) illustrates a cooperative network with $N = 4$ nodes. Each solid line represents a transmission link from a source node to its next node, and each dashed line represents a link from a source to a relay. In addition, Figure 14.10(b) shows a power-allocation matrix $\mathbf{P}$ that corresponds to the cooperative network in Figure 14.10(a). Each nonzero diagonal element of $\mathbf{P}$ represents a transmit power of a source node. In Figure 14.10(a), node 1 helps relay information of node 2 and 3 to their intended destination. Therefore P_{12} and P_{13} in the first row of $\mathbf{P}$ contain nonzero elements, which represent power that node 1 helps node 2 and node 3, respectively. Similarly, P_{41} is a nonzero element because node 4 helps forward information of node 1.

Suppose node j acts as a source (or a helped node) and node i acts as a relay (or a helping node). When node j sends information to n_j in Phase 1, the received signal at n_j is given in (14.20). However, the received signal at the helping node i is given by

$$y_{ji} = \sqrt{P_j} h_{ji} x_j + w_{ji}, \tag{14.24}$$

where h_{ji} denotes a channel coefficient from node j to node i, and w_{ji} represents an additive noise. In Phase 2, the relay (node i in this case) forwards the information of node j to n_j only if the symbol is correctly decoded [298]. The received signal at n_j can be expressed as [298]

$$y_{in_j} = \sqrt{\tilde{P}_{ij}} h_{in_j} x_j + w_{in_j}, \tag{14.25}$$

where $\tilde{P}_{ij} = P_{ij}$ if the relay correctly decodes the symbol, and $\tilde{P}_{ij} = 0$ otherwise. In (14.25), h_{in_j} and w_{in_j} are modeled as $CN(0, \sigma_{in_j}^2)$ and $CN(0, N_o)$, respectively. After that, the destination (n_j in this case) combines the directly received signal from the source in Phase 1 with that from the relay in Phase 2 by MRC. Assuming that x_j has unit energy; then an instantaneous SNR at the MRC output of n_j is

$$\gamma_{n_j} = \frac{P_j |h_{jn_j}|^2 + \tilde{P}_{ij} |h_{in_j}|^2}{N_0}. \tag{14.26}$$

By taking into account the decoding result at the relay and averaging the conditional BER over the Rayleigh distributed random variables, the average BER in the case of MPSK modulation can be expressed as [298]

$$\begin{aligned}
\text{BER}_j &= \frac{1}{\log_2 M} F\left(1 + \frac{b P_j \sigma_{jn_j}^2}{N_0 \sin^2 \theta}\right) F\left(1 + \frac{b P_j \sigma_{ji}^2}{N_0 \sin^2 \theta}\right) \\
&\quad + \frac{1}{\log_2 M} F\left[\left(1 + \frac{b P_j \sigma_{jn_j}^2}{N_0 \sin^2 \theta}\right)\left(1 + \frac{b P_{ij} \sigma_{in_j}^2}{N_0 \sin^2 \theta}\right)\right] \\
&\quad \times \left[1 - F\left(1 + \frac{b P_j \sigma_{ji}^2}{N_0 \sin^2 \theta}\right)\right],
\end{aligned} \tag{14.27}$$

where $F[x(\theta)] = \frac{1}{\pi} \int_0^{(M-1)\pi/M} [x(\theta)]^{-1} d\theta$ and b is defined in (14.22). The first term on the right-hand side of (14.27) corresponds to incorrect decoding at the relay, whereas the second term corresponds to correct decoding at the relay. By assuming that all channel links are available, i.e., $\sigma_{jn_j}^2 \neq 0$ and $\sigma_{ji}^2 \neq 0$, the BER upper bound of (14.27) can be obtained by removing the negative term and all one's in (14.27), and we have [298]

$$\text{BER}_j \leq \frac{N_0^2}{b^2 \log_2 M} \cdot \frac{A^2 P_{ij} \sigma_{in_j}^2 + B P_j \sigma_{ji}^2}{P_j^2 P_{ij} \sigma_{jn_j}^2 \sigma_{ji}^2 \sigma_{in_j}^2}, \tag{14.28}$$

where $A = (M-1)/(2M) + \sin(2\pi/M)/(4\pi)$ and $B = 3(M-1)/(8M) + \sin(2\pi/M)/(4\pi) - \sin(4\pi/M)/(32\pi)$. We can see from (14.28) that cooperative transmission obtains a diversity order of two as indicated in the power of N_o. Hence, with cooperative diversity, the total power required at the source and the relay is less than that required for noncooperative transmission in order to obtain the same BER performance. Therefore, by properly allocating the transmit power at the source (P_j) and the transmit power at the relay (P_{ij}), the lifetime of the source can be significantly increased, whereas the lifetime of the relay is slightly decreased. Note that, for multihop relay networks, the signal model in [263] can also be applied to the proposed framework in a similar way. In addition, a practical way to perform symbol-by-symbol detection at the relay is to use a simple threshold test on the received signal as proposed in [1, 175, 286].

Here, we aim to maximize the minimum device lifetime among all cooperative nodes in the network. First, we formulate the lifetime maximization problem. Then, an analytical solution is provided for a network with two cooperative nodes. After that, based on the solution for the two-node network, a fast suboptimal algorithm is developed to solve a problem for a network with multiple cooperative nodes.

As shown previously, the cooperative scheme requires less power to achieve the same performance as the noncooperative scheme, and thus it can be used to improve the minimum device lifetime. Note that different nodes may have different remaining energy, and they may contribute to different performance improvement because of their different locations. As a result, the nodes with energy or location advantages can help forward information of other energy-depleting nodes. In what follows, we formulate an optimization problem to determine which node should be a helping node and how much power should be allocated to efficiently increase the minimum device lifetime.

Let us first determine the device lifetime in a noncooperative network. From previous analysis, the noncooperative device lifetime of node j is given by

$$T_j = \frac{\kappa \varepsilon \sigma_{jn_j}^2 E_j}{(\kappa \varepsilon \sigma_{jn_j}^2 \lambda_j P_s + N_0 \sum_{l=1}^{N} \lambda_{lj})}, \tag{14.29}$$

where $\kappa = 4b \log_2 M$ and P_s represents a processing power. From (14.29), we can see that the lifetime of each node depends on its initial energy and its geographical location. Intuitively, the node whose energy is small and location is far away from its next node tends to have small device lifetime.

In the case of a cooperative network, the overall transmit power of each node is a summation of the power that the node needs for transmitting its own information and the power that the node needs for cooperatively helping forward information of other nodes. Let P_r be a processing power at each relay node, i.e., a power that the relay uses for decoding and forwarding information. From the power-allocation matrix $\mathbf{P}$, the overall transmit power of the cooperating node i is $P_i \sum_{l=1}^{N} \lambda_{li} + \sum_{j=1, j \neq i}^{N} P_{ij}(\sum_{l=1}^{N} \lambda_{lj})$, and the overall processing power of node i is $\lambda_i P_s + \sum_{j=1, j \neq i}^{N} P_r \text{sgn}(P_{ij})(\sum_{l=1}^{N} \lambda_{lj})$, where $\text{sgn}(P_{ij})$ represents the sign function that returns 1 if $P_{ij} > 0$, and 0 otherwise. Therefore the lifetime of the cooperative node i can be written as

$$T_i(\mathbf{P}) = \frac{E_i}{\lambda_i P_s + P_i \sum_{l=1}^{N} \lambda_{li} + \Lambda(P_r, P_{ij}, \lambda_{lj})}, \tag{14.30}$$

where $\Lambda(P_r, P_{ij}, \lambda_{lj}) = \sum_{j=1, j \neq i}^{N} [P_r \text{sgn}(P_{ij}) + P_{ij}] (\sum_{l=1}^{N} \lambda_{lj})$, and E_i is an initial energy of node i. Obviously, the lifetime of node i is reduced if node i helps transmit information of other nodes. On the other hand, the more the power P_{ij} that node i helps forwarding information of node j, the longer the lifetime of node j. Therefore it is crucial to properly design the power-allocation matrix $\mathbf{P}$ such that the minimum device lifetime is maximized.

With an objective to maximize the minimum device lifetime under the BER constraint on each node, the optimization problem can be formulated as

$$\max_{\mathbf{P}} \min_{i} T_i(\mathbf{P}) \tag{14.31}$$

$$\text{s.t.} \begin{cases} \text{Performance: } \text{BER}_i \leq \varepsilon, \ \forall i; \\ \text{Power: } 0 < P_i \leq P_{\max}, \ \forall i; \\ \text{Power: } 0 \leq P_{ij} \leq P_{\max}, \ \forall j \neq i, \end{cases}$$

where ε denotes a BER requirement. In (14.31), the first constraint is to satisfy the BER requirement as specified in (14.27), the second constraint guarantees that each node has information to be transmitted and the transmit power is no greater than $P_{\max}$, and the third constraint ensures that all the allocated power is nonnegative and no greater than $P_{\max}$. Because of its assignment and combinatorial nature, the formulated problem is NP-hard [158]. Even though each source–destination route is already known, the proposed work needs to optimize the pairing between each source and its relay. This problem of choosing relay is an assignment problem.

To gain some insight on the formulated problem, we provide a closed-form analytical solution in a high SNR scenario for a network with two cooperative nodes ($N = 2$). Each node transmits its information directly to the destination d. In this two-node network, there are three possible transmission strategies, namely, (1) each node transmits noncooperatively, (2) one node helps forward information of the other, and (3) both nodes help forward information of each other. Next, we maximize the minimum device lifetime for each strategy. Without loss of generality, we assume that a transmit power required for a noncooperative transmission is less than $P_{\max}$.

Based on the previous discussion, the optimum power allocation for the noncooperative case is $P_j = N_0/(\kappa \varepsilon \sigma_{jd}^2)$ for $j = 1, 2$, and $P_{ij} = 0$ for $i \neq j$. Using (14.29), the optimum device lifetime for this transmission strategy is given by

$$T_{\text{noncoop}}^* = \min \left[\frac{\kappa \varepsilon \sigma_{1d}^2 E_1}{\lambda_{11}(\kappa \varepsilon \sigma_{1d}^2 P_s + N_0)}, \frac{\kappa \varepsilon \sigma_{2d}^2 E_2}{\lambda_{22}(\kappa \varepsilon \sigma_{2d}^2 P_s + N_0)} \right]. \tag{14.32}$$

Without loss of generality, we provide a solution for a case in which node i helps relay information of node j to the destination. In this case, the lifetimes of node i and node j are given by $T_i = E_i/[\lambda_i(P_s + P_i) + \lambda_j(P_r + P_{ij})]$ and $T_j = E_j/[\lambda_j(P_s + P_j)]$, respectively. Hence, appropriately choosing P_i, P_{ij}, and P_j can improve the minimum device lifetime while maintaining a specified BER requirement.

For node i to satisfy the BER requirement ε, the optimum transmit power of node i is $P_i = N_0/(\kappa \varepsilon \sigma_{id}^2)$. To determine P_j and P_{ij}, we first note that, according to the BER upper bound in (14.28), P_j and P_{ij} must satisfy

$$\frac{N_0^2 A^2 P_{ij} \sigma_{id}^2 + N_0^2 B P_j \sigma_{ji}^2}{b^2 \log_2(M) P_j^2 P_{ij} \sigma_{jd}^2 \sigma_{ji}^2 \sigma_i^2} = \varepsilon. \tag{14.33}$$

Then we can express P_{ij} in terms of P_j as

$$P_{ij} = \frac{P_j}{C_{ij} P_j^2 - D_{ij}} = f(P_j), \tag{14.34}$$

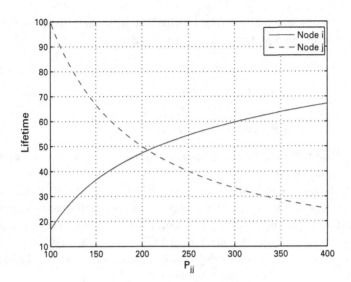

Figure 14.11 Lifetimes of the two cooperative nodes as functions of the transmit power of the helped node (P_{22}).

where $C_{ij} = (\varepsilon \sigma_{id}^2 \sigma_{jd}^2 b^2 \log_2 M)/(BN_0^2)$ and $D_{ij} = (A^2 \sigma_{id}^2)/(B\sigma_{ji}^2)$. With P_i and P_{ij}, we have

$$T_i = \frac{E_i}{\lambda_i(P_s + \frac{N_0}{\kappa \varepsilon \sigma_{id}^2}) + \lambda_j[P_r + f(P_j)]}, \tag{14.35}$$

which is a function of P_j. Therefore, optimization problem (14.31) is simplified to

$$T_{i-\text{helps}-j}^* = \max_{P_j} \left[\min \left(\frac{E_i}{\lambda_i \left(P_s + \frac{N_0}{\kappa \varepsilon \sigma_{id}^2} \right) + \lambda_j[P_r + f(P_j)]}, \frac{E_j}{\lambda_j(P_s + P_j)} \right) \right]. \tag{14.36}$$

As an illustrated example, Figure 14.11 plots lifetimes T_i and T_j as functions of P_j for a specific set of parameters. For an unconstrained optimization of (14.36), Figure 14.11 shows that the optimum power P_j in (14.36) is the one that results in $T_i = T_j$. Therefore the optimum device lifetime in the case that node i helps node j is

$$T_{i-\text{helps}-j}^* = \frac{E_i}{\lambda_i(P_s + \frac{N_0}{\kappa \varepsilon \sigma_{id}^2}) + \lambda_j[P_r + f(P_j^*)]} = \frac{E_j}{\lambda_j(P_s + P_j^*)}, \tag{14.37}$$

where P_j^* is the solution to $C_{ij} E_i \lambda_j P_j^3 + KC_{ij} P_j^2 - (D_{ij}\lambda_j E_i + \lambda_j E_j)P_j - \Upsilon D_{ij} = 0$ in which $\Upsilon = E_i \lambda_j P_s - E_j \lambda_i P_s - \lambda_j E_j P_r + (\lambda_i N_0 E_j)/(\kappa \varepsilon \sigma_{id}^2)$. Accordingly, we can find $P_{ij}^* = f(P_j^*)$ from (14.34).

If the resulting P_{ij}^* is not larger than $P_{\max}$, then (14.37) is the optimum device lifetime for this scenario. Otherwise, let $P_{ij}^* = P_{\max}$ and find P_j^* that satisfies the BER requirement (14.33). After some manipulations, we have

$$P_j^* = \frac{-Q_1 + \sqrt{Q_1^2 + Q_2 Q_3 P_{\max}^2}}{Q_2 P_{\max}}, \tag{14.38}$$

where $Q_1 = B\sigma_{ji}^2 N_0^2/(b^2 \log_2 M)$, $\quad Q_2 = 2\varepsilon\sigma_{id}^2\sigma_{ji}^2\sigma_{jd}^2$, $\quad$ and $\quad Q_3 = 2A^2\sigma_{id}^2 N_0^2/$ $(b^2 \log_2 M)$. Therefore the lifetimes of node i and node j are $T_i^* = E_i/$ $[\lambda_i(P_s + P_i) + \lambda_j(P_r + P_{\max})]$ and $T_j^* = E_j/[\lambda_j(P_s + P_j^*)]$, respectively. Hence the optimum device lifetime when $P_{ij}^* > P_{\max}$ is the minimum among T_i^* and T_j^*. As a result, the optimum device lifetime when node i helps node j can be summarized as follows:

$$T_{i-\text{helps}-j}^* = \begin{cases} \frac{E_j}{\lambda_j(P_s + P_j^*)}, & P_{ij}^* \le P_{\max} \\ \min\{T_i^*, T_j^*\}, & T_{ij}^* > P_{\max} \end{cases}. \tag{14.39}$$

Under this cooperation strategy, P_{ij} and T_i are given in (14.34) and (14.35), respectively. The optimum device lifetime in this case can be obtained by finding P_i^* and P_j^* that maximize T_i (or T_j) under the condition $T_i = T_j$; we have

$$T_{\text{both-help}}^* = \frac{E_i}{\lambda_i(P_s + P_i^*) + \lambda_j\left[P_r + \frac{P_j^*}{C_{ij}(P_j^*)^2 - D_{ij}}\right]}$$

$$= \frac{E_j}{\lambda_j(P_s + P_j^*) + \lambda_i\left[P_r + \frac{P_i^*}{C_{ji}(P_i^*)^2 - D_{ji}}\right]}, \tag{14.40}$$

where P_i^* and P_j^* are the solutions to

$$\arg\max_{P_i, P_j} \frac{E_i}{\lambda_{id}(P_s + P_i) + \lambda_{ji}(P_r + \frac{P_j}{C_{ij}P_j^2 - D_{ij}})} \tag{14.41}$$

$$\text{s.t.} \begin{cases} \frac{[(\lambda_i P_s + \lambda_j P_r + \lambda_i P_i)(C_{ij}P_j^2 - D_{ij}) + \lambda_j P_j](C_{ji}P_i^2 - D_{ji})}{[(\lambda_j P_s + \lambda_i P_r + \lambda_j P_j)(C_{ji}P_i^2 - D_{ji}) + \lambda_i P_i](C_{ij}P_j^2 - D_{ij})} = \frac{E_i}{E_j}; \\ P_j > \sqrt{\frac{D_{ij}}{C_{ij}}}, \forall j \ne i, \end{cases}$$

in which the first constraint ensures that $T_i = T_j$, and the second constraint guarantees that $P_{ij} = P_j/(C_{ij}P_j^2 - D_{ij}) > 0$.

If $P_{ij}^* \le P_{\max}$ and $P_{ji}^* \le P_{\max}$, then the solution to (14.40) is the optimum device lifetime for this transmission strategy. Otherwise, the optimization problem is separated into two subproblems. First, we let $P_{ij}^* = P_{\max}$ and find P_j^* from (14.38). Then both T_i and T_j are functions of P_i. Therefore the optimum device lifetime for this subproblem is the maximal $\min\{T_i, T_j\}$ over P_i. Second, we let $P_{ji}^* = P_{\max}$ and find P_i^* from (14.38). In this case, T_i and T_j are functions of P_j. The optimum device lifetime for the second subproblem is the maximal $\min\{T_i, T_j\}$ over P_j. Finally, the optimum device lifetime when $P_{ij}^* > P_{\max}$ or $P_{ji}^* > P_{\max}$ is the maximum among these two solutions. Therefore the optimum device lifetime when both nodes help each other can be summarized as follows:

$$T_{\text{both-help}}^* = \begin{cases} T_i, & P_{ij}^* \text{ and } P_{ji}^* \le P_{\max} \\ \max[T(ii), T(jj)], & P_{ij}^* \text{ or } P_{ji}^* > P_{\max} \end{cases}, \tag{14.42}$$

where $T_i = E_i/(\lambda_i(P_s + P_i^*) + \lambda_j[P_r + \frac{P_j^*}{C_{ij}(P_j^*)^2 - D_{ij}}])$. In (14.42), we denote $T(ii) = \max_{P_{ii}} \min\{T_i, T_j\}$, and $T(jj) = \max_{P_{jj}} \min\{T_i, T_j\}$.

Finally, the optimum device lifetime for the two-node cooperative network is

$$T_D^* = \max\left\{T_{\text{noncoop}}^*, T_{1-\text{helps}-2}^*, T_{2-\text{helps}-1}^*, T_{\text{both-help}}^*\right\}, \tag{14.43}$$

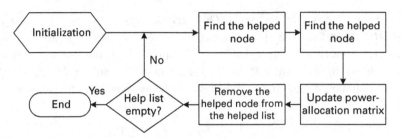

Figure 14.12 A flowchart to illustrate the proposed suboptimal algorithm.

in which T_D^* is the maximal lifetime among these four possible transmission strategies. Although the optimum solution can be obtained through full searching, it is computationally expensive for a large cooperative network. To reduce complexity of the problem, we propose next a suboptimal greedy algorithm to determine the power allocation and the corresponding device lifetime.

The basic idea of the greedy suboptimal algorithm is to find a node to be helped and a helping node step by step. In each step, the algorithm selects a node with a minimum lifetime and that has never been helped by others. Then, the algorithm chooses a helping node as the one that maximizes the minimum device lifetime after the helped node has been served. In this way, the minimum device lifetime can be increased step by step. The iteration stops when the device lifetime cannot be significantly improved or all cooperative nodes have been helped. A flowchart that summarizes the proposed algorithm is shown in Figure 14.12. Note that the proposed greedy suboptimal approach can be applied to any multinode cooperation strategy.

In what follows, we first maximize the minimum device lifetime for a given pair of helped and helping nodes, and then we describe our proposed algorithm in detail. For a given pair of helped and helping nodes, their transmit power and the corresponding lifetime can be determined in a similar way to those for the two-node network. Specifically, we consider a two-node cooperation strategy. The optimum device lifetime when node i helps node j can be obtained by solving

$$T_{i\text{-helps}-j}^* = \max_{P_j} \left[\min \left(\frac{E_i}{\Psi_i + [P_r + f(P_j)]\Sigma_{l=1}^N \lambda_{lj}}, \frac{E_j}{\Psi_j + P_j \Sigma_{l=1}^N \lambda_{lj}} \right) \right], \quad (14.44)$$

where

$$\Psi_i = \lambda_i P_s + P_i \Sigma_{l=1}^N \lambda_{li} + \Sigma_{k=1,\, k\neq i,j}^N [P_r \text{sgn}(P_{ik}) + P_{ik}](\Sigma_{l=1}^N \lambda_{lk}), \quad (14.45)$$

and

$$\Psi_j = \lambda_j P_s + \Sigma_{k=1,\, k\neq j}^N [P_r \text{sgn}(P_{jk}) + P_{jk}](\Sigma_{l=1}^N \lambda_{lk}), \quad (14.46)$$

in which Ψ_i and Ψ_j are constants that do not depend on P_j. Using the equality $T_i = T_j$ and after some manipulations, we can find that

$$T_{i\text{-helps}-j}^* = \frac{E_j}{\Psi_j + P_j^* \Sigma_{l=1}^N \lambda_{lj}}, \quad (14.47)$$

where P_j^* is the solution to

$$C_{ij} E_i \Sigma_{l=1}^N \lambda_{lj} P_j^3 + G C_{ij} P_j^2 - (D_{ij} E_i + E_j)(\Sigma_{l=1}^N \lambda_{lj}) P_j - G D_{ij} = 0, \qquad (14.48)$$

in which $G = E_i \Psi_j - E_j \Psi_i - E_j P_r \Sigma_{l=1}^N \lambda_{lj}$. If the resulting $P_{ij}^* = f(P_j^*)$ is larger than P_{max} then the same calculation steps (as previously mentioned) can be used to determine $T_{i-\text{helps}-j}^*$. This formulation is used to find the device lifetime at each step in the proposed algorithm.

Initially, the power-allocation matrix $\mathbf{P}$ is assigned as a diagonal matrix with its diagonal component $P_j = N_0/(\kappa \varepsilon \sigma_{jd}^2)$, i.e., the initial scheme is the noncooperative transmission scheme. The corresponding lifetime of node j is $T_j = E_j/(\lambda_j P_s + P_j \Sigma_{l=1}^N \lambda_{lj})$. We construct a helped list that is a list of all possible nodes to be helped: $H_{\text{list}} = \{1, 2, \ldots, N\}$. First, the algorithm finds a helped node from the helped list by choosing the node that has minimum lifetime, i.e., the helped node $\hat{j}$ is given by

$$\hat{j} = \arg \min_{j \in H_{\text{list}}} T_j. \qquad (14.49)$$

Second, the algorithm finds a node to help node $\hat{j}$ from all nodes i, $i = 1, 2, \ldots, N$ and $i \neq \hat{j}$. For each possible helping node i, the algorithm uses (14.44) to find power allocation for the helping node i and the helped node $\hat{j}$. Then the algorithm determines $T_D^*(i)$ as the minimum lifetime among cooperative nodes after node i finishes helping node $\hat{j}$. The obtained $T_D^*(i)$ from all possible helping nodes are compared, and then the algorithm selects node $\hat{i} = \arg \max_i T_D^*(i)$ as the one who helps node $\hat{j}$. Next, the algorithm updates $\mathbf{P}$ and the helped list by removing node $\hat{j}$ from the helped list. Then the algorithm goes back to the first step. The iteration continues until all nodes have been helped, i.e., the helped list is empty or the device lifetime cannot be significantly increased. The resulting $\mathbf{P}$ is the optimum power allocation that gives an answer to these questions: Which node should help which node and how much power should be used for cooperation? The detailed algorithm is summarized in Table 14.3.

Note that the proposed algorithm is suboptimal. Even though the algorithm is based on a cooperation strategy with only one relay ($K = 1$), the proposed algorithm significantly improves the device lifetime, and this will be confirmed by simulation results. In terms of complexity of the proposed algorithm, it increases only quadratically with the number of cooperative nodes. In addition, the minimum device lifetime can be further improved by cooperation with more than one relay; nevertheless, such lifetime improvement trades off with higher complexity. Note also that all necessary computations can be performed off-line. Once the algorithm is executed, each cooperative node follows the determined power-allocation and cooperation strategy. Because the proposed algorithm allocates power based on the average channel realizations, the algorithm is updated only when the network topology changes considerably. Furthermore, additional overhead for the cooperation assignment is required only at the beginning of the transmission. In (14.44), it is obvious that the helped node and helping node should be close to each other. According to this observation, we can further reduce the complexity of the proposed algorithm by searching for a helping node among cooperative nodes that are in the vicinity of the helped node. In this way, only local information is needed to compute the

Table 14.3 Suboptimal algorithm for maximizing the minimum device lifetime of wireless network with multiple cooperative nodes

---

Initialization: $P_j = N_0/(\kappa \varepsilon \sigma_{jd}^2)$, $T_j = E_j/\lambda_j(P_s + P_j)$, $T_D^* = \min T_j$, and $H_{list} = \{1, 2, \ldots, N\}$.
Iteration:

1. Select the helped node with the minimum lifetime from the helped list: $\hat{j} = \arg\min_{j \in H_{list}} T_j$, where $T_j = E_j/(\lambda_j P_s + P_j \sum_{l=1}^N \lambda_{lj})$.
2. Select the helping node from $\phi_{\hat{j}} = \{1, 2, \ldots, N\} - \{\hat{j}\}$.
   - For each $i \in \phi_{\hat{j}}$, solve (14.36) for T_i and $T_{\hat{j}}$, and then find the corresponding minimum device lifetime $T_D^*(i)$.
   - Select $\hat{i}$ that results in maximum of minimum device lifetime, $\hat{i} = \arg\max_{i \in \phi_{\hat{j}}} T_D^*(i)$, as the helping node.
3. Update power-allocation matrix $\mathbf{P}$ and helped list H_{list}. Go to 1.

End if the helped list is empty: $H_{list} = \emptyset$, or the device lifetime cannot be significantly increased. return $\mathbf{P}$.

---

power-allocation matrix. Although this may lead to some performance degradations, we will show through computer simulations that such a performance loss is insignificant.

Next, we improve the device lifetime by exploiting cooperative diversity through a deployment of cooperative relays in an energy-depleting network. Not all of these relays have information to be transmitted; however, they help forward information of all energy-depleting nodes. The relay deployment reduces the need of frequent battery changing for each node, which in turn helps reduce maintenance cost. In addition, the relay deployment does not require any modification in the cooperative nodes. An additional implementation cost is the installation cost of the relays. By using a proper number of cooperative relays and placing these relays in appropriate locations, the device lifetime can be greatly increased while the overall cost is minimized. In the following, we determine the location of each cooperative relay in the network with an objective to maximize the minimum device lifetime.

We consider a wireless network with N randomly located nodes, K cooperative relays, and a destination. The cooperative nodes are denoted as nodes $1, 2, \ldots, N$, and the cooperative relays are represented by $R_1, R_2, \ldots, R_K$. Because there is no cooperation among the cooperative nodes, the power-allocation matrix $\mathbf{P}$ is an $N \times N$ diagonal matrix whose diagonal element, P_j, represents a power that node j transmits information to its next node n_j. We assume that all cooperative nodes have information to be transmitted, i.e., $P_j > 0$ for all j. Hence the lifetime of node j is given by

$$T_j(\mathbf{P}) = \frac{E_j}{\lambda_j P_s + P_j \Sigma_{l=1}^N \lambda_{lj}}. \tag{14.50}$$

In addition, we also define a $K \times N$ relay power-allocation matrix $\hat{\mathbf{P}}$ whose element, $\hat{P}_{ij}$, represents a power that the relay R_i helps the node j. We assume that each relay does not have its own information to transmit; it helps transmit information of only other

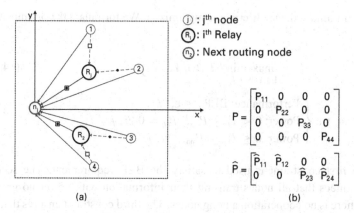

$$P = \begin{bmatrix} P_{11} & 0 & 0 & 0 \\ 0 & P_{22} & 0 & 0 \\ 0 & 0 & P_{33} & 0 \\ 0 & 0 & 0 & P_{44} \end{bmatrix}$$

$$\hat{P} = \begin{bmatrix} \hat{P}_{11} & \hat{P}_{12} & 0 & 0 \\ 0 & 0 & \hat{P}_{23} & \hat{P}_{24} \end{bmatrix}$$

Figure 14.13 Cooperative wireless network with relay deployment: (a) one cluster with four nodes, two relays, and one destination; (b) the corresponding power-allocation matrices (**P**) and ($\hat{\mathbf{P}}$) for the nodes and the relays, respectively.

cooperative nodes. By denoting E_{R_i} as an initial energy of a relay R_i, the lifetime of R_i is

$$T_{R_i}(\hat{\mathbf{P}}) = \frac{E_{R_i}}{\sum_{j=1}^{N}[P_r \operatorname{sgn}(\hat{P}_{ij}) + \hat{P}_{ij}](\sum_{l=1}^{N} \lambda_{lj})}. \tag{14.51}$$

As an example, a wireless network with four cooperative nodes and two cooperative relays is depicted in Figure 14.13(a). In the figure, the solid line represents a link from a node (source j or relay) to the next node n_j, and the dashed line represents a link from a source to a relay. Figure 14.13(b) shows the power-allocation matrix **P** and the relay power-allocation matrix $\hat{\mathbf{P}}$ that correspond to the wireless network in Figure 14.13(a). Because all four cooperative nodes transmit their information to n_j, then all diagonal elements of **P** are nonzeros. As shown in Figure 14.13(a), solid lines with square and circle represent the cases in which relay R_1 helps transmit information of node 1 and node 2, respectively. Accordingly, $\hat{P}_{11}$ and $\hat{P}_{12}$ are nonzero elements in the first row of $\hat{P}$. Similarly, relay R_2 helps transmit information of node 3 and node 4 to the next node n_j; $\hat{P}_{23}$ and $\hat{P}_{24}$ are nonzero elements in the second row of $\hat{P}$.

We denote x_j and y_j as locations of node j on the x axis and the y axis, respectively. Then we represent a location of node j in a vector form as $\bar{\mathbf{D}}_j = [\mathbf{x_j} \ \mathbf{y_j}]^T$. Accordingly, the channel variance between node j and its next node n_j is given by $\sigma_{jn_j}^2 = \eta \|\bar{\mathbf{D}}_j - \bar{\mathbf{D}}_{\mathbf{n}_j}\|^{-\alpha}$ where $\| \cdot \|$ denotes the Frobenius norm [136]. Locations of the cooperative relays are specified by a $2 \times K$ matrix $\mathbf{D_R} = [\bar{\mathbf{D}}_{\mathbf{R}_1} \ \bar{\mathbf{D}}_{\mathbf{R}_2} \cdots \bar{\mathbf{D}}_{\mathbf{R}_K}]$ in which the ith column indicates the location of relay R_i, i.e., $\bar{\mathbf{D}}_{\mathbf{R}_i} = [\mathbf{x_{R_i}} \ \mathbf{y_{R_i}}]^T$ is the location vector of the relay R_i. Then the channel variance between R_i and node n_j is $\sigma_{R_i,n_j}^2 = \eta \|\bar{\mathbf{D}}_{\mathbf{R}_i} - \bar{\mathbf{D}}_{\mathbf{n}_j}\|^{-\alpha}$, and the channel variance between node j and R_i is $\sigma_{j,R_i}^2 = \eta \|\bar{\mathbf{D}}_j - \bar{\mathbf{D}}_{\mathbf{R}_i}\|^{-\alpha}$. If node j is helped by R_i, then the BER of node j has a form similar to that of (14.27) with P_{ij} and σ_{ji}^2 replaced with $\hat{P}_{ij}$ and σ_{j,R_i}^2, respectively. Our objective is used to determine $\mathbf{D_R}$, **P**, and

$\hat{\mathbf{P}}$ such that the minimum device lifetime is maximized. We formulate the optimization problem as

$$\max_{\mathbf{D_R},\mathbf{P},\hat{\mathbf{P}}} \min_{i,j} \left\{ T_j(\mathbf{P}), \mathbf{T_{R_i}}(\hat{\mathbf{P}}) \right\}, \tag{14.52}$$

$$\text{s.t.} \begin{cases} \text{Performance: BER}_j \leq \varepsilon, \ \forall j; \\ \text{Power: } 0 < P_i \leq P_{\max}, \ P_{ij} = 0 \ \forall i, j \neq i; \\ \text{Power: } 0 \leq \hat{P}_{ij} \leq P_{\max}, \ \forall i, j. \end{cases}$$

In (14.52), the first constraint is used to satisfy the BER requirement. The second constraint guarantees that all nodes transmit their information with power no greater than $P_{\max}$ and there is no cooperation among nodes. The third constraint ensures that the power for each cooperative relay is nonnegative and not greater than $P_{\max}$. Because of the assignment and combinatorial nature of the formulated problem, the problem in (14.52) is NP-hard [158]. Because it is computationally expensive to obtain the optimum solution to (14.52), a fast suboptimal algorithm is proposed to solve the formulated problem.

The basic idea of the proposed algorithm is to add one cooperative relay at a time into the network. Each time the optimum location of the added relay is chosen as the location that maximizes the minimum device lifetime, among all possible locations. The algorithm stops when the device lifetime improvement is insignificant after another cooperative relay is added or when the maximum number of relays is reached. In the following, we first describe the algorithm to determine the device lifetime in each step, and then we describe the proposed algorithm in detail. To maximize the minimum device lifetime when the number of relays and their locations are given, we use the step algorithm similar to the one described in Figure 14.12. Initially, all nodes are sorted in ascending order according to their noncooperative lifetimes, as specified in (14.29), and then the algorithm registers them in a helped list H_{list}. In each iteration, first, select the first node in the helped list as the one to be helped. Second, determine the minimum device lifetime after all of the cooperative relay R_i's $(i = 1, 2, \dots, K)$ finish helping the selected node, and then choose the relay $R_{\hat{i}}$, where $\hat{i}$ is the relay that maximizes the minimum device lifetime to help the selected node. Next, update the power-allocation matrices $\mathbf{P}$ and $\hat{\mathbf{P}}$ and remove the selected node in the first step from the helped list. The iteration continues until all nodes have been helped and the helped list is empty or until the device lifetime improvement is insignificant.

The algorithm used to find an optimum location of each cooperative relay is described as follows. We denote $K_{\max}$ as the maximum number of cooperative relays and denote Φ_D as a set of all possible relay locations. Initially, the number of relays is set to zero. In each iteration, the number of relays is increased by one, and the optimum relay location $\hat{\mathbf{D}}$ is determined by use of one of the heuristic search methods (e.g., local search or simulated annealing) together with the algorithm in Table 14.4. The location $\hat{\mathbf{D}}$ that results in the maximum of the minimum device lifetime is selected as the optimum relay location. Then the device lifetime is updated. Finally, the algorithm goes back to the first step. The algorithm stops if the device lifetime improvement is insignificant or the number of relays reaches $K_{\max}$. The detailed algorithm is presented in Table 14.5.

Table 14.4 Suboptimal algorithm to determine device lifetime when relay locations are fixed

Initialization: $P_j = \frac{N_0}{\kappa \varepsilon \sigma_{jd}^2}$, $T_j = \frac{E_j}{\lambda_j(P_s+P_j)}$, $T_D^* = \min T_j$, Sort N nodes by their lifetimes in
   ascending order list in H_{list}.

Iteration:
1. Select the first node in the H_{list} as the helped node.
2. Select the helping relay $R_{\hat{i}}$ from the set of K relays.
   • For each i, use the heuristic algorithm to maximize the minimum device lifetime, $T_D^*(i)$.
   • Select $R_{\hat{i}}$ that results in maximum of minimum device lifetime to help the node $\hat{j}$.
3. Update $P_{\hat{j}\hat{j}}$ in **P** and update $\hat{P}_{\hat{i}\hat{j}}$ in $\hat{\mathbf{P}}$.
   Set $\hat{P}_{ij} = 0$ for all $i \neq \hat{i}$ and set $T_D^* = T_D(\hat{i})$.
   Remove node $\hat{j}$ from the helped list H_{list}. Go to 1.

End: If the helped list is empty: $H_{\text{list}} = \emptyset$, or the device lifetime cannot be significantly
   increased. Return **P**, $\hat{\mathbf{P}}$, T_D^*.

Table 14.5 Algorithm to determine relay locations

Initialization: $q = 0$

Iteration:
1. Increase number of relays: $q = q+1$
2. For each location $\mathbf{D_l} \in \Phi_\mathbf{D}$. Set $\mathbf{D_{R_q}} = \mathbf{D_l}$.
   Find T_D^*, **P** and $\hat{\mathbf{P}}$ using the algorithm in Table 14.4
   Denote the obtained results by $T_D^*(l)$, $\mathbf{P}(l)$ and $\hat{\mathbf{P}}(l)$
3. Find the relay location R_q: $\mathbf{D_{R_q}} = \mathbf{D_{l^*}}$, where $l^* = \arg\max_l T_D^*(l)$
4. Update **P**, $\hat{\mathbf{P}}$, and T_D^*. Go to 1.

End: If the device lifetime improvement is insignificant, or $q = K_{\max}$. Return **P**, $\hat{\mathbf{P}}$, T_D^*.

Note that the proposed algorithm allows, at most, one relay to help each node. Although the algorithm is suboptimal, simulation results show that the proposed algorithm
significantly improves the device lifetime. In addition, all of the required computations
can be performed off-line. Moreover, the problems and algorithms are closely related to
the goal of extending the device lifetime by exploiting cooperative diversity. The cooperative diversity is exploited by cooperation among devices. However, cooperative relays
are deployed, and the cooperative diversity is exploited by cooperation between each
device and one of these additional cooperative relays. Therefore the basic algorithms
for finding **P** and finding $\hat{\mathbf{P}}$ are similar. The search process of **P** and $\hat{\mathbf{P}}$ is done under the
given locations of relays (i.e., for a fixed **D**); the process is not affected by the method
of finding **D**.

In all simulations, BPSK modulation is used in the system, the propagation-loss
factor is $\alpha = 3$, $\eta = 1$, and $\varepsilon = 10^{-3}$ (unless stated otherwise). The processing power
of each node (P_s) is set at 25% of the transmit power of the node whose location is at

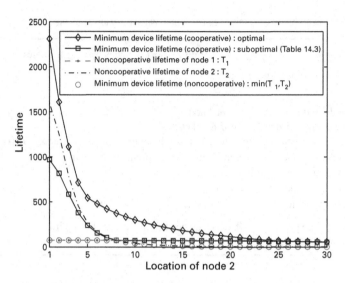

Figure 14.14 Device lifetime in a two-node wireless network.

(10 m, 0) [277]. The processing power of each relay (P_r) is set at 50% of P_s. All nodes are equipped with equal initial energy of $E_j = 10^5$. The noise variance is set at $N_0 = 10^{-2}$. The nodes are randomly distributed based on uniform distribution, and the destination is located in the center of the area. Each node sends information to the destination via a route that is determined by Dijkstra's algorithm.

In Figure 14.14, we consider a two-node wireless network in which the destination is located at coordinate (0, 0). Node 1 is fixed at coordinate (0, 8 m). The location of node 2 varies from (0, 1) to (0, 30 m). We can see that the minimum device lifetime of the noncooperative scheme is determined by the lifetime of the node that is located farther from the destination, as shown by a curve with circle ("∘"). Under cooperative transmission, the minimum device lifetime is significantly increased, especially when node 2 is located close to the destination. The reason is that node 2 requires small transmit power to reach the destination. After node 2 helps node 1, the transmit power of node 2 slightly increases, while the transmit power of node 1 greatly reduces because of the cooperative diversity. With the proposed suboptimal algorithm (Table 14.3), the minimum device lifetime is improved to almost the same as the lifetime of the node that is closer to the destination (see a curve with rectangular). By using the optimum power allocation, the minimum device lifetime can be further increased (see a curve with diamonds) because both nodes take advantage of the cooperative diversity while using a smaller amount of their transmit power.

Figure 14.15 depicts the minimum device lifetime according to density of cooperative nodes in a square area. The number of randomly located nodes varies from 20 to 50 over an area of 100 m × 100 m. In the simulation, we normalize the transmission rate to be the same for all network sizes. For local search, the helping node is chosen among the nodes whose distances from the source node are less than 20 m. From the figure, we can see that the minimum device lifetime of the cooperative network is higher than that of

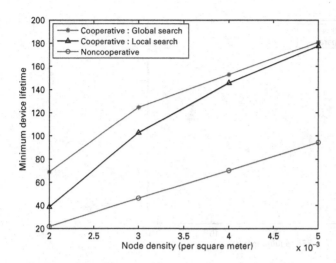

Figure 14.15 Minimum device lifetime with different numbers of randomly located nodes.

the noncooperative network for all network sizes. For example, the cooperative network doubles the minimum device lifetime compared with that of noncooperative network when there are 50 nodes in the network. Note that the performance gain is calculated as $T_{\text{coop}}/T_{\text{noncoop}}$, where T_{coop} and T_{noncoop} represent the lifetime of cooperative and noncooperative networks, respectively. From Figure 14.15, the cooperative scheme with local search yields a performance similar to the one with global search, especially when the node density is high. This confirms our expectation that the helping node is chosen as the one that is located close to the helped node. Note that the minimum device lifetimes for noncooperative and cooperative networks increase with the number of nodes because the chance of being helped by a node with good location and high energy increases.

In Figure 14.16, we consider improvement of the minimum device lifetime according to different BER requirements. We assume in the simulation that there are 30 randomly located cooperative nodes in an area of size 100 m $\times$ 100 m. We can see from the figure that the minimum device lifetime is small at a BER requirement of 10^{-6} under both noncooperative and cooperative networks. This is because each node requires large transmit power to satisfy such a small BER requirement. As the BER constraint increases, the minimum device lifetime also increases since the transmit power required to satisfy the BER constraint decreases. Note that the cooperative network achieves longer device lifetime than that for the noncooperative network over the entire range of the BER requirement. For example, the cooperative network achieves $125/49 = 2.6$ times longer lifetime than the noncooperative network at a BER requirement of 10^{-3}. However, both cooperative and noncooperative networks yield almost the same device lifetime at a BER constraint of 10^{-2}. The reason is that the transmit power required to satisfy the BER of 10^{-2} is much smaller than the processing power. The effect of processing power on the device lifetime dominates that of transmit power in this case.

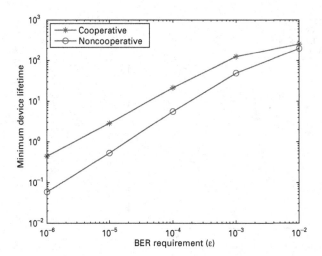

Figure 14.16 Minimum device lifetime improvement according to different BER targets.

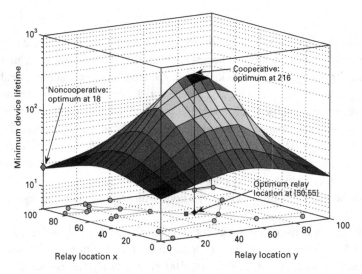

Figure 14.17 Minimum device lifetime for a cooperative network with a cooperative relay.

Figure 14.17 shows the minimum device lifetime for different relay locations. We consider a case in which there are 20 randomly located nodes and a relay with initial energy of $E_{R_i} = 10^6$ in area of 100 m × 100 m. In the figure, a node with a circle represents a randomly located node, and a node with a rectangle shows the location of the destination. In the simulation, we vary the relay location in a grid area of 100 m × 100 m. From the figure, the minimum device lifetime of the noncooperative network is the same for all possible relay locations (as indicated by a point with "◇"). However, the minimum device lifetime of the cooperative network gradually increases when the relay moves closer to the destination. Specifically, the minimum device lifetime is the same as that of noncooperative network when the relay is far away from the destination. But

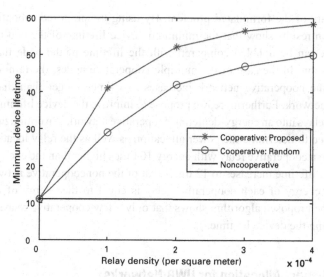

Figure 14.18 Minimum device lifetime according to different numbers of deployed cooperative relays.

the minimum device lifetime further improves to $216/18 = 12$ times longer than that of the noncooperative network when the relay is close to the center of the area. This is because the node that is nearest to the destination tends to drain out its battery first, and its lifetime can be greatly improved by placing the relay near the destination.

Figure 14.18 shows the minimum device lifetime according to the density of cooperative relays (i.e., the number of relays per square meter). We consider a cooperative network with 20 randomly located nodes in an area of 100 m × 100 m. The initial energy of each relay is 10^5. The minimum device lifetime of the cooperative network with one randomly added cooperative relay is about $28/11 = 2.55$ times longer than that of the noncooperative network (as shown by a curve with circle "o"). If the relay is placed at its optimum location, the minimum device lifetime (a curve with star "∗") can be improved to $42/11 = 3.83$ times longer than that of noncooperative network. Furthermore, when two to four relays are added into the network, the minimum device lifetime can be further increased under a case with optimally placed relays as well as a case with randomly placed relays. However, the minimum device lifetime is almost saturated when more than two relays are deployed in the network.

Overall, we propose the lifetime maximization by cooperative node employment and relay deployment in wireless networks. By introducing cooperation protocol among nodes, both energy advantage and location advantage can be explored such that the device lifetime is improved. First, the DF cooperation protocol is employed among nodes. We determine which nodes should cooperate and how much power should be allocated for cooperation. An optimization problem is formulated with an aim to maximize the minimum device lifetime under a BER constraint. An analytical solution for a two-node cooperative network is provided. In the case of a multiple-node scenario, it turns out that the formulated problem is NP-hard. A suboptimal algorithm is developed to reduce the complexity of the formulated problem. By using the proposed suboptimal algorithm, simulation results show that the minimum device lifetime of the two-node cooperative network can be doubled compared with the lifetime of the node that is closer to the destination. In the case of the multiple cooperative nodes, the minimum device lifetime of the cooperative network increases two times longer than that of the noncooperative network. Furthermore, we propose to improve the device lifetime by adding cooperative relays into an energy-depleting cooperative network. An optimization problem is formulated to determine the power allocation as well as the relay locations. By optimally placing a cooperative relay with energy 10 times higher than the energy of the nodes, the device lifetime increases to 12 times that of the noncooperative network. Furthermore, when energy of each cooperative relay is equal to the energy of each cooperating node, the proposed algorithm shows that only a few cooperative relays are required for improving the device lifetime.

14.4   Power Control/Channel Allocation for UWB Networks

UWB is an emerging technology that offers great promise to satisfy the growing demand for low-cost and high-speed digital wireless home networks. In 1998, the FCC mandated that UWB radio transmission can legally operate in the range from 3.1 to 10.6 GHz, at a transmit power of -41.3 dBm/MHz [371]. Traditional UWB technology is based on single-band systems [73, 334] that directly modulate data into a sequence of pulses which occupy the available bandwidth of 7.5 GHz. Recently, innovative multiband UWB schemes were proposed in [17, 78]. Instead of using the entire UWB frequency band to transmit information, a multiband technique divides the spectrum into several subbands. Each subband occupies a bandwidth of at least 500 MHz in compliance with the FCC limit [371]. By interleaving the transmitted symbols across subbands, multiband UWB systems can still maintain the average transmit power as if the large GHz bandwidth were used. The advantage is that the information can be processed over much smaller bandwidths, thereby reducing overall design complexity, as well as improving spectral flexibility and worldwide compliance. The current leading proposal for the IEEE 802.15.3a WPAN standard [375] is based on multiband OFDM, which utilizes a combination of OFDM and time–frequency interleaving [17]. The OFDM technique is efficient at collecting multipath energy in highly dispersive channels, as is the case for most UWB channels [17]. Time–frequency interleaving allows the OFDM symbols to be transmitted on different subbands. A frequency synthesizer can be utilized to perform frequency hopping. By using proper time–frequency codes, the multiband UWB system provides not only diversity, but also multiple-access capability [17].

Because many applications enabled by UWB are expected to be in portable devices, low power consumption becomes a fundamental requirement. The low transmit power of UWB emissions ensures a long lifetime for the energy-limited devices. In addition, UWB systems are expected to support an integration of multimedia traffic, such as voice, image, data, and videostreams. This requires a cross-layer algorithm that is able to allocate the available resources to a variety of users with different service rates in an effective way. An overview of resource allocation in UWB communications is provided in [370]. In [54], a joint rate and power-assignment problem that is central in multiuser UWB networks is considered, and a radio resource-sharing mechanism that performs a handshaking procedure to establish a communication link is proposed. In [244], a joint scheduling, routing, and power-allocation problem is discussed with an objective to maximize the total utility of a UWB system.

Most of the existing resource-allocation schemes for UWB systems (see [244, 370] and references therein) are based on single-band impulse radio technology. On the other hand, most research efforts on multiband UWB systems have been devoted to the physical layer issues [17, 215, 258]. Some of the key issues in multiband UWB systems that remain largely unexplored are resource allocations such as power control and channel allocation. The current multiband proposal divides the subbands into groups, each comprising two to three subbands. A set of certain time–frequency codes is used to interleave the data within each band group [17]. This strategy lacks the ability to allocate subbands optimally because the available subbands are not assigned to each user according to its channel condition. Moreover, in the multiband proposal [17], the transmit power of each user is equally distributed among its assigned subbands without any power adaptation to the channel variations. As a result, adaptive optimization of the subband assignment and power control can greatly improve the system performances of multiband UWB systems.

We propose a novel cross-layer channel-allocation scheme for multiband UWB wireless networks (e.g., a piconet, as in the IEEE 802.15.3 standard). By efficiently allocating the subbands, transmit power, and data rates among all users, the proposed scheme enables the multiband UWB system to operate at a low transmit power level, while still achieving desired performance. First, we formulate a subband assignment and power-allocation problem as an optimization problem whose goal is to minimize the overall transmit power provided that all users achieve their requested data rates and desired PER, whereas the power spectral density complies with the FCC limit [371]. To take into account the fact that users in the multiband UWB system may have different data rates that in turn imply different channel-coding rates, frequency-spreading gains, and/or time-spreading gains, our formulated problem considers not only the limitation on transmit power level, but also band hopping for users with different data rates. It turns out that the formulated problem is an integer programming problem whose complexity is NP-hard. Then to reduce the complexity of the formulated problem, we propose a fast suboptimal algorithm that can guarantee to obtain a near-optimal solution, but requires low computational complexity. To ensure the system feasibility in variable channel conditions, we further develop a joint rate assignment and power-controlled channel-allocation algorithm that is able to allocate resources to the users according to three different system optimization goals, namely maximizing overall rate, achieving proportional fairness, and reducing maximal rate. Simulation results based on the UWB channel model specified in the IEEE 802.15.3a standard [77] show that the proposed algorithm achieves up to 61% of transmit power saving compared with that of the standard multiband scheme [17]. Moreover, the proposed algorithm can also find feasible solutions adaptively when the initial system is not feasible for the rate requirements of the users.

Table 14.6 Rate-dependent parameters

Data rate (Mbps)	Modulation	Coding rate	Conjugate symmetric inputs to IFFT	Time-spreading factor
53.3	QPSK	1/3	Yes	2
55	QPSK	11/32	Yes	2
80	QPSK	1/2	Yes	2
106.7	QPSK	1/3	No	2
110	QPSK	11/32	No	2
160	QPSK	1/2	No	2
200	QPSK	5/8	No	2
320	QPSK	1/2	No	1
400	QPSK	5/8	No	1
480	QPSK	3/4	No	1

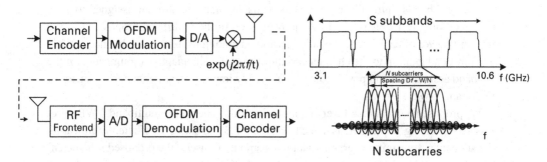

Figure 14.19 Multiband UWB system illustration.

We consider a UWB system using multiband OFDM that has been proposed for the IEEE 802.15.3a WPAN standard [375]. As shown in Figure 14.19, the available UWB spectrum, from 3.1 to 10.6 GHz, is divided into S subbands. Each subband occupies a bandwidth of at least 500 MHz in compliance with FCC regulations. The UWB system employs OFDM with N subcarriers, which are modulated using QPSK. At each OFDM symbol period, the modulated symbol is transmitted over one of the S subbands. These symbols are time interleaved across subbands. Different data rates are achieved by using different channel-coding, frequency-spreading, or time-spreading rates [17]. The frequency-domain spreading is obtained by choosing conjugate symmetric inputs to the IFFT. Specifically, $N/2$ complex symbols are transmitted in the first half of the subcarriers, and their conjugate symmetric symbols are transmitted in the second half of the subcarriers. Time-domain spreading is achieved by transmitting one OFDM symbol followed by a permutation of that OFDM symbol, i.e., transmitting the same information twice during two OFDM symbol periods. The receiver combines the information transmitted via different times or frequencies to increase the SNR of received data.

As listed in Table 14.6, the multiband UWB system provides data rates ranging from 53.3 to 480 Mbps. For rates not higher than 80 Mbps, both time and frequency spreadings are performed, yielding the overall spreading gain of four. For the rates between 106.7 and

200 Mbps, only time-domain spreading is utilized, which results in the overall spreading gain of two. The system with data rates higher than 200 Mbps exploits neither frequency nor time spreading, and the overall spreading gain is one. Forward error-correction codes with coding rates of 1/3, 11/32, 1/2, 5/8, or 3/4 are employed to provide different channel protections with various data rates.

In what follows, we describe the channel model and system model for the multiuser, multiband UWB system.

The channel model specified in the IEEE 802.15.3a standard [77] is a modified version of the SV model for indoor channels [264]. The channel impulse response for the kth user at time t can be represented by

$$h_k(t) = X_k \sum_{m=0}^{M_k} \sum_{l=0}^{L_k} \alpha_k(m, l)\delta[t - T_k(m) - \tau_k(m, l)], \qquad (14.53)$$

where X_k is the log-normal shadowing, and $\alpha_k(m, l)$ denotes the gain of the lth multipath component in the mth cluster. The time duration $T_k(m)$ represents the delay of the mth cluster, and $\tau_k(m, l)$ is the delay of the lth path in the mth cluster relative to the cluster-arrival time. The cluster arrivals and the path arrivals within each cluster can be modeled as Poisson distribution with rate Λ_k and rate λ_k ($\lambda_k > \Lambda_k$), respectively. In (14.53), M_k is the number of clusters, and L_k is the number of paths within each cluster. The path amplitude $|\alpha_k(m, l)|$ follows the log-normal distribution, whereas the phase $\angle\alpha_k(m, l)$ is a uniform random variable over $[0, 2\pi)$. The model parameters corresponding to several scenarios are provided in [77]. With the choice of cyclic prefix length greater than the duration of the channel impulse response, OFDM allows for each UWB subband to be divided into a set of N orthogonal narrowband channels. The baseband frequency response at the nth ($n = 0, 1, \ldots, N - 1$) subcarrier is

$$H_k(n) = X_k \sum_{m=0}^{M_k} \sum_{l=0}^{L_k} \alpha_k(m, l)e^{-j2\pi n\Delta f[T_k(m)+\tau_k(m,l)]}, \qquad (14.54)$$

where $j = \sqrt{-1}$ and Δf is the frequency separation between two adjacent subcarriers. It is worth noting that, for most WPAN applications, the transmitter and receiver are stationary [210]. As a result, the UWB channel is very slowly fading. The standard channel model assumes that the channel stays either completely static or is time invariant during the transmission of each packet [77, 210]. We assume that the channel-state information is known at both the transmitter and the receiver.

We consider a multiuser multiband UWB scenario in which K users simultaneously transmit their information. The kth user has the data rate R_k, which can be any value specified in Table 14.6. As shown in Table 14.6, if the rate is higher than 200 Mbps, there is no time spreading; otherwise, the time-domain-spreading operation is performed with a spreading factor of two. In this case, any time–frequency code with a period of two time slots can guarantee that each user will achieve the additional diversity by transmitting the same information over two OFDM blocks. The time–frequency codes with a period longer than two can also be used to improve the multiple-access capability for asynchronous UWB wireless networks [17]. To simplify the problem formulation, we consider a multiband UWB system employing time–frequency codes of length two. The extension to UWB systems with longer time–frequency codes is straightforward.

To specify in which subbands each user can transmit its information, we define a $K \times S$ assignment matrix $\mathbf{A}$, whose (k, s)th element is denoted by a_{ks}, for $k = 1, 2, \ldots, K$ and $s = 1, 2, \ldots, S$. This a_{ks} represents the number of OFDM symbols that user k is allowed to transmit on the sth subband during two OFDM symbol periods. Assuming that each user utilizes one subband per transmission, a_{ks} can take any value from the set $\{0, 1, 2\}$. However, when the kth data rate of the user is less than or equal to 200 Mbps, we need to ensure that the band hopping is performed to obtain the diversity from time spreading. In this case, a_{ks} is restricted to $a_{ks} \in \{0, 1\}$. Thus the element of assignment matrix satisfies [281, 282]

$$a_{ks} \in \phi(R_k) = \begin{cases} 0, 1, & R_k \leq 200 \text{ Mbps} \\ 0, 1, 2, & R_k > 200 \text{ Mbps} \end{cases}. \tag{14.55}$$

During each OFDM symbol period, one user will occupy one subband. Because we consider the duration of two OFDM blocks, the assignment strategy needs to satisfy

$$\sum_{s=1}^{S} a_{ks} = 2, \quad k = 1, 2, \ldots, K. \tag{14.56}$$

In addition, to minimize multiple-access interference, each subband is assigned to a specific user at a time, and hence each subband can be used at most twice during two OFDM symbol periods. Therefore the subband assignment also follows

$$\sum_{k=1}^{K} a_{ks} \leq 2, \quad s = 1, 2, \ldots, S. \tag{14.57}$$

Let $P_k^s(n)$ denote the transmit power of the kth user at subcarrier n of the sth subband. Accordingly, the SNR of user k at the sth subband and the nth subcarrier is given by

$$\Gamma_k^s(n) = \frac{P_k^s(n) G_k^s(n)}{\sigma_k^2}, \tag{14.58}$$

where $G_k^s(n)$ is the corresponding channel gain. We can express $G_k^s(n)$ as

$$G_k^s(n) = |H_k^s(n)|^2 \left(\frac{4\pi d_k}{\lambda_k^s} \right)^{-\nu}, \tag{14.59}$$

in which $H_k^s(n)$ is the channel frequency response at subband s and subcarrier n, ν is the propagation-loss factor, d_k represents the distance between the transmitter and receiver, $\lambda_k^s = 3 \times 10^8 / f_{c,k}^s$ is the wavelength of the transmitted signal, and $f_{c,k}^s$ is the center frequency of the waveform. In (14.58), σ_k^2 denotes the noise power at each subcarrier, defined as

$$\sigma_k^2 = 2 \times 10^{[-174 + 10 \log_{10}(R_k) + N_F]/10}, \tag{14.60}$$

where R_k is the data rate of the kth user and N_F is the received noise figure referred to as the antenna terminal [17]. As in the multiband standard, we assume that the noise power σ_k^2 is the same for every subcarrier within each subband. We assume ideal band hopping such that the signal transmitted over different subbands undergoes independent fading, and there is no multiple-access interference.

Because of the consideration for the simple transceiver of UWB, the current standard assumes that there is no bit loading and the power is equally distributed across subcarriers within each subband. Similarly, we assume that $P_k^s(n) = P_k^s(n')$ for any $0 \leq n, n' \leq N - 1$. Denote

$$P_k^s(n) = P_k^s, \quad n = 0, 1, \ldots, N - 1; \tag{14.61}$$

then the $K \times S$ power allocation matrix can be defined as $[\mathbf{P}]_{ks} = P_k^s$, in which the (k, s)th component represents the transmit power of the kth user in subband s.

In the multiband frequency band plan [17], the subband center frequencies span a wide range from 3.43 to 10.3 GHz. Consequently, different subbands tend to undergo different fading and propagation loss. Additionally, the channel condition for a specific subband may be good for more than one user. Therefore, to efficiently reduce the power consumption, we need to optimize the subband assignment matrix $\mathbf{A}$ and power-allocation matrix $\mathbf{P}$ under some practical constraints.

First, we derive a generalized SNR expression for various UWB transmission modes. Second, we provide a necessary condition for the SNR so as to satisfy the PER requirement. Then we propose a problem formulation to minimize the overall transmit power provided that all users achieve their requested data rates and desired PER, while the transmit power level is below the FCC limitation and rate parameters are in accord with the standard proposal given in Table 14.6. We develop a fast suboptimal scheme to solve the proposed problem. Finally, to ensure the system feasibility, we develop a joint rate adaptation, subband assignment, and power-allocation algorithm.

Assuming that the channel-state information is perfectly known at the receiver, the receiver employs MRC to combine the information transmitted via different times or frequencies. As a result, the average SNR at the output of MRC depends not only on the channel-coding rate, but also on the time- and frequency-spreading factors. The following proposition provides a generalized expression of the average SNR for any data rates.

PROPOSITION 4 *Assume maximum ratio combining and $P_k^s(n) = P_k^s$ for all subcarriers n, then the average SNR of the kth user is given by*

$$\bar{\Gamma}_k = \sum_{s=1}^{S} a_{ks} P_k^s F_k^s, \tag{14.62}$$

where

$$F_k^s = \frac{b_k}{N\sigma_k^2} \sum_{n=0}^{N-1} G_k^s(n), \tag{14.63}$$

and b_k is a constant that depends on the data rate of the kth user as follows:

$$b_k = \begin{cases} 2, & R_k \leq 80 \ Mbps; \\ 1, & 80 < R_k \leq 200 \ Mbps; \\ 1/2, & R_k > 200 \ Mbps. \end{cases} \tag{14.64}$$

Proof 9 Recall that when R_k is not higher than 80 Mbps, the information is spread across both time and frequency with the overall spreading gain of four. Consequently, the total SNR for the kth user at subcarrier n, $n = 0, 1, \ldots, N/2 - 1$, is

$$\Gamma_k(n) = \sum_{s=1}^{S} a_{ks} \left[\Gamma_k^s(n) + \Gamma_k^s(n + N/2) \right]. \tag{14.65}$$

Note that the SNR in (14.65) is based on the assumptions of no multiuser interference and no correlation among the data bits; it leads to an upper bound on the performance. Average (14.65) over $N/2$ subcarriers, resulting in the average SNR

$$\bar{\Gamma}_k = \frac{1}{N/2} \sum_{n=0}^{N/2-1} \Gamma_k(n) = \frac{1}{N/2} \sum_{n=0}^{N-1} \sum_{s=1}^{S} a_{ks} \Gamma_k^s(n). \tag{14.66}$$

By substituting (14.58) into (14.66) and assuming $P_k^s(n) = P_k^s$, we obtain

$$\bar{\Gamma}_k = \frac{2}{N} \sum_{n=0}^{N-1} \sum_{s=1}^{S} a_{ks} P_k^s \frac{G_k^s(n)}{\sigma_k^2}$$

$$= \sum_{s=1}^{S} a_{ks} P_k^s \left[\frac{2}{N\sigma_k^2} \sum_{n=0}^{N-1} G_k^s(n) \right]. \tag{14.67}$$

When R_k is between 106.7 and 200 Mbps, only time spreading is performed, and hence the total SNR at subcarrier n, $n = 0, 1, \ldots, N - 1$, becomes

$$\Gamma_k(n) = \sum_{s=1}^{S} a_{ks} \Gamma_k^s(n) = \sum_{s=1}^{S} a_{ks} \frac{P_k^s(n) G_k^s(n)}{\sigma_k^2}. \tag{14.68}$$

Thus the average SNR can be obtained from (14.68) as

$$\bar{\Gamma}_k = \frac{1}{N} \sum_{n=0}^{N-1} \Gamma_k(n) = \sum_{s=1}^{S} a_{ks} P_k^s \left[\frac{1}{N\sigma_k^2} \sum_{n=0}^{N-1} G_k^s(n) \right]. \tag{14.69}$$

For R_k higher than 200 Mbps, there is no spreading and the average SNR of the kth user is simply the average of $\Gamma_k^s(n)$ over N subcarriers and two subbands, i.e.,

$$\bar{\Gamma}_k = \frac{1}{2N} \sum_{n=0}^{N-1} \sum_{s=1}^{S} a_{ks} \Gamma_k^s(n)$$

$$= \sum_{s=1}^{S} a_{ks} P_k^s \left[\frac{1}{2N\sigma_k^2} \sum_{n=0}^{N-1} G_k^s(n) \right]. \tag{14.70}$$

Express (14.67), (14.69) and (14.70) in terms of F_k^s defined in (14.63) leading to the results in (14.62).   **QED**

A common performance requirement of UWB systems is to offer packet transmission with an error probability less than a desired threshold value. The PER metric is directly related to the BER performance, which in turn depends on the SNR at the output of the MRC. By keeping the SNR level higher than a specific value, the PER can be ensured to be lower than the PER threshold. Next, we provide a necessary condition for the average SNR so as to satisfy the PER requirement.

Suppose the maximum PER is ε and the packet length is L bits, then the bit error probability after the channel decoder for the kth user, P_k, needs to satisfy

$$1 - (1 - \mathrm{P}_k)^L \leq \varepsilon. \tag{14.71}$$

By the assumptions of the use of convolutional coding and Viterbi decoding with perfect interleaving, P_k is given by [239]

$$\mathrm{P}_k \leq \sum_{d=d_{\text{free}}}^{\infty} a_d \mathrm{P}_k(d), \tag{14.72}$$

where d_{free} is the free distance of the convolutional code, a_d denotes the total number of error events of weight d, and $\mathrm{P}_k(d)$ represents the probability of choosing the incorrect path with distance d from the correct path. With hard-decision decoding, $\mathrm{P}_k(d)$ is related to the average BER, $\bar{B}_k$, as [239]

$$\mathrm{P}_k(d) = \sum_{l=(d+1)/2}^{d} \binom{d}{l} \bar{B}_k^l (1 - \bar{B}_k)^{d-l}, \tag{14.73}$$

when d is odd, and

$$\mathrm{P}_k(d) = \sum_{l=\frac{d}{2}+1}^{d} \binom{d}{l} \bar{B}_k^l (1 - \bar{B}_k)^{d-l} + \frac{1}{2} \binom{d}{\frac{d}{2}} \bar{B}_k^{\frac{d}{2}} (1 - \bar{B}_k)^{\frac{d}{2}}, \tag{14.74}$$

when d is even.

The average BER $\bar{B}_k$ can be obtained by averaging the conditional BER over the PDF of the SNR at the output of MRC. With Γ_k denoting the instantaneous SNR at the MRC output, the conditional BER is [239]

$$B_k(\Gamma_k) = Q\left(\sqrt{\Gamma_k}\right), \tag{14.75}$$

where $Q(\cdot)$ is the Gaussian error function. From (14.71) and (14.72), we can see that, for a given value of PER threshold ε, a corresponding BER threshold can be obtained. Because the error probability P_k in (14.72) is related to the coding rate through the parameters d_{free} and a_d, the BER requirement depends not only on the value of ε, but also on the data rate R_k. This implies that the SNR threshold is also a function of both ε and R_k. Let $\gamma(\varepsilon, R_k)$ be the minimum SNR of the kth user that is required to achieve the data rate R_k with a PER less than ε. Then, the necessary condition for the average SNR [defined in (14.62)] to satisfy the PER requirement is given by

$$\bar{\Gamma}_k = \sum_{s=1}^{S} a_{ks} P_k^s F_k^s \geq \gamma(\varepsilon, R_k). \tag{14.76}$$

The optimization goal is to minimize the overall transmit power subject to the PER, rate, and FCC regulation constraints. Recall from (14.55) that the assignment matrix $\mathbf{A}$

has $a_{ks} \in \phi(R_k)$, $\forall k, s$. We can formulate the problem as follows:

$$\min_{\mathbf{A},\mathbf{P}} \quad P_{\text{sum}} = \sum_{k=1}^{K} \sum_{s=1}^{S} a_{ks} P_k^s, \qquad (14.77)$$

$$\text{s.t.} \begin{cases} \text{Rate and PER: } \sum_{s=1}^{S} a_{ks} P_k^s F_k^s \geq \gamma(\varepsilon, R_k), \ \forall k; \\ \text{Assignment (14.56): } \sum_{s=1}^{S} a_{ks} = 2, \ \forall k; \\ \text{Assignment (14.57): } \sum_{k=1}^{K} a_{ks} \leq 2, \ \forall s; \\ \text{Power: } P_k^s \leq P_{\text{max}}, \ \forall k, s, \end{cases}$$

where the first constraint in (14.77) is used to ensure rate and PER requirements. The second and third constraints are described later. The last constraint is related to the limitation on the transmit power spectral density of -41.3 dBm/MHz, according to FCC, Part 15, rules [371]. Here, P_{max} is the maximum transmit power taking into consideration the effects such as the peak-to-average-power ratio (PAPR), i.e., P_{max} is the maximum power considering both the average maximum power allowed by the FCC and the PAPR of OFDM signals.

If the elements in the assignment matrix $\mathbf{A}$ are binary, the problem defined in (14.77) can be viewed as a generalized form of a generalized assignment problem [158] that is NP-hard. Because the components of $\mathbf{A}$ can be 0, 1, or 2, the problem is an even harder integer programming problem. As a result, the existing channel-assignment approaches, e.g. in [335], are not applicable in (14.77). Although the optimal solution can be found through full search, it is computationally expensive. To overcome the complexity issue, we propose next a fast suboptimal scheme, which is near optimal but has very low computational complexity.

The basic idea is a greedy approach to assign a_{ks} for a user step by step, so that the power consumption is minimized. The initialization is to set $\mathbf{A} = \mathbf{0}_{K \times S}$. We define the user optimization list $K_{\text{live}} = \{1, 2, \ldots, K\}$, and define the subband optimization list $S_{\text{live}} = \{1, 2, \ldots, S\}$. First, each user makes a hypothesis that it can assign its transmission into different subbands regarding the absence of other users. For each hypothesis, a dummy overall transmission power P_{dummy}^k is calculated by finding the minimum power among all possible subbands such that the BER performance requirement of user k is satisfied. The user with the highest dummy overall transmit power to achieve its rate will be assigned first, so that the best channel is assigned to the user that can most reduce the overall power. Then this user is removed from the optimization list K_{live}. Because each subband can accommodate only one user per symbol period, and we consider two OFDM symbol periods, when a subband is assigned twice this subband is removed from the optimization list S_{live}. Then we go to the first step for the remaining users to assign their transmissions into the remaining subbands. This iteration is continued until all users are assigned with their subbands, i.e., $K_{\text{live}} = \emptyset$. Finally, the algorithm checks if the maximum power is larger than the power limitation. If yes, an outage is reported; otherwise, the final values of $\mathbf{A}$ and $\mathbf{P}$ are obtained. The proposed algorithm can be described as follows:

Initialization: $a_{ks} = 0$, $\forall k, s$, $K_{\text{live}} = \{1, \ldots, K\}$, $S_{\text{live}} = \{1, \ldots, S\}$

Iteration: Repeat until $K_{\text{live}} = \emptyset$ or $S_{\text{live}} = \emptyset$

1. For $k \in K_{\text{live}}$
   $$P_{\text{dummy}}^k = \min \sum_{s=1}^{S} a_{ks} P_k^s \text{ s.t. } a_{ks} \in S_{\text{live}}$$
   End
2. Select k' with the maximal P_{dummy}^k, $\forall k$, assign the corresponding $a_{k's}$ to **A**, and update **P**.
3. $K_{\text{live}} = K_{\text{live}} \backslash k'$
4. If $\sum_{k=1}^{K} a_{ks'} = 2$, $S_{\text{live}} = S_{\text{live}} \backslash s'$, $\forall s'$.

End: If $(\max(\mathbf{P}) > P_{\max})$ or $(S_{\text{live}} = \emptyset$ and $K_{\text{live}} \neq \emptyset)$, an outage is reported. Otherwise, return **A** and **P**.

The complexity of the proposed algorithm is only $O(K^2 S)$. Although the algorithm is suboptimal, simulation results shows that the proposed fast suboptimal algorithm has performances very close to the optimal solutions obtained by full search. Another complexity issue is that, for the proposed scheme, power control is needed for each subband.[1] This will increase the system complexity slightly, but from the simulation results, we can see that the performance improvement is significant. Moreover, the proposed algorithm can be implemented by the master node to manage the power and subband usages of all users in a UWB picocell system, as adopted in the IEEE 802.15.3a standard [375]. The signaling information needed to be broadcasted at the master node includes the band-hopping sequence of each user and the corresponding transmit power. The algorithm is updated when a new user joins the network or when the channel link quality of each user changes considerably. In each update, the algorithm requires the channel-state information for all considered subbands (instead of three subbands as in the standard multiband scheme) between all transmitters and the receiver. Such an update does not frequently occurs thanks to the small size of the picocell and the stationary nature of most transceivers in WPAN applications.

Because the transmit power in each subband is limited by maximal power $P_{\max}$, solutions to (14.77) may not exist in some situations, such as when the requested rates of the users are high but the channel conditions are poor. Under such conditions, some desired rates of the users cannot be satisfied, and we say that the system is infeasible. When the system is not feasible, the requested rates need to be reduced. Here, we develop a joint rate-assignment and power-control channel-allocation algorithm that is able to obtain the feasible solutions adaptively when the initial system is not feasible for the rate requirements of the users. Basically, the proposed algorithm comprises two main stages, namely resource-allocation and rate-adaptation stages. Figure 14.20 shows the flowchart of the proposed algorithm.

At the initialization step, the data rate of the kth user, R_k, $k = 1, 2, \ldots, K$, is set to its requested rate. After the initial setting, the first stage is to perform the subband and power allocation using the algorithm described previously. If there is a solution for this

---

[1] But no power control or bit loading for subcarriers within each subband.

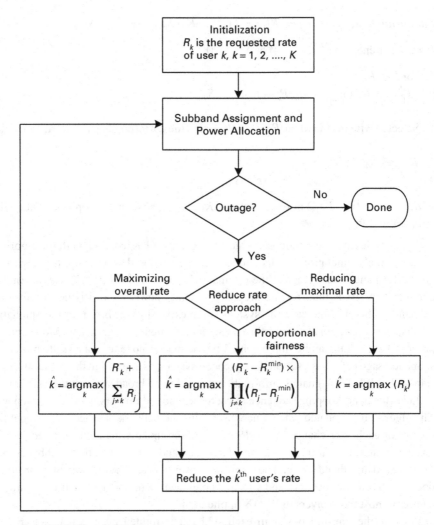

Figure 14.20 Flowchart of the proposed algorithm.

assignment, then it is done. Otherwise, an outage will be reported, indicating that the requested rates of the users are too high for the current channel conditions. In this case, we proceed to the second stage, in which the rate adaptation is performed.

In the rate-adaptation stage, the algorithm chooses only one user and reduces its rate to the next lower rate as listed in Table 14.6. To specify which user is to be selected we consider three different goals, namely maximizing overall rate, achieving proportional fairness[2] [159], and reducing maximal rate. In particular, given the data rate of the kth user, R_k, we denote its one-step reduced rate by R_k^-. For instance, from Table 14.6, the

---

[2] Note that proportional fairness is achievable when the utility is a log function. We have discrete and nonconvex cases, so the same product form is used as the system performance goal instead of the log function. From the simulations, this goal achieves trade-off of performances and fairness between the maximal rate goal and reducing the maximal rate goal.

reduced rate R_k^- corresponding to a rate $R_k = 320$ Mbps is $R_k^- = 200$ Mbps. Note that when the rate R_k reaches the minimum allowable rate of 53.3 Mbps, we let $R_k^- = R_k$, i.e., the rate R_k is not further reduced. Then the user $\hat{k}$ whose rate will be reduced can be determined according to the performance goals as follows:

- Maximizing overall rate: $\arg\max_k R_k^- + \sum_{j=1, j \neq k}^{K} R_j$.
- Proportional fairness: $\arg\max_k \prod_{j=1, j \neq k}^{K} (R_j - R_j^{\min}) \times (R_k^- - R_k^{\min})$.
- Reducing maximal rate: $\hat{k} = \arg\max_k(R_k)$,

where $R_k^{\min}$ denotes a minimal rate requirement for user k. With maximizing the overall rate approach, the overall system rate is maximized in every reduction step. In the case of the proportional fairness approach, the product of rates minus minimal rate requirements [159] is maximized. For reducing the maximal rate approach, the highest rate in the system will be reduced. Note that if there is still no solution to the assignment after the rates of all users are reduced to the minimum allowable rate, then an outage is reported. This indicates that the system under the current channel conditions cannot support the transmission of all K users at the same time. The proposed joint resource-allocation and rate-adaptation algorithm is summarized as follows.

Initialization: Iteration index $n' = 0$, $R_k(0) =$ requested rate of user k, $k = 1, 2, \ldots K$

Iteration:

1. Given $R_k(n')$, solve subband assignment and power-allocation problem in (14.77).
2. If (14.77) has a solution, the algorithm ends. Otherwise,
   - If $R_k(n') = R_k^-(n')$, $\forall k$, then an outage is reported and the algorithm ends.
   - Determine $\hat{k}$.
   - Update the rates:

$$R_k(n' + 1) = \begin{cases} R_k^-(n'), & k = \hat{k} \\ R_k(n'), & \text{otherwise} \end{cases}. \qquad (14.78)$$

   - Set $n' = n' + 1$.

To illustrate the performance of the proposed schemes, we perform simulations for multiband UWB systems with $N = 128$ subcarriers, $S = 14$ subbands, and a subband bandwidth of 528 MHz. Following the IEEE 802.15.3a standard proposal [17], we utilize the subbands with center frequencies $2904 + 528 \times n_b$ MHz, $n_b = 1, 2, \ldots, 14$. The OFDM symbol is of duration $T_{\text{FFT}} = 242.42$ ns. After adding the cyclic prefix of length $T_{\text{CP}} = 60.61$ ns and the guard interval of length $T_{\text{GI}} = 9.47$ ns, the symbol duration becomes $T_{\text{SYM}} = 312.5$ ns. As in [17], the convolutional encoder with a constraint length of 7 is used to generate different channel coding rates. The maximum transmit power is -41.3 dBm/MHz, and the PER is maintained such that PER $< 8\%$ for a 1024-byte packet. The average noise power follows (14.60) with $N_F = 6.6$ dB and the propagation loss factor is $\nu = 2$.

We consider a multiuser scenario in which each user is located at a distance of less than 4 m from the central base station. The performance is evaluated in multipath channel

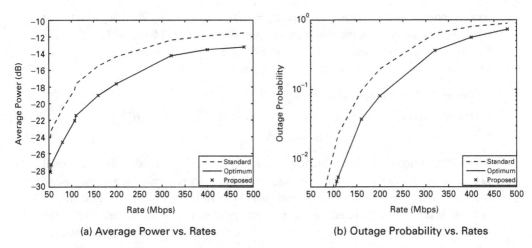

(a) Average Power vs. Rates   (b) Outage Probability vs. Rates

Figure 14.21 Performances of three-user system with random locations.

environments specified in the IEEE 802.15.3a channel-modeling subcommittee report [77]. We employ channel models 1 and 2, which are based on channel measurements over the range of 0–4 m. The simulated channels were constant during the transmission of each packet, and independent from one packet to another. In each simulation, we averaged over a minimum of 50,000 channel realizations.

Next, we present the average transmit power and the outage probability curves for multiband UWB systems. Here, the outage probability is the probability that the requested rate cannot be supported under the constraints in (14.77). We compare the performances of the proposed scheme with those of the current multiband scheme in the standards proposal [17].

For Figures 14.21(a) and 14.21(b), the number of users is fixed to $K = 3$, and each user is randomly located at the distance of 1 to 4 m from the base station. In Figure 14.21(a), we illustrate the average transmit power as a function of the data rates for a standard multiband scheme, the proposed fast suboptimal scheme, and the optimal scheme obtained by full search. It is apparent that the proposed algorithm can achieve almost the same performance as the optimal scheme. In addition, the proposed algorithm greatly reduces the average transmit power compared to that in the standard proposal. The performance gain in terms of power reduction that is achieved by the proposed scheme, compared with the standard scheme, can be computed as $(P_{\text{standard}} - P_{\text{proposed}})/P_{\text{standard}}$, where P_{standard} and P_{proposed} denote the average power for the standard and the proposed schemes, respectively. The results in Figure 14.21(a) show that both the fast suboptimal and optimal approaches can reduce about 60% of average transmit power at low rates (53.3–200 Mbps) and up to 35% at high rates (320–480 Mbps). Notice that the curves are not smooth because of the discrete nature of the problem. Figure 14.21(b) shows the outage probability versus the transmission rates. We can see that the proposed scheme achieves lower outage probability than that of the standard multiband scheme for any rates. For instance, at 110 Mbps, the outage probability of the proposed scheme is 5.5×10^{-3}, whereas that of the standard multiband scheme is 2.3×10^{-2}.

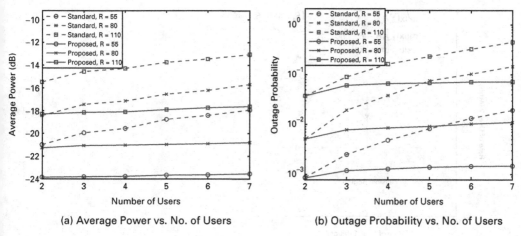

(a) Average Power vs. No. of Users   (b) Outage Probability vs. No. of Users

Figure 14.22 Performances of multiple-user system.

We also consider a multiuser system with different numbers of users, each located at a fixed position of about 4 m from the base station. Specifically, the distance between the kth user and the base station is specified as $d_k = 4 - 0.1(k - 1)$ for $k = 1, 2, \ldots, K$. In Figures 14.22(a) and 14.22(b), we show the average transmit power and outage probability as functions of the number of users for data rates of 55, 80, and 110 Mbps. In both figures, we use the standard multiband scheme and the proposed scheme. We can observe from Figure 14.22(a) that the transmit power increases with the number of users. This results from the limited available subbands with good channel conditions. When the number of users is large, some users have to occupy the subbands with worse channel conditions. Comparing the proposed algorithm with the standard multiband approach, we can see that the proposed scheme achieves lower transmit power for all the rate requirements.

Figure 14.22(b) shows that the outage probability increases with the number of users. As the number of users increases, the system is more crowded and may not be feasible to support all the users at all times. Observe that, at any rate, the performance of the standard multiband scheme degrades as the number of users increases. On the other hand, when the proposed scheme is employed, the effect of the number of users to the outage probability is insignificant when the rates are not higher than 110 Mbps. As we can see, the proposed algorithm achieves smaller outage probabilities than those of the standard scheme under all conditions.

Next, we illustrate the performances of the proposed joint rate-assignment and resource-allocation algorithm for a multiband system. We consider a multiuser system with different numbers of users. Each user is randomly located at the distance of 1 to 4 m from the base station. The requested rates of users are also randomly selected from the set {200, 320, 400, 480} Mbps, and the minimum rate requirement is $R_k^{\min} = 50$ Mbps $\forall k$ for the proportional fairness goal. The joint rate-assignment and resource-allocation algorithm proposed is performed for each set of requested rates and channel conditions.

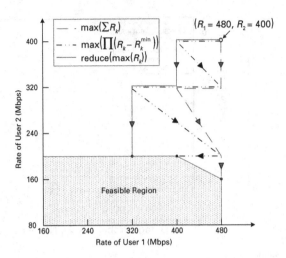

Figure 14.23 One realization of rate adaptation for two-user system.

Figure 14.23 illustrates one realization of rate adaptation for a two-user system with three different goals. The shaded area represents the feasible range for R_1 and R_2 in the current channel conditions. In this example, the requested rates are $R_1 = 480$ and $R_2 = 400$ Mbps, and both users are located at about 4 m from the base station. We can observe from Figure 14.23 that the reducing maximal rate approach has the lowest overall rate in every adaptation step. This is because the highest rate in the system can always be reduced. On the other hand, the maximizing overall rate approach tends to reduce the lower rate because most low rates have a smaller decreasing step size than high rates have. Although the maximizing overall rate approach always yields superior system performance, it is unfair to those applications with low data rates. The proportional fairness goal provides the performance that is between the maximizing overall rate approach and reducing maximal rate approach.

Figures 14.24(a) and 14.24(b) show the average system performance versus the number of users. In Figure 14.24(a), we present the performances in terms of the average data rates of the users. We can see that the average rates of all three approaches decrease when the number of users increases. This is due to the limited subbands with good channel conditions. Comparing the performances of three approaches, we can see that the proportional fairness yields a slightly lower average rate than that of the maximizing overall rate approach, and both proportional fairness and maximizing overall rate approaches achieve much higher rates than that of the reducing maximal rate approach.

In Figure 14.24(b), we show the standard deviations of the data rates of the users for three approaches. Here the standard deviation represents the fairness of allocation among users. We can observe that the standard deviation for every scheme increases with the number of users because the larger the number of users, the higher the variation of the rates. At any fixed number of users, the reducing maximal rate approach results in the smallest standard deviation, and its standard deviation slightly increases with the number of users. This is because the feasible rates obtained from the reducing maximal

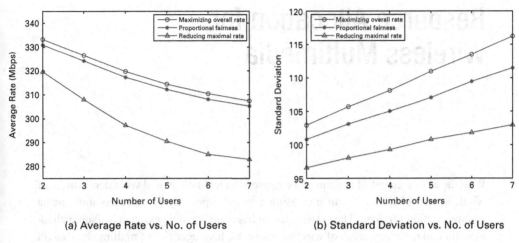

(a) Average Rate vs. No. of Users   (b) Standard Deviation vs. No. of Users

Figure 14.24 Average rate and standard deviation of multiple-user systems.

rate approach are close to each other. In contrast, the maximizing overall rate scheme can yield the feasible rates of around 100 to 480 Mbps at the same time. Thus its standard deviation increases much faster with the number of users. The standard deviation of the proportional fairness approach is between those of the other two schemes. As a result, the proportional fairness approach is a trade-off between the maximal rate approach and reducing maximal rate approach for both performances and fairness.

On the whole, low power consumption is one of the key elements in making multiband UWB technology the solution for future indoor wireless communications. We propose an efficient cross-layer algorithm for allocating subband and power among users in a multiband UWB system. The proposed scheme aims to reduce power consumption without compromising performance. We propose a general framework to minimize the overall transmit power under the practical implementation constraints. The formulated problem is NP-hard; however, with the proposed fast suboptimal algorithm, we can reduce the computational complexity to only $O(K^2 S)$, where K is the number of users and S is the number of subbands. Simulation results show that the proposed algorithm achieves performances comparable with those of the expensive optimal full search algorithm, and can save up to 61% of power consumption compared to the standard multiband scheme. Moreover, the proposed algorithm can obtain the feasible solutions adaptively when the initial system is not feasible for the rate requirements of the users. Among three different system optimization goals used in the proposed rate adaptation algorithm, the proportional fairness approach turns out to be a trade-off between the maximal rate approach and reducing maximal rate approach for both performance and fairness.

15 Resource Allocation for Wireless Multimedia

With the advancement of multimedia compression technology and wide deployment of wireless networks, there is an increasing demand especially for wireless multimedia communication services. The system design has many challenges, such as fading channels, limited radio resources of wireless networks, heterogeneity of multimedia content complexity, delay and decoding dependency constraints of multimedia, mixed-integer optimization, and trade-offs among multiuser service objectives. To overcome these challenges, dynamic resource allocation is a general strategy used to improve the overall system performance and ensure individual QoS. Specifically in this chapter, we consider two aspects of design issues: *cross-layer optimization* and *multiuser diversity*. We study how to optimally transmit multiuser multimedia streams, encoded by current and future multimedia codecs, over resource-limited wireless networks such as 3G cellular systems, WLANs, 4G cellular systems, and future WLAN/WMANs.

15.1   Introduction

Over the past few decades, wireless communications and networking have experienced an unprecedented growth. With the advancement in multimedia coding technologies, transmitting real-time encoded multimedia programs over wireless networks has become a promising service for such applications as video-on-demand and interactive video telephony. In most scenarios, multiple multimedia programs are transmitted to multiple users simultaneously by sharing resource-limited wireless networks.

The challenges for transmitting multiple compressed multimedia payloads (such as videos) over wireless networks in real time lie in several factors. First, wireless channels are impaired by detrimental effects such as fading and CCIs. Second, there are limited radio resources, such as bandwidth and power, in the wireless networks. In addition, unlike generic data and voice transmission, the rates of compressed videos can be highly bursty because of the differences in video contents and intracoding/intercoding modes. This fact complicates the video rate allocation. Moreover, the optimizations in different layers are cross related and mostly with parameters of mixture of continuous and integer values, so that the formulated problem is often NP-hard. Further, handling multiple multimedia streams over a wireless system involves several important service objectives, such as system efficiency and individual fairness, but there are inherent trade-offs among those objectives.

To overcome the aforementioned design challenges, dynamic resource allocation integrates the system's parameters and utilizes the limited system resources optimally in the source and communication layers for multiple users. We address two major aspects, cross-layer design and multiuser diversity, to optimize resource allocation, so as to accommodate a large number of users with acceptable received QoS. Specifically, we discuss in this chapter the multiuser video system in current 3G cellular networks, WLANs, 4G cellular networks, and future WLAN/WMANs from a resource-allocation point of view. Notice that we have already discussed the voice payloads in Part II of this book as varieties of examples. We first present the generic framework, the wireless network components, and the state-of-art video source codec for multiuser wireless video systems. We then investigate some major design methodologies for multiuser video communication with cross-layer design. The goal is to provide a broad overview and perspective of the recent and future advances in multiuser wireless video transmissions.

This chapter is organized as follows: In Section 15.2, we propose a framework for multimedia over wireless networks and study two major topics. Then we illustrate the basic resource-allocation techniques and explain several current and future wireless networks. In Section 15.3, we address how to transmit video such as MPEG4-FGS over current and 3G CDMA system. In Section 15.3, we investigate videoconferencing over WLAN/WiFi. In Section 15.5, we examine future videostreaming over future wireless networks like 4G and WiMAX. Finally, a summary is given in Section 15.6.

15.2   Framework of Multimedia Over Wireless Networks

In this section, we first present an overview of multiuser cross-layer systems for transmitting video streams over wireless networks. We then briefly review the communication subsystem and video source-coder subsystem. For each subsystem, we discuss the available resources, study how to control parameters to achieve the desired goals, and analyze the corresponding constraints in the practical implementation.

Figure 15.1 depicts a generic framework [296] for video transmission over wireless network with a total of N communication links. There are four major subsystems, namely, the video source-codec subsystem, the communication subsystem, the receiver system, and the resource-allocation subsystem. The resource-allocation subsystem first collects the necessary information from the video source subsystem, the communication subsystem, and the receiver system. In the video source-coding subsystem, each video program is encoded in real time. The encoders compress the incoming video frames and send the corresponding R-D information to the resource-allocation subsystem. The communication subsystem provides the available resources in the network, MAC, and physical layer, and supplies the channel information obtained via feedback from the receivers. After gathering the information, the resource-allocation subsystem executes optimization algorithms and allocates system resources to different links so as to achieve the system optimization objectives. We consider the following two design aspects for optimizing the framework:

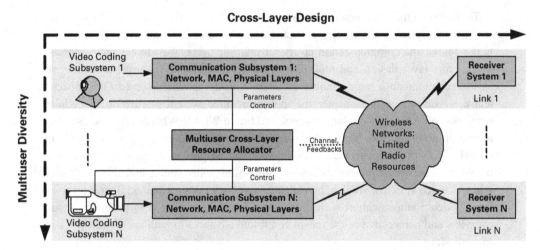

Figure 15.1 Framework of multiuser cross-layer video transmission system over wireless networks.

1. **Cross-layer optimization:**
   Traditionally, wireless data networks are designed in layers. According to Shannon's Separation Theorem, source and channel coding can be designed separately while still achieving the optimality, if an arbitrarily long delay is allowed. However, in video wireless networks, because the packets must be transmitted with delay constraints, separating the layers is no longer an optimal design. As a result, it is critical to design the video transmission system in a cross-layer fashion to achieve system-wide optimality as opposed to local optima within individual layers.

2. **Multiuser diversity:**
   System resources are limited and shared to transmit multiple videostreams. In a multiuser system, the channel conditions experienced by different users differ and vary from time to time. In addition, the transmission rates of different compressed video bit streams vary among users and change over time. By exploring multiuser diversity and allocating system resources dynamically, we can improve the network performances and achieve the baseline QoS satisfactions for individual users.

15.2.1   Current and Future Wireless Network Paradigms

Because the available radio resources are limited, modern wireless networks often adopt the following resource-allocation methods to adaptively improve spectrum utilization.

1. **Power control**: The gains and phases of wireless channels fluctuate over time generally. To maintain the link quality, the SINR should be dynamically controlled to meet a threshold known as the *minimum protection ratio*. The objective of power control is to guarantee certain link quality and reduce CCI.

2. **Adaptive modulation and coding (AMC)**: In current and future wireless communications, adaptive modulation is applied to have better spectrum utilization. To combat

different levels of channel-introduced errors, adaptive forward error coding (FEC) is widely used in wireless transceivers. Further, joint consideration of adaptive modulation and adaptive FEC provides each user with the ability to adjust the transmission rate and to achieve the desired error protection level, thus facilitating the adaptation to various channel conditions.

3. **Channel assignment**: The channel used here is a general concept that represents the smallest unit of radio resources a user can be assigned to transmit data, such as frequency band and time slot. Considering the different channel conditions and users' transmission requirements, dynamic channel assignment can improve the utilization of system resources by exploring the multiuser diversities, time diversity, and frequency diversity. Because of the discrete nature of channel assignment, the optimization problem is mostly NP-hard.

4. **Scheduling/random access**: Scheduling and random access are two types of schemes for multiple users to take turns sharing the limited radio resources. Scheduling has a centralized control to decide which user can transmit at a specific time. Random access can reduce transmission delay in light-loaded networks and avoid conflict of resource usage in an autonomous way. Scheduling is usually applied in cellular networks, whereas random access is implemented in ad hoc networks such as WLAN.

In the following, we discuss the current and future broadband communication networks that can support real-time video transmission.

1. **3G cellular networks: CDMA/scheduling**
   3G wireless communication systems employ CDMA. CDMA uses unique spreading codes to spread the baseband data before transmission. The signal occupies a much broader bandwidth than narrowband transmissions and is transmitted with a power density below the noise level. The receiver uses a correlator to despread the signal of interest, which is passed through a narrow-bandpass filter to remove unwanted interferences. This brings many benefits, such as immunity to narrowband interferences, jamming, and multiuser access.

   Figure 15.2(a) shows the CDMA system with a scheduler to allocate resources to users. Widely adopted in 3G networks, the scheduler allocates a different number of CDMA codes or CDMA codes with various spreading factors to users at different times according to the channel conditions, QoS types, bandwidth requirements, and buffer occupancies.

2. **WLAN: OFDM, CDMA/random access**
   WLAN can provide a higher transmission rate within local areas. There are two major current standards for WLAN, namely, IEEE 802.11b and IEEE 802.11g. IEEE 802.11b uses CDMA technology and supports up to 11 Mbps; and IEEE 802.11g uses OFDM technology and supports up to 54 Mbps. OFDM splits a high-rate data stream into a number of lower-rate streams and transmits them over a number of frequency subcarriers simultaneously. In addition, guard time with cyclical extension is inserted in each OFDM symbol. Thus intersymbol and intercarrier interference are almost eliminated in OFDM systems.

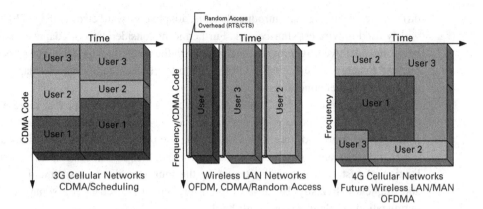

Figure 15.2 Illustration of multiuser resource allocation for 3G, WLAN, 4G, future WLAN/WMAN.

Figure 15.2(b) illustrates the multiple access in current IEEE 802.11 standard. IEEE 802.11 MAC protocol supports two access methods: distributed coordination function (DCF) and point coordination function (PCF). At each time slot, both functions only allow one user to occupy all radio resources. The DCF is the basic random-access mechanism using CSMA/CA or ready-to-send (RTS) / clear-to-send (CTS). In contrast, the PCF is based on polling controlled by a point coordinator.

3. **4G cellular networks and future WLAN/WMAN: OFDMA**
   In current OFDM systems, all subcarriers are assigned to a single user at each moment, and multiple users are supported through time division. However, for a given subcarrier, different users experience different channel conditions and the probability for all users to have deep fades in the same subcarrier is very low. OFDMA allows multiple users to transmit simultaneously on the different subcarriers, and each subcarrier is assigned to the user who is experiencing a good channel condition.

   Figure 15.2(c) shows the multiuser resource-allocation strategy for the OFDMA system. Users' transmission can be allocated to different time–frequency slots. By doing this, the multiuser, time, and frequency diversity can be fully explored to improve the system performance.

15.2.2   Current and Future Video-Coding Paradigms

Owing to the perceptual characteristics of human vision, the received video can tolerate a certain level of quality degradation. Trading in lossless reconstructed quality by lossy compression can substantially reduce the required bit rate and still maintain acceptable visual quality. Most current standardized video codecs, such as H.261/3/4 and MPEG-1/2/4, adopt block-based motion-compensated prediction and block discrete-cosine-transform (DCT) coding with quantization to remove temporal and spatial redundancy [279]. Researchers have been also exploring three-dimensional (3D) wavelet coding to simultaneously remove the spatiotemporal redundancies.

The potential application for video technology has evolved from precompressed files in predistributed storage, such as DVDs, to real-time encoded bit streams over wireless networks, such as videoconferencing. However, there are still many remaining design challenges for real-time video compression and transmission. We summarize them as follows:

1. **Perceptual quality control**: Unlike throughput as a major concern in data transmission systems, video systems concern video quality in terms of either subjective quality assessment or objective distortion measurement such as MSE or PSNR. We need to control the source-coding parameters to obtain acceptable video quality. To adapt to the variation of the scene complexity [301], the communication module needs to either dynamically adjust bandwidth or employ buffers to smoothen traffic.

2. **Rate/delay control**: Compressed video bit streams have decoding dependency on the previous coded bit streams, owing to the spatial and temporal prediction. Therefore transmitting videostreams in real time has a strict delay constraint that belated video data are useless for its corresponding frame and will cause error propagation for the video data that are predictively encoded using that frame as reference. We need to adjust the coding parameters to control the rate such that the bit stream can arrive at the receiver and be decoded in time.

3. **Error control**: Because of decoding dependency, video bit streams are also vulnerable to bit error, as bit error may cause the following bit stream to be decoded incorrectly. A wireless video system should take channel error into account [291]. Error-resilient tools, which increase the robustness of video bit streams, and error-concealment schemes, which utilize the received bit stream to conceal the damaged bit stream, can be integrated together to improve the end-to-end video quality [327].

4. **Scalability**: A scalable video codec provides a new coding paradigm, whereby the video is encoded once and can be transmitted and decoded in many targeted rates according to the channel conditions or users' needs. Several technologies, such as fine granularity scalability (FGS) coding and scalable video coding (SVC), have been proposed in MPEG-4 to provide spatial, temporal, and quality scalability.

15.2.3   Design Principles of Multiuser Cross-Layer Resource Allocation

For video communication systems designed in layers, different layers have their own resources with practical constraints such as the feasible ranges or finite sets of discrete values that are associated with the resource parameters. For a system with cross-layer design, the allocation of system resources is constrained *vertically across layers*. For example, bandwidth consumption for use in the application layer should not exceed the achievable capacity by the physical layer. Unlike in a single-user system, network resources are shared by multiple users in a multiuser wireless video system. Allocating these resources to one user would affect the performances of the other users because of the limited amount of resources or interference of simultaneous usage. In other words, the allocation of system resources is further constrained *horizontally among users*. Owing to the time heterogeneity of the video source and time-varying characteristics of the channel condition, the allocation of system resources should be performed *dynamically along time*. Moreover, real-time video transmission has additional delay constraints such as playback deadline.

A multiuser video transmission system should consider not only the video quality of each individual user but also different perspectives from a network-level point of view. In general, we can formulate the resource-allocation problem to optimize the network objective by allocating the resources across layers and among users subject to system constraints. Two essential network objectives, *efficiency* and *fairness*, are often considered. Efficiency concerns how to attain the highest overall video quality using the available system resources, and fairness concerns the video quality deviation among users who subscribe the same QoS. There is a trade-off between efficiency and fairness.

The resource-allocation problem often has to deal with resources having both continuous and integer-valued parameters. Systems may also have nonlinear or/and nonconvex constraints and many local optima may exist in the feasible range. Thus obtaining the optimal solution is often NP-hard. General approaches for solving the problem are to reduce the search space by some bounds or to adaptively find the solution close to optimum. Some engineering heuristics can be employed for certain network scenarios. To allocate system resources, the resource allocators require some level of up-to-date information about available resources. Depending on the communication/computation cost and existence of central authority, systems can employ centralized or distributed algorithms. In general, resource allocators with more information can have better performance but require more communication overhead for accurate information.

From these design principles, we investigate in the next sections three major design methodologies for various video transmission applications over different types of wireless networks.

15.3   Transmitting Video Over 3G Cellular Networks

Transmitting encoded videos to users through cell phone networks has become an emerging service. Video telephony and TV on demand are two such examples. In this section, we consider a downlink scenario of sending multiple real-time encoded video programs to multiple mobile users over CDMA systems [119, 292]. Figure 15.3 shows the block diagram of the wireless video server located at the base station, which transmits multiple real-time encoded video programs to multiple mobile users. There are three major subsystems in the system: the video source-coding subsystem, the multicode (MC) CDMA subsystem, and the resource-allocation subsystem for managing distortion.

The resource-allocation subsystem first collects the necessary information from the video source coders and MC-CDMA subsystem. In the video source-coding subsystem, each video program is encoded by a FGS encoder in real time. These FGS encoders compress the incoming video frames and send the corresponding R-D information to the resource-allocation subsystem. The downlink channel information is obtained via feedback from mobiles.

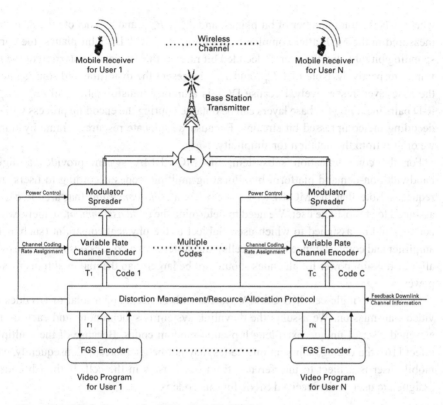

Figure 15.3 Block diagram for FGS over MC-CDMA.

To facilitate the rate/quality adaptation, MPEG-4 FGS is adopted for the video-coding part. The encoder first generates a nonscalable base layer at a low bit rate using a large quantization step and computes the residues between the original frame and the base layer. These residues are transformed into the DCT domain, and the bit planes of DCT coefficients are then encoded sequentially as an enhance-ment layer and form the FGS layer. The decoder can decode any truncated seg-ment of the FGS bit stream corresponding to each frame. The more bits the de-coder receives and decodes, the higher the reconstructed perceptual quality of video we obtain. The video codec has a *rate constraint*: The transmitted rate for each video frame needs to be between the base-layer rate and the maximal-available-FGS rate.

Previous studies in [360, 361] show that a piecewise linear function is a good ap-proximation to the R-D curve of the FGS video at the frame level. This piecewise linear function model can be summarized as

$$D_{n,j}(r_{n,j}) = M_{n,j}^k(r_{n,j} - R_{n,j}^k) + E_{n,j}^k, \; k = 0, \ldots, p-1, \qquad (15.1)$$

$$\text{with } M_{n,j}^k = \frac{E_{n,j}^{k+1} - E_{n,j}^k}{R_{n,j}^{k+1} - R_{n,j}^k}, \; R_{n,j}^k \le r_{n,j} \le R_{n,j}^{k+1},$$

where p is the total number of bit planes, and $E_{n,j}^k$, $R_{n,j}^k$, and $r_{n,j}$ denote the distortion measured in the MSE after completely decoding the first k DCT bit planes, the corresponding bit rate, and the overall decoded bit rate for the jth user's distortion of the nth frame, respectively. Note that $E_{n,j}^0$ and $R_{n,j}^0$ represent the distortion and source rate of the base layer, respectively. Because DCT is a unitary transformation, all $(R_{n,j}^k, E_{n,j}^k)$ R-D pairs including the base layers can be obtained during the encoding process without decoding the compressed bit streams. Because we allocate resources frame by frame, we omit n from the notation for simplicity.

For the communication subsystem, the MC-CDMA system provides a digital bandwidth-on-demand platform by allocating multiple codes according to users' rate requests. Note that the MC-CDMA system has a *code constraint* that a code can be assigned to at most one user. We need to determine the *code assignment*, namely, which code should be assigned to which user. Subject to the physical constraint (such as the amplifier and cointerference to other cells), the system has a *power constraint*: The overall transmission power for all codes should not be larger than the maximal transmission power.

Consider a single-cell MC-CDMA system with N users and a total of C codes for video transmission. We assume the downlink system is synchronous and each user is assigned a set of unique fixed-length pseudo-random codes. Because of the multipath effect [16], the orthogonality among codes may not be guaranteed. Consequently, each mobile user is subject to interference from other users in the cell. If the ith code is assigned to user j, the received SINR for this code is

$$\Gamma_i^j = \frac{W}{R} \frac{P_i G_j}{G_j \sum_{k=1, k \neq i}^{C} \alpha_{ki} P_k + \sigma^2}, \tag{15.2}$$

where W is the total bandwidth and is fixed, R is the transmission rate, P_i is the transmission power from the base station for code i, α_{ki} is orthogonality factor between the kth and ith codes and can be estimated statistically [16], G_j is the jth user's path loss, which is assumed to be stable within each video frame and can vary from frame to frame, and σ^2 is the thermal-noise level and assumed to be the same at all mobile receivers. The ratio W/R is the processing gain. In this section, we use BPSK modulation for simplicity.

To protect the video bit stream from bit error during transmission, we use a rate-compatible punctured convolutional (RCPC) code [101], which provides a wide range of channel-coding rates and is relatively simple to implement. A family of RCPC codes is described by the mother code of rate $1/M$. The output of the coder is punctured periodically following a puncture table. The puncturing period Q determines the range of channel coding rates $r = \frac{Q}{Q+l}$, $l = 1, \ldots, (M-1)Q$, which are between $\frac{1}{M}$ and $\frac{Q}{Q+1}$ with decreasing channel-error-protection ability.

The goal of channel coding is to provide sufficiently low BER such that the end-to-end video quality is controllable. For the MPEG-4 video codec, the degradation of video quality is negligible if we enable the error-resilient functions and error-concealment mechanism as well as keep the BER below a threshold [119, 292] set around 10^{-6}. To achieve the BER requirement, the received SINR should not be less than a targeted SINR. Our study shows that this targeted SINR can be approximated by an exponential function of the channel coding rate of RCPC as

$$\gamma_i = 2^{AT_i + B}, \tag{15.3}$$

where γ_i is the required targeted SINR, T_i is the channel-coding rate with a discrete value in the range $[T_{min}, T_{max}]$, and A and B are the parameters related to the channel coding that can be experimentally determined. Their values depend on the BER requirements, fading conditions, and channel-code family being used.

Although a CDMA code with a higher channel-coding rate can carry more source bits to improve the video quality, the required power to meet the BER requirement is higher. The received SINR for each code is subject to interference from other users' codes in the cell because of nonorthogonality among codes caused by multipath fading. Because the overall power is limited, we need to determine the *channel-coding rate assignment* of each code to achieve the optimal video quality subject to the power constraint and interference.

Overall, the key issue is how to jointly perform the rate adaptation, code allocation, and power control to achieve the required perceptual qualities of received video. We denote $a_{ij} \in \{0, 1\}$ as an indicator to specify whether the ith code is assigned to user j. The total available power for video transmission is P_{max}, and each user's throughput, r_j, should be larger than the base-layer rate R_j^0 to guarantee the baseline quality and should be smaller than the maximum source rate R_j^P. The transmission rate for each CDMA code is R. Each code will carry information of rate RT_i, where T_i is the channel coding rate for the ith code. We formulate this problem so as to minimize the overall users' distortion by determining the code assignment and channel-coding rate assignment, subject to the constraints on CDMA codes, power, and video-coding rate:

$$\min_{T_i, a_{ij}} \sum_{j=1}^{N} D_j(r_j), \tag{15.4}$$

$$\text{s.t.} \begin{cases} \text{code constraint: } \sum_{j=1}^{N} a_{ij} \leq 1, a_{ij} \in \{0, 1\}, \forall i; \\ \text{power constraint: } P_{sum} = \sum_{i=1}^{C} P_i \leq P_{max}; \\ \text{rate constraint: } R_j^0 \leq r_j = R \sum_{i=1}^{C} a_{ij} T_i \leq R_j^P, \forall j. \end{cases}$$

This problem is a mixed-integer programming problem, which is NP-hard and requires a formidably high amount of computational complexity to obtain an optimal solution. In searching for an effective and efficient practical solution, we have found an important heuristic on balancing code and power usage. More specifically, the code and power resource should be used in a balanced way to avoid exhausting one resource first while having the other resource left, which leads to low system performance.

We have developed a distortion management algorithm by balancing the code and power usages. There are two main stages in the algorithm, which are illustrated in Figure 15.4. In the first stage, we allocate the resources for delivering the base-layer data to

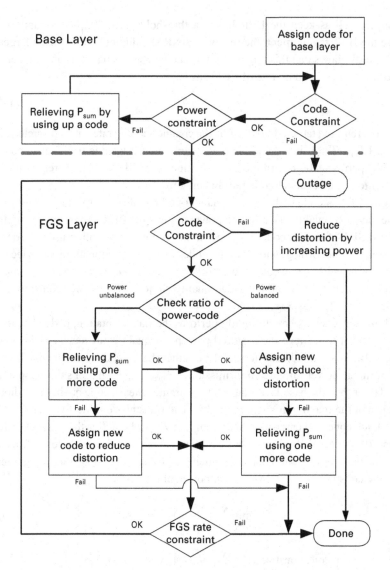

Figure 15.4 Distortion management algorithm for base layer and FGS layer.

provide the baseline video quality for each user. Some FGS enhancement-layer data are then delivered to reduce the overall distortion in the second stage.

We define the power-to-code usage ratio as the ratio of the current transmission power to the number of assigned CDMA codes. We assign one code at a time to a user and apply different algorithms for the code assignment according to the current power-to-code usage ratio. If the current ratio is larger than the ratio of the maximal transmission power over the total number of CDMA codes, the system is in a power-unbalanced state, i.e., consuming higher than average power per code. In this case, a new code is assigned to relieve the transmitted power usage and keep the same overall distortion, such that

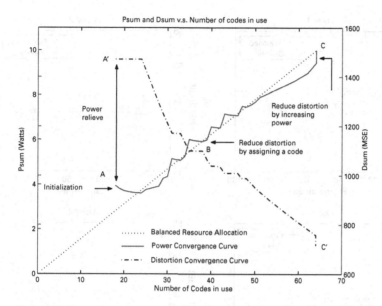

Figure 15.5 Power convergence and distortion convergence vs. the number of assigned codes.

the power-to-code usage ratio is reduced. Otherwise, the system is in a power-balanced state, and a new code that consumes more power can be assigned to reduce the overall distortion.

We use Figure 15.5 to illustrate how the scheme balances the code and power limitation to fully utilize the system resources. The figure shows the convergence trace of the overall power and video distortion with respect to the number of assigned codes. After initialization (shown at position A), a total of 18 codes are assigned to deliver the base layer of all users. The overall visual distortion (shown by position A') is large because only the base layer is transmitted. We can see that, at this point, the system is power unbalanced, i.e., the operating point A is above the balanced resource-allocation line. Whenever power unbalance occurs, we apply the power-relief algorithm to reduce the power while keeping the distortion fixed. When the system is no longer power unbalanced (such as position B), we assign codes to reduce distortion until all the codes are used up. Note that, by doing so, the required power is increased. Finally, we perform a round of refinement algorithm to further reduce distortion (shown at position C') by using the remaining power quota (shown at position C). At the end, all available power and code resources are fully utilized.

We compare the algorithm with a modified greedy approach [111]. This modified approach is similar to our framework, but uses a greedy approach for the code assignment in the FGS layer. For the base layer, the greedy algorithm executes the same procedure as our algorithm. For each iteration in the FGS layer, this greedy scheme will favor the users close to the base station and with simple video content complexity.

Figure 15.6 shows the frame-by-frame PSNR results in a four-user system in which users 1–4 are located at 700, 400, 600, and 20 m, respectively. Users 1–4 receive 100-frame video sequence of *Claire*, *Coastguard*, *Grandmother*, and *Akiyo*, respectively.

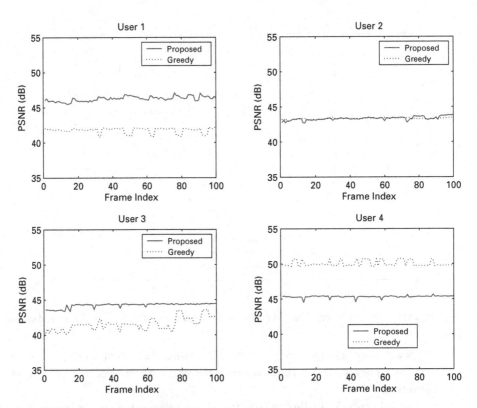

Figure 15.6 Frame-by-frame PSNR results for user 1 to user 4 vs. frame index.

The first three users receive better or similar video qualities using our algorithm. The greedy algorithm assigns codes to the users who can use the least power to obtain the largest decreased distortion, which is the fourth user in this example. Compared with our scheme, the greedy scheme cannot effectively reduce other three users' distortions.

Figure 15.7 shows the number of users versus the average of the total distortion D_{sum}. The content program for each user is 100 frames and starts from a randomly selected frame of the concatenated testing video. The location for each user is uniformly distributed within the cell with a radius from 20 to 1000 m. We repeat the simulations 300 times. The simulation results demonstrate that the average D_{sum} of our algorithm outperforms that of the greedy algorithm by 21%–32%. The reason for this gain is that the greedy algorithm ignores the balance between power and code usage and thus depletes one resource while wasting other resources. In other words, only one system constraint becomes active after several iterations, which leaves no room to improve the overall system performance even if other resources are still available. Our scheme shows performance improvement by fully utilizing both power and code resources.

In summary, we propose a framework for transmitting multiple videostreams over wireless communication networks, which is a promising service in current wireless networks. Specifically, we have developed a system to transmit multiple real-time

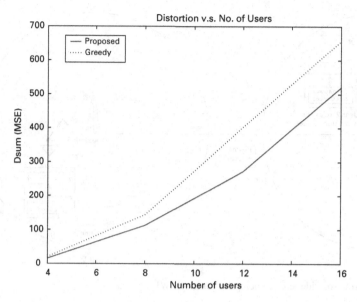

Figure 15.7 Performance comparison of the schemes.

MPEG-4 FGS video programs over downlink MC-CDMA networks. The resource allocation is formulated as an optimization problem to minimize the overall received distortion of all users subject to the baseline video-quality requirement, the maximal transmission power, and the number of codes constraints. To fully utilize the limited resources, we propose a distortion management algorithm to jointly allocate source-coding rates, channel coding rates, CDMA codes, and transmission power. The scheme balances the constraints of codes and power so that no resources are wasted. We also derive optimal solutions for the demand-limited and code-limited case and a performance upper bound for the power-limited case.

Experimental results show that our approach provides an efficient solution for sending videos over downlink MC-CDMA systems. The system can fully utilize the available radio resources by balancing the power and code limits. The scheme is a promising solution for real-time multiple video transmissions in current CDMA networks or the next-generation CDMA networks like CDMA2000.

15.4   Videoconferencing Over WLANs

With the wide deployment of WLAN, transmitting real-time video such as videoconferencing to mobile users is becoming an attractive service [198, 270]. In this section, we present a real-time interactive videoconferencing framework to support multiple video-conferencing pairs [293]. In this system, for each conversation, there are two video-streams exchanged between a conversation pair, which can be located in the same cell or in different cells, and each videostream is transmitted through an uplink and a down-link. The transmitted packets of each videostream often experience different channel

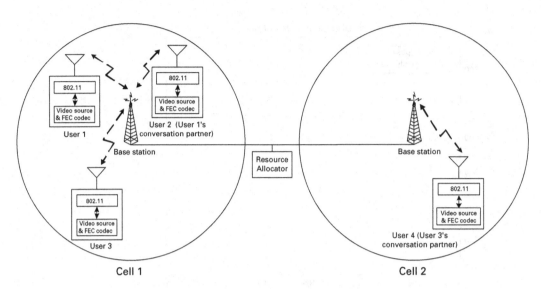

Figure 15.8 System block diagram of videoconferencing for multicell WLAN case.

conditions in both links. Our framework explores the diversity of video content, the heterogeneity of uplink and downlink channel conditions experienced by different users, and multiuser diversity over the resource-limited wireless networks.

Figure 15.8 shows the framework. Without loss of generality, we use a system with two cells as an example. Single-cell and multiple-cell cases can be investigated in a similar way. We assume the wired channel connecting both cells is reliable without any packet loss and the bandwidth is large enough to transmit all packets. We also assume that the coherent time of the channel condition is much larger than the propagation delays induced by the wired link. In addition to an intracell conversation with a user within the same cell (e.g., the conversation between users 1 and 2 in Figure 15.8), a user can have an intercell conversation with a user located in another cell (e.g., conversation between users 3 and 4). The resource allocator needs to first gather R-D information of all videostreams and channel information of all links, and then perform centralized distortion control. Note that, in this multicell system, there is only one user who can send data at any time in each cell. In other words, there are two users who can transmit video packets simultaneously in this two-cell system. The major tasks of the resource allocator are how to jointly consider the traffic load in both cells and how to allocate system resources to each user in each link such that the maximal distortion among all users is minimized.

The IEEE 802.11a PHY layer provides eight operation modes using different modulation schemes and convolutional coding rate. Different PHY modes have different BER performances and data rates ranging from 6 to 54 Mbps. In this system, we need to determine the *PHY mode assignment*, namely, selecting which PHY mode of uplink and downlink for each user can achieve the optimal video quality.

The IEEE 802.11 MAC protocol allows only one user to occupy the entire bandwidth at each time. Each individual user's transmission time can be controlled by either PDF or enhanced DCF. For the real-time video transmission scenario, all users should transmit their video frames within each frame sampling period. Consequently, the system has a *transmission time constraint* that the overall transmission time of all users is bounded by this period. In this chapter, we study how to optimize the amount of transmission time allocated to each user under the transmission time constraint.

The transmission errors can be detected by checking the CRC bits of a packet. If the server finds the packet is corrupted by errors, this packet will not be forwarded to its destination, which leads to packet loss. With the transmission index, we can identify which packet is lost in the receiver. Thus we can model this channel as an erasure channel. Applying application layer FEC across packets, such as using Reed–Solomon (RS) codes, has been shown to be an effective solution to alleviate the problem caused by packet loss. For the source-coding part, we adopt MPEG-4 FGS to facilitate rate adaptation. To jointly determine source coding and channel protection, we need to determine the *packet assignment*, namely, the number of source packets and parity check packets for each video frame.

In this system, we consider fairness among users by choosing the PHY mode assignment and packet assignment subject to the transmission time constraint. Searching the optimal setting is NP-hard. A two-stage scheme is developed to obtain a near-optimal solution. First, for each user, we map the required system resources and the corresponding expected distortion into a function relating to the transmission time and the expected distortion. We call this function the T-D function. By doing this, the traditional R-D function in a single-user transmission evolves to resource-distortion function in the multiuser scenario. Second, a round of bisection search on all users' T-D functions is performed to find the solutions.

We compare the preceding scheme with a traditional scheme that allocates equal transmission time budgets for both uplink and downlink and sequentially optimizes these two links. More specifically, the scheme first allocates the optimal configuration based on only the uplink channel information and then optimizes the downlink configuration based on the packets received successfully by the server. We consider a four-user system, in which users 1–4 are located 91, 67, 71, and 20 m away from the server, respectively. 90-frame QCIF (176 × 144) video sequence, *Akiyo*, *carphone*, *Claire*, and *foreman*, are delivered to users 1–4. Figure 15.9 shows the frame-by-frame PSNR results. As shown, our scheme can provide higher minimal and average PSNR, more uniform video quality among all users, and lower-quality fluctuation along each received video sequence than the sequential optimization scheme. The performance gain is attributed to the dynamic bandwidth allocation by our scheme to users in uplink and downlink transmission paths. Note that the sequential optimization scheme allocates fixed $T/2$ for all uplinks and another $T/2$ for all downlinks. Because of the asymmetric channel conditions along uplink and downlink for each videostream and the time heterogeneity of video content, the sequential optimization scheme lacks the freedom to dynamically adjust the time budget for uplink and downlink to attain better video quality.

Four performance criteria are used to evaluate our scheme and the traditional scheme. Let $PSNR_{i,n}$ denote the PSNR of the received video frame n for user i. Because the service objective is to minimize the maximal distortion, our first performance metric is

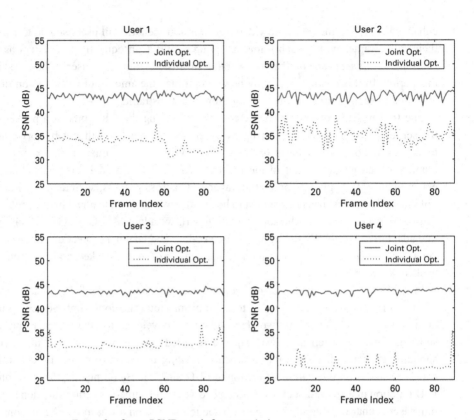

Figure 15.9 Frame-by-frame PSNR result for users 1–4.

the worst received video quality among all users. We measure the minimal PSNR among all users at frame n as $\mathrm{minPSNR}_n = \min_i \{\mathrm{PSNR}_{i,n}\}$ and take the average of the minimal PSNRs' over M video frames :

$$\mathrm{minPSNR} = \frac{1}{M} \sum_{n=1}^{M} \mathrm{minPSNR}_n. \tag{15.5}$$

The second metric is the average video quality received by all users, averaged over M frames:

$$\mathrm{avePSNR} = \frac{1}{M} \sum_{n=1}^{M} \mathrm{PSNR}_n, \tag{15.6}$$

where PSNR_n is the average received video quality of all users' nth video frame. The higher the avePSNR is, the higher the system efficiency in terms of overall video quality we have.

The third metric measures the fairness through examining the deviation of video qualities received by users. If users pay the same price for certain video quality, the received qualities for these users should be similar. To quantify the fairness, we calculate the standard deviation for all users' nth video frame and take the average along the whole M-frame video, i.e.,

$$\text{stdPSNR} = \frac{1}{M} \sum_{n=1}^{M} \left\{ \frac{1}{N} \sum_{i=1}^{N} (\text{PSNR}_{i,n} - \text{PSNR}_n)^2 \right\}^{\frac{1}{2}}. \tag{15.7}$$

The lower the stdPSNR is, the fairer the video quality each user receives.

The fourth metric concerns the quality fluctuation. Significant quality differences between consecutive frames can bring irritating flickering and other artifacts to viewers even when the average video quality is acceptable. To quantify the fluctuation of quality between nearby frames, we use the mean absolute difference of consecutive frames' PSNR, madPSNR, to measure the perceptual fluctuation along each video sequence and take the average over N users:

$$\text{madPSNR} = \frac{1}{N} \sum_{i=1}^{N} \left\{ \frac{1}{M-1} \sum_{n=2}^{M} \left| \text{PSNR}_{i,n} - \text{PSNR}_{i,n-1} \right| \right\}. \tag{15.8}$$

For the multicell case, without loss of generality, we simulate a two-cell system in which there are 8, 12, and 16 users. For each simulation profile, each user is randomly located in either cell, the distance from each user to his/her cell's access point is randomly selected between 20 and 100 m, and each user's first video frame is also randomly picked from the testing video sequence. We repeat the simulation using 100 different profiles and average the results to evaluate the performance. Figures 15.10(a) and 15.10(b) show the minPSNR and avePSNR using both schemes for different numbers of users in this system, respectively. The joint uplink and downlink optimization scheme outperforms the sequential uplink and downlink optimization scheme by 4.92–10.50 dB for the minimal PSNR and by 3.04–7.43 dB for the average PSNR. Because there are three different types of videoconferencing calls in this system, namely, intercell call between cells one and two, intracell call within cell one, and intracell call within cell two. We compare the stdPSNR for each type of call separately. As revealed by Figure 15.10(c), our algorithm can provide lower quality deviation for all three types of calls. Figure 15.10(d) shows the quality fluctuation along each received video sequence for both schemes, suggesting that our scheme provides lower quality fluctuation than the sequential optimization scheme. In summary, our scheme can provide higher minPSNR, higher avePSNR, lower stdPSNR, and lower madPSNR, which again demonstrates the superiority of joint uplink and downlink optimization.

In summary, we have constructed a videoconferencing framework for multiple conversation pairs within IEEE 802.11 networks. The framework dynamically performs multidimensional resource allocation by jointly exploring cross-layer error protection, multiuser diversity, and the heterogeneous channel conditions in all paths. With the transformation of system resources into T-D functions, the whole system is formulated to a min–max optimization problem to provide satisfactory video quality for all users. A fast algorithm is constructed to obtain the transmission configuration for each user in both single-cell and multicell scenarios.

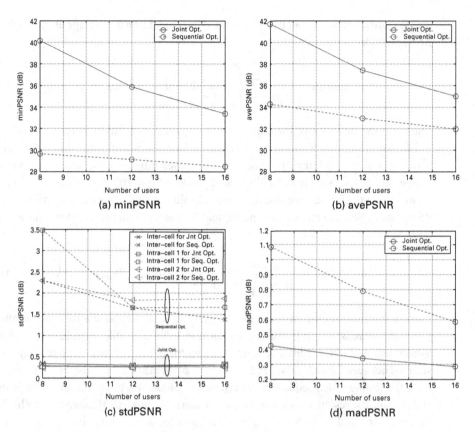

Figure 15.10 PSNR results for different numbers of users for two-cell case.

15.5   Next-Generation Video Over Multiuser OFDMA Networks

In this section, we discuss the next-generation wireless video transmission both in terms of the wireless communication networks as well as the video source encoders. We consider the downlink scenario of streaming multiple embedded video bit streams over a single-cell multiuser OFDMA system [294, 295].

For the video source coding, we use a 3D embedded wavelet video (EWV) as an example of a scalable video codec. The EWV encoder collects a group of frames (GOF) as a coding unit and performs 3D wavelet transform to obtain wavelet spatiotemporal subbands (or "subbands" in short). The encoder will encode each subband into several coding that passes. The more consecutive coding passes of each subband a receiver receives, the higher decoded video quality we have. The coding that passes among all subbands can be further grouped into several quality layers such that the received video quality can be refined progressively. In the video-coding subsystem, we need to determine *coding-pass assignment*, namely, which coding pass should be assigned to which quality layer for each user.

Assume the channel condition is stable within a period of time called the transmission interval. We can divide the maximal allowed time duration (for meeting the delay constraint) to transmit an encoded GOF bit stream into L transmission intervals with equal length. For each video, we dynamically construct a quality layer from the coded coding passes and transmit the quality layer in each transmission interval. The total allocated rate for each quality layer has a *rate constraint* in that it cannot exceed the available bandwidth.

To maintain the coding efficiency, the R-D curve in each subband should be convex [308]. Some coding passes in a subband cannot serve as feasible truncation points to maintain the convexity, and they will be pruned from the truncation point list. To facilitate the discussion, we call all the coding passes between two truncation points coding pass clusters.

Consider now a total of B subbands for the kth user and the subband b has $T_k^{b,\max}$ coding-pass clusters. We can measure the rate and the corresponding decrease in normalized mean-squared distortion of the tth coding-pass cluster in subband b for the kth user [308] and denote them as $\Delta r_{t,b,k}$ and $\Delta d_{t,b,k}$, respectively. We divide the whole duration for transmitting a total of L quality layers into L transmission intervals with equal length. The lth quality layer is transmitted at the lth transmission interval. The received distortion D_k^l and rate R_k^l for quality layers 0 to l can be expressed as

$$D_k^l = D_k^{\max} - \sum_{b=0}^{B-1} \sum_{t=0}^{T_k^{b,l}-1} \Delta d_{t,b,k}, \tag{15.9}$$

$$R_k^l = \sum_{q=0}^{l} \Delta R_k^q. \tag{15.10}$$

Here $D_k^{\max}$ is the distortion without decoding any coding-pass cluster,

$$\Delta R_k^l = \sum_{b=0}^{B-1} \sum_{t=T_k^{b,l-1}}^{T_k^{b,l}-1} \Delta r_{t,b,k}, \tag{15.11}$$

and $T_k^{b,l}$ is the total number of coding-pass clusters of subband b in the quality layers 0 to l, which satisfies

$$0 \le T_k^{b,l-1} \le T_k^{b,l} \le T_k^{b,\max}, \quad \forall b \text{ and } 0 < l < L. \tag{15.12}$$

Define the number of coding-pass clusters for subband b in quality layer l as $\Delta T_k^{b,l} = T_k^{b,l} - T_k^{b,l-1}$ and for all subbands

$$\Delta \mathbf{T}_k^l = [\Delta T_k^{0,l}, \Delta T_k^{1,l}, \ldots, \Delta T_k^{B-1,l}]. \tag{15.13}$$

We also define a matrix $\Delta \mathbf{T}^l$ whose kth row is $\Delta \mathbf{T}_k^l$. Thus, in each transmission interval l, the source-coding part of our system determines the coding-pass cluster assignment $\Delta \mathbf{T}_k^l$ and packetizes them as a quality layer for each user. We use Figure 15.11 to illustrate the relationship among coding pass, subband, and quality layer. Note that, owing to different content complexities and motion activities shown in video sources, the R-D

Figure 15.11 Illustration of the relationship among coding pass, subband, and quality layer.

information should be evaluated for each GOF of each user to capture the characteristics of the corresponding bit stream.

We consider a downlink scenario of a single-cell multiuser OFDM system in which there are K users randomly located. The system has N subcarriers and each subcarrier has bandwidth of W. We use an indicator $a_{kn} \in \{0, 1\}$ to represent whether the nth subcarrier is assigned to user k. Note that, in a single-cell OFDM system, each subcarrier can be assigned to at most one user, i.e., $\sum_{k=0}^{K-1} a_{kn} \in \{0, 1\}, \forall n$. The overall subcarrier-to-user assignment can be represented as a matrix $\mathbf{A}$ with

$$[\mathbf{A}]_{kn} = a_{kn}. \tag{15.14}$$

Let r_{kn} be the kth user's transmission rate at the nth subcarrier, and the total rate for the kth user can be expressed as $\sum_{n=0}^{N-1} a_{kn} r_{kn}$. The overall rate allocation can also be represented as a matrix $\mathbf{R}$ with $[\mathbf{R}]_{kn} = r_{kn}$.

In mobile wireless communication systems, signal transmission suffers from various impairments such as frequency-selective fading that is due to multipath delay [247]. The continuous complex baseband representation of user k's wireless channel impulse response is expressed as

$$g_k(t, \tau) = \sum_i \upsilon_{k,i}(t) \delta(\tau - \tau_{k,i}), \tag{15.15}$$

where $\upsilon_{k,i}(t)$ and $\tau_{k,i}$ are the gain and the delay of path i for user k, respectively. In Rayleigh fading, the sequence $\upsilon_{k,i}(t)$ is modeled as a zero-mean circular symmetric complex Gaussian random variable with variance $\sigma_{v_{k,i}}^2$ proportional to $d^{-\alpha}$, where d is the distance and α is the propagation loss factor. All $v_{k,i}(t)$ are assumed to be independent for different paths. The RMS delay spread is the square root of the second central moment of the power-delay profile:

$$\sigma_{k,\tau} = \sqrt{\overline{\tau_k^2} - (\bar{\tau}_k)^2}, \qquad (15.16)$$

where $\overline{\tau_k^2} = \dfrac{\sum_i \sigma_{v_{k,i}}^2 \tau_{k,i}^2}{\sum_i \sigma_{v_{k,i}}^2}$ and $\bar{\tau}_k = \dfrac{\sum_i \sigma_{v_{k,i}}^2 \tau_{k,i}}{\sum_i \sigma_{v_{k,i}}^2}$.

After sampling at the receiver, the channel gain of OFDM subcarriers can be approximated by the discrete samples of the continuous channel frequency response as

$$G_{kn}^h = \int_{-\infty}^{\infty} g_k(t, \tau) e^{-j2\pi f\tau} d\tau \big|_{f=nW, t=hT_f}, \qquad (15.17)$$

where T_f is the duration of an OFDM symbol and h is the sampling index. This approximation does not consider the effect of the smoothing filter at the transmitter and the front-end filter at the receiver.

We assume a slow fading channel in which the channel gain is stable within each transmission interval.[1] The resource-allocation procedure will be performed in each transmission interval. To facilitate the presentation, we omit h in the channel-gain notation. The channel parameters from different subcarrier of different users are assumed perfectly estimated, and the channel information is reliably fed back from mobiles users to the base station in time for use in the corresponding transmission interval. Denote Γ_{kn} as the kth user's SNR at the nth subcarrier as

$$\Gamma_{kn} = P_{kn} G_{kn} / \sigma^2, \qquad (15.18)$$

where P_{kn} is the transmission power for the kth user at the nth subcarrier and σ^2 is the thermal-noise power that is assumed to be the same for each subcarrier of different users. Further, let $[\mathbf{G}]_{kn} = G_{kn}$ be the channel-gain matrix and $[\mathbf{P}]_{kn} = P_{kn}$ the power-allocation matrix. For the downlink system, because of the practical constraints in implementation, such as the limitation of power amplifier and consideration of cochannel interferences to other cells, the overall power is bounded by P_{max}, i.e., $\sum_{k=0}^{K-1} \sum_{n=0}^{N-1} a_{kn} P_{kn} \leq P_{max}$.

The goal of the framework is to provide good subjective video quality of the reconstructed video. Because the distortion introduced by channel error is typically more annoying than the distortion introduced by source lossy compression, the system should keep the channel-induced distortion at a negligible portion of the end-to-end distortion so that the video quality is controllable by the source-coding subsystem. This can be achieved when we apply an appropriate amount of channel coding to keep the BER after the channel coding below some targeted BER threshold (10^{-6} in our system). In addition, joint consideration of adaptive modulation, adaptive channel coding, and power control

---

[1] In practice, the duration of a transmission interval can be adjusted shorter enough so that the channel gain is stable within a transmission interval.

Table 15.1 Required SNRs and transmission rates using adaptive modulation and convolutional coding rates [12]

k	Rate v_k(W)	Modulation	Convolutional coding rate	SNR ρ_k (dB) for BER $\leq 10^{-6}$	SNR (dB) for BER $\leq 10^{-5}$
1	1	QPSK	1/2	4.65	4.09
2	1.33	QPSK	2/3	6.49	5.86
3	1.5	QPSK	3/4	7.45	6.84
4	1.75	QPSK	7/8	9.05	8.44
5	2	16QAM	1/2	10.93	10.04
6	2.66	16QAM	2/3	12.71	12.13
7	3	16QAM	3/4	14.02	13.29
8	3.5	16QAM	7/8	15.74	15.01
9	4	64QAM	2/3	18.50	17.70
10	4.5	64QAM	3/4	19.88	18.99
11	5.25	64QAM	7/8	21.94	21.06

can provide each user with the ability to adjust each subcarrier's data transmission rate r_{kn} to control video quality while meeting the required BER.

We focus our attention on MQAM modulation and convolutional codes with bit interleaved coded modulation (BICM) as they provide high spectrum efficiency and strong forward error protection, respectively. We list the required SNRs and the adopted modulation with convolutional coding rates to achieve different supported transmission rates under different BER requirements in Table 15.1 based on the results in [12]. Given a targeted BER, there is a one-to-one mapping between the selected transmission rate and the chosen modulation scheme with the convolutional coding rate when the required SNR is satisfied. In this case, to obtain r_{kn} is equal to determine the modulation and channel coding rate. For each rate allocation $[\mathbf{R}]_{kn}$, the corresponding power allocation $[\mathbf{P}]_{kn}$ should maintain the SNR in (15.18) larger than the corresponding value listed in Table 15.1 to achieve the BER requirement. To facilitate our discussion, we define the feasible set of transmission rates in Table 15.1 as $v = \{v_0, v_1, v_2, \ldots, v_Q\}$ and the corresponding set of the required SNRs for BER $\leq 10^{-6}$ as $\rho = \{\rho_0, \rho_1, \rho_2, \ldots, \rho_Q\}$. Here, $v_0 = 0$ and $\rho_0 = 0$, and Q represents the number of combinations for different modulation with convolutional coding rates (11 in our case). All transmission rates r_{kn} should be selected from the feasible rate set v.

For a single-cell OFDMA system, there is a *subcarrier constraint* that each subcarrier can be assigned to at most one user. For a given subcarrier, users may experience different channel conditions. To achieve the same SNR, users in good channel conditions require less power than users in bad channel conditions. In this system, we maintain a BER below 10^{-6}. The supported rate of a subcarrier is a function of the required SNR for such a BER through AMC. A subcarrier with a higher power level can have a higher transmission rate. As the overall transmission power for all subcarriers is limited, this system design has a *power constraint*. Therefore we need to jointly determine which subcarrier should be assigned to which user (*subcarrier assignment*) and the subcarrier rate in each subcarrier (*rate assignment*) subject to power constraint.

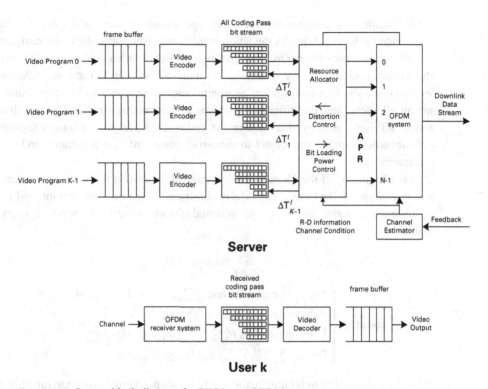

Figure 15.12 System block diagram for EWV over OFDMA.

The block diagram of the wireless video system is shown in Figure 15.12. The upper and lower parts show the modules located on the server side and the mobile user side, respectively. For the server side, the server buffers each user's incoming video frames in the user's frame buffer. For each video, a GOF with H frames is moved to a wavelet video encoder and compressed as a coding-pass bit stream. The selected coding-pass clusters will be transmitted during the next GOF transmission time of H/F seconds long, where F is the video frame sampling rate. The resource allocator obtains the channel condition information from the channel estimator and the R-D information from the source coder. With the estimated channel conditions, the resource allocator can predict how many data rates with BER $\leq 10^{-6}$ the wireless networks can support in the next transmission interval. With the R-D information, the resource allocator can estimate the qualities of the reconstructed videos after decoding at each mobile terminal. By jointly considering the R-D information and the estimated channel conditions, the resource allocator performs resource optimization and distributes video and radio resources to each video stream in each transmission interval. According to the allocated resources, the source-coding subsystem will group the selected coding-pass clusters into a quality layer for each user and pass them to the transmission system; and the multiuser OFDM subsystem will load the video data to be transmitted to different subcarriers at a controlled amount of power. On the mobile user side, an OFDM receiver buffers the received data until the end of the current GOF transmission time. Then those data are decoded by a wavelet video decoder and the decoded frames are sent for display.

We consider two essential service objectives, namely, fairness and efficiency. Fairness concerns the video-quality deviation among users who subscribe to the same QoS, and efficiency relates to how to attain the highest overall video quality using the available system resources. There is an inherent trade-off between fairness and efficiency. We formulate the fairness problem as minimizing the maximal end-to-end distortion among all users, and the efficiency problem as minimizing total end-to-end distortion of all users, both by determining the coding-pass assignment, rate assignment, and subcarrier assignment, subject to subcarrier constraint, rate constraint, and power constraint.

At the beginning of the *l*th transmission interval, according to the channel information and subject to the transmission delay constraint as one transmission interval long, the resource allocator minimizes the maximal distortion received among all users as follows:

$$\min_{\mathbf{A},\mathbf{R},\Delta\mathbf{T}^l} \max_k w_k f(D_k^l) \tag{15.19}$$

subject to
$$\begin{cases}
\text{Subcarrier assignment: } \sum_{k=0}^{K-1} a_{kn} \leq 1, a_{kn} \in \{0, 1\}, \forall n; \\
\text{Subcarrier rate: } r_{kn} \in v, \quad \forall k, n; \\
\text{User rate: } 0 \leq \Delta R_k^l \leq \sum_{n=0}^{N-1} a_{kn} r_{kn}, \forall k; \\
\text{Power: } \sum_{k=0}^{K-1} \sum_{n=0}^{N-1} a_{kn} P_{kn} \leq P_{\max};
\end{cases}$$

where w_k is the quality weighting factor and $f(\cdot)$ is the perceptual distortion function. We solve this optimization problem by selecting the values of subcarrier assignment matrix **A** in (15.14), rate-assignment matrix **R** (15.10), and coding pass cluster assignment $\Delta\mathbf{T}^l$ in (15.13) subject to four constraints: The first constraint is on the subcarrier assignment, that a subcarrier can be assigned to, at most, one user; the second one restricts the subcarrier rate to be selected from only the feasible rate set v; the third constraint states that the user's overall assigned rate in (15.11) should be no larger than the overall assigned subcarrier rate; and the fourth one is on the maximal power available for transmission. Note that the system can provide differentiated service by setting $\{w_k\}$ to different values according to the quality levels requested by each user. As a proof of this concept, we consider the case of $w_k = 1$, $\forall k$ and $f(D_k^l) = D_k^l$ for providing uniform quality among all users here. The problem in (15.19) is a multidimension generalized assignment problem, which is an NP-hard problem [201].

We formulate the efficiency problem to minimize the overall (weighted) end-to-end distortion among all users subject to constraints on subcarrier assignment, subcarrier rate, user rate, and power:

$$\min_{\mathbf{A},\mathbf{R},\Delta\mathbf{T}^l} \sum_{k=0}^{K-1} w_k f(D_k^l), \tag{15.20}$$

subject to
$$\begin{cases}
\text{Subcarrier assignment: } \sum_{k=0}^{K-1} a_{kn} \leq 1, a_{kn} \in \{0, 1\}, \forall n; \\
\text{Subcarrier rate: } r_{kn} \in v, \quad \forall k, n; \\
\text{User rate: } 0 \leq \Delta R_k^l \leq \sum_{n=0}^{N-1} a_{kn} r_{kn}, \forall k; \\
\text{Power: } \sum_{k=0}^{K-1} \sum_{n=0}^{N-1} a_{kn} P_{kn} \leq P_{\max}.
\end{cases}$$

Similar to the fairness case, the delay constraint is implicitly imposed in problem (15.20) so the transmission delay is restricted within a transmission interval. The problem in (15.20) is also NP-hard.

We develop two fast algorithms, a fairness algorithm and an efficiency algorithm, for the real-time requirement. Both algorithms have the following two steps. First, the resource allocator obtains the operational R-D function of all unsent coding passes for the current GOF of each user. The R-D function specifies the required data rates for reducing a certain amount of distortion. In the second step, the resource allocator can search the R-D functions and find the highest required rate that can be supported by the OFDMA system.

To achieve a trade-off between fairness and efficiency, for each GOF, we propose to apply the fairness algorithm in the first x transmission intervals to ensure the baseline fairness, and then apply the efficiency algorithm in the remaining $y = L - x$ transmission intervals to improve the overall video quality. We denote this class of strategies as $F_x E_y$ algorithms.

The simulations are set up as follows. The OFDM system has 32 subcarriers over a total 1.6-MHz bandwidth. The delay spread in RMS is 3×10^{-7} s. An additional 5-μs guard interval is used to avoid intersymbol interference that is due to channel delay spread. This results in a total block length as 25 μs and a block rate as 40k per second. The Doppler frequency is 10 Hz and the transmission interval is 33.33 ms. The mobile is uniformly distributed within the cell with radius of 50 m, and the minimal distance from mobile to the base station is 10 m. The noise power is 5×10^{-9} W, and the maximal transmission power is 0.1 W. The propagation loss factor is 3 [247]. The video sampling rate is 30 frames per second with CIF resolution (352 $\times$ 288). The GOF size is 16 frames, and each GOF is encoded using the codec [141] with four-level temporal decomposition.

Figure 15.13 shows the fairness, measured in terms of PSNR deviation, and efficiency, measured in terms of average of PSNR, for our algorithms. We first compare the performances for the $F_x E_y$ algorithm family. We see that the $F_{16} E_0$ algorithm achieves the lowest PSNR deviation among all algorithms but has the lowest average PSNR; and the $F_0 E_{16}$ algorithm achieves the highest average PSNR but has the highest PSNR deviation. We can have different tradeoffs on these two criteria by selecting a different number of transmission intervals for the fairness algorithm. Next, we compare the $F_x E_y$ algorithm family with a TDMA algorithm. Instead of allowing subcarriers in a transmission interval to be allocated among multiple users, the TDMA algorithm assigns all subcarriers in one transmission interval to only one user whose current distortion is the largest. As shown in Figure 15.13, for achieving the same average PSNR, the $F_x E_y$ algorithm family is about 1 dB lower in PSNR deviation than the TDMA algorithm. When achieving the same PSNR deviation, our algorithm can have higher average PSNR than the TDMA algorithm. In other words, our algorithm provides fairer and better quality than the TDMA algorithm. This is because our scheme exploits additional diversity in frequency and multiuser.

Figure 15.14 shows the average value of the worst received PSNR among all users from 10 different terminals' locations for the different numbers of users in the system. We can see that our algorithm, $F_{16} E_0$, can improve the minimal PSNR better than that

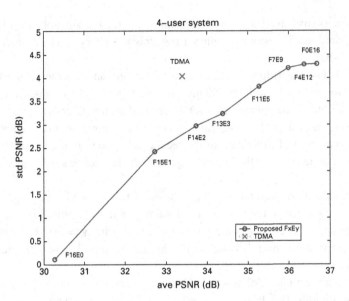

Figure 15.13 Efficiency and fairness results of the F_xE_y algorithm family and TDMA algorithm.

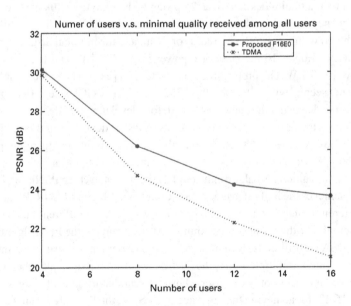

Figure 15.14 Performance comparison for the worst quality received among all users using our algorithm and TDMA algorithm.

of the TDMA scheme. There is about 0.5–3 dB gain for different numbers of users. The performance gap increases when the number of users increases owing to the multiuser diversity.

We first use a four-user system with uniform quality among users to demonstrate our algorithm in achieving pure fairness ($F_{16}E_0$) among all users. Each user receives

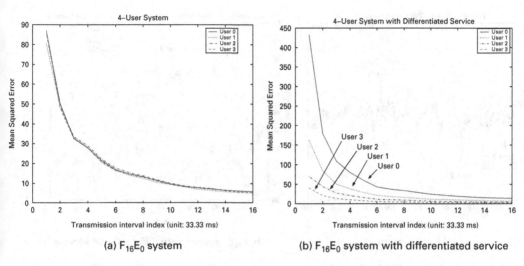

(a) $F_{16}E_0$ system

(b) $F_{16}E_0$ system with differentiated service

Figure 15.15 Comparison for the $F_{16}E_0$ system providing uniform quality and differentiated service.

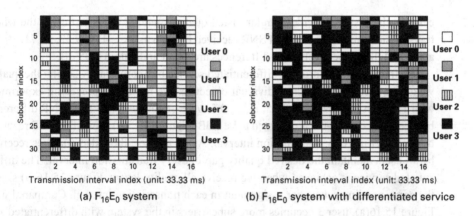

(a) $F_{16}E_0$ system

(b) $F_{16}E_0$ system with differentiated service

Figure 15.16 Subcarrier assignment for the $F_{16}E_0$ system in each transmission interval.
(a) Uniform quality. The system assigns more subcarriers to user 0 at most intervals due to the required rate of video sequence 0 to achieve the same quality is higher than other sequences.
(b) Differentiated service. The system assigns more subcarriers to user 3 because of the highest requested quality.

one GOF from one of the four video sequences, *Foreman, Hall Monitor, Mother and daughter,* and *Silent,* respectively. Figure 15.15(a) shows the received video quality in terms of mean squared error in every transmission interval. As we can see, all four users have similar video quality in each transmission interval and the received video quality is improved by receiving more quality layers till the last transmission interval. The corresponding subcarrier assignment in each transmission interval is shown in Figure 15.16(a). As the source-coding rate of each user is allocated in different time and frequency slots according to the channel conditions and source characteristics, the frequency diversity, time diversity, and multiuser diversity are jointly exploited. Figure 15.17(a) shows the frame-by-frame PSNR for all users by receiving 10 GOFs. As we

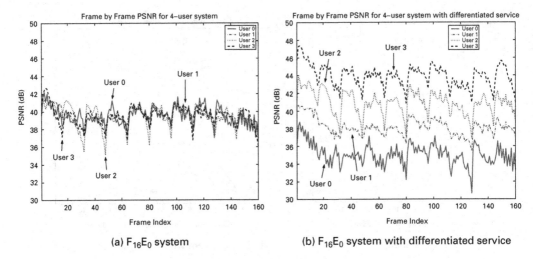

(a) $F_{16}E_0$ system   (b) $F_{16}E_0$ system with differentiated service

Figure 15.17 Frame-by-frame PSNR for the $F_{16}E_0$ system without and with differentiated service.

can see, the system provides similar visual qualities among all users during the whole transmission time. The average PSNR received by each users 1–4 is 39.52, 39.71, 39.46, and 39.54 dB, respectively. The differences are small and within 0.25 dB.

The framework can provide differentiated service by appropriately setting the quality weighting factor $\{w_k\}$ to the individual objective. We repeat the preceding experiment with a new set $\{w_k\}$ as $w_0 = 0.25$, $w_1 = 0.5$, $w_2 = 1$, and $w_3 = 2$. The PSNR difference between user i and $i + 1$ is expected to be 3 dB. Figure 15.15(b) shows the MSE received by each user in every transmission interval. As we can see, the video qualities received by all users maintain the desired quality gap in every transmission interval. The differentiated service is achieved when we receive all quality layers. Figure 15.16(b) shows the corresponding subcarrier assignment in each transmission interval. Compared with Figure 15.16(a), user 3 occupies more subcarriers in the system with differentiated service than in the system with uniform quality. Figure 15.17(b) shows the frame-by-frame PSNR for all users by receiving 10 GOFs. As we expected, user 3 has the highest received video quality and user 0 has the lowest PSNR. The average PSNR received by each user is 35.06, 38.16, 40.91, and 43.92 dB, respectively. The PSNR differences between user i and $i + 1$ for $i = 0, 1, 2$ are 3.10, 2.75, 3.01 dB, respectively, which is close to the design goal of 3-dB differentiated service.

In summary, we have constructed a framework sending multiple scalable video programs over multiuser OFDM networks. By leveraging the frequency, time, and multiuser diversity of the OFDM system and the scalability of the 3D EWV codec, the framework can allocate system resources to each videostream to achieve desired video quality. We address two service objectives: fairness and efficiency. For the fairness problem, we formulate the system to achieve fair quality among all users as a min–max problem. For the efficiency problem, we formulate the system to attain the lowest overall video distortion as a minimization problem. To satisfy the real-time requirement, two fast algorithms are constructed to solve the preceding two problems.

15.6   Summary

We have discussed the multiuser video transmission over wireless networks from the resource-allocation point of view. Unlike the traditional schemes, we consider two aspects of design issues:

1. Cross-layer optimization: To achieve system-wide optimality that is due to the delay constraints of video transmissions.
2. Multiuser diversity: To exploit multiuser diversity and allocating system resources dynamically so as to improve the network performances and achieve the baseline QoS satisfactions for individual users.

Then a general resource-allocation *framework* is constructed to control the qualities of multiple video transmissions that share the limited resources over wireless networks. We review several schemes transmitting videos over 3G, 4G, and WLAN/WMAN systems and demonstrate that the multiuser cross-layer design can provide better video quality than that of the traditional schemes. Specifically, we provide three solutions as follows:

1. Multiuser distortion management of layered video over resource-limited downlink MC-CDMA,
2. Joint uplink and downlink optimization for real-time multiuser videoconferencing over WLANs,
3. Scalable multiuser framework for video over OFDM networks with fairness and efficiency concerns.

Bibliography

[1] A. Adinoyi and H. Yanikomeroglu, "Multi-antenna aspects of parallel fixed wireless relays," in *IEEE* Wireless Communications and Networking Conference *(WCNC'06)* (2006).

[2] J. M. Aein, "Power balancing in system employing frequency reuse," *COMSAT Tech. Rev.,* Communication Satellite Corporation (1973), Vol. 3.

[3] A. Ahluwalia, E. Modiano, and L. Shu, "On the complexity and distributed construction of energy-efficient broadcast trees in static ad hoc networks," in *Proceedings of the 36th Annual Conference on Information Sciences and Systems* (2002), pp. 807–813.

[4] I. F. Akyildiz, W. Y. Lee, M. C. Vuran, and S. Mohanty, "Next generation/dynamic spectrum access/cognitive radio wireless networks: A survey," *Computer Networks: The Int. J. Comput. Telecommun. Network.* **50**, 2127–2159 (2006).

[5] I. F. Akyildiz, D. A. Levine, and I. Joe, "A slotted CDMA protocol with BER scheduling for wireless multimedia networks," *IEEE/ACM Trans. Network.* **7** (1999).

[6] M. Almgren, H. Andersson, and K. Wallstedt, "Power control in a cellular system," in *Proceedings of the IEEE Vehicular Technology Conference (VTC)*, New York, IEEE (1994).

[7] M. S. Alouini and A. J. Goldsmith, "Adaptive modulation over Nakagami fading channels," *Wireless Personal Commun.* **13**, 119–143 (2000).

[8] M. S. Alouini and A. J. Goldsmith, "A unified approach for calculating error rates of linearly-modulated signals over generalized fading channels," *IEEE Trans. Commun.* **47**, 1324–1334 (1999).

[9] E. Altman, A. A. Kherani, P. Michiardi, and R. Molva "Non-cooperative forwarding in ad-hoc networks," in *The 15th IEEE International Symposium on Personal, Indoor and Mobile Radio Communications (PIMRC)*, IEEE, New York (2004).

[10] E. Altman, A. A. Kherani, P. Michiardi, and R. Molva, "Non-cooperative forwarding in ad-hoc networks," *INRIA*, Tech. Rep. No. 5116 (2004).

[11] L. Anderegg and S. Eidenbenz, "Ad Hoc-VCG: A truthful and cost-efficient routing protocol for mobile ad hoc networks with selfish agents," in *Proceedings of the ACM Annual International Conference on Mobile Computing and Networking (MobiCom)* (2003).

[12] E. Armanious, D. D. Falconer, and H. Yanikomeroglu, "Adaptive modulation, adaptive coding, and power control for fixed cellular broadband wireless systems: some new insights," in *IEEE WCNC* (2003).

[13] B. S. Atal, V. Cuperman, and A. Gersho, *Speech and Audio Coding for Wireless and Network Application*, Boston: Kluwer Academic (1993).

[14] R. J. Aumann, "Subjectivity and correlation in randomized strategy," *J. Math. Econ.* **1**, 67–96 (1974).

[15] R. J. Aumann, "Correlated equilibrium as an expression of Bayesian rationality," *Econometrica* **55**, 1–18 (1987).

[16] O. Awoniyi, N. B. Mehta, and L. J. Greenstein, "Characterizing the ortogonality factor in WCDMA downlinks," *IEEE Trans. Wireless Commun.* **2**, 621–625 (2003).

[17] A. Batra, J. Balakrishnan, G. R. Aiello, J. R. Foerster, and A. Dabak, "Design of a multiband OFDM system for realistic UWB channel environments," *IEEE Trans. Microwave Theor. Tech.* **52**, 2123–2138 (2004).

[18] A. Batra, J. Balakrishnan, A. Dabak, R. Gharpurey, J. Lin, P. Fontaine, J.-M. Ho, S. Lee, M. Frechette, S. March, and H. Yamaguchi, "Multi-band OFDM physical layer proposal for IEEE 802.15 task group 3a," IEEE P802.15-03/268r3 (March 2004).

[19] M. S. Bazaraa, *Nonlinear Programming: Theory and Algorithms* (2nd ed.), New York: Wiley (1993).

[20] M. S. Bazaraa, H. D. Sherali, and C. M. Shetty, *Nonlinear Programming: Theory and Algorithms* (2nd ed.), New York: Wiley (1993).

[21] R. Beard, D. Kingston, M. Quigley, D. Snyder, R. Christiansen, W. Johnson, T. Mclain, and M. Goodrich, "Autonomous vehicle technologies for small fixed wing UAVs," *AIAA J. Aerospace Comput. Inf. Commun.* **2**, 92–108 (2005).

[22] R. Beard, T. McLain, D. Nelson, and D. Kingston, "Decentralized cooperative aerial surveillance using fixed-wing miniature UAVs," *IEEE Proceedings: Special Issue on Multi-Robot Systems*, technical report available at $https : //dspace.byu.edu/handle/1877/60$.

[23] J. J. van de Beek, P. O. Börjesson, M. -L. Boucheret, D. Landström, J. M. Arenas, P. Ödling, C. Östberg, M. Wahlgvist and S. K. Wilson. "A time and frequency synchronization scheme for multiuser OFDM," *IEEE J. Sel. Areas Commun.* **17**, 1900–1914 (1999).

[24] J. F. Benders, "Partitioning procedures for solving mixed-variables programming problems," *Numer. Math.* **4**, 238–252 (1962).

[25] C. Berrou, A. Glavieux, and P. Thitimajshima, "Near Shannon limit error-correcting coding and decoding: Turbo-codes," in *Proceedings of the International Conference on Communication (ICC93)* (1993), pp. 1064–1070.

[26] D. P. Bertsekas, *Dynamic Programming and Optimal Control*, Athena Scientific (1995).

[27] D. P. Bertsekas, *Nonlinear Programming* (2nd ed.), Athena Scienific (1999).

[28] G. Bersekas and R. Gallager, *Data Network* (2nd ed.), Englewood Cliffs, NJ: Prentice- Hall (1992).

[29] P. Bhagwat, P. Bhattacharya, A. Krishna, and S.K. Tripathi, "Enhancing throughput over wireless LANs using channel state dependent packet scheduling," in *Proceedings of the IEEE 27th Conference on Computer Communications (IEEE Infocom)* (1996).

[30] M. Bhardwaj, T. Garnett, and A. P. Chandrakasan, "Upper bounds on the lifetime of sensor networks," in *Proceeding of the International Conference on Communication* (2001), Vol. 3, pp. 785–790.

[31] V. Bharghavan, S. Lu, and T. Nandagopal, "Fair queueing in wireless networks: Issues and approaches," *IEEE Personal Commun. Mag.* **6** (1), 44–53 (February 1999).

[32] A. Bletsas, A. Lippman, and D. P. Reed, "A simple distributed method for relay selection in cooperative diversity wireless networks based on reciprocity and channel measurements," in *VTC* (2005), Vol. 3, pp. 1484–1488.

[33] S. Borst, "User-level performance of channel-aware scheduling algorithms in wireless data networks," in *IEEE Infocom* (2003).

[34] S. Boyd and L. Vandenberghe, *Convex Optimization*, New York: Cambridge University Press (2006). (http://www.stanford.edu/~boyd/cvxbook.html).

[35] J. Boyer, D. D. Falconer, and H. Yanikomeroglu, "Multihop diversity in wireless relaying channels," *IEEE Trans. Commun.* **52**, 1820–1830 (2004).

[36] D. G. Brennan, "Linear diversity combining techniques," *Proc. IEEE* **91**, 331–356 (2003).

[37] S. Buchegger and J-Y. Le Boudec, "Performance analysis of the confidant protocol (cooperation of nodes—fairness in dynamic ad-hoc networks)," *in Proceedings of the 3rd ACM Internation Symposium on Mobile Ad Hoc Networking and Computing (MobiHoc'02)*, New York: Association for Computing Machinery (2002), pp. 80–91.

[38] L. Buttyan and J. P. Huaux, "Stimulating cooperation in self-organizing mobile ad hoc networks," *ACM J. Mobile Networks*, special issue on mobile ad hoc networks (2002).

[39] G. Caire, G. Taricco, and E. Biglieri, "Bit-interleaved coded modulation," *IEEE Trans. Inf. Theory* **44**, 927–946 (1998).

[40] T. Camp, J. Boleng, and V. Davies, "A survey of mobility models for ad hoc network research," *Wireless Commun. Mobile Comput.* **2**, 483–502 (2002).

[41] Y. Cao and V. O. K. Li, "Scheduling algorithms in broad-band wireless networks," *Proc. IEEE* **89**, 76–87 (2001).

[42] D. W. Casbeer, D. B. Kingston, R. W. Beard, T. W. McLain, S. Li, and R. Mehra, "Cooperative forest fire surveillance using a team of small unmanned air vehicles," *International Journal of Systems Sciences*, Technical Report available at https://dspace.byu.edu/handle/1877/55.

[43] N. H. L. Chan and P. T. Mathiopoulos, "Efficient video transmission over correlated Nakagami fading channels for IS-95 CDMA systems," *IEEE J. Sel. Areas Commun.* **18**, 996–1011 (2000).

[44] Y. S. Chan and J. W. Modestino, "A joint source coding-power control approach for video transmission over CDMA networks," *IEEE J. Sel. Areas Commun.* **21**, 1516–1525 (2003).

[45] J.-H. Chang and L. Tassiulas, "Maximum lifetime routing in wireless sensor networks," *IEEE/ACM Trans. Network.* **12**, 609–619 (2004).

[46] C.-C. Chiang, H.-K. Wu, W. Liu, and M. Gerla, "Routing in clustered multihop, mobile wireless networks with fading channel," in *Proceedings of the IEEE Singapore International Conference on Networks (SICON 97)*, New York: IEEE (1997), pp. 197–211.

[47] C. Cho, H. Zhang, and M. Nakagawa "A UWB repeater with a short relaying-delay for range extension," *IEEE WCNC* (2004), Vol. 3, pp. 1436–1441.

[48] J. C.-I. Chuang, "Autonomous adaptive frequency assignment for TDMA portable ratio systems," *IEEE Trans. Veh. Technol.* **VT- 40**, 627–635 (1991).

[49] S. T. Chung and A. J. Goldsmith, "Degrees of freedom in adaptive modulation: A unified view," *IEEE Trans. Communun.* **49**, 1561–1571 (2001).

[50] S. T. Chung, S. J. Kim, and J. M. Cioffi, "A game-theoretic approach to power allocation in frequency-selective Gaussian interference channels," in *Proceedings of the IEEE International Symposium on Information Theory (ISIT '03)* (2003).

[51] M. W. Cooper and K. Farhangian, "Multicriteria optimization for nonlinear integer-variable problems," *Large Scale Syst.* **9**, 73–78 (1985).

[52] T. Cover and J. Thomas, *Elements of Information Theory*, New York: Wiley (1991).

[53] T. M. Cover and A. El Gamal, "Capacity theorems for the relay channel," *IEEE Inf. Theory* **25**, 572–584 (1979).

[54] F. Cuomo, C. Martello, A. Baiocchi, and F. Capriotti "Radio resource sharing for ad hoc networking with UWB," *IEEE J. Sel. Areas Commun.* **20**, 1722–1732 (2002).

[55] L. A. DaSilva, D. W. Petr, and N. Akar, "Static pricing and quality of service in multiple service networks," in *Proceedings of the 5th Joint Conference on Information Sciences (JCIS'00)* (2000), Vol. 1, pp. 355–358.

[56] G. B. Dantzig, "Programming of interdependent activities. II. Mathematical model," *Econometrica* **17**, 200–211 (1949).

[57] B. Deep and W.-C. Feng, "Adaptive code allocation in multicode-CDMA for transmitting H.263 video," in *IEEE WCNC* (1999).

[58] Z. Ding and Y. Li, *Blind Equalization and Identification*, Signal Processing and Communications Series, Marcel Dekker, 2001.

[59] R. Dube, C. D. Rais, K.-Y. Wang, and S. K. Tripathi, "Signal stability-based adaptive routing (SSA) for ad-hoc mobile networks," *IEEE Personal Commun. Mag.*, 36–45 (February 1997).

[60] D. A. Eckhardt and P. Steenkiste, "Effort-limited fair (ELF) scheduling for wireless networks," in *IEEE Infocom* (2000).

[61] B. Eklundh, "Channel utilization and blocking probability in a cellular mobile telephone system with directed retry," *IEEE Trans. Commun.* **COM-34**, 530–535 (1989).

[62] C. Eklunk, R. B. Marks, K. L. Stanwood, and S. Wang, "IEEE standard 802.16: A technical overview of the WirelessMAN air interface for broadband wireless access," *IEEE Commun. Mag.* 98–107 (June 2002).

[63] S. A. El-Dolil, W. C. Wong, and R. Steele, "Teletraffic performance of highway microcells with overlay macrocell," *IEEE J. Sel. Areas Commun.* **7**, 71–78 (1989).

[64] T. Elbatt and A. Ephremides, "Joint scheduling and power control for wireless ad hoc networks," in *IEEE Infocom* (2002).

[65] W. H. R. Equitz and T. M. Cover, "Successively refinement of information," *IEEE Trans. Inf. Theory* **37**, 269–274 (1991).

[66] N. C. Ericsson, "Adaptive modulation and scheduling over fading channels," in *Proceedings of the IEEE Global Communication Conference (Globecom99)* (1999), pp. 2668–2672.

[67] R. Etkin, A. Parekh, and D. Tse, "Spectrum sharing for unlicenced bands," in *Proceedings of The First IEEE International Symposium on New Frontiers in Dynamic Spectrum Access Networks (DySPAN 2005)*, New York: IEEE (2005), pp. 251–258.

[68] S. Falahati, A. Svensson, T. Ekman, and M. Sternad, "Adaptive modulation systems for predicted wireless channels," in *Proceedings of the 7th Wireless World Research Forum Workshop* (2002).

[69] S. Falahati, A. Svensson, T. Ekman, and M. Sternad, "Effect of prediction errors on adaptive modulation systems for wireless channels," in *Proceedings, Radiovetenskap och Kommunikation* (2002), pp. 234–238.

[70] L. M. Feeney and M. Nilsson, "Investigating the energy consumption of a wireless network interface in an ad hoc networking environment," in *IEEE Infocom 2001* (2001), Vol. 3, pp. 1548–1557.

[71] M. Felegyhazi, L. Buttyan and J. P. Hubaux "Equilibrium analysis of packet forwarding stratiegies in wireless ad-hoc networks – the static case," in *Personal Wireless Communications 2003* (2003).

[72] N. Feng, S. Mau, and N. Mandayam, "Pricing and power control for joint network-centric and user-centric radio resource management," in *Proceedings of the 39th Allerton Conference on Communication, Control, and Computing*, (2001).

[73] Z. Feng and T. Kaiser, "On the evaluation of channel capacity of multi-antenna UWB indoor wireless systems," in *Proceedings of the IEEE International Symposium on Spread Spectrum Technology and Applications*, New York: IEEE (2004), pp. 525–529.

[74] M. L. Fisher, "The Lagrangian method for solving integer programming problems," Management Sci. **27**, 1–18 (1981).

[75] P. Floreen, P. Kaski, J. Kohonen, and P. Orponen, "Lifetime maximization for multicasting in energy-constrained wireless networks," *IEEE J. Sel. Areas Commun. (Special Issue on Wireless Ad Hoc Networks)* **23**, 117–127 (2005).

[76] J. R. Foerster, "The performance of a direct-sequence spread ultra wideband system in the presence of multipath, narrowband interference, and multiuser interference," in *IEEE Proceedings of the Conference on Ultra Wideband Systems and Technologies*, New York: IEEE (2002), pp. 87–91.

[77] J. Foerster et al., "Channel Modeling Sub-committee Report Final," IEEE802.15-02/490, Nov. 18, 2003.

[78] J. R. Foerster et al., "Intel CFP presentation for a UWB PHY," IEEE P802.15-03/109r1, March 3, 2003.

[79] G. J. Foschini and Z. Miljanic, "A simple distributed autonomous power control algorithms and its convergence." *IEEE Trans. Veh. Technol.* **40**, 641–646 (1993).

[80] D. Fudenberg and J. Tirole, *Game Theory*, Cambridge, MA: MIT Press (1991).

[81] A. E. Gamal, C. Nair, B. Prabhakar, E. U. Biyikoglu, and S. Zah, "Energy efficient scheduling of packet transmissions over wireless networks," in *IEEE Infocom* (2002).

[82] F. R. Gantmacher, *The Theory of Matrices*, New York: Chelsea (1990), Vol. 2.

[83] A. Gersho and R. M. Gray, *Vector Quantization and Signal Compression*, Boston: Kluwer Academic (1992).

[84] A. Goldsmith and S. G. Chua, "Adaptive coded modulation for fading channels," *IEEE Trans. Commun.* **46**, 595–602 (1998).

[85] G. Golub, *Matrix Computations* (3rd ed.), Baltimore: Johns Hopkins University Press (1996).

[86] J. Gomez, A.T. Campbell, and H. Morikawa, "The Havana framework for supporting appliaction and channel dependent QoS in wireless networks," in *Proceedings of the IEEE International Conference on Network Protocols (ICNP)* (1999).

[87] R. E. Gomory, "Outline of an algorithm for integer solution to linear programs," *Bull. Am. Math. Soc.* **64**, 275–278 (1958).

[88] D. J. Goodman, R. A. Valenzula, K. T. Gayliard, and B. Ramamurthi, "Packet reservation multiple access for local wireless communications," *IEEE Trans. Commun.* **37**, 885–890 (1989).

[89] V. K. Goyal, "Multiple description coding: Compression meets the network," *IEEE Sig. Process. Mag.* **18**(5), 74–93 (September 2001).

[90] I. S. Gradshteyn and I. M. Ryzhik, *Table of Integrals, Series, and Products* (5th ed.), New York: Academic (2000).

[91] S. Grandhi, R. Vijayan, D. J. Goodman, and J. Zander, "Centralized power control for cellular radio systems" *IEEE Trans. Veh. Technol.* **42**, (1993).

[92] S. A. Grandhi, R. Vijayan, and D. J. Goodman. "Distributed power control in cellular radio systems," *IEEE Trans. Commun.* **42** (1994).

[93] S. A. Grandhi, R. Vijayan, and D. J. Goodman, "A distributed algorithm for power control in cellular radio systems," in *30th Allerton Conference* (1992).

[94] S. A. Grandhi, J. Zander, and R. Yates, "Constrained power control," *Wireless Personal Commun.* **2** (1995).

[95] D. Grosu, A. T. Chronopoulos, and M.Y. Leung, "Load balancing in distributed systems: an approach using cooperative games," in *IPDPS 2002* (2002), pp. 52–61.

[96] M. Grotschel, "Discrete mathematics in manufacturing," in *Proceedings of the Second International Conference on Industrial and Applied Mathematics*.

[97] D. L. Gu, H. Ly, X. Hong, M. Gerla, G. Pei, and Y. Lee, "C-ICAMA, a centralized intelligent channel assigned multiple access for multi-layer ad-hoc wireless networks with UAVs," in *IEEE WCNC* (2000), Vol. 2, pp. 879–884.

[98] D. L. Gu, G. Pei, H. Ly, M. Gerla, B. Zhang, and X. Hong, "UAV aided intelligent routing for ad-hoc wireless network in single- area theater," in *IEEE WCNC* (2000), Vol. 3, pp. 1220–1225.

[99] M. Guignard and S. Kim, "Lagrangian decomposition: a model yielding stronger Lagrangian bounds," *Math. Program.* **39**, 215–228 (1987).

[100] D. Gunduz and E. Erkip, "Joint source-channel cooperation: Diversity versus spectral efficiency," in *ISIT* (2004), p. 392.

[101] J. Hagenauer, "Rate compatible punctured convolutional (RCPC) codes and their applications," *IEEE Trans. Commun.* **36**, 389–399 (1988).

[102] Z. Han, F. R. Farrokhi, and K. J. Ray Liu, "Joint power control and blind beamforming over wireless networks: A cross layer approach," *EURASIP J. Appl. Sig. Process.* **5**, 751–761 (2004) (special issue on MIMO Communications and Signal Processing).

[103] Z. Han, F. R. Farrokh, Z. Ji, and K. J. Ray Liu, "Capacity optimization using subspace method over multicell OFDMA networks," in *WCNC* (2004), Vol. 4, pp. 2393–2398.

[104] Z. Han, T. Himsoon, W. Siriwongpairat, and K. J. Ray Liu, "Energy efficient cooperative transmission over multiuser OFDM networks: Who helps whom and how to cooperate," in *WCNC* (2005), Vol. 2, 1030–1035.

[105] Z. Han, Z. Ji, and K. J. Ray Liu, "Power minimization for multi-cell OFDM networks using distributed non-cooperative game approach," in *Proceedings of the IEEE Global Telecommunications Conference*, New York: IEEE (2004).

[106] Z. Han, Z. Ji, and K. J. Ray Liu, "A referee-based distributed scheme of resource competition game in multi-cell multi-user OFDMA networks," *IEEE J. Sel. Areas Commun.* (special issue on noncooperative behavior in networking), **53**(10) (2007).

[107] Z. Han, Z. Ji, and K. J. Ray Liu, "Fair multiuser channel allocation for OFDMA networks using nash bargaining solutions and coalitions," *IEEE Trans. Commun.* **53**, 1366–1376 (2005).

[108] Z. Han, Z. Ji, and K. J. Ray Liu, "Low-complexity OFDMA channel allocation with Nash bargaining solution fairness," in *IEEE Globecom* (2004).

[109] Z. Han, Z. Ji, and K.J. Ray Liu, "Dynamic distributed rate control for wireless networks by optimal cartel maintenance strategy," in *IEEE Globecom* (2004).

[110] Z. Han, A. Kwasinski, K. J. Ray Liu, and N. Farvardin, "A near-optimal joint source-channel speech resource allocation scheme over downlink CDMA networks," *IEEE Trans. Wireless Commun.* **54**, 1682–1692 (2006).

[111] Z. Han, A. Kwasinski, K. J. Ray Liu, and N. Farvardin, "Pizza party algorithm for real-time distortion management in downlink single-cell CDMA systems," in *41st Allerton Conference* (2003).

[112] Z. Han and K. J. R. Liu, "Adaptive SINR threshold allocation for joint power control and beamforming over wireless network," in *IEEE VTC* (2001), pp. 1548–1552.

[113] Z. Han and K. J. R. Liu, "Power minimization under constant throughput constraint in wireless networks with beamforming," in *IEEE VTC* (2002).

[114] Z. Han and K. J. Ray Liu, "Throughput maximization using adaptive modulation in wireless networks with fairness constraint," in *IEEE WCNC* (2003).

[115] Z. Han and K. J. Ray Liu, "Non-cooperative power control game and throughput game over wireless networks," *IEEE Trans. Commun.* **53** (2005).

[116] Z. Han and K. J. Ray Liu, "Power minimization under throughput management over wireless networks with antenna diversity," *IEEE Trans. Wireless Commun.* **3** (2004).

[117] Z. Han and K. J. Ray Liu, "Joint adaptive link quality and power management with fairness constraint over wireless networks," *IEEE Trans. Veh. Technol.* **53**, 1138–1148 (2004).

[118] Z. Han and K. J. Ray Liu, "Power minimization under throughput management over wireless networks with antenna diversity," *IEEE Trans. Wireless Commun.* **3**, 2170–2181 (2004).

[119] Z. Han, X. Liu, Z. J. Wang, and K. J. Ray Liu, "Delay sensitive scheduling schemes with short-term fairness for heterogeneous QoS over wireless network," *IEEE Trans. Wireless Commun.* **6** (2007).

[120] Z. Han, C. Pandana, and K. J. R. Liu, "A self-learning repeated game framework for optimizating packet forwarding networks," in *IEEE WCNC* (2005), pp. 2131–2136.

[121] Z. Han, C. Pandana, and K. J. Ray Liu, "Distributive opportunistic spectrum access for cognitive radio using correlated equilibrium and no-regret learning," in *IEEE WCNC* (2007).

[122] Z. Han, G. Su, A. Kwasinski, M. Wu, and K. J. Ray Liu, "Multiuser distortion management of layered video over resource limited downlink multi-code CDMA," *IEEE Trans. Wireless Commun.* **5**, 3056–3067 (2006).

[123] Z. Han, A. Lee Swindlehurst, and K. J. Ray Liu, "Smart deployment/movement of unmanned air vehicle to improve connectivity in MANET," in *IEEE WCNC* (2006).

[124] Z. Han, Z. Jane Wang, and K. J. Ray Liu, "A resource allocation framework with credit system and user autonomy over heterogeneous wireless networks," in *Proceedings of IEEE Globecom* (2003).

[125] S. V. Hanly, "An algorithm for combined cell-site selection and power control to maximize cellular spread spectrum capacity," *IEEE J. Sel. Areas Commun.* **13**, 1332–1340 (1995).

[126] L. Hanzo, P. Cherriman, and J. Steit, *Wireless Video Communications, Second to Third Generation Systems and Beyond*, New York: IEEE (2001).

[127] L. Hanzo, S. X. Ng, T. Keller, and W. T. Webb, *Single and Multicarrier Quadrature Amplitude Modulation: From Basics to Adaptive Trellis-Coded, Turbo-Equalised and Space-Time Coded OFDM, CDMA and MC-CDMA Systems*, New York: Wiley (2004).

[128] P. Harley, "Short distance attenuation measurements at 900 MHz and 1.8 GHz using low antenna heights for microcells," *IEEE Trans. Sel. Areas Commun.* **37**, 220–222 (1988).

[129] S. Hart and A. Mas-Colell, "A simple adaptive procedure leading to correlated equilibrium," *Econometrica* **68**, 1127–1150 (2000).

[130] S. Haykin, *Adaptive Filter Theory* (3rd ed.), Englewood Cliffs, NJ: Prentice-Hall (1996).

[131] S. Haykin, "Cognitive radio: Brain-empowered wireless communications," *IEEE J. Sel. Areas Commun.* **23**, 201–220 (2005).

[132] T. Himsoon, W. Pam Siriwongpairat, Z. Han, and K. J. Ray Liu, "Lifetime maximization framework by cooperative nodes and relay deployment in Wireless Networks," *IEEE J. Sel. Areas Commun.* (special issue on cooperative communications and networking), **25**(2), 306–317 (2007).

[133] T. Himsoon, W. Siriwongpairat, Z. Han, and K. J. Ray Liu, "Lifetime maximization with cooperative diversity in wireless sensor networks," in *IEEE WCNC* (2006).

[134] K. J. Hole, H. Holm, and G. E. Oien, "Adaptive multidimensional coded modulation over flat fading channels," *IEEE J. Sel. Areas Commun.* **18**, 1153–1158 (2000).

[135] D. Hong and S. S. Rappaport, "Traffic model and performance analysis for cellular mobile radio telephone systems with prioritized and nonprioritized handoff procedures," *IEEE Trans. Veh. Technol.* **VT-35**, 448–461 (1986).

[136] R. A. Horn and C. R. Johnson, *Matrix Analysis*, New York: Cambridge University Press (1985).

[137] A. Host-Madsen, "Upper and lower bounds for channel capacity of asynchronous cooperative diversity networks," *IEEE Trans. Inf. Theory* **50**, 3062–3080 (2004).

[138] A. Host-Madsen, "A new achievable rate for cooperative diversity based on generalized writing on dirty paper," in *IEEE ISIT* (2003), p. 317.

[139] Y. T. Hou, Y. Shi, H. D. Sherali, and S. F. Midkiff, "On energy provisioning and relay node placement for wireless sensor networks," *IEEE Trans. Commun.* **4**, 2579–2590 (2005).

[140] Z. Hu and B. Li, "On the fundamental capacity and lifetime limits of energy-constrained wireless sensor networks," *Proceedings of the 10th IEEE Real-Time and Embedded Technology and Applications Symposium*, New York: IEEE (2004), pp. 2–9.

[141] J. Hua, Z. Xiong, and X. Wu, "High-performance 3-D embedded wavelet video (EWV) coding," in *Proceedings of the IEEE Fourth Workshop on Multimedia Signal Processing*, New York: IEEE (2001).

[142] J. Huang, R. A. Berry, and M. L. Honig, "Spectrum sharing with distributed interference compensation," in *DySPAN* (2005), pp. 88–93.

[143] M. R. Hueda, C. Rodriguez, and C. Marques, "Enhanced-performance video transmission in multicode CDMA wireless systems using a feedback error control scheme," in *IEEE Globecom 2001* (2001), pp. 619–626.

[144] X. L. Hung and B. Bensaou, "On max-min fairness and scheduling in wireless ad-hoc networks: analytical framework and implementation," in *MobiHoc* (2001).

[145] T. E. Hunter and A. Nosratinia, "Performance analysis of coded cooperation diversity," in *ICC03* (2003), Vol. 4, pp. 2688–2692.

[146] T. E. Hunter, S. Sanayei, and A. Nosratinia, "Outage analysis of coded cooperation," *IEEE Trans. Inf. Theory.* **52**, 375–391 (2006).

[147] A. S. Ibrahim, A. K. Sadek, W. Su, and K. J. R. Liu, "Cooperative communications with partial channel state information: when to cooperate?" in *Globecom'05* (2005), Vol. 5, pp. 3068–3072.

[148] W. C. Jakes, *Microwave Mobile Communications* (2nd ed.), New York: IEEE (1994).

[149] A. Jalali, R. Padovani, and R. Pankaj, "Data throughput of CDMA-HDR a high eciency-high data rate personal communication wireless system," in *VTC* (2000), Vol. 3.

[150] A. Jamalipour, T. Wada, T. Yamazato, "A tutorial on multiple access technologies for beyond 3G mobile networks," *IEEE Commun. Mag.* (Feb. 2005).

[151] R. Jantti and S.-L. Kim, "Second-order power control with asymptotically fast convergence," *IEEE J. Sel. Areas Commun.* **SAC-18**, 447–457 (2000).

[152] N. S. Jayant and P. Noll, *Digital Coding of Waveforms*, Englewood Cliffs, NJ: Prentice-Hall (1984).

[153] D. B. Johnson, "Routing in ad hoc networks of mobile hosts," in *Proceedings of the Workshop on Mobile Computing Systems and Applications*, Washington, DC: IEEE Computer Society (1994), pp. 158–163.

[154] S. Kandukuri and S. Boyd, "Optimal power control in interference-limited fading wireless channels with outage-probability specifications," *IEEE Trans. Wireless Commun.* 46–55 (2002).

[155] J. Karlsson and B. Eklundh, "A cellular mobile telephone system with load sharing-an enhancement of directed retry," *IEEE Trans. Commun.* **COM-37**, 530–535 (1989).

[156] N. Karmarkar, "A new polynomial-time algorithm for linear programming," *Combinatorica* **4**, 373–395 (1984).

[157] A. K. Katsaggelos, Y. Eisenberg, F. Zhai, R. Berry, and T. N. Pappas, "Advances in efficient resource allocation for packet-based real-time video transmission," *Proc. IEEE* **93**, 135–147 (2005).

[158] H. Kellerer, U. Pferschy, and D. Pisinger, *Knapsack Problems*, New York: Springer (2004).

[159] F. Kelly, "Charging and rate control for elastic traffic," *European Trans. Telecommun.* **8**, 33–37 (1997).

[160] F. P. Kelly, A. K. Maulloo and D. K. H. Tan, "Rate control in communication networks: shadow prices, proportional fairness and stability," *J. Operat. Res. Soc.* **49**, 237–252 (1998).

[161] L. G. Khachian, "A polynomial algorithm in linear programming." *Dokl. Akad. Nauk SSSR* **244**, 1093–1096 (1979). English translation in *Sov. Math. Dokl.* **20**, 191–194 (1979).

[162] A. E. Khandani, E. Modiano, L. Zheng, and J. Abounadi, "Cooperative routing in wireless networks," in *Advances in Pervasive Computing and Networking*, B. K. Szymanski and B. Yener (eds.), Boston: Kluwer Academic (2004).

[163] M. A. Khojastepour, A. Sabharwal and B. Aazhang, "On the capacity of 'cheap' relay networks," in *Proceedings of the 37th Annual Conference on Information Sciences and Systems* (2003).

[164] S. Kim, Z. Roseberg, and J. Zander, "Combined power control and transmission rate selection in cellular networks," in *IEEE VTC* (1999), pp. 1653–1654.

[165] L. Kleinrock, *Queueing Systems, Volume II: Computer Applications*, New York: Wiley (1976).

[166] I. Koutsopoulos and L. Tassiulas, "Carrier assignment algorithms in wireless broadband networks with channel adaptation," in *IEEE ICC'01*, (2001), pp. 1401–1405.

[167] V. Krishna, *Auction Theory*, New York: Academic (2002).

[168] B. Krishnamachari, *Networking Wireless Sensors*, New York: Cambridge University Press (2005).

[169] H. W. Kuhn, "The Hungarian method for the assignment problem," *Nav. Res. Logistics*, Quarterly 2 (1955).

[170] A. Kwasinski, Z. Han, and K. J. R. Liu, "Joint source coding and cooperation diversity for multimedia communications," *Proceedings of the 2005 IEEE Workshop on Signal Processing Advances in Wireless Communication* (*SPAWC*), New York: IEEE (2005), pp. 129–133.

[171] A. Kwasinski, Z. Han, and K. J. R. Liu, "Cooperative multimedia communications: Joint source coding and collaboration," in *Proceedings of IEEE Global Telecommunications Conference*, New York: IEEE (2005).

[172] A. Kwasinski, Z. Han, K. J. Ray Liu, and Nariman Farvardin, "Dynamic real time distortion management over multimedia downlink CDMA," in *IEEE WCNC* (2004), Vol. 2, pp. 650–654.

[173] R. J. La and V. Anantharam, "Optimal routing control: Repeated game approach," *IEEE Trans. Autom. Control.* 437–450 (2002).

[174] I. Lambadaris, P. Narayan, and I. Viniotis, "Optimal service allocation among two heterogeneous traffic types with no queueing," in *Proceedings of the 26th IEEE Conference on Decision and Control*, New York: IEEE (1987), pp. 1496–1498.

[175] J. N. Laneman, D. N. C. Tse, and G. W. Wornell, "Cooperative diversity in wireless networks: efficient protocols and outage behavior," *IEEE Trans. Inf. Theory* **50**, 3062–3080 (2004).

[176] J. N. Laneman and G. W. Wornell, "Distributed space-time coded protocols for exploiting cooperative diversity in wireless networks," *IEEE Trans. Inf. Theory* **49**, 2415–2525 (2003).

[177] J.-Y. Le Boudec and M. Vojnovic, "Perfect simulation and stationarity of a class of mobility models, in *IEEE Infocom* (2005).

[178] J. Lee, *A First Course in Combinatorial Optimization*, New York: Cambridge University Press (1996).

[179] K. K. Leung and L. C. Wang, "Controlling QoS by integrated power control and link adaptation in broadband wireless networks," *Eur. Trans. Telecommun.* No. 4, 383–394 (2000).

[180] F. Li, K. Wu, and A. Lippman, "Energy-efficient cooperative routing in multi-hop wireless ad hoc networks," in *Proceedings of the IEEE International Performance, Computing, and Communications Conference*, New York: IEEE (2006), pp. 215–222.

[181] Q. Li and M. van der Schaar, "Providing adaptive QoS to layered video over wireless local area networks through real-time retry limit adaptation," *IEEE Trans. Multimedia* **6**, 278–290 (2004).

[182] Y. Li and K. J. R. Liu, "Adaptive blind multi-channel equalization for multiple signal separation," *IEEE Trans. Inf. Theory* **44**, 2864–2876 (1998).

[183] Y. C. Liang, F. P. S. Chin, and K. J. R. Liu, "Downlink beamforming for DS-CDMA mobile radio with multimedia services," *IEEE Trans. Communun.* **49**, 1288–1298 (2001).

[184] H. Liu and G. Xu, "Closed-form blind symbol estimation in digital communications," *IEEE Trans. Sig. Process.* **43**, 2714–2723 (1995).

[185] X. Liu, E. K. P. Chong, and N. B. Shroff, "A framework for opportunistic scheduling in wireless networks," *Comput. Networks* **41**, 451–474 (2003).

[186] X. Liu, E. K. P. Chong, and N. B. Shroff, "Opportunistic transmission scheduling with resource-sharing constraints in wireless networks," *IEEE J. Sel. Areas Commun.* **19**, 2053–2064 (2001).

[187] X. Liu, E. K. P. Chong, N. B. Shroff, "Transmission scheduling for efficient wireless utilization," in *IEEE Infocom* (2001).

[188] X. Liu and S. Shankar, "Sensing-based opportunistic channel access," *ACM J. Mobile Networks* **11**, 577–591 (2006).

[189] X. Liu and W. Wang, "On the characteristics of spectrum-agile communication networks," in *IEEE DySPAN* (2005), pp. 214–223.

[190] Y. Liu and E. Knightly, "Opportunistic fair scheduling over multiple wireless channels," in *IEEE Infocom* (2003).

[191] S. Lu, V. Bharghavan, and R. Srikant, "Fair scheduling in wireless packet networks," *IEEE/ACM Trans. Network.* **7** (1999).

[192] S. Lu, T. Nandagopal, and V. Bharghavan, "A wireless fair service algorithm for packet cellular networks," in *ACM MobiCom* (1998).

[193] M. G. Luby, M. Mitzenmacher, M. A. Shokrollahi, and D. A. Spielman, "Improved low-density parity-check codes using irregular graphs," *IEEE Trans. Inf. Theory* **47** (2001).

[194] H. Luo, S. Lu, and V. Bharghavan, "A new model for packet scheduling in multihop wireless networks," in *ACM MobiCom* (2000).

[195] J. Luo, R. S. Blum, L. J. Greenstein, L. J. Cimini, and A. M. Haimovich, "New approaches for cooperative use of multiple antennas in ad hoc wireless networks," in *IEEE VTC* (2004), Vol. 4, pp. 2769–2773.

[196] A. B. MacKenzie and S. B. Wicker, "Game theory and the design of self-configuring, adaptive wireless networks," *IEEE Commun. Mag.* 126–131 (November 2001).

[197] A. B. MacKenzie and S. B. Wicker, "Stability of multipacket slotted Aloha with selfish users and perfect information," in *IEEE INFOCOM* (2003).

[198] A. Majumda, D. G. Sachs, I. V. Kozintsev, K. Ramchandran, and M. M. Yeung, "Multicast and unicast real-time video streaming over wireless LANs," *IEEE Trans. Circuits Syst. Video Technol.* **12**, 524–534 (2002).

[199] I. Maric and R. D. Yates, "Cooperative multicast for maximum network lifetime," *IEEE J. Sel. Areas Commun.* **23**, 127–135 (2005).

[200] R. J. Marks, A. K. Das, and M. El-Sharkawi, "Maximizing lifetime in an energy constrained wireless sensor array using team optimization of cooperating systems," in *Proceedings of the International Joint Conference on Neural Networks* (2002).

[201] S. Martello and P. Toth, *Knapsack Problems*, New York: Wiley (1990).

[202] S. Meguerdichian, F. Koushanfar, M. Potkonjak, and M. B. Srivastava, "Coverage problems in wireless ad hoc sensor networks," *in IEEE INFOCOM* (2001).

[203] A. Mercado and K. J. Ray Liu, "NP-hardness of the stable matrix in unit interval family problem in discrete time," *Sys. Control Lett.*, 261–265 (2001).

[204] J. F. Mertens and A. Neyman, "Stochastic games," *Int. J. Game Theory* **10**, 53–66 (1981).

[205] M. Mesbahi, M. G. Safonov, and G. P. Papavassilopoulos, "Bilinearity and complementarity in robust control," in *Advances in Linear Matrix Inequality Methods in Control*, Philadelphia: Society for Industrial and Applied Mathematics (2000).

[206] P. Michiardi and R. Molva, "A game theoretical approach to evaluate cooperation enforcement mechanisms in mobile ad hoc networks," *IEEE/ACM Workshop WiOpt 2003*, New York: IEEE (2003).

[207] P. Michiardi and R. Molva, "CORE: a collaborative reputation mechanism to enforce node cooperation in mobile ad hoc networks," *in Proceedings of the International Federation of Information Processing Communication and Multimedia Security Conference* (2002).

[208] M. Milano, *Constraint and Integer Programming: Toward a Unified Methodology*, Boston: Kluwer Academic (2004).

[209] D. Mitra, "An asynchronous distributed algorithm for power control in cellular radio systems," in *Fourth WINLAB Workshop on Third Generation Wireless Information Networks*, New Brunswick, NJ: Rutgers University (1993).

[210] A. F. Molisch, J. R. Foerster, and M. Pendergrass, "Channel models for ultrawideband personal area networks," *IEEE Wireless Commun.* **10**(6), 14–21 (2003).

[211] R. A. Monzingo and T. W. Miller, *Introduction to Adaptive Arrays*, New York: Wiley (1980).

[212] S. Murthy and J. J. Garcia-Luna-Aceves, "An efficient routing protocol for wireless networks," *ACM Mobile Networks and Appl. J., Special Issue on Routing in Mobile Communication Networks*, 183–197 (1996).

[213] R. B. Myerson, *Game Theory: Analysis of Conflict* (5th ed.), Cambridge, MA: Harvard University Press (2002).

[214] A. F. Naguib, A. Paulraj, and T. Kailath, "Capacity improvement with base-station antenna arrays in cellular CDMA," *IEEE Trans. Veh. Technol.* **43**, (1994).

[215] Y. Nakache, P. V. Orlik, W. M. Gifford, A. F. Molisch, I. Ramachandran, G. Fang, and J. Zhang, "Low-complexity ultrawideband transceiver with compatibility to multiband-OFDM," Technical Report, Mitsubishi Electronic Research Laboratory. Internet: www.merl.com/reports/docs/TR2004-051.pdf.

[216] T. Nandagopal, S. Lu, and V. Bharghavan, "A unified architecture for the design and evaluation of wireless fair queueing algorithms," in *Proceedings of Mobicom '99*, (1999), pp. 132–142.

[217] J. Nash, "Equilibrium points in n-person games," *Proc. Natl. Acad. Sci.* **36**, 48–49 (1950).

[218] S. G. Nash and A. Sofer, *Linear and Nonlinear Programming*, New York: McGraw-Hill (1995).

[219] J. Neel, J. Reed, R. Gilles. "Game models for cognitive radio algorithm analysis," in *Proceedings of the SDR Forum Technical Conference* (2004), pp. 15–18.

[220] R. W. Nettleton and H. Alavi, "Power control for spread spectrum cellular mobile radio system," in *VTC* (1983).

[221] T. S. Eugene Ng, I. Stoica, and H. Zhang, "Packet fair queueing algorithms for wireless networks with location-dependent errors," in *IEEE Infocom 98* (1998).

[222] T. Nguyen and Z. Ding, "Blind CMA beamforming for narrowband signals with multipath arrivals," *Int. J. Adaptive Control Sig. Process.* **12**, 157–172 (1998).

[223] N. Nie and C. Comaniciu, "Adaptive channel allocation spectrum etiquette for cognitive radio networks," *ACM Mobile Networks and Appl., special issue on Reconfigurable Radio Technologies in Support of Ubiquitous Seamless Computing*, **11**, 779–797 (2006).

[224] D. Niyato and E. Hossain, "Call admission control for qos provisioning in 4g wireless networks: issues and approaches," *IEEE Network*, 5–11 (September/October 2005).

[225] P. Ormeci, X. Liu, D. L. Goeckel, and R. D. Wesel, "Adaptive bit-interleaved coded modulation," *IEEE Trans. Commun.* **49**, 1572–1581 (2001).

[226] B. Ottersten, R. Roy, and T. Kailath, "Signal waveform estimation in sensor array processing," in *Proceedings of the 23rd Asilomar Conference on Signals, Systems, and Computers* (1989).

[227] G. Owen, *Game Theory* (3rd ed.), New York: Academic (2001).

[228] L. Ozarov, "On a source coding problem with two channels and three receivers," *Bell Syst. Tech. J.* **59**, 1909–1921 (1980).

[229] C. Pandana and K. J. Ray Liu, "Near-optimal reinforcement learning framework for energy-aware sensor communications," *IEEE J. Sel. Areas Commun.* **23**, 259–268 (2005).

[230] V. D. Park and M. S. Corson, "A highly adaptive distributed routing algorithm for mobile wireless networks," in *IEEE Infocom* (1997).

[231] D. C. Parkes, "Iterative combinatorial auctions: Achieving economic and computational efficiency," Ph.D. dissertation, the University of Pennsylvania (2001).

[232] N. Passas, L. Merakos, and D. Skyrianoglou, "Traffic scheduling in wireless ATM networks," in *Proceedings of the IEEE ATM Workshop*, New York: IEEE (1997).

[233] C. Perkins, "Highly dynamic destination-sequenced distance-vector routing (DSDV) for mobile computers," in *Proceedings of the ACM SIGCOMM'94, Conference on Communications Architectures, Protocols and Applications*, London: Association for Computing Machinery (1994).

[234] C. Perkins and E. Royer, "Ad-hoc on-demand distance vector routing," *Proceedings of the 2nd IEEE Workshop on Mobile Computing Systems and Applications*, New York: IEEE (1999), pp. 90–100.

[235] W. W. Peterson and E. J. Weldon, *Error-Correcting Codes* (2nd ed.), Cambridge, MA: MIT Press (1972).

[236] S. Pietrzyk and G. J. M. Janssen, "Multiuser subcarrier allocation for QoS provision in the OFDMA systems," in *IEEE VTC* (2002).

[237] L. Ping and K. Y. Wu, "Concatenated tree codes: A low-complexity, high-performance approach," *IEEE Trans. Inf. Theory* **47** (2001).

[238] R. H. Porter, "Optimal cartel trigger price strategies," *J. Econ. Theory* **29**, 313–318 (1983).

[239] J. G. Proakis, *Digital Communications* (3rd ed.), New York: McGraw-Hill (1995).

[240] R. Puri, K.-W. Lee, K. Ramchandran, and V. Bharghavan, "An integrated source transcoding and congestion control paradigm for video streaming in the internet," *IEEE Trans. Multimedia* **3**(1), 18–32 (2001).

[241] X. Qiu and K. Chawla, "On the performance of adaptive modulation in cellular systems," *IEEE Trans. Commun.* **47**, 884–895 (1999).

[242] L. R. Rabiner, *Fundamentals of Speech Recognition*, Englewood Cliffs, NJ: Prentice-Hall (1993).

[243] H. M. Radha, M. van der Schaar, and Y. Chen, "The MPEG-4 fine-grained scalable video coding method for multimedia streaming over IP," *IEEE Trans. Multimedia* **3**(1), 53–68 (2001).

[244] B. Radunovic and J.-Y. Le Boudec, "Optimal power control, scheduling, and routing in UWB networks," *IEEE J. Sel. Areas Commun.* **22**, 1252–1270 (2004).

[245] S. Ramakrishna and J. M. Holtzman, "A scheme for throughput maximization in a dual-class CDMA system," *IEEE J. Sel. Areas Commun.* **16**, 830–844 (1998).

[246] P. Ramanathan and P. Agrawal, "Adapting packet fair queueing algorithms to wireless networks," in *ACM/IEEE MobiCom'98* (1998).

[247] T. S. Rappaport, *Wireless Communications: Principle and Practice* (2nd ed.), Englewood Cliffs, NJ: Prentice-Hall (2002).

[248] F. Rashid-Farrokhi, K. J. R. Liu, and L. Tassiulas, "Transmit beamforming and power control for cellular wireless systems," *IEEE J. Sel. Areas Commun.* **16**, 1437–1449 (1998).

[249] F. Rashid-Farrokhi, K. J. R. Liu, and L. Tassiulas, "Transmit and receive diversity and equalization in wireless networks with fading channels," in *Globecom* (1997).

[250] F. Rashid-Farrokhi, L. Tassiulas, and K. J. R. Liu, "Joint optimal power control and beamforming in wireless network using antenna arrays," *IEEE Trans. Communun.* **46**, 1313–1323 (1998).

[251] R. Rezaiifar, A. M. Makowski, and S. Kumar, "Stochastic control of handoffs in cellular networks," *IEEE J. Sel. Areas Commun.* **13**, 1348–1362 (1995).

[252] W. Rhee and J. M. Cioffi, "Increase in capacity of multiuser OFDM system using dynamic subchannel allocation," in *IEEE VTC'00* (2000), pp. 1085–1089.

[253] J. Ribas-Corbera and S. Lei, "Rate control in DCT video coding for low-delay communications," *IEEE Trans. Circuits Syst. Video Technol.* **9**, 172–185 (1999).

[254] T. J. Richardson and R. L. Urbanke, "The capacity of low-density parity-check codes under message-passing decoding," *IEEE Trans. Inf. Theory* **47** (2001).

[255] T. J. Richardson and R. L. Urbanke, "Efficient encoding of low-density parity-check codes," *IEEE Trans. Inf. Theory* **47** (2001).

[256] Z. Rosberg, P. Varaiya, and J. Walrand, "Optimal control of service in tandem queues," *IEEE Trans. Autom. Control.* **AC-27**, 600–610 (1982).

[257] I. Rubin, A. Behzad, H. Ju, R. Zhang, X. Huang, Y. Liu, and R. Khalaf, "Ad hoc wireless networks with mobile backbones," in *IEEE PIMRC* (2004), Vol. 1, pp. 566–573.

[258] E. Saberinia, J. Tang, A. H. Tewfik, and K. K. Parhi, "Design and implementation of multi-band pulsed-OFDM system for wireless personal area networks," in *IEEE ICC* (2004), Vol. 2, pp. 862–866.

[259] E. Saberinia and A. H. Tewfik, "Multi-user UWB-OFDM communications," in *IEEE Proceedings on Pacific Rim Conference on Communication, Computers and Signal Processing* (2003), Vol. 1, pp. 127–130.

[260] A. K. Sadek, Z. Han, and K. J. R. Liu, "A distributed relay-assignment algorithm for cooperative communications in wireless networks," in *IEEE ICC* (2006).

[261] A. K. Sadek, Z. Han, and K. J. Ray Liu, "Relay-assignment for cooperative communications in cellular networks to extend coverage area," in *IEEE WCNC* (2006).

[262] A. K. Sadek, W. Su, and K. J. R. Liu, "A class of cooperative communication protocols for multi-node wireless networks," in *Proceedings of the IEEE International Workshop on Signal Processing Advances in Wireless Communications (SPAWC)*, New York: IEEE (2005).

[263] A. K. Sadek, W. Su, and K. J. R. Liu, "Multinode cooperative resource allocation to improve coverage area in wireless networks," in *IEEE Globecom* (2005).

[264] A. A. M. Saleh and R. A. Valenzuela, "A statistical model for indoor multipath propagation," *IEEE J. Sel. Areas Commun.* **5**, 128–137 (1987).

[265] A. Sampath and J. M. Holtzman, "Access control of data in integrated voice/data CMDA systems: benefits and tradeoffs," *IEEE J. Sel. Areas Commun.* **15**, 1511–1526 (1997).

[266] B. Sampath, K. J. R. Liu, and Y. G. Li, "Deterministic blind subspace MIMO equalization," *EURASIP J. Appl. Sig. Process.* **2002**, 538–551 (2002).

[267] S. Sampei, S. Komaki, and N. Morinaga, "Adaptive modulation/TDMA scheme for large capacity personal multi-media communication systems," *IEICE Trans. Commun.* **77**, 1096–1103 (1994).

[268] C. U. Saraydar, N. B. Mandayam, and D. J. Goodman, "Efficient power control via pricing in wireless data networks," *IEEE Trans. Commun.* **50**, 291–303 (2002).

[269] A. H. Sayed, A. Tarighat, and N. Khajehnouri, "Network-based wireless location," *IEEE Sig. Process. Mag.* **22**, 24–40 (2005).

[270] M. van der Schaar, S. Krishnamachari, S. Choi, and X. Xu, "Adaptive cross-layer protection strategies for robust scalable video transmission over 802.11 WLANs," *IEEE J. Sel. Areas Commun.* **21**, 1752–1763 (2003).

[271] M. van der Schaar and H. M. Radha, "A hybrid temporal-SNR fine-granular scalability for internet video," *IEEE Trans. Circuits Syst. Video Technol.* **11**, 318–331 (2001).

[272] A. Sendonaris, E. Erkip, and B. Aazhang, "User cooperation diversity, Part I: System description," *IEEE Trans. Commun.* **51**, 1927–1938 (2003).

[273] A. Sendonaris, E. Erkip, and B. Aazhang, "User cooperation diversity, Part II: Implementation aspects and performance analysis," *IEEE Trans. Commun.* **51**, 1939–1948 (2003).

[274] S. Shakkottai and R. Srikant, "Scheduling real-time traffic with deadlines over a wireless channel," *ACM/Baltzer Wireless Networks J.* **8**, 13–26 (2002).

[275] S. Shakkottai and A. L. Stolyar, "Scheduling algorithms for mixture of real-time and non-real-time data in HDR," in *Proceedings of the 17th International Teletraffic Congress (ITC-17)* (2001).

[276] L. S. Shapley, "Stochastic games," *Proc. Natl. Acad. Sci. USA* **39**, 1095–1100 (1953).

[277] V. Shnayder, M. Hempstead, B.-R. Chen, G. W. Allen, and M. Welsh, "Simulating the power consumption of large-scale sensor network applications," in *Proceedings of the 2nd International Conference on Embedded Networked Sensor Systems* (2004), pp. 188–200.

[278] M. Sikora, J. N. Laneman, M. Haenggi, D. J. Costello, and T. E. Fuja, "Bandwidth- and power-efficient routing in linear wireless networks," *IEEE Trans. Inf. Theory* **52**, 2624–2633 (2006).

[279] T. Sikora, "Trends and perspectives in image and video coding," *Proc. IEEE* **93**, 6–17 (2005).

[280] M. K. Simon and M. S. Alouini, *Digital Communication Over Fading Channels: A Unified Approach to Performance Analysis*, New York: Wiley (2000).

[281] W. P. Siriwongpairat, Z. Han, and K. J. R. Liu, "Energy-efficient channel allocation for multiuser multiband UWB system," in *IEEE WCNC* (2005), pp. 813–818.

[282] W. Pam Siriwongpairat, Z. Han, and K. J. Ray Liu, "Power controlled channel allocation for multiuser multiband UWB," *IEEE Trans. Wireless Commun.* **6** (2007).

[283] W. Siriwongpairat, W. Su, Z. Han, and K. J. Ray Liu, "Coverage enhancement for multiband UWB systems using cooperative communications," in *IEEE WCNC* (2006).

[284] W. P. Siriwongpairat, W. Su, and K. J. R. Liu, "Characterizing performance of multi-band UWB systems using poisson cluster arriving fading paths," in *IEEE SPAWC* (2005), pp. 264–268.

[285] W. P. Siriwongpairat, W. Su, M. Olfat, and K. J. R. Liu, "Space–time-frequency coded multiband UWB communication systems," in *IEEE WCNC* (2005), Vol. 1, pp. 426–431.

[286] W. Pam Siriwongpairat, T. Himsoon, W. Su, and K. J. R. Liu, "Optimum threshold-selection relaying for decode-and-forward cooperation protocol," in *IEEE WCNC'06* (2006).

[287] V. Srinivasan, P. Nuggehalli, C. F. Chiasserini, and R. R. Rao, "Cooperative in wireless ad hoc networks," in *IEEE Infocom'03* (2003).

[288] V. Srivastava, J. Neel, A. MacKenzie, R. Menon, L. A. DaSilva, J. Hicks, J. H. Reed, and R. Gilles, "Using game theory to analyze wireless ad hoc networks," *IEEE Commun. Surv. Tutor.* **7**(4), 46–56 (2005).

[289] V. M. Stankovic, R. Hamzaoui, and Z. Xiong, "Real-time error protection of embedded codes for packet erasure and fading channels," *IEEE Trans. Circuits Syst. Video Technol.* **14**, 1064–1072 (2004).

[290] R. Steele and M. Nofal, "Teletraffic performance of microcellular personal communication networks," *IEE Proc.* **139**(4) (1992).

[291] K. Stuhlmuller, N. Farber, M. Link, and B. Girod, "Analysis of video transmission over lossy channels," *IEEE J. Sel. Areas Commun.* **18**, 1012–1032 (2000).

[292] G.-M. Su, Z. Han, A. Kwasinski, M. Wu, K. J. R. Liu, and N. Farvardin, "Distortion management of real-time MPEG-4 FGS video over downlink multicode CDMA networks," in *IEEE ICC* (2004), Vol. 5, pp. 3071–3075.

[293] G.-M. Su, Z. Han, M. Wu, and K. J. R. Liu, "Joint uplink and downlink optimization for video conferencing over wireless LAN," in *IEEE International Conference on Acoustics, Speech, and Signal Processing, (ICASSP)*, New York: IEEE (2005).

[294] G.-M. Su, Z. Han, M. Wu, and K. J. R. Liu, "Dynamic distortion control for 3-D embedded wavelet video over multiuser OFDM networks," in *IEEE Globecom* (2004), Vol. 2, pp. 650–654.

[295] G.-M. Su, Z. Han, M. Wu, and K. J. Ray Liu, "A scalable multiuser framework for video over OFDM networks: Fairness and efficiency," *IEEE Trans. Circuit Syst. Video Technol.* **16**, 1217–1231 (2006).

[296] G.-M. Su, Z. Han, M. Wu, and K. J. Ray Liu, "Multiuser cross-layer resource allocation for video transmission over wireless networks," *IEEE Network Mag., Special Issue on Multimedia over Broadband Wireless Networks*, 21–27 (March/April, 2006).

[297] H. J. Su and E. Geraniotis, "Adaptive closed-loop power control with quantized feedback and loop filtering," *IEEE Trans. Wireless Commun.* 76–86 (2002).

[298] W. Su, A. K. Sadek, and K. J. R. Liu, "SER performance analysis and optimum power allocation for decode-and-forward cooperation protocol in wireless networks," in *IEEE WCNC* (2005).

[299] B. Suard, A. Naguib, G. Xu, and A. Paulraj, "Performance analysis of CDMA mobile communication systems using antenna array," in *ICASSP' 93* (1994), Vol. IV, pp. 153–156.

[300] G. Sun, J. Chen, W. Guo, and K. J. Ray Liu, "Signal processing techniques in network-aided positioning: A survey," *IEEE Sig. Process. Mag.* **22**(4), 12–23 (2005).

[301] M. T. Sun and A. Reibman, *Compressed Video Over Networks,* Signal Processing Series, New York: Marcel Dekker (2001).

[302] R. S. Sutton and A. G. Barto, *Reinforcement Learning: An Introduction*, Cambridge, MA: MIT Press (1998).

[303] S. Talwar and A. Paulraj, "Blind separation of synchronous co-channel digital signals using an antenna array. Part II: Performance analysis," *IEEE Trans. Sig. Process.* **45**, 706–718 (1997).

[304] S. Talwar, M. Viberg, and A. Paulraj, "Blind separation of synchronous co-channel digital signals using an antenna array. Part I: Algorithms," *IEEE Trans. Sig. Process.* **44**, 1184–1197 (1996).

[305] A. S. Tanenbaum, *Computer Networks*, Englewood Cliffs, NJ: Prentice-Hall (1981).

[306] X. Tang, M. S. Alouini, and A. J. Goldsmith, "Effect of channel estimation error on M-QAM BER performance in Rayleigh fading," *IEEE Trans. Commun.* **47**, 1856–1864 (1999).

[307] L. Tassiulas1 and S. Sarkar, "Maxmin fair scheduling in wireless networks," in *IEEE Infocom* (2002).

[308] D. S. Taubman, "High performance scalable image compression with EBCOT," *IEEE Trans. Image Process.* **9**, 1158–1170 (2000).

[309] C. K. Toh, "Associativity-based routing for ad hoc mobile networks," *Wireless Personal Commun.: An Intl. J.* **4**, 103–139 (1997).

[310] L. Tong, G. Xu, and T. Kailath, "Blind channel identification based on second order statistics: a frequency domain approach," *IEEE Trans. Inf. Theory* **41**, 329–334 (1995).

[311] J. Torrance, D. Didascalou, and L. Hanzo, "The potential and limitations of adaptive modulation over slow Rayleigh fading channels," in *IEE Colloquium on the Future of Mobile Multimedia Communications*, London: Institution of Electrical Engineers (1996).

[312] J. M. Torrance and L. Hanzo, "Adaptive modulation in a slow Rayleigh fading channel," in *IEEE PIMRC* (1996), pp. 497–501.

[313] T. Ue, S. Sampei, and N. Morinaga, "Symbol rate and modulation level controlled adaptive modulation/TDMA/TDD for personal communication systems," in *IEEE VTC* (1995), pp. 306–310.

[314] S. Ulukus and R. Yates, "Stochastic power control for cellular radio systems," *IEEE Trans. Commun.* **46**, 784–798 (1998).

[315] G. Ungerboeck, "Channel coding with multilevel/phase signals," *IEEE Trans. Inf. Theory* **IT-28**, 55–67 (1982).

[316] N. H. Vaidya, P. Bajl, and S. Gupta, "Distributed fair scheduling in a wireless LAN," in *ACM Mobicom* (2000).

[317] M. C. Valenti and B. Zhao, "Distributed turbo codes: Towards the capacity of the relay channel," in *IEEE VTC* (2003), Vol. 1, pp. 322–326.

[318] S. Verdu, *Multiuser Detection*, New York: Cambridge University Press (2001).

[319] P. Viswanath, D. Tse, and V. Anantharam, "Asymptotically optimal waterfilling in vector multiple access channels," *IEEE Trans. Inf. Theory* **12**, 241–267 (2001).

[320] P. Viswanath, D. N. C. Tse, and R. Laroia, "Opportunistic beamforming using dumb antennas," *IEEE Trans. Inf. Theory* **48**, 1277–1294 (2002).

[321] H. Viswanathan, X. Li, and R. Krishnamoorthy, "Adaptive coded modulation over slow frequency-selective fading channels," in *IEEE VTC* (1999).

[322] J. P. Walters, Z. Liang, W. Shi, and V. Chaudhary, "Wireless sensor network security: a survey," in *Security in Distributed, Grid, and Pervasive Computing*, New York: Auerbach (2006).

[323] B. Wang, Z. Han, and K. J. Ray Liu, "Distributed relay selection and power control for multiuser cooperative communication networks using buyer / seller game," in *Infocom* (2007).

[324] B. Wang, Z. Han, and K. J. Ray Liu, "Stackelberg Game for Distributed Resource Allocation over Multiuser Cooperative Communication Networks," in *IEEE Globecom* (2006).

[325] H. S. Wang and N. Moayeri, "Finite-state markov channel – a useful model for radio communication channel," *IEEE Trans. Veh. Technol.* **44**, 163–170 (1995).

[326] L. Wang and A. H. Aghvami, "Optimal power allocation for a multi-rate packet CDMA system with multi-media traffic," in *IEEE WCNC99* (1999).

[327] Y. Wang, J. Ostermann, and Y.-Q. Zhang, *Video Processing and Communications*, Englewood Cliffs, NJ: Prentice-Hall (2001).

[328] J. H. Wen, W. J. Chen, and S. Y. Lin, "Time-slot selection algorithms for quasi-fixed frequency assignment TDMA cellular systems," *J. Chinese Inst. Electr. Eng.* **5**(3), 223–233 (1998).

[329] J. H. Wen, W. J. Chen, S. Y. Lin, and K. T. Huang, "Performance evaluation of LIBTA/hybrid time-slot selection algorithm for cellular systems," in *IASTED International Conference on Modeling, Simulation and Optimization* (1997), pp. 178–181.

[330] J. H. Wen, W. J. Chen, and J. K. Ho, "Time-slot selection algorithm with directed retry/load sharing for personal communication systems," *Proc. NSC, Part A: Physical Sci. Eng.* **21**, 631–636 (1997).

[331] H. Weyl, "Elementare theorie der konvexen polyheder," *Comm. Math. Helv.* **7**, 290 (1935) (translated in *Contributions to the Theory of Games* **1**, 3, 1950).

[332] S. Wicker, *Error Control Systems for Digital Communication and Storage*, Englewood Cliffs, NJ: Prentice-Hall (1995).

[333] J. E. Wieselthier, G. D. Nguyen, and A. Ephremides, "On the construction of energy-efficient broadcast and multicast trees in wireless networks," in *IEEE Infocom* (2000), Vol. 2, pp. 585–594.

[334] M. Z. Win and R. A. Scholtz, "Impulse radio: How it works," *IEEE Commun. Lett.* **2**(2), 36–38 (1998).

[335] C. Y. Wong, R. S. Cheng, K. B. Letaief, and R. D. Murch, "Multiuser OFDM with adaptive subcarrier, bit, and power allocation," *IEEE J. Sel. Areas Commun.* **17**, 1747–1758 (1999).

[336] C. Y. Wong, C. Y. Tsui, R. S. Cheng, and K. B. Letaief, "A real-time sub-carrier allocation scheme for multiple access downlink OFDM transmission," in *IEEE VTC'99* (1999), pp. 1124–1128.

[337] M. K. Wood and G. B. Dantzig, "Programming of interdependent activities. I. General Discussion," *Econometrica* **17**, 193–199 (1949).

[338] D. Wu and R. Negi, "Effective capacity: A wireless link model for support of quality of service," *IEEE Trans. Wireless Commun.* **2**, 630–643 (2003).

[339] Y. Xiao, "IEEE 802.11e: QoS provisioning at the MAC layer," *IEEE Trans. Wireless Commun.* **11**, 72–79 (2004).

[340] H. Xie and S. Kuek, "Priority handoff analysis," in *IEEE VTC' 93* (1993), pp. 855–858.

[341] G. Xu, Y. Cho, A. Paulraj, and T. Kailath, "Maximum likelihood detection of co-channel communication signals via exploitation of spatial diversity," in *Proceedings of the 26th Asilomar Conference on Signals, Systems, and Computers* (1992).

[342] K. Xu, X. Hong, M. Gerla, H. Ly, and D. L. Gu, "Landmark routing in large wireless battlefield networks using UAVs," in *Proceedings of the IEEE Military Communications Conference (MILCOM 2001)* (2001), Vol. 1, pp. 230–234.

[343] H. Yaiche, R. R. Mazumdar, and C. Rosenberg, "A game theoretic framework for bandwidth allocation and pricing in broadband networks," *IEEE/ACM Trans. Network.* **8**, 667–678 (2000).

[344] X. Yang and N. H. Vaidya, "Priority scheduling in wireless ad hoc networks," *Proceedings of the 3rd ACM International Symposium on Mobile Ad Hoc Networking and Computing* (2002).

[345] Y. Yang and T. S. P. Yum, "Delay distributions of slotted ALOHA and CSMA," *IEEE Trans. Commun.* **51**, 1846–1857 (2003).

[346] Z. Yang, J. Liu, and A. Host-Madsen, "Cooperative routing and power allocation in ad-hoc networks," in *IEEE Globecom* (2005).

[347] R. Yates, "A framework for uplink power control in cellular radio systems," IEEE *J. Sel. Areas Commun.* **13**, 1341–1348 (1995).

[348] R. Yates and C. Y. Huang, "Integrated power control and base station assignment," *IEEE Trans. Veh. Technol.* **44**, 638–644 (1995).

[349] H. Yin and H. Liu, "An efficient multiuser loading algorithm for OFDM-based broadband wireless systems," in *IEEE Globecom* (2000), Vol. 1, pp. 103–107.

[350] W. Yu and J. M. Cioffi, "FDMA capacity of Gaussian multi-access channels with ISI," *IEEE Trans. Commun.* **50**, 102–111 (2002).

[351] J. Zander, "Performance of optimum transmitter power control in cellular radio systems," *IEEE Trans. Veh. Technol.* **41**, 57–62 (1992).

[352] J. Zander, "Distributed cochannel interference control in cellular radio systems," *IEEE Trans. Veh. Technol.* **41**, 305–311 (1992).

[353] J. Zander, "Transmitter power control for co-channel interference management in cellular radio systems," in *WINLAB* (1993).

[354] J. Zander, "Performance bounds for joint power control and link adaption for NRT bearers in centralized (bunched) wireless Networks," in *PIMRC 99* (1999).

[355] J. Zander and S. L. Kim, *Radio Resource Managment for Wireless Networks*, Norwood, MA: Artech House (2001).

[356] E. Zehavi, "8-PSK trellis codes for a Rayleigh channel," *IEEE Trans. Communun.* **40**, 873–884 (1992).

[357] Q. Zeng and D. P. Agrawal, "Handoff in wireless mobile networks," in *Handbook of Wireless Networks and Mobile Computing* (I. Stojmenovic, ed.), New York: Wiley (2002).

[358] Q.-A. Zeng, K. Mukumoto, and A. Fukuda, "Performance analysis of mobile cellular radio systems with two-level priority reservation handoff procedure," *IEICE Trans. Commun.*, **E80-B**, 598–604 (1997).

[359] Q. Zhang, Z. Ji, W. Zhu, and Y.-Q. Zhang, "Power-minimized bit allocation for video communication over wireless channels," *IEEE Trans. Circuits Syst. Video Technol.* **12**, 398–410 (2002).

[360] X. M. Zhang, A. Vetro, Y. Q. Shi, and H. Sun, "Constant quality constrained rate allocation for FGS-coded videos," *IEEE Trans. Circuits Syst. Video Technol.* **13**, 121–130 (2003).

[361] L. Zhao, J. Kim, and C.-C. J. Kuo, "MPEG-4 FGS video streaming with constant-quality rate control and differentiated forwarding," *SPIE, Conf. on Visual Comm. and Image* (2002).

[362] Q. Zhao, P. Cosman, and L. Milstein, "Optimal allocation of bandwidth for source coding, channel coding, and spreading in CDMA systems," *IEEE Trans. Commun.* **52**, 1797–1808 (2004).

[363] Q. Zhao, P. Cosman, and L. Milstein, "On the optimal allocation of bandwidth for source coding, channel coding and spreading in a coherent DS-CDMA system employing an MMSE receiver," in *European Wireless Conference Proceedings* (2002), Vol. 2, pp. 663–668.

[364] Q. Zhao, L. Tong, and A. Swami, "Decentralized cognitive MAC for dynamic spectrum access," in *DySPAN* (2005), pp. 224–232.

[365] S. Zhao, Z. Xiong, and X. Wang, "Optimal resource allocation for wireless video over CDMA networks," in *ICME* (2003).

[366] H. Zheng and L. Cao, "Device-centric spectrum management," in *DySPAN* (2005), pp. 56–65.

[367] L. Zheng and D. N. C. Tse, "Diversity and multiplexing: A fundamental tradeoff in multiple-antenna channels," *IEEE Trans. Inf. Theory* **49**, 1073–1096 (2003).

[368] S. Zhong, R. Y. Yang, and J. Chen, "Sprite: a simple, cheat-proof, credit-based system for mobile ad-hoc networks," in *IEEE Infocom* (2003).

[369] C. Zhou, M. L. Honig, and S. Jordan, "Two-cell power allocation for wireless data based on pricing," in *Proceedings of the Allerton Conference on Communication, Control and Computing* (2001).

[370] W. Zhuang, X. Shen, and Q. Bi, "Ultra-wideband wireless communications," *Wireless Commun. Mobile Computing (Special Issue on Ultra-Broadband Wireless Communications for the Future, Invited paper)*, **3**, 663–685 (2003).

[371] Federal Communications Commission Report FCC 98-153 "Revision of part 15 of the commission's rules regarding ultra-wideband transmission systems, first report and order," Washington, DC: FCC (Feburary 14, 2002).

[372] IEEE Computer Society, *Achieving wireless broadband with WiMAX*, June 2004.

[373] 3rd Generation Partnership Project, Technical Specification Group RAN, Working Group 2 (TSG RAN WG2), "Radio resource management strategies," 3G TR 25.922, V2.0.0, December 1999.

[374] IEEE P802.11e/Draft 6.0, "Draft amendment to IEEE std 802.11, 1999 edition, medium access control enhancements for quality of service," November 2003.

[375] IEEE 802.15WPAN High Rate Alternative PHY Task Group 3a (TG3a). Internet: www.ieee802.org/15/pub/ TG3a.html

[376] IEEE Standards Department, "Wireless LAN medium access control (MAC) and physical layer (PHY) specifications," ANSI/IEEE Standard 802.11, Aug. 1999.

[377] IEEE Standard for Local and Metropolitan Area Networks Part 16: Air Interface for Fixed Broadband Wireless Access Systems IEEE Std 802.16-2004 (Revision of IEEE Std 802.16-2001) 2004 Page(s): 27–28.

[378] IEEE 802.22 Working group, WRAN ReferenceModel, Doc Num. 22-04-0002-12-0000.

[379] Wikipedia, The Free Encyclopedia.

[380] Wikipedia, $http://en.wikipedia.org/wiki/Stackelberg\_game$

[381] Wikipedia, the free encyclopedia, $http://en.wikipedia.org/$.

[382] $http://iris.gmu.edu/ \sim khoffman/papers/newcomb1.html$

[383] $http://www.wimaxforum.org/home/$

[384] $http://www.bluetooth.com/bluetooth/$

[385] $http://w3.antd.nist.gov/wahn\_ssn.shtml$

[386] $http://www.umtsworld.com$

[387] $http://www.cdg.org$

[388] $http://www.gametheory.net$

Index

Printed in the United States
By Bookmasters